A GUIDE TO THE *ANOLIS* LIZARDS (ANOLES)

of Mainland Central and South America

A GUIDE TO THE *ANOLIS* LIZARDS (ANOLES)

of Mainland Central and South America

STEVEN POE

Princeton University Press
Princeton and Oxford

For Beeling and Camilo and Pascal and Ernest Williams

Copyright © 2025 by Princeton University Press

Princeton University Press is committed to the protection of copyright and the intellectual property our authors entrust to us. Copyright promotes the progress and integrity of knowledge created by humans. By engaging with an authorized copy of this work, you are supporting creators and the global exchange of ideas. As this work is protected by copyright, any reproduction or distribution of it in any form for any purpose requires permission; permission requests should be sent to permissions@press.princeton.edu. Ingestion of any IP for any AI purposes is strictly prohibited.

Published by Princeton University Press
41 William Street, Princeton, New Jersey 08540
99 Banbury Road, Oxford OX2 6JX
press.princeton.edu

GPSR Authorized Representative: Easy Access System Europe – Mustamäe tee 50, 10621 Tallinn, Estonia, gpsr.requests@easproject.com

All Rights Reserved
ISBN (pbk.) 978-0-691-19287-1
ISBN (e-book) 978-0-691-26819-4
Library of Congress Control Number: 2025937733

British Library Cataloging-in-Publication Data is available

Editorial: Robert Kirk and Megan Mendonça
Production Editorial: Karen Carter
Text Design: D & N Publishing, Wiltshire, UK
Jacket/Cover Design: Wanda España
Production: Steven Sears
Publicity: Caitlyn Robson-Iszatt and Matthew Taylor
Copyeditor: Annie Gottlieb

Cover image: Nature Picture Library / Alamy Stock Photo

This book has been composed in Minion Pro

Printed in China

10 9 8 7 6 5 4 3 2 1

CONTENTS

Foreword viii

Acknowledgments x

Introduction 1

CHAPTER 1 Biology of Mainland Anoles 17
CHAPTER 2 Collecting: How to Find and Catch Mainland Anoles 27
CHAPTER 3 How to Identify Mainland Anoles 35
CHAPTER 4 Species Accounts 49

Plate Set 1: Dewlaps 50

aeneus 66
aequatorialis 67
agassizi 68
allisoni 69
altae 70
alvarezdeltoroi 71
amplisquamosus 72
anatoloros 73
anchicayae 74
anisolepis 75
annectens 76
anoriensis 77
antioquiae 78
antonii 80
apletophallus 81
apollinaris 82
aquaticus 83
arenal 85
auratus 86
baleatus 87
barkeri 88
beckeri 89

bellipeniculus 90
benedikti 91
bicaorum 92
binotatus 93
biporcatus 95
biscutiger 96
boettgeri 97
bombiceps 98
boulengerianus 99
brasiliensis 100
brianjuliani 101
brooksi 102
caceresae 103
calimae 104
callainus 105
campbelli 106
capito 107
caquetae 109
carlostoddi 110
carolinensis 111
carpenteri 112
casildae 113

charlesmyersi 114
chloris 115
chrysolepis 116
cobanensis 117
compressicauda 118
concolor 119
crassulus 120
cristatellus 121
cristifer 123
cupreus 124
cuprinus 125
cuscoensis 126
cusuco 127
cybotes 128
cymbops 129
danieli 130
datzorum 131
dissimilis 132
distichus 133
dollfusianus 134
dracula 136
duellmani 137

CONTENTS

dunni 138
elcopeensis 139
equestris 140
eulaemus 141
euskalerriari 142
extremus 143
fasciatus 144
festae 146
fitchi 147
fortunensis 148
fraseri 149
frenatus 150
fungosus 152
fuscoauratus 153
gadovii 154
gaigei 155
garmani 157
gemmosus 158
ginaelisae 159
gorgonae 160
gracilipes 161
graniliceps 162
gruuo 163
heterodermus 164
heteropholidotus 166
hobartsmithi 167
huilae 168
humilis 169
hyacinthogularis 170
ibanezi 171
immaculogularis 172
inderenae 173
insignis 174
jacare 175
johnmeyeri 176

kathydayae 176
kemptoni 178
kreutzi 179
kunayalae 180
laevis 181
laeviventris 182
lamari 183
latifrons 185
lemurinus 186
limifrons 187
limon 188
lineatus 189
liogaster 190
lionotus 191
lososi 192
loveridgei 193
lynchi 194
macrinii 195
macrolepis 196
macrophallus 197
maculigula 199
maculiventris 200
magnaphallus 201
maia 202
mariarum 204
marmoratus 205
marsupialis 206
matudai 207
medemi 208
megalopithecus 209
megapholidotus 210
menta 211
meridionalis 212
microlepidotus 214
microtus 215

milleri 216
mirus 217
monteverde 218
morazani 219
muralla 220
nasofrontalis 221
naufragus 223
neblininus 224
nebuloides 225
nebulosus 226
neglectus 227
nemonteae 228
nicefori 229
notopholis 230
ocelloscapularis 231
omiltemanus 232
onca 233
orcesi 234
ortonii 235
otongae 237
oxylophus 238
pachypus 239
parilis 240
parvauritus 241
parvicirculatus 242
pentaprion 243
peraccae 244
peruensis 245
petersii 247
peucephilus 248
phyllorhinus 249
pijolense 250
pinchoti 251
planiceps 252
podocarpus 253

poecilopus 254
poei 255
polylepis 256
princeps 257
proboscis 258
propinquus 259
pseudokemptoni 260
pseudopachypus 261
pseudotigrinus 262
punctatus 263
purpurescens 265
purpurgularis 266
purpuronectes 267
pygmaeus 268
quercorum 269
quimbaya 270
richteri 272
riparius 273
rivalis 274
roatanensis 275
robinsoni 276
rodriguezii 277
rubiginosus 278

rubribarbaris 280
ruizii 281
sagrei 282
salvini 283
santamartae 284
savagei 285
schiedii 286
scypheus 287
sericeus 289
serranoi 290
sminthus 291
soinii 292
solitarius 293
squamulatus 294
subocularis 295
sulcifrons 296
tandai 297
taylori 298
tenorioensis 299
tequendama 301
tetarii 302
tigrinus 303
tolimensis 304

townsendi 305
trachyderma 306
transversalis 307
trinitatis 308
triumphalis 309
tropidogaster 310
tropidolepis 311
tropidonotus 313
uniformis 314
urraoi 315
ustus 316
utilensis 317
vanzolinii 318
vaupesianus 319
ventrimaculatus 321
vicarius 322
villai 323
vittigerus 324
wermuthi 325
williamsmittermeierorum 326
woodi 327
yoroensis 328
zeus 330

Plate Set 2: Bodies 331

APPENDIX 1: Evaluations of Validity for Some Species of *Anolis* 349

APPENDIX 2: Species Lists of *Anolis* by Country, Region, and Gestalt 356

Notes 362

References 370

Photographic Credits 409

Index 410

FOREWORD

Anoles! For an increasing number of biologists, these small to medium-sized lizards are the most fascinating organisms on planet Earth. They certainly have been the focus of copious and diverse studies. Beyond their aesthetic appeal, there are good reasons why so much research has focused on anoles. Anoles are active by day, and many are conspicuous, particularly when they signal with their flashy dewlaps ("throat fans") and climb on tree trunks and other vertical surfaces aided by adhesive pads on their digits ("fingers" and "toes"). Those two features alone invite investigation, but anoles have also speciated and diversified extensively to form a clade approaching 450 currently recognized species, distributed throughout the Neotropics. There are numerous variations on the anole theme, including many cases of convergent evolution. Moreover, some anoles adapt well to laboratory conditions, making them amenable to both lab and field studies. That fact, along with the accumulating information about the phylogeny and natural history of anoles, has facilitated deep dives into anole biology, such as genomics and developmental biology.

For a long time, the West Indian anole species have been much better known than their mainland relatives, which comprise more than half of the currently recognized species. That's not to say that all mainland species were poorly known. In fact, two of the most extensively studied species, *Anolis carolinensis* and *A. sagrei*, are mainland species, although both likely colonized the mainland of North America from Cuba (one naturally, the other with human assistance). Overall, however, there has been a much higher proportion of mainland species about which very little was known, particularly in Central and South America, including many species previously unknown to scientists. This situation likely resulted at least in part from the much larger area over which the mainland species occur, including many places that are still difficult to access, as well as from the seeming greater scarcity, or at least lesser conspicuousness, of anoles of many mainland species.

But the situation has recently been changing. The mainland anole fauna is rapidly becoming better known, not only in terms of taxonomy and natural history but also in the diverse other areas of biology (from behavior and community ecology to phenotypic plasticity and physiology) that are being studied using anoles. Although many biologists are responsible for this change, arguably no one has contributed more to mainland anole taxonomy and natural history than Steven Poe, Professor of Biology at the University of New Mexico. I've known Steve since I served as his advisor for a Smithsonian Predoctoral Fellowship (1997–1998) at the National Museum of Natural History, where I am a Research Zoologist and Curator of the amphibian and reptile collections. I've had the good fortune of advising many excellent predoctoral and postdoctoral fellows, and Steve Poe is among a select few with whom I am most proud to be associated. Steve spent his yearlong fellowship assembling a large morphological dataset for a comprehensive phylogenetic study of anoles. We later conducted fieldwork on anoles together in Cuba (in the company of several other now-prominent anole biologists) and have coauthored articles on the phylogeny and evolution of anoles (and the philosophy of phylogenetic inference). Through these shared experiences, I became well acquainted with Steve's ardor for anoles and his passion for fieldwork.

Steven Poe has most likely become the world's leading authority on mainland anoles in total. Steve has been studying anoles for over thirty years, since his undergraduate days at Harvard University (A.B. 1993), where he was mentored by the first godfather of anole biology, Ernest Williams. The full extent of his career has been devoted to anoles, through graduate school at the University of Texas at Austin, a postdoctoral fellowship at the University of California at Berkeley, and his current positions as a professor at the University of New Mexico and a curator in the Museum of Southwestern Biology. Steve has conducted fieldwork on anoles throughout the Neotropics, in both the West Indies and the mainland of North, Central, and South America, including Bolivia, Belize, Panama, Peru, Colombia, Costa Rica, Ecuador, Honduras, Guatemala, and Mexico, collecting thousands of specimens now available for further study in natural history museums. He has observed anoles belonging to 181 of the mainland species in their natural habitats, and he has discovered and named 17 species of anoles (with more to come). Steve, his graduate students, and other collaborators (from various anole-inhabited countries) have investigated diverse aspects of anole biology, including systematics (species

FOREWORD

delimitation and phylogeny), taxonomy and nomenclature (of both species and clades), convergent evolution, naturalization, genomics, signal evolution, adaptive radiation, and biogeography.

So here we have Steven Poe's long-anticipated opus on the anoles of mainland Central and South America. There hasn't been a comprehensive taxonomic treatment of mainland anoles since 1934, when Thomas Barbour published a briefly annotated list of the then-known species. Poe's book goes far beyond that, not only in covering more than twice as many species, but also in its depth of coverage. In addition to introductory chapters on general anole biology, methods for finding and capturing anoles, and how to identify anoles, Poe provides a detailed account for each species, including a photograph, distribution map, reference for the species name, type specimen(s), type locality, description of its organismal morphology, comparisons with similar species, verbal description of the geographic distribution, and information on its natural history.

This work promises to be a landmark resource for current and future anole biologists, not only in summarizing the current state of taxonomy and natural history for mainland anoles, but also in helping anole biologists find and identify mainland anoles, thereby promoting further studies on these fascinating lizards about which there is still much to be learned.

Kevin de Queiroz
Research Zoologist and Curator of Amphibians and Reptiles
National Museum of Natural History
Smithsonian Institution

ACKNOWLEDGMENTS

The following readers gave helpful reviews of chapters: Kevin de Queiroz (Introduction), Jonathan Losos (Biology), Bee Armijo (Introduction, Collecting), Fernando Ayala-Varela (species accounts), Rafael Moreno (species accounts), Luke Mahler (species accounts).

The following workers kindly contributed photos. Each photo is credited as it appears (photos directed and financed by me, including those of laboratory specimens and those taken on field trips [usually dewlaps], may not be credited to the photo-taker): Esteban Alzate, César Barrio Amorós, Chris Anderson, Tito Barros, David Bejarano, Alejandro Calzada, Rosario Castañeda, Colin Donihue, Marco Antonio de Freitas, Rich Glor, Diego Gómez, José Luis Pérez González, Freddy Grisales, Anthony Herrel, Juan Pablo Hurtado Gómez, Roberto Langstroth, David Laurencio, Edgar Lehr, José Daniel Lara-Tufiño, Ian Latella, Jonathan Losos, D. Luke Mahler, Guido F. Medina-Rangel, Aurelien Miralles, Andrew Odum, Luis A. Rodriguez, Miguel Trefaut Rodrigues, Fabio Olmos, Janson Jones, Gilson Rivas, Paolo Sampaio, Ivan Prates, Mason Ryan, Santiago Ron, Omar Torres, Hermes Vega, Pablo Venegas, Oliver Komar, Rick Stanley, Rich Sajdak, Bob Powell, Juan Daza, John Sullivan, Adan Bautista-del Moral, Josue Ramos Galdamez, Michael Kielb, Alberto Nadal, Josh Vandermeulen, Gunther Köhler, Delmer Jonathan, Zsombor Károlyi. Special thanks to Tom Kennedy for his numerous excellent photos and exemplary fieldwork that contributed to the production of this book.

Numerous researchers helped with fieldwork and observations that contributed to this work: Bee Armijo, Juan Carlos Chaparro, Caleb Hickman, Mason Ryan, Eric Schaad, Julian Davis, Erik Hulebak, Chris Anderson, Joseph Barnett, Devon Graham, Levi Gray, Jenny Hollis, Sofia Nuñez, Ian Latella, Andrés Quintero-Angel, Julie Ray, Juan Salvador, Andrew Crawford, Martha Calderón-Espinosa, José Luis Pérez González, Adrian Nieto, Omar Torres, Brad Truett, Fernando Ayala-Varela, Beto Rueda-Solano, Carlos Vasquez, Julian Velasco, Christian Yañez-Miranda, Alvaro Juan Aguilar Kirigin, Rafael Moreno-Arias, Miguel Angel Méndez Galeano, UNM Herpetology classes, Juan Daza, Natalie Blea, Simon Scarpetta, Heather MacInnes.

Museum workers provided specimens and hosted visits that benefited this book: Jose Rosado, Jonathan Losos, Joe Martinez, Chris Phillips, Dan Wylie, Toby Hibbits, Lee Fitzgerald, Alan Resetar, Rafe Brown, Federico Bolaños, Roberto Ibañez, James Aparicio, Juan Daza, Kevin de Queiroz, David Kizirian, Neftali Camacho, David Wake, Marvalee Wake. Special thanks to Luke Mahler, who provided invaluable information from his reviews of Colombian anole collections and other type material.

Funding for collection of some of the data that appear in this book was provided by the National Science Foundation and the University of New Mexico.

Thanks to Karen Carter, Robert Kirk, and Megan Mendonca at Princeton University Press for their help and patience. Thanks to Annie Gottlieb for her sterling copyediting and to Jen Burton for creating the index.

This work would not have been possible without the help of Joseph Barnett and Chris Anderson, who prepared maps and compiled literature, respectively, that were essential to the book. My parents, Sally and David Poe, have been extraordinarily supportive throughout my life and provided critical childcare during book completion. Thanks especially to Bee Armijo for her help and support throughout this work.

INTRODUCTION

— SCOPE AND ORGANIZATION —

This book is intended to serve as both an identification guide and a biological and taxonomic reference for *Anolis* lizards, commonly called anoles, found in mainland America.

Chapter 1 describes the anatomy, ecology, and evolution of mainland anoles. This chapter is necessarily brief in its coverage, as a comprehensive treatment of mainland anole biology is well outside the scope of this book. Jonathan Losos's (2009) excellent summary remains the starting point for any student wishing to investigate anole biology in depth.

Chapter 2 gives a detailed treatment on how to find and catch anoles. I cover daytime and nighttime techniques, but focus heavily on night work, as I have found nighttime searching to be most efficient for securing the greatest number and diversity of anoles, especially rare species. This chapter draws mainly from my own experience with 275 (181 mainland) species of anole in nature. My lab at the University of New Mexico has long concentrated our research on anoles, so we have tailored our fieldwork specifically toward finding anoles.

Chapter 3 describes approaches to identifying anoles in hand. I describe both field and laboratory techniques and include some discussion of the relative utility of competing approaches and the diagnostic potential of alternative traits. As in chapter 2, I rely heavily on traits and techniques found to be most useful in my own field and lab research, with exceptions made for species for which I lack experience. I emphasize practicality over infallibility. For example, some trait of internal anatomy may be foolproof for species identification of some form, but this trait may not be easily accessed.

Chapters 1 through 3 may be read as stand-alone treatments or as background information for the species accounts of chapter 4. Each account is composed of biological and taxonomic information for a species of mainland anole, including sections covering the original description of the species, morphology, range, natural history, and tips for distinguishing the species from potentially sympatric anole species. Although not informationally comprehensive, these accounts may serve as a point of departure for any study of the biology of a given mainland anole species. In addition to summarizing species biology, the accounts are intended to provide enough information for accurate identification of mainland anole species in the field. A more detailed presentation of the material in chapter 4 is at the end of this Introduction.

— THE FAUNA: MAINLAND ANOLES —

The faunal purview of this book includes all anole species with established breeding populations on mainland North, Central, and South America and immediately surrounding islands. I treat "naturalized" species such as *Anolis sagrei* in Florida no differently than "native" mainland forms.

The anoles that inhabit the American mainland share a suite of traits that make assignment to *Anolis* straightforward. Males and some females of all mainland anole species display an extensible signaling organ under the neck called a dewlap (fig. 0.1a). This dewlap is anatomically distinctive and easily distinguished from similar structures in iguanas and other lizards. Also, all individuals of almost all species of mainland anole possess sets of laterally expanded specialized scales under the digits called toepads (fig. 0.1b). Although easily differentiated from cohabitating lizard species, mainland anoles are not a natural group. That is, there are mainland anole species that are more closely related to Caribbean island forms than to some other mainland species. In fact, there are two independent evolutionary radiations composed mainly of mainland anoles, formally designated *Dactyloa*[1] for the clade[2] that dispersed from South America northward as far as Costa Rica and *Draconura* for the clade that invaded Central America from the Caribbean and migrated south (Poe et al. 2017a; see further discussion in chapter 1).[3] Individuals of these radiations live in sympatry[4] throughout most of Central and South America. Although there is no external character that consistently differentiates *Dactyloa* and *Draconura* anoles, some trends are evident. *Dactyloa* includes more large species (>100 mm body length) and more species that possess large dewlaps in females as well as males. Well-developed female dewlaps are uncommon in *Draconura*, and body length is seldom greater than 80 mm.[5]

INTRODUCTION

FIG. 0.1 a) *Anolis fraseri* displaying dewlap (photo by Tom Kennedy). b) Underside of anole foot showing expanded lamellae on toes.

Given the lack of evolutionary unity in mainland anoles, why study them as a single group? The obvious answers to this question are taxonomic tradition and geographic convenience, but there are more substantive reasons. First, the morphological similarity coupled with geographic proximity of *Draconura* and *Dactyloa* anoles begs for a common identification reference for these groups. Second, the ecological similarity of anoles in these two groups indicates shared niches, or ecological roles, in the environment. Third, scientific research on mainland anoles tends to treat draconuran and dactyloan anoles together, a point to which I return in chapter 1.

So if mainland anoles are so similar to each other and each large mainland lineage has closer relatives in the Caribbean, then why treat the mainland anoles separately from the island forms? Setting aside practical geographic reasons, there do appear to be biological differences between mainland and island anole faunas. In particular, mainland species tend to be less abundant (Buckley and Jetz 2007), some morphological and ecological types tend to be restricted to either mainland or island environments,[6] and anole assemblages[7] tend to be less morphologically diverse on the mainland relative to the islands (Poe and Anderson 2019). The study of faunal differences in mainland and island environments is ongoing in anoles and other organisms.

Many scientists, including myself, do study both mainland and island anoles. But island anoles have been much more intensively studied than mainland forms, especially for ecological and comparative multispecies work. I will return to the study of mainland vs. island anoles in chapter 1. For now, I note that this research imbalance was one of the impetuses for this book. Some outstanding resources notwithstanding,[8] there remains a need for a comprehensive guide to the mainland anoles, especially with regard to the complex anole faunas of Mexico, Panama, and Colombia. Hopefully this book will help fill that void.

— THE UNITS OF ANALYSIS: SPECIES —

The first step in constructing a guide to some set of species is to decide which species to include. This decision requires either accepting whatever public list seems most reliable, or evaluating each purported species of the studied group based on an operationalization of some species concept. Being an evolutionary biologist, a longtime student of anole taxonomy, and a frequent arbiter of anole species decisions via journal reviews and species lists, I have chosen the latter option. My adopted species concept is that which is most accepted among theoretical systematists today: the idea of species as evolutionary lineages. Originating with George Gaylord Simpson (1951), fleshed out by Edward Wiley (1978), and expanded by Kevin de Queiroz (1998), this Evolutionary Species Concept (ESC)[9] views vertebrate species as groups of individuals that live and breed with some spatial independence and temporal continuity relative to other such groups. Put another way, species are groups of individuals that undergo evolution as a unit.

You might notice that this well-accepted species concept gives no prescription for actually deducing, in practice, which groups of individuals form species and which do not.[10] There is a vast and ever-expanding literature on approaches to species identification (see, e.g., Flot 2015), but much of it addresses near-ideal conditions of broad geographic sampling and multigene datasets. The practice of species identification is advancing rapidly, and nowadays many published species descriptions include copious and diverse molecular data. But the fact remains that nearly all species currently recognized were identified without reference to the molecular tools that are justifiably popular today.

In this book, I have generally accepted species as valid if they are geographically continuous and supported by two relatively independent lines of evidence. I consider morphological structural traits (e.g., scale counts), dewlap color, hemipenes structure, and genetic similarity to be roughly independent lines of data. Thus, a hypothesized anole species with a substantial genetic distance[11] from other species and a uniquely colored dewlap is accepted as valid, whereas one with a slight difference in scale structure relative to nearby populations but no other distinguishing traits is not. These principles are adopted primarily in an attempt to reflect evolutionary reality, but also with an eye toward practicality in the field. That is, for example, if two geographically contiguous purported species that are diagnosed by some minor genetic distance are indistinguishable to herpetologists in the field, there seems little point in separating those "species." Still, as the goal of identifying species is to represent evolutionary reality rather than to convenience humans, these criteria are not embraced rigidly. In special cases I have made exceptions to these rules,[12] especially when molecular data present compelling evidence of "cryptic" species.[13]

This approach results in the absorption of some commonly recognized species within other species—that is, the taxonomic "sinking" of one name as a junior synonym of another. In many of these cases, the sunken name is likely to survive upon detailed and geographically broad taxonomic studies if the type localities[14] of named forms are geographically distant. This being the case, why not just leave things be? First, these species names are being sunk in many cases at least in part because the current conception of the species is indistinguishable from some other species.[15] Second, newer uses of sunken names may not resemble previous conceptions of them. For example, the name *Anolis cryptolimifrons* (type locality in Bocas del Toro, Panama) may eventually apply to all *limifrons*-like anoles on the Caribbean versant of Panama. However, that usage would differ significantly from the current conception of *A. cryptolimifrons*, i.e., as a small coastal population externally indistinguishable from *A. limifrons* that interbreeds with *A. limifrons* and is geographically surrounded by it.

— THE REGION: MAINLAND LATIN AMERICA —

The prehuman range of anoles is from the southeastern United States and northern Mexico south to Paraguay[16] (fig. 0.2). This area inhabited by anoles is vast and physiographically complex, including the world's longest river and mountain chain and multiple regions of endemism. Virtually all parts of this range that are not too high or too dry are home to anoles (fig. 0.3). Some pockets of semixeric and/or temperate habitation notwithstanding, nearly the entire anole range is considered tropical both in terms of location between the Tropic of Cancer and the Tropic of Capricorn and with regard to the warm and humid areas within this region to which anoles are restricted.

According to Köppen-Geiger classification[17] (Beck et al. 2018), the anole range is dominated by Tropical Rainforest and Tropical Monsoon climates, with some Tropical Savannah occupied as well (fig. 0.4). In each of these climates, temperatures are high and relatively constant across seasons, so local temperature differences are mainly due to elevation. Rainfall is plentiful in each Tropical climate, but seasonality and volume of rainfall differs. Tropical Rainforests are largely aseasonal, whereas Tropical Savannahs have a pronounced dry season. Tropical Monsoon climates display seasonal rainfall patterns but are wetter than Tropical Savannahs. Many anole species are restricted to particular climate regions, an effect that is especially evident in Central America, where several otherwise widespread lowland species are absent from Tropical Savannah environments.

It is convenient to discuss the physiography of the mainland anole range separately for South and Middle America.

INTRODUCTION

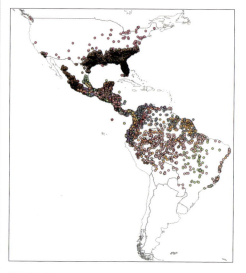

FIG. 0.2. Distribution map for *Anolis* based on museum records, iNaturalist, and my observations. Points in the northern United States are observations from iNaturalist that are not likely to represent breeding populations.

BELOW: FIG. 0.3. Prime roadside anole habitat: pasture near Mindo, Ecuador (upper left); undisturbed forest near Cusuco National Park (upper right). Prime trailside anole habitat: Cerro Dantas, Costa Rica (lower left); El Copé park, Panama (lower right). All photos by Tom Kennedy.

4

— THE REGION: MAINLAND LATIN AMERICA

SOUTH AMERICA

The geography of South America within the anole range is dominated by the Andes mountains to the west and the Amazon and Orinoco river basins centrally and to the east (fig. 0.5). These features provide opposite biogeographic effects, with the topographic complexity of the Andes serving to separate numerous vicariant[18] anole lineages and the uniformity of Amazonia harboring a relatively depauperate complement of anole species, each with a broad distribution.

The Andes extend from northern Venezuela south and west along the Pacific side of the continent, rising above 4000 meters in some areas to form an effective biogeographic barrier for anoles and other organisms. Indeed, the anole faunas of Amazonia and the Pacific Coast are nearly entirely distinct. River valleys within the Andes create additional topography and fertile biogeographic grist

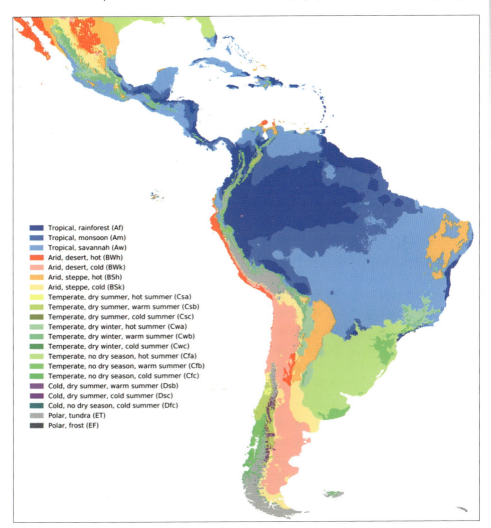

FIG. 0.4. Köppen-Geiger temperature/rainfall categorizations of the region occupied by *Anolis*. Anoles are found in Tropical Rainforest, Tropical Monsoon, and Tropical Savannah habitats, i.e., the blue areas of the map.

5

INTRODUCTION

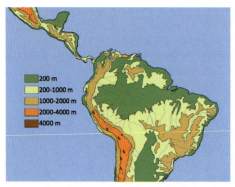

FIG. 0.5. Elevation map of Middle and South America (mapsland.com).

for speciation. The southwestern Cauca and northeastern Magdalena River valleys split the Colombian Andes into Cordilleras Oriental to the east, Central between the valleys, and Occidental to the west. The geographic complexity of this region undoubtedly contributes to Colombia's country record of 74 anole species.[19] Much of the remaining intra-Andean topography to the south, such as the Peruvian-Bolivian altiplano, is too high to support anoles, but several drainages from the Andes engender biogeographic patterns in anoles.[20]

The physiographic complexity of the Colombian Andes fades to the south in Andean Ecuador, Peru, and Bolivia, where the mountain range forms a single barrier with faunally distinct east ("cis," Amazonian) and west ("trans," Pacific) sides. The dry Pacific regions of Bolivia and Peru lack anoles, but the tropical forested Pacific slope of Ecuador harbors some of the most species-rich anole assemblages, including spectacular forms such as *Anolis proboscis* and *A. fraseri*. The scientific field station Rio Palenque, an 87-hectare plot of forest at 177 meters elevation surrounded by kilometers of pasture and sap fields in central Pacific Ecuador, is home to at least 11 sympatric anole species.

The higher anole-hospitable elevations of the Amazonian slope of the Andes from Colombia to southern Peru include a standard complement of geographically replacing species. There generally is a large green or complexly patterned form (e.g., *Anolis cuscoensis* in central Peru, *A. soinii* in southern Ecuador and northern Peru), one or two "twig"[21] anole species (e.g., *A. williamsmittermeierorum* and *A. peruensis* in central Peru, *A. lososi* and *A. hyacinthogularis* in southern Ecuador) and, at lower elevations, some additional common Amazonian species such as *A. fuscoauratus*.

North and east of the Andes, respectively, are the other major South American mountain formations, the Sierra Nevada de Santa Marta[22] (SNSM) of extreme northern Colombia and the Guiana Highlands, part of the Guiana Shield of Venezuela. Each of these highland regions includes its own endemic[23] anole fauna. The SNSM includes the highest peak in Colombia and is home to three apparently parapatric[24] "solitary"[25] anole species, *Anolis solitarius*, *A. menta*, and *A. santamartae*.[26] The tepuis, isolated "table-top" mountains of the Guiana Highlands, harbor endemic anoles *A. neblininus* (Neblina tepui), *A. carlostoddi* (Abacapá tepui), and *A. bellipeniculus* (Cerro Yavi).

The only other area of the South American anole range with significant relief is the Brazilian Highlands. Covering most of the southern part of Brazil, this network of plateaus and craggy outcrops averages about 1000 meters in height but includes peaks approaching 3000 meters. The savannahs and shrublands that make up the Brazilian Highlands harbor few anoles, but Amazonian forms are found at the north, west, and east fringes of these areas, and the Atlantic coastal forest of Brazil is home to three endemic species with apparently small ranges: *Anolis nasofrontalis*, *A. pseudotigrinus*, and *A. neglectus*.

The vast Amazon basin extends from the foothills of the Andes to the Atlantic Ocean and south to Bolivia, covering over a third of the South American continent. This massive land area is climatically and elevationally relatively homogeneous tropical rainforest. Consequently, localities in Amazonia contain a fairly consistent complement of sympatric anole species: *Anolis punctatus*, *A. transversalis*, *A. ortonii*, *A. fuscoauratus*, and one or two species of anole similar to *A. chrysolepis*.[27] At some localities, one additional rare species may be found: *A. caquetae*, *A. phyllorhinus*, or *A. dissimilis*.

Finally, numerous offshore islands are home to species complements with South American roots. Distant Pacific rocky Malpelo supports only the curious island endemic *Anolis agassizi*, whereas Gorgona, 30 km from the mainland near Buenaventura Colombia, harbors an anole assemblage of two endemic (*A. gorgonae*, *A. medemi*) and three more widespread (*A. parvauritus*, *A. princeps*, *A. purpurescens*) species.

MIDDLE AMERICA

Much of Middle America is characterized by central highlands creating a geographic barrier between wet Pacific and wetter Caribbean forests (fig. 0.5). This geography creates relatively distinct Caribbean, Pacific, and highland anole faunas in Mexico and much of Central America. Peaks in Central America are lower than those in the Andes and northern Mexico, and mountain ranges are not continuous. Consequently, lowland anole endemism in Central America is not as extreme as in northern Mexico or in Peru, Ecuador, and southern Colombia, where the number of shared Pacific and Amazonian anole species is zero. As in South America, highland anole species in Middle America tend to have restricted ranges, and lowland forms tend to be widely distributed, in some cases throughout nearly the entire region. *Anolis capito* ranges from at least Darién, Panama, to central Mexico, on both Caribbean and Pacific slopes in Panama and Costa Rica, and only on the Caribbean slope from Honduras north to Mexico.[28] These general patterns notwithstanding, Middle America is physiographically diverse, with each country displaying its own idiosyncratic topography and anole fauna.

The narrow land bridge of Panama connects South and Central America from east to west in a flattened S-shape. A central depression at the Panama Canal separates the high Cordillera Central

FIG. 0.6. Ten species of anole found within 200 meters of the field station over two nights at Omar Torrijos park in western Panama. *Anolis ibanezi* (large green, top left), *A. capito* (large brown, top center), *A. biporcatus* (large green, top right), *A. frenatus* (largest green, left center), *A. lionotus* (unpatterned brown, upper center), *A. elcopeensis* (small brown, leftmost), *A. pentaprion* (banded gray-brown, middle), *A. kunayalae* (blue-green, right center), *A. limifrons* (small brown, bottom left), *A. humilis* (small brown, bottom center).

7

INTRODUCTION

that extends into Costa Rica in the west from lower-elevation ranges to the east. The Darién range runs along the Caribbean coast northwest from Colombia, the Majé range runs parallel to the western vestiges of the Darién range just east of the canal, and the Jungurudó range occurs southwest of the highest parts of the Darién range near the Colombian border. There is some evidence for evolutionary differentiation associated with these lower eastern ranges (e.g., Batista et al. 2015), but most anole species present in eastern Panama occur across all surveyed elevations.[29]

The low Isthmus region of Panama does not appear to have been a major biogeographic barrier for anoles. We[30] (Poe et al. 2017a) estimated 18–20 evolutionary crossings of the Isthmus by anole lineages, including some as early as 30 million years ago—earlier than any seriously entertained estimate of Isthmus closure.[31]

Western Panama is dominated by the Cordillera Central. The significant elevation of this range, up to 1874 meters at Volcán Barú, creates a distinct anole fauna at its upper reaches and diverse assemblages at middle elevations. Omar Torríjos (El Copé) park in west-central Panama at 900 meters elevation has, at 13 species (fig. 0.6), the highest documented local diversity of any mainland site, and possibly of any anole site. Relatively low mountain saddles such as the Fortuna pass apparently preclude species-level differentiation between Pacific and Caribbean anole faunas in Panama, as many lowland species are found on both sides of the Cordillera Central.[32] The distinct highland anole fauna of western Panama includes several species similar to *Anolis tropidolepis*, some of which are sympatric (Lotzkat et al. 2011; personal observation), as well as spectacular giant anoles such as *A. ginaelisae* and *A. kathydayae*.

Southern Costa Rica displays a relatively straightforward geography of a Central Cordillera, the Talamanca range, extending into Panama and separating Pacific and Caribbean lowlands. Northern Costa Rica is more geologically complex, with active and dormant volcanoes punctuating discontinuous highlands. From south to north these northern ranges are the Cordillera Central, which is nearly continuous with the Talamanca range; the smaller Cordillera Tilarán; and the isolated volcanic peaks of the Cordillera de Guanacaste.[33]

Anole species in Costa Rica may be distributed exclusively in the highlands (e.g., species similar to *Anolis tropidolepis*), throughout the lowlands (e.g., *A. humilis*), restricted to the Caribbean versant (e.g., *A. frenatus*), or restricted to the Pacific versant (e.g., *A. polylepis*). Some species that previously were considered to inhabit the entire lowlands have been found to constitute different species on each side of the cordillera. For example, the Caribbean version of *A. pentaprion* was separated as *A. charlesmyersi* (Köhler 2010), and southwestern *A. savagei* was previously considered part of *A. insignis* (Poe and Ryan 2017). Other widely distributed lowland species may be similarly divided (see, e.g., the molecular divergences between populations of *A. humilis* in Phillips et al. [2015]). Localities for *A. insignis* near the Nicaraguan border mark the northern extent of the *Dactyloa* clade of anoles.

North of Costa Rica, Nicaragua is composed of a broad eastern plain, northern low mountains, and a narrow strip of Pacific lowlands. The eastern plain, which includes the southern aspect of the narrow Miskito (Mosquito) Coast, is mostly tropical forest but includes open savannahs to the northeast. The Pacific Coastal region includes the two largest lakes in Central America, Lago de Managua and Lago de Nicaragua, and a string of volcanoes extending from the Golfo de Fonseca at the country's northern border south to Lago Nicaragua. The central highlands are continuous with the central mountains of Honduras but are lower, seldom reaching above 2000 meters.

Excluding the island form *Anolis villai*, Nicaragua shares its entire anole fauna with Costa Rica, Honduras, or both countries. The only exclusively highland forms in Nicaragua are *A. laeviventris* and *A. wermuthi*. Other anole species are found countrywide or only absent from the northwest.[34]

The region of Honduras and El Salvador is composed of a dominant central highland area and Pacific and Caribbean lowlands that include the northern extent of the Moskitia. The central highlands are divided by a depression running north to south from the Caribbean to the Gulf of Fonseca. Several peaks over 2000 meters are separated by numerous low valleys to create a complex stage for faunal differentiation. Indeed, Honduras is home to numerous endemic anole species known only from particular peaks or ranges in the country. The *crassulus*-group anoles especially follow this pattern, including *Anolis muralla*, known only from La Muralla National Park; *A. rubribarbaris* from Montaña de Santa Bárbara National Park; *A. morazani* from Montaña de Yoro National Park; and *A. amplisquamosus* from Cusuco National Park.

Several anole species are found in Honduras only in the eastern region, including northern Central American forms such as *Anolis rodriguezii*, which reaches its southern range limit in northeast Honduras, as well as predominantly southern species such as *A. oxylophus* found in Honduras only in the southeast Moskitia region. The Pacific lowland anole fauna of this region includes two species with ranges that extend farther north along the coast, *A. serranoi* and *A. macrophallus*, and a species of the *A. sericeus* complex. The Bay islands off northern Caribbean Honduras harbor island endemic forms *A. bicaorum*, *A. roatanensis*, *A. utilensis*, and *A. allisoni*,[35] as well as the more widespread *A. sericeus*, *A. lemurinus*, and *A. sagrei*.

North of Honduras, the Yucatan peninsula and the area along the Gulf of Mexico north to the Isthmus of Téhuantépec form a biogeographically continuous lowland inhabited by a series of broadly distributed anole species. Among these, *Anolis biporcatus*, *A. capito*, *A. beckeri*, and *A. uniformis* are absent from the dry sandy regions of Yucatan and Campeche States forming the extension of the peninsula, whereas xeric-tolerant forms *A. lemurinus*, *A. rodriguezii*, *A. ustus*, and *A. tropidonotus* are found throughout this region north of the highlands of Guatemala. *Anolis sagrei* inhabits coastal areas here.

Central Guatemala is composed of roughly parallel east-west mountain ranges that extend into southern Mexico, the western Cuchumatanes range and eastern Minas range to the north, and the southern Sierra Madre. Some anole species appear restricted to one of these ranges, such as *Anolis cobanensis* and *A. campbelli* in the northern range. Others, such as *A. laeviventris*, are widely distributed throughout the Guatemalan highlands and north into Mexico.[36] The southern Sierra Madre slopes to the coast, where a largely regionally endemic anole fauna is shared across Guatemala and southern Mexico. Pacific lowland endemics *A. serranoi*, *A. cristifer*, *A. macrophallus*, and, at higher elevations, *A. dollfusianus* coexist with more widespread forms *A. sericeus* and, at higher elevations, *A. laeviventris* and *A. petersii*.

The eastern mountains of Guatemala extend west into Mexico, becoming the Chiapas Central Highlands to the north and the Sierra Madre of Chiapas in the south. These ranges surround a central depression to the east and fragment with low valleys to the west, terminating in the mountainous Chimalapas region at the eastern edge of the Isthmus of Téhuantépec. The high areas of this region include several geographically separated members of the *Anolis schiedii* group: *Anolis campbelli* in Montebello National Park, *A. hobartsmithi* on the northernmost slopes of the Central Highlands in Chiapas, *A. matudai* in the Sierra Madre, *A. parvicirculatus* north of Ocozocuautla, and *A. cuprinus* in the Chimalapas. Several ecologically specialized species inhabit low to moderate localities in the Chimalapas-Isthmus region: cave specialist *A. alvarezdeltoroi* on the eastern Chimalapas slope, semiaquatic form *A. purpuronectes* west and north of the Chimalapas, ground anoles *A. compressicauda* and *A. pygmaeus* from east of the Chimalapas across the Isthmus, and semiaquatic anole *A. barkeri* from north of the Isthmus to Los Tuxtlas. Other species in this species-rich area are restricted to the Pacific lowlands (see above) or are widespread highland (e.g., *A. laeviventris*) or Caribbean lowland forms (see above).

Mexico north of the Isthmus of Téhuantépec includes distinct Pacific and Caribbean anole faunas. The Caribbean fauna is composed of the widespread forms discussed above as part of the Yucatan fauna, *Anolis duellmani*, and high-elevation species similar to *A. schiedii* (from south to north) *A. rubiginosus*, *A. milleri*, *A. cymbops*, *A. schiedii*, and *A. naufragus*. Excluding *A. carolinensis*, *A. naufragus* represents the northernmost prehuman extent of *Anolis* along the Caribbean coast.

Excepting widespread *Anolis sericeus*, all *Anolis* of Pacific Mexico west of the Isthmus of Téhuantépec are endemic to that region. These include *Anolis boulengerianus* and *A. immaculogularis* near the Isthmus and *A. dunni*, *A. gadovii*, *A. liogaster*, *A. macrinii*, *A. megapholidotus*, *A. microlepidotus*, *A. nebuloides*, *A. nebulosus*, *A. omiltemanus*, *A. peucephilus*, *A. quercorum*, *A. subocularis*, and *A. taylori* farther north.[37] *Anolis nebulosus* currently has a particularly large range, extending from at least the middle of Oaxaca north into Sonora to reach the northernmost extent of Pacific *Anolis* in Mexico.

INTRODUCTION

— HISTORY —

The study of mainland anoles is nearly as old as the study of lizards. Carl Linnaeus and some pre-Linnaean workers seem to have studied *Anolis carolinensis* and/or similar anoles, but the species identities of their study subjects are murky (see Vance 1991). The first scientific work that clearly discussed anoles was volume 4 of Francois Daudin's (1802) *Histoire naturelle, générale et particulière, des reptiles* (fig. 0.7).[38] This work gave a morphological diagnosis of the genus and divided anoles into two groups based on tail shape. Among mainland-associated species still recognized today, *Anolis lineatus* was placed in the *Première Section* of species with compressed tails, while *A. auratus* and *A. punctatus* were grouped with the cylindrical-tailed species in the *Deuxième Section*.

Daudin's work was followed by additional descriptive work on mainland anoles. The early history of mainland anole alpha taxonomy[39] mirrors that of vertebrate taxonomy in general. The primary anole describers of the 1800s and early 1900s—A.M.C. Duméril, W. Peters, E. D. Cope, G. A. Boulenger—each described hundreds of species of reptiles, mammals, birds, and fish, in addition to anoles (fig. 0.8). These authors, each associated with major natural history museums,[40] are responsible for the early spike in newly described anoles in the second half of the nineteenth century (fig. 0.9, fig. 0.10).

Although many species descriptions resulted from the personal fieldwork of these authors,

FIG. 0.7. Page from Daudin (1802) showing initial description of *Anolis*.

FIG. 0.8. Prolific describers of *Anolis* species include, from left to right, Ernest Williams (from the Ernst Mayr Library and Archives of the Museum of Comparative Zoology, Harvard University), Edward Cope, George Boulenger, and Gunther Köhler.

especially Cope and Peters, many more were due to collections from throughout the range of *Anolis*, donated or sold to these authors' museums.[41] Thus, for example, George Boulenger, curator of the British Museum, described *Anolis tropidolepis* from the mountains of Costa Rica, *A. wattsi* from the Lesser Antilles, and *A. notopholis* from Pacific coastal Colombia. Much material for these descriptions came from collectors on expeditions supported by natural history museums.[42] Anole specimens also resulted from surveys associated with human development such as the creation of the Panama Canal (Schmidt 1933), as well as from field research in other disciplines such as archaeology (e.g., Stuart 1958).

In the latter part of the twentieth century, much of the next phase of mainland anole research came from workers specializing in particular regions such as Hobart Smith[43] in Mexico, Edward Taylor in Costa Rica, Ernest Williams in South

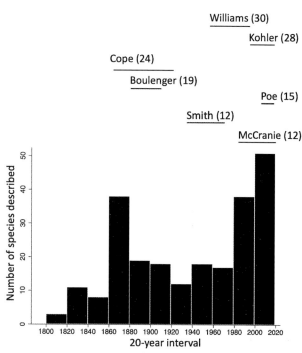

ABOVE: FIG. 0.9. Number of mainland anole species described in each 20-year period that are still valid today, including number of species described by some authors.

BELOW: FIG. 0.10. Cumulative number of mainland anole species described per year that are valid today.

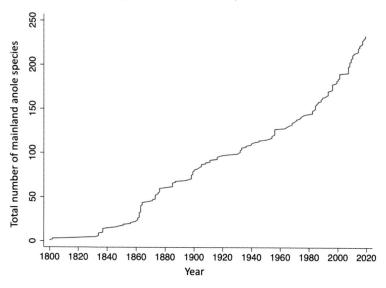

INTRODUCTION

America,[44,45] James McCranie in Honduras, and Fernando Castro and Stephen Ayala in Colombia. This work led to regional checklists and monographs summarizing *Anolis* faunas either exclusively (e.g., Stuart [1955]: Guatemala; McCranie and Köhler [2015]: Honduras; Ugueto et al. [2007, 2009]: Venezuela) or as components of broader herpetological treatments (e.g., Taylor [1956], Savage [2002]: Costa Rica; Smith and Taylor [1950]: Mexico; Ayala [1986]: Colombia; Avila-Pires [1995]: Amazonia; Peters and Donoso-Barros [1970]: Americas). Williams (1976a) listed all known South American anole species, and Köhler (2003, 2008) summarized Central American forms.

As with any other animal group, ecological and multispecies studies of mainland anoles only became practicable after taxonomy stabilized. Aside from some observations of *Anolis carolinensis* (e.g., Trowbridge 1937) and the inclusion of the occasional *Anolis* species in broader studies (e.g., Park 1938), ecological and comparative studies of mainland anoles did not materialize until the second half of the twentieth century.[46] Early ecological research on mainland anoles was undertaken by Owen Sexton and collaborators on *A. limifrons* and other common species in Panama (Sexton et al. 1963; Sexton 1967; Ballinger et al. 1970); Thomas Jenssen (1970a, b) on *A. nebulosus* in Mexico; Henry Fitch and collaborators on several anole species in Costa Rica and other mainland sites (Echelle et al. 1971, Fitch 1972); Robin Andrews and collaborators on *A. polylepis* (Andrews 1971) and other Central American anoles (Andrews and Rand 1974; Scott et al. 1976; Andrews 1979b); Craig Guyer (1986a, b; 1988a, b) on *A. humilis*; and Stan Rand and Stephen Humphrey (1968) on a lizard community including anoles in Brazil. Laurie Vitt (e.g., Vitt et al. 1995) led a series of papers describing the autecology of common mainland anoles. Andrews (1976, 1979a) was the first to compare the natural history of mainland and island anoles.

The past forty years has produced an explosion of research on mainland anoles. The recent increase in species descriptions of mainland anoles (figs. 0.9, 0.10) likely is due to the greater number of practicing herpetologists, more accessible museums and natural areas, and the development of new data and analytic tools for identifying "cryptic" species. Ernest Williams, the deceased longtime Curator of the Museum of Comparative Zoology, who worked mainly in South America[44] (and the Caribbean), and Gunther Köhler, current Curator at the Senckenberg Museum, concentrating in Central America, and their students and collaborators have been instrumental in increasing the number of species of mainland anole (fig. 0.8).

Evolutionary, ecological, and distributional studies of mainland anole species continue unabated, with much of this work spearheaded by Latin American herpetologists. Rafael Moreno-Arias and Martha Calderón-Espinosa (2015) of the National University of Colombia applied the ecomorph concept of correlated ecology and morphology to mainland anoles, following early work by Pounds (1988) and Irschick et al. (1997), and have contributed other important works on mainland anoles (e.g., Méndez-Galeano and Calderón-Espinosa 2017). Julián Velasco of the Universidad Nacional Autónoma de México (UNAM), previously of University of Valle in Colombia, has employed mainland anoles as a study system in his cutting-edge comparative macroevolutionary work (e.g., Velasco et al. 2020). Fernando Castro of the University of Valle and Jhon Tailor Rengifo of the Technological University of Chocó have greatly increased our knowledge of the ecology and distribution of Colombian anoles (e.g., Castro 1988; Rengifo et al. 2015).[47] Adrian Nieto Montes de Oca of UNAM continues to elucidate taxonomic and phylogenetic issues in Mexican anoles (e.g., Nieto Montes de Oca 2001; Nieto Montes de Oca et al. 2013). Ivan Prates, Teresa Ávila-Pires, Miguel Trefaut Rodrigues, and Annelise D'Angiolella produce model phylogenetic work incorporating Brazilian anoles (e.g., Prates et al. 2015, 2018; D'Angiolella et al. 2016). Gilson Rivas, Tito Barros, and collaborators continue to expand our knowledge of Venezuelan species (e.g., Ugueto et al. 2009; Rivas et al. 2012). The reptile section of the Pontificia Católica University in Quito, led by Omar Torres and Fernando Ayala, has become the center for anole research in Ecuador (e.g., Torres-Carvajal et al. 2019; Ayala-Varela et al. 2014). Roberto Ibañez has performed and facilitated anole research in Panama through his position at the Smithsonian Tropical Research Institute (e.g., Arosemena and Ibañez 1993; Hulebak et al. 2007), as has Federico Bolaños at the University of Costa Rica (e.g., Márquez-Baltán et al. 2005).

Research on mainland anoles is still in its incipience. A comprehensive estimate of the species diversity of mainland anoles is a distant goal, and basic ecological information is lacking for most known species. New species will be discovered by molecular analyses of widespread forms (e.g., Hofmann and Townsend 2017) and by exploration of poorly surveyed areas. Many widespread and morphologically variable forms are likely to harbor cryptic species (Hofmann et al. 2018), including, for example, *Anolis biporcatus* (Armstead and Poe 2017; Phillips et al. 2019), *A. auratus* (Calderón-Espinosa et al. 2014), *A. tropidolepis* (Köhler et al. 2014c), and *A. limifrons* (Hulebak 2008). Ecological studies of local anole assemblages, as in Moreno-Arias et al. (2020), are warranted across the mainland anole range, as are autecological studies such as those by Vitt and collaborators (e.g., Vitt et al. 2002) for the majority of mainland anole species.

— DATA —

The data that make up the bulk of the information in this book were taken from several sources, including my own lab research and field experience. Here I describe sources and procedures for obtaining the data used in the species accounts and elsewhere in the book.

MORPHOLOGY Variation in anatomical structure of each anole species is compiled in the species accounts, and general mainland anole anatomy is described in chapter 1. Material for the species accounts is gleaned from primary literature, photos, field observations and, mainly, from my examinations of several thousand preserved anole specimens. I have scored preserved individuals of 237 of the 240 mainland anole species for a standard set of characters. These characters are described in Williams (1995), Poe (2004), and species descriptions for which I am a coauthor (e.g., Poe et al. 2015a). Several of these characters are described in chapter 3. Anatomical information for chapter 1 is taken from primary research literature, supplemented by my own observations.

ECOLOGY Variation in the ecology of each anole species is compiled in the species accounts under "Natural History," and research in mainland anole ecology is described in chapter 1. In the species accounts I have included information from both my own field observations and reliable literature reports. The summary information on ecology in chapter 1 is taken almost exclusively from peer-reviewed literature with only a sprinkling of personal observations.

MAPS Production of the anole species range maps has been among the more challenging and time-consuming tasks in the completion of this book. Range maps are the product of locality data—the places where an individual organism was found—for collected or observed individuals. As such, they depend on accurate inferences regarding species identity and location. Ideally these two sources of error, identification and location, would be vetted for each individual lizard whose data contribute to the resulting map. This twin vetting is the approach aspired to in most species descriptions and in some excellent regional monographs such as Lee (1996) and Savage (2002). The obvious difficulty with this ideal approach is that it is prohibitively time-consuming for larger regions or groups. According to VertNet, there are at least 4617 specimens of *Anolis limifrons* distributed across at least 29 natural history museums. Imagine examining each of these specimens for accurate identification, georeferencing each associated locality, then evaluating the authenticity of this information. Now imagine repeating this process for the other 239 mainland *Anolis* species. Worthy as such an approach would be, it is not feasible for mainland anoles at this time.

In lieu of the formidable task of complete vetting of all hypothesized mainland anole localities, I have adopted the following approach. The starting point for each species range map is a set of locality points taken from the Global Biodiversity Information Facility (GBIF) database, Velasco et al. (2019), or a combination of these two sources. I mapped out localities from these sources for each species on Google Earth and eliminated those points that were likely errors according to vetted primary literature, my own knowledge of anole distributions, and my (re)georeferencing of questionable points.[48] For example, anole specimens from west of the Andes that are identified as the Amazonian species *Anolis fuscoauratus* are likely actually to be *A. maculiventris*, and an observation of high-

INTRODUCTION

elevation form *Anolis crassulus* at sea level is not credible. Obviously, there is subjectivity in this step. I relied heavily on reliable primary sources such as Lee (1996) for the Yucatan peninsula, McCranie and Köhler (2015) for Honduras, Savage (2002) for Costa Rica, the Pontificia Universidad Católica website (Torres-Carvajal et al. 2019) and Arteaga et al. (2023) for Ecuador, and Ribeiro Junior (2015) for Amazonia, as well as personal experience. In addition to verifying localities, I added locality points listed in these works. I accepted iNaturalist observations if I was able to confirm species identification based on the photographic evidence accompanying the observation. I vetted most points at the spatial and elevational limits of each species, and generally accepted points that were well within my inferred range limits for each species. For example, I accepted all observations of *A. kemptoni* near Boquete, Panama, as that area is a well-known locality for this form, but scrutinized reports of this species in Costa Rica.

Elevational ranges were determined as in the locality vetting described above, with lower elevational limits rounded down to the nearest 50 meters and upper elevational limits rounded up to the nearest 50 meters. For example, a species known from a single locality of elevation 1672 meters was listed as being found from 1650 to 1700 meters. For small islands where anoles have been found islandwide (e.g., San Andrés, Malpelo), the maximum elevation of the island was used. Additionally, elevations were taken and localities were georeferenced for elevation from original species descriptions.

— SPECIES ACCOUNTS —

The species accounts of chapter 4 form the principal substance of this book. Each account includes the following components:

DESCRIPTION, a reference for the publication in which each species was named. I include reference information for the entire publication if the species description was the sole focus of the paper (example title: "A new species of anole from Panama"). If the species was described in the context of some larger work, I give reference for the pages on which the species is described (example title: "Notes on a collection of lizards by Dr. Popovich in Colombia"). For many recently described species, the publication initially describing the species is the most comprehensive source of information for it. For species described decades or centuries ago, the description is mainly of academic interest.

HOLOTYPE specimen for each species. The holotype is an actual preserved lizard specimen considered to be the "name bearer" of the species. The definitive word on identification of a specimen is the type specimen with which its population is best associated. That is, if specimen X is shown to be a member of the species for which type specimen Y is itself constituent, then specimen X is assigned the same species name as specimen Y. Other statements of specimen status used in the species accounts include neotype (a specimen designated as the new name bearer of a species after the original holotype is lost or destroyed), syntype (in earlier literature, multiple individuals could carry the status of a holotype; these are syntypes), lectotype (a specimen selected from syntypes to be the single name-bearer of the species), and paratype (individuals included in the species description to help characterize variation in the species). Museum acronyms for specimen numbers follow Sabaj (2020).

TYPE LOCALITY of each species. The type locality is the place where the holotype specimen was collected. It is critically important information for taxonomic work. Any specimen of the same species as the holotype that is collected at the type locality is called "topotypical." Recent descriptions almost invariably provide precise type locality information, including GPS coordinates. Descriptions published before the common use of GPS frequently suffer from one or both of two problems.

First, many very old descriptions mention type locality information that is so vague as to be useless. For example, the listed type locality of *Anolis punctatus* is "South America." In such cases, it may be possible to narrow the topotypical possibilities by reconstructing the trip upon which the holotype specimen was collected.[49] In some cases I have attempted such reconstructions, but in others I have surrendered to the realities of historical imprecision.

Second, many descriptions in which the author(s) recognized the importance of type locality precision nevertheless predated widespread use of mapping coordinates. In some of such cases, workers failed to appreciate that local place names may not be findable on any map available to future workers. For example, the listed type locality of *Anolis parvicirculatus* is "El Suspiro, Chiapas, 1200m." According to the website fallingrain.com, an online gazetteer, there are nine places called "El Suspiro" in Chiapas, Mexico. Which is to say, additional information beyond the location name is needed if one wishes to visit the type locality of *A. parvicirculatus* (and thereby be certain of catching true *A. parvicirculatus*). In many such cases it is possible to pinpoint the exact location of a given type locality using additional resources, such as older maps and travelogues and gazetteers such as Paynter's (1997) *Ornithological Gazetteer of Colombia*. I have attempted, not nearly successfully, to assign a GPS point to the type locality of every species of mainland *Anolis*. In my discussions of the elucidation of these points I have tried to communicate the thought process and therefore the uncertainty involved with the suggested points so future workers can assess my inferences more evidentially. The format of this section is a verbatim listing of the type locality from the species description followed, if not clear from this listing, by my estimate of the actual location where the type specimen was collected.

SIMILAR SPECIES comparisons distinguish anoles that may be confused with the species in question. This section offers practical advice for species identification, given the location and gestalt of the individual. For example, if you are in the Panama Canal area and have a small brown anole in hand, you may be uncertain whether you are holding *Anolis apletophallus*, *A. auratus*, *A. gaigei*, *A. humilis*, or *A. elcopeensis*. The Similar Species section for any or each of these species may be consulted for convenient means to distinguish these forms.

Comparisons focus on concision and easily scoreable traits, and generally compare only similar anole species that are potentially sympatric. I do not attempt genus or countrywide diagnoses, or to distinguish species that are differentiable with a glance or that differ vastly in elevational range. That is, I assume that a worker will not confuse *Anolis apletophallus*, which is always brown and has a body length of about 40 mm, with *A. biporcatus*, which is almost always bright green and has an adult body length of about 85 mm. And I see no need to describe differences between lowland form *A. gaigei* and highland form *A. solitarius*. Although both species are found in the region of the Sierra Nevada de Santa Marta in northern Colombia, their elevational paths do not cross. This geographical aspect of the identification process is important. Quite simply, there are too many similar mainland anoles to try to distinguish some small brown species from all other possibilities without geographic information.[50]

In the spirit of the above concerns, I classify anole species according to color, size, proportions, and, if distinctive, ecology. I categorize each anole species as *small brown*, *large brown*, *small green*, *large green*, *giant*, *semiaquatic*, or/and *twig*. Description and discussion of the utility of these categories is in chapter 3.

MORPHOLOGY sections give information on a standard set of traits: body size in males (M) and females (F); presence or absence of ventral keeling; condition of middorsal scales; relative size of interparietal and tympanum; number of scales across the snout at the second canthals; number of toe lamellae; contact of subocular and supralabial scale rows; contact of supraorbital semicircles; condition of supraocular scales; anterior extent of adpressed hindlimb relative to landmarks on the head; overall dorsal body color pattern; dewlap color pattern. Additional traits are noted if these are expected to be especially definitive. The characters and associated terms used here are defined in chapter 3.

RANGE describes the geographic scope of the species, including elevation. The range maps give graphic and more precise locality information.

INTRODUCTION

NATURAL HISTORY sections begin with a short, one- to two-sentence statement on the detectability and basic habitat use of the species, taken from my reading of the literature and my personal experience. For detectability, I label a species *very common* if one is likely to find that species without focusing effort toward finding it.[51] For example, if you eat at an outdoor restaurant in Boquete, Panama, on a sunny day near any vegetation, you are likely to see the very common species *Anolis polylepis*. I consider a species *common* if you are likely to find it in one evening when targeting your search for it. That is, if you go to a locality where you are certain *A. latifrons* has been found, you are likely to find *A. latifrons* if you know where and how to look for it (at night, sleeping high on twigs, leaves, or vines) and direct your efforts accordingly. I consider a species *rare* if even a few nights of targeting that species are unlikely to result in its procurement. For example, I know *A. brooksi* may be found on the short loop trail outside the visitor center at Parque Omar Torrijos in Panama because I have found it there. But finding it is a rare treat. One of the best anole-finding graduate students in my lab walked that loop for 28 consecutive nights and never saw *A. brooksi*. Note that my detectability categories are largely but not completely based on personal experience and do not take range size into account. That is, a species may be restricted to some tiny geographic area or specialized habitat, but if it is possible to get to that area or habitat and find individuals easily then I still consider that species to be *common*. Several species have only been found at a single locality. By my conventions, such a species may or may not be *rare*, depending on its detectability at that locality. Note further that these detectability comments are less statements of the ecology of a species[52] and more rough categorizations of the difficulty expected in finding that particular species of anole,[53] and that the degree of such difficulty is almost certainly to be influenced by locality. I have found *A. brooksi* to be rare in Parque Omar Torrijos and elsewhere, but perhaps there is a glorious honeyhole of habitat elsewhere in Panama where the *brooksi* are as many as the *sagrei* on planted Florida hotel grounds.

Note that the natural history of most anole species would be coarsely but accurately described as "Common, inhabits forest, eats insects, diurnally active arboreally, sleeps on leaves and twigs at night." My brief summaries of natural history will be most useful, then, when they deviate from this paradigm. *Anolis auratus*, for example, tends to be found in open areas rather than forest, *A. taylori* forgoes vegetation for activity on large boulders, and *A. tropidonotus* tends to sleep in leaf litter rather than arboreal vegetation. The best way to find most anoles is to survey twigs and leaves in forest and forest edge at night, but exceptions such as these species require alternative approaches.

The summary statement of natural history is followed by a list of references that give more detailed treatments of ecology, behavior, and life history. I have not attempted to summarize all information in each listed reference, even by topic,[54] but I have tried (and certainly failed in many cases) to include all pertinent references related to ecology of species in wild environments.[55] I generally exclude the original species description from these lists unless it is the only reference for a species, even though many descriptions contain ecological information. It can be assumed that any modern species description contains at least some ecological information, such as where in the habitat specimens of the type series were collected. The references in the list may be consulted for further help in finding a species or for deeper understanding of its ecology.

COMMENTS are offered mainly on taxonomic status but may include any information deemed important that was not included in the other sections.

CHAPTER 1
BIOLOGY OF MAINLAND ANOLES

In conclusion, I may say that Anolis *is a very difficult genus and I cannot pretend to any knowledge which satisfies me at all.*
—E. R. DUNN (1930:22)

Anole research comes in three flavors: island, mainland, and *Anolis carolinensis*. Like fruit flies, white mice, and zebrafish, the Green Anole, *A. carolinensis*, is what is known as a "model organism." Model organisms are species that are subjects of intense research aimed at understanding basic biological phenomena, usually with an eye toward application to humans. *Anolis carolinensis* is a mainland form, being found in the southeastern United States and naturalized in many mainland areas, but it is often associated with island forms because its evolutionary roots are on Cuba. If you have studied just one species of anole, chances are that species is *A. carolinensis*. This species was the first anole to have its anatomy described (Cope 1892) and the first to have its genome sequenced (Alföldi et al. 2011). It has been at the forefront of research advances on diverse topics such as the neural bases of behavior (e.g., Greenberg 1977), the structure and capability of the visual system (e.g., Kawamura and Yokoyama 1998), and the genetic basis of tissue regeneration (e.g., Hutchins et al. 2014). A book could be written on research on *A. carolinensis* alone, and that book would cover a significant percentage of what is known of anole biology.

The other two categories of anole study are geographic but not natural evolutionary units (see Phylogeny section below). It is convenient to treat these geographic groups differently in terms of research for three reasons. First, comparative studies generally have treated mainland and island anoles separately. Most studies of evolutionary convergence in anoles, for example, have focused on convergence between islands, not between islands and mainland (e.g., Losos et al. 1998). Second, there is an imbalance in research effort that has been applied to these groups. Many more researchers, and many more scientific publications, have addressed scientific questions using island rather than mainland anoles. Finally, research foci differ between the two groups. For example, the ecomorph concept of ecological and morphological correlations (Williams 1972) has been a fruitful and ongoing research avenue for some fifty years in the Caribbean. Only recently have attempts been made to apply those ideas to the mainland (e.g., Moreno-Arias et al. 2016). Conversely, detailed autecological studies of mainland anole species may exceed those of island species.[1]

Below I will draw from each of these research sources, *carolinensis*, island, and mainland, to briefly summarize what is known about mainland anoles. For some topics I describe particular studies of interest, but in most cases I give single example citations as support for statements. Which is to say I make no pretension of a comprehensive literature review. More extensive citation lists are found in the species accounts of chapter 4.

— ANATOMY —
INTEGUMENT

As in other lizards, the external covering of anoles is composed of distinct layers of epidermal cells. A thin (~ 1 μm) layer of cornified cells (the Oberhäutchen) provides a scaly, impermeable outer surface. This external layer is bordered sequentially medially by b-keratin, mesos, and a-keratin layers above a stratum germinativum that generates cells and a deep, nerve- and blood vessel–rich dermis (e.g., Alexander and Parakkal 1969). These layers correspond to different cell types, that is, cells with proteins and/or structural configurations that differ. The hard b-keratin component provides stiffness and protection while the a-keratin layer is a more pliable barrier. Forty beta-proteins and 41 alpha-proteins have been recorded in *Anolis carolinensis*, with the density of protein type related to the structure and function of associated regions. For example, claws are composed of a higher concentration of more corneous proteins than are toe lamellae (Alibardi 2016). The cells that compose the Oberhäutchen of anoles are flattened and juxtaposed, and microornamented with spinules. The

FIG. 1.1 Scaly dorsal and nuchal crests and snout extension in *Anolis proboscis* (top). Erectable nuchal and dorsal crests in *A. kunayalae* (middle). Tail crest in *A. cristatellus* (bottom). Photos by Tom Kennedy

structure and arrangement of these cells vary between anoles and other lizard lineages. Spinule arrangement and structure varies between anole species (Peterson 1983, 1984).

The above construction presents as a scaly exterior over nearly the entire anole surface. Anole species vary in scale size, structure, and arrangement, with mostly uniform scales on the body (especially ventrally) and a set of heterogeneous scales on the head (especially dorsally). Individual scales may be keeled or smooth, and sets of scales in a particular body region generally are juxtaposed or overlapping. The surface of the dewlap is exceptional in displaying a mix of scattered or regularly arranged scales punctuating exposed skin. Scale characters are useful for species identification, and many conditions such as ventral keeling usually are invariant within species. Caudal, nuchal, and middorsal crests, as well as scaly snout extensions (fig. 1.1) have evolved in anoles. See chapter 3 for description and figures of scale variation and scale traits used in this book.

Two aspects of the *Anolis* integument are unusual among lizards: expanded digital scales that form a toepad and color-changing ability.[2] Many species of anole have the ability to change color from shades of green to shades of brown under stress or as a thermoregulatory mechanism (e.g., Jenssen et al. 1995). Studies of *Anolis carolinensis* have demonstrated three layers of pigment cells, including an outer layer of xanthophores and deeper reflecting iridophores responsible for green coloration, as well as basal melanophores that extend distally and laterally to overlay the iridophore and xanthophore layers. Color change is accomplished by the hormone-triggered movement of melanin granules from positions near the melanophore nucleus to the distal cellular extensions. In this dorsal position, the melanin granules prevent light from reaching xanthophores and iridophores, thereby dulling the animal's color from green to brown (e.g., Taylor and Hadley 1969).

Most mainland anole species possess laterally expanded scales on their digits, a trait that seems an obvious adaptation for climbing.[3] In particular, toepads formed by these expanded scales allow effective climbing on smooth surfaces such as leaves and boulders. Claws supplement climbing ability on many substrates, and both claw and toepad size are correlated with vertical habitat use in some anoles (Macrini et al. 2003; Crandell et al. 2014). Anole toepad scales are covered with densely packed microscopic hairs, or setae, that function to increase clinging ability (Ruibal and Ernst 1965). The similarity in gross structure of gecko and anole toepads (Williams and Peterson 1982) suggests that the van der Waals adhesion forces operating in geckos (Autumn et al. 2002) occur in anoles as well. In fact, anoles have become a model system for understanding gecko toepads and their potential human applications (Garner et al. 2019).

— ANATOMY

MUSCULOSKELETAL SYSTEM

Early studies of lizard osteology that included or focused on anoles were systematic surveys (e.g., Etheridge 1959) or purely descriptive works (e.g., Stimie 1964). Cope's (1892: 199–201) treatment is the first osteological study of which I am aware that includes a separate, detailed section on *Anolis*. This work drew mainly from skeletons of *Anolis carolinensis*, but Cope did note some interspecies variation in abdominal ribs, a quality that later was foundational in Etheridge's (1959) landmark systematic study.

The known variable osteological traits in mainland anoles are described in Etheridge (1959), Poe (2004), and other sources.[4] These works focused on head and trunk skeletons. Etheridge (1959) documented interspecific variability in rib counts and structure, interclavicle structure, number of vertebrae in different body regions, presence or absence of fracture planes in tail vertebrae,[5] and structure of caudal vertebrae[6] among anoles. I (Poe 1998, 2004) and others[4] documented variation in the structure and presence of skull bones in anoles, including, for example, the presence or absence of the postfrontal bone and the shape of the parietal bone. The appendicular skeleton of mainland anoles has been studied by Ríos-Orjuela et al. (2020) and Feiner et al. (2021; also see island studies by Myers [1997], Herrel et al. [2008]). Studies of Caribbean anoles have discovered some variation in bone and muscle structure to be correlated with ecology (Tinius et al. 2020).

Anolis musculature mainly has been studied in *Anolis carolinensis* and island forms (but see Ríos-Orjuela et al. [2020] for a mainland example), and for systems of specific interest in *Anolis*. For example, there is a large amount of literature on the musculature of dewlap extension (e.g., Bels 1990) and on comparison of limb musculature in different ecomorphological types (e.g., Anzai et al. 2014). Comparative studies of jaw musculature, often coupled with evaluations of bite force, have included some mainland species (e.g., Wittorski et al. 2016).

Perhaps the most unusual musculoskeletal innovation in anoles is the support for the dewlap, an extensible flap of skin located in the gular area.[7] This flap incorporates a distal sheath around an elongate central (i.e., second) process of the hyoid skeleton.[8] Specialized *M. ceratohyoideus* muscles pull the lateral aspects of the hyoid posteriorly in the floor of the mouth, which rotates the anterior part of the hyoid dorsally, thereby bowing the long central process ventrally and extending the dewlap (e.g., Font and Rome 1990).

NERVOUS, ENDOCRINE, AND SENSORY SYSTEMS

Anolis possess the typical vertebrate brain structure of an anterior telencephalon including paired cerebral hemispheres and elongate, paired olfactory bulbs; an anteroventral diencephalon composed of thalamus and hypothalamus; paired optic lobes forming the midbrain; and posterior pons and medulla oblongata forming the hindbrain leading to the spinal cord (Northcutt 2002). Huber and Crosby (1933) described the diencephalon and midbrain, Greenberg (1982) described the forebrain, and Willard (1915) described the cranial nerves of *A. carolinensis*. Armstrong et al. (1953) described the telencephalon of three species of Jamaican anole. Compared to other lizards, anoles share gross structural aspects with close phylogenetic relatives (Northcutt 2013), have relatively large brains, and share elements of brain proportion with other visually oriented arboreal species (Macri et al. 2019). For example, anoles possess a large cerebrum and cerebellum and reduced olfactory anatomy (Armstrong 1953). The acceptance of these generalizations should be tempered by the fact that almost all anole brain studies are of a single species, that is, *A. carolinensis*. However, the few comparative interspecific studies of *Anolis* brains (Powell and Leal 2012; Armstrong et al. 1953) have found little variation between species. Minor seasonal and sexual variation in the size of reproductively oriented brain regions has been recorded (e.g., O'Bryant and Wade 2002).

The sensory experience of mainland anoles appears to be dominated by visual input. Evidence for this contention comes from multiple sources, including the excellent color vision and complex eye structure of anoles, as well as their keen ability to discriminate signals from background noise (Fleishman 1992). Like some visually oriented birds, anoles have a second fovea—a region of high photoreceptor density—in the retina (Underwood 1970), and ultraviolet vision (Fleishman et al. 1993). The importance of elaborate and colorful visual behavioral cues for anole biology, especially dewlap displays, further supports the idea that mainland anoles are largely visual animals.

Mainland anole ear structure and hearing acuity do not differ greatly from other iguanian lizards (Wever 1978), with best hearing occurring at an optimal temperature (Werner 1972). Mainland anoles have a vertically oval external tympanic membrane that may be large and obvious (e.g., *Anolis alvarezdeltoroi*) or barely exposed (e.g., *A. dissimilis*). Sound is conducted from this membrane to the oval window of the cochlea of the inner ear via a columella and extracolumella in the air-filled middle ear, thence through the inner ear to the basilar papillae, or auditory sensory cells. There is some variation in the size and structure of the basilar papillae, both relative to nonanole iguanians and between studied anole species (Wever 1978), although this variation does not necessarily translate into differences in hearing ability (Manley 2002). Vocalization has been documented in at least 16 anole species including the mainland forms *A. biporcatus*, *A. purpurescens*, and *A. salvini* (formerly known as *A. vociferans*).

The olfactory apparatus of studied anoles is poorly developed relative to other lizards (e.g., Pratt 1948), and chemoreception has not been found to play a role in prey detection or reproduction (e.g., Curio and Mobius 1978; Orrell and Jenssen 2003; but see Baeckens et al. 2016).

REPRODUCTIVE SYSTEM

Anolis carolinensis has been shown to share basic characteristics of reproductive anatomy with most studied lizard species (e.g., Rheubert et al. 2014). That is, males possess testes where sperm develop in seminiferous tubules, testicular ducts for delivery of sperm, a sexual segment of the kidney, a cloaca, and hemipenes. Paired hemipenes are stored posterior to the cloacal opening in the tail. One of the two hemipenial structures folds inside out from the cloacal opening via male-specific muscles for intromittance during copulation. Females have paired ovaries and oviducts that alternate development of a single egg.[9] Each oviduct is composed of a posterior vagina that enters the cloaca and an anterior uterus wherein an egg develops. Sperm storage tubules at the utero-vaginal transition may store sperm for several months and allow multiple paternity (e.g., Calsbeek et al. 2007). Activity of male and female reproductive structures is controlled by hormones, with morphological and physiological changes in these structures occurring during maturation and with cyclical (often seasonal) changes in reproductive behavior as adults (Connor and Crews[10] 1980; Lovern et al. 2004; Rheubert et al. 2014; and references therein).

A few studies have addressed interspecific variation in reproductive anatomy among anole species. Johnson et al. (2014) examined the size and evolution of several morphological reproductive traits in nine anole species, including two mainland forms. They found significant interspecific variation in the size of five of these traits, including the seminiferous tubules and the retractor penis magnus muscle (RPM), with the size of two traits (RPM, hemipenes size) correlated with copulatory frequency. Kahrl et al. (2019) found variation in sperm structure and testis size and in the rates of evolution of these traits across 26 species of anoles including 6 mainland species. Extreme variation occurs in hemipenial structure among mainland anole species (see species descriptions by Gunther Köhler and collaborators). Rate of evolution of hemipenial structure may be rapid relative to other morphological traits (Klaczko et al. 2015).

OTHER ANATOMY

To my knowledge, the circulatory, digestive, excretory, and respiratory systems of mainland anoles have not been studied comparatively. *Anolis carolinensis* and/or *A. sagrei* have been included in some studies of these systems as a lizard representative, but little or no variation between anole species or between anoles and other lizard species has been documented. See Gans and Parsons (1977) and Gans and Gaut (1998)[11] for helpful reviews of lizard soft anatomy.

Like other lizards, mainland anoles possess a heart with paired separate dorsal atria and partially separated ventricles, with a disproportionately large right side that receives venous blood from the body and shunts it to the lungs (White 1968; Jensen et al. 2014). Lungs are paired organs composed of compliant air sacs that exchange oxygen and carbon dioxide during negative pressure ventilation. That is, the thoracic cavity is expanded using trunk muscles that also function in lateral movement to create passive air flow from the environment into the lungs (Perry 1998). Presumably this reliance on trunk musculature for both breathing and movement hinders or precludes breathing while running in anoles, as in other lizards (Carrier 1987).

The digestive system of *Anolis carolinensis* is typical of other small carnivorous iguanian lizards, with comparable digestive efficiency (Licht and Jones 1967). Food is procured by biting capture, often after leading with a protrusion of the muscular tongue. Taste buds are present throughout the tongue and on the roof of the mouth (Willard 1915; Schwenk 1985). Teeth are unicuspid anteriorly and tricuspid (or rounded, in a few species) posteriorly in all mainland anole skulls I have examined. Food items usually are swallowed whole, or with some brief mastication, and food proceeds down the esophagus to a slightly expanded stomach and then an elongate small intestine wherein most digestion occurs. Food then proceeds through a caecum to the colon, where water and vitamins are resorbed and nutrient extraction is finalized. The colon empties into the cloaca, which is also the termination point for the ureter of the kidney and the urinary bladder. A urinary bladder has been observed in *A. carolinensis* and several close relatives of *Anolis* (Gabe and Saint Girons 1965; Beauchat 1986) and so probably is present in all anole species. The cloaca empties gastrointestinal and urogenital waste from the body.

— ECOLOGY —

DISTRIBUTION AND ABUNDANCE

Mainland anoles are found from the southeastern United States and continuously from northern Mexico south through Central America to Paraguay and southern Brazil (fig. 0.2). Most tropical and subtropical areas in this region are occupied by anoles, with dense concentrations of species at mid elevations along the central cordillera in Central America and the Andes of South America. A particular mainland area may be occupied by a single anole species (called a "solitary" species; see Williams et al. 1970), or multiple anole species may live in sympatry (e.g., Fitch 1975).

At the microhabitat level, mainland anole species may live terrestrially or occupy virtually any arboreal or saxicolous surface. However, as occurs famously in the Caribbean (e.g., Williams 1983), certain mainland species appear to concentrate on particular microhabitats, specializing in narrow or broad arboreal perches, trees or rocks, or the ground (e.g., Irschick et al. 1997). For example, *Anolis taylori* inhabits boulders (Smith and Spieler 1945), *A. lionotus* is semiaquatic (Campbell 1973), and *A. tandai* is commonly on the ground (Vitt et al. 2001). Arboreal forms may further specialize on particular surfaces such as tree trunks (e.g., Miyata 2013). Within species, different sexes and age-classes may differ in habitat use and locomotor approach (e.g., Pounds 1988). Species may change microhabitat use seasonally (e.g., Lister and Garcia Aguayo 1992) or hourly (e.g., Miyata 2013). Some species specialize on particular habitats at all known localities (see above examples), but others use whatever habitat is available at different localities where they are found (e.g., Irschick et al. 2005). Habitat partitioning among sympatric species is evident (e.g., Talbot 1979), but at least some species do not appear to alter their habitat use based on number of sympatric congeners (e.g., Vitt et al. 2002). Particular assemblages may include both habitat specialists and generalists (e.g., Moreno-Arias et al. 2020).

Anole abundances generally are much lower on the mainland relative to Caribbean islands (Buckley and Jetz 2007). However, mainland anole abundance may rival Caribbean norms in some areas. Disturbed areas in particular may harbor large numbers of species such as *Anolis auratus* and *A. sagrei*. Interspecific differences in abundance occur (e.g., Miyata 2013). Within species, mainland anole populations may vary in abundance seasonally (e.g., Fleming and Hooker 1975), across sites (e.g., Mesquita et al. 2007), and between sexes (e.g., D'Cruze and Stafford 2006). Abundance may be affected by the presence of birds, which in addition to preying on anoles (see below) may compete with them for food (Wright 1979).

LIFE HISTORY AND BEHAVIOR

Individuals of most studied mainland anole populations tend to grow faster, reach sexual maturity earlier, and die younger than their island counterparts (Andrews 1976). These differences may result from greater competition for food in island forms and greater predation risk for mainland forms

(Andrews 1979a). Mainland anoles seldom live longer than one year (but see Bock et al. 2010[12]), with degree of survivorship potentially varying by species, sex, and site (e.g., Andrews and Nichols 1990). Sexual maturity is reached in two to nine months, and females lay a single egg or occasionally two eggs every one to two weeks during periods of active breeding (e.g., Andrews 1976).

Breeding in mainland anoles occurs roughly year-round in wet tropical areas and seasonally at northern distributional limits and regions with long dry seasons (e.g., Watling et al. 2005). Most studied species have been found to be territorial, with a male territory encompassing multiple female territories (e.g., Stamps 1977). Males generally have been found to defend territories against other males, although subadult males may be tolerated in a dominant male anole's territory in some cases (e.g., Fleishman 1988). Females also may defend territories, but at lower rates than males (e.g., Nunez et al. 1997). Territory size for males and females is smaller when food is more abundant (e.g., Stehle et al. 2017). Males engage in courtship via species-specific displays involving extension and retraction of the dewlap, head bobs, pushups, and other movements (e.g., Greenberg and Noble 1944). Females may prefer particular males or mate randomly (e.g., Andrews 1985), and may mate with multiple males (e.g., Tokarz 1998). Copulation proceeds via male insertion of one hemipenis into the cloaca of the female, as in other squamates.

Anole dewlap displays are complex multifunctional behaviors (e.g., Tokarz et al. 2003). Particular displays are often specific to species and function, but may vary within and between individuals and populations of a species (e.g., Jenssen 1971). Male dewlap use has been studied in several species of mainland anole (e.g., Echelle et al. 1971) and shown to function in attracting a mate, deterring predators, and defending territory. Female dewlap use has been poorly studied, but work on *Anolis carolinensis* and some Caribbean species suggest use in deterring unwanted males or in intrasexual interactions (Hicks and Trivers 1983; Orrel and Jenssen 2003).

Mainland anole activity levels may vary seasonally and ontogenetically (e.g., Lister and Garcia-Aguayo 1992). All species are diurnal (although some forms will engage in nocturnal activity around artificial lights; e.g., Badillo-Saldana et al. 2016), and some display peaks of activity and dormancy during the day (e.g., Vitt et al. 2003a). Most species sleep nocturnally on twigs and/or leaves, but a few sleep on rock surfaces or near the ground in leaf litter (see examples in the species accounts). Daily activities of anoles have been roughly categorized as either territorial or foraging, with foraging the commonest behavior among females and territoriality more common in males (e.g., Andrews 1971). Mean home range size of adult mainland anoles ranges from 3.7 m^2 in female *Anolis limifrons* (Andrews and Rand 1983) to 804.6 m^2 in male *A. frenatus* (Losos et al. 1991). Home range size varies between males, females, and juveniles within species, with males having the largest ranges (e.g., Verwaijen and Van Damme 2008). Type of activity (i.e., foraging vs. territorial) may be correlated with perch height (e.g., Andrews 1971). Neuroendocrinological correlates for some behaviors have been documented (e.g., Crews 1980).

Anoles are extraordinarily well studied for thermal biology, even forming the basis for an attempted standardization of field research protocols for small ectotherms (Hertz et al. 1993). But with notable exceptions (e.g., Ballinger et al. 1970; Vitt 2001), most of this work focuses on island forms (see Huey 1982). These studies have shown variation in thermal tolerance and thermoregulatory behavior (e.g., basking) across species, populations, times of day, seasons, regions, elevations, and habitats. Traits of thermal biology tend to evolve quickly (Hertz et al. 2013), and mainland species have lower optimal body temperatures than island species (Salazar et al. 2019)

FEEDING AND PREDATION

Although there is some variation in mainland anole approaches to feeding, most studied species are roughly categorized as "sit and wait" predators (e.g., Magnusson et al. 1985). That is, individuals statically survey an area for food items and strike when suitable prey is nearby. Detailed studies of stomach contents have been published for several mainland anole species (e.g., Vitt et al. 2003a). Consistently across these studies, the most common anole prey items are orthopterans (grasshoppers, crickets), coleopterans (beetles), ants, dipterans (flies), spiders, and insect larvae. Vertebrates are taken occasionally, including anoles (e.g., Gerber and Echternacht 2000), as are invertebrates including

— EVOLUTION

FIG. 1.2. *Anolis cristatellus* being swallowed by a snake on Culebra Island, Puerto Rico.

noxious forms (Odendaal et al. 1997: *Battus* butterflies). Several species of island *Anolis* have been shown to eat fruit (e.g., Herrel et al. 2004). Among mainland species, *A. heterodermus* has been observed to eat fruit (Rafael Moreno-Arias, personal communication 2021). Even comparably sized sympatric species tend to consume different prey complements (e.g., Vitt and Zani 1998), and there is variation in prey consumption between sexes and age classes (e.g., Perry 1995) but not between seasons (Sexton et al. 1972). Goodman (1971) found that *A. polylepis* preferred living katydids whereas *A. aquaticus* would eat living or dead individuals.

Birds (e.g., Poulin et al. 2001) and snakes (e.g., Henderson 1982; fig. 1.2) are common predators of mainland anoles. Other vertebrates such as fish (Pianka and Vitt 2003: 73), monkeys (Mitchell 1989), and nonsnake lizards (e.g., Gerber and Echternacht 2000) also have been observed to prey on mainland anoles. Invertebrates such as spiders (e.g., Nyffeler et al. 2017; Reyes-Olivares et al. 2020) and whip scorpions (e.g., Oliver 2020) have been observed eating anoles. Studies of island *Anolis* have shown the use of displays to deter predators (e.g., Leal 1999) and the ability to distinguish harmful bird predators via auditory cues (Cantwell and Forrest 2013).

— EVOLUTION —
PHYLOGENY

Although mainland anoles were implicitly or explicitly included in several early treatments of lizard evolution (e.g., Camp 1923), the modern study of anole evolutionary history, or phylogeny, began with Richard Etheridge's (1959) University of Michigan PhD dissertation. Titled "The relationships of the anoles (Reptilia: Sauria: Iguanidae): an interpretation based on skeletal morphology," this work employed a handful of osteological characters and before-its-time cladistic[13] reasoning to divide *Anolis* into two large *sections*, termed alpha and beta, and several smaller groups Etheridge called

BIOLOGY OF MAINLAND ANOLES

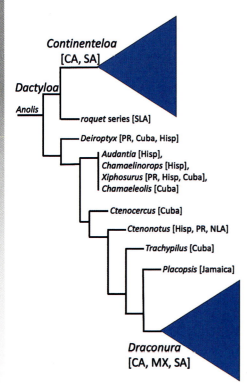

FIG. 1.3. Anole phylogeny. Relationships and named groups mainly refer to Poe et al. (2017). Parenthetical terms give geographic range for each clade. Each clade includes several species. CA = Central America, SA = South America, SLA = Southern Lesser Antilles, PR = Puerto Rico, Hisp = Hispaniola, MX = Mexico. Large blue triangles indicate clades that are the focus of this book.

series. In spite of its limited character sampling, this work has held up remarkably well. For example, subsequent studies have repeatedly found the anoles in Etheridge's beta section to be close evolutionary relatives.

Etheridge's work was followed by Williams's (1976a) influential taxonomic treatment of South American anoles, wherein he identified "sections," "series," and "species groups." These informal groups formed the units in several taxonomic and comparative evolutionary studies in the decades that followed. The emergence of molecular data in systematics brought reorganization of Williams's groups (e.g., Gorman 1973) as well as molecular phylogenetic analyses of many subgroups of island *Anolis* (e.g., Gorman et al. 1983; Hedges and Burnell 1990; Schneider et al. 2001; Glor et al. 2004). Mainland anoles remained understudied, with only a few mainland species included in genus-wide treatments of *Anolis* (Jackman et al. 1999; Poe 2004; Alföldi et al. 2011; Nicholson et al. 2012; Gamble et al. 2014). Eventually, subgroups of mainland species were analyzed (e.g., Castañeda and de Queiroz 2013; Hofmann and Townsend 2017) and greater numbers of species were included in larger phylogenetic analyses. Our study (Poe et al. 2017a) included all valid species of *Anolis* known at that time.

Most large-scale phylogenetic studies of *Anolis* have converged on the conclusion of two predominately mainland clades, which we (Poe et al. 2017a) called *Dactyloa* (see Guyer and Savage [1986] for initial designation of this group) and *Draconura*, respectively. Note that these clades are not closest relatives among anoles (fig. 1.3). That is, draconurans are more closely related to northern Caribbean anoles than to dactyloans, and dactyloans are more closely related to southern Caribbean anoles than to draconurans. Biogeographically, draconurans appear to have invaded Central America from the northern Caribbean and dispersed southward to the southern limits of *Anolis* in South America. Dactyloans originated in South America and spread northward as far as Costa Rica (Nicholson et al. 2005; Poe et al. 2017a).

COMPARATIVE EVOLUTIONARY BIOLOGY OF ISLAND AND MAINLAND ANOLES

Any study that analyzes more than one species may fall under the rubric of Comparative Biology. As such studies are implicitly or explicitly of evolution, they may be more precisely categorized as Comparative Evolutionary Biology[14,15] (CEB), especially if they account for some evolutionary covariance between species. *Anolis* has been a model system in CEB, but the focus in anoles until recently usually has been on island forms. In the past few years there has been expansion of anole comparative studies to mainland lineages. In some cases such studies extend or essentially duplicate island studies (e.g., Moreno-Arias and Calderón-Espinosa 2016) or include a few mainland species in analyses that are predominately island-centered (e.g., Ingram et al. 2016). Other studies address new questions focused on mainland anoles (e.g., Gray et al. 2020) or on the entirety of *Anolis* with ample mainland sampling (e.g., Velasco et al. 2020).

Large-scale taxonomically comprehensive comparisons of mainland and island lineages[16] of anoles have found comparable rates of speciation, morphological convergence, and morphological evolution among mainland and island lineages[17] (Pinto et al. 2008; Poe et al. 2018; Poe and Anderson 2019). However, island anoles occupy narrower niche breadths[18] than mainland anoles (Velasco et al. 2016), and island and mainland lineages may show different patterns of limb modularity[19] (Feiner et al. 2021). Furthermore, although both mainland and island assemblages[20] tend to include close evolutionary relatives (Anderson and Poe 2018), island assemblages include a wider range of morphological diversity than mainland assemblages (Poe and Anderson 2019). Morphological convergence occurs between mainland and island anoles and between mainland lineages, but the degree to which this convergence involves the classic Greater Antillean ecomorphs is debatable, and clearly many "morphotypes" are unique to either islands or mainland (Pinto et al. 2008; Schaad and Poe 2010; Poe and Anderson 2019; Patton et al. 2021; Huie et al. 2023). Semiaquatic, rock, and terrestrial specialists have evolved on both islands and the mainland (Leal et al. 2002; Huie et al. 2023; Anderson 2024). Patton et al. (2021) found that mainland interactions between the island-derived *Draconura* clade and the South America-sourced *Dactyloa* clade "favored" draconurans, and that these clades evolved their ecological and morphological diversity in different ways.[21] As in Caribbean anoles (Schoener 1969), mainland anole species that live without sympatric congeners exhibit increased sexual size dimorphism and uniform body size. Both these characteristics evolved exaptively[22] rather than adaptively on the mainland (Velasco et al. 2019), which contrasts with the island pattern of sexual size dimorphism evolving adaptively (Poe et al. 2007).

Caribbean anoles have been an undeniably productive model system for comparative evolutionary biology, especially evolutionary ecology. But mainland anole lineages have become integral subject matter for diverse comparative studies beyond mainland–island comparisons. For example, Tollis et al. (2018) compared entire genomes of four mainland anole species and found evidence for accelerated evolution and positive selection in genes involved in traits associated with adaptive radiation and species reinforcement.[23] Prates et al's (2018) comparative genomic study found a broad spectrum of candidate genes showing genome-environment associations in *Anolis punctatus*. Mainland anoles have been valuable subjects for understanding mainland speciation (e.g., Hofmann et al. 2019) as well as Central American (e.g., Phillips et al. 2018) and South American (e.g., Prates et al. 2017) biogeography. And of course, high-profile comparative work comparing mainland anole *Anolis carolinensis* to other vertebrate lineages continues to flourish in genomics[24] (e.g., Yuasa et al. 2015). There is an unlimited store of comparative research projects on mainland anoles to be mined.

— CONSERVATION —

No mainland anole species is known to have gone extinct during human history,[25] and no anole species is mentioned in the same resigned breath as vaunted conservation poster species such as polar bears and bald eagles. This is not to say that no mainland anoles are perceived to be at risk. The International Union for Conservation of Nature (IUCN)[26] has evaluated the status of 78 mainland anole species[27] and deemed that 1 species (*Anolis vanzolinii*) is Critically Endangered, 9 are Endangered, 7 are Vulnerable, 4 are Near Threatened, 38 are Least Concern (meaning they are in no danger of decline or extinction), and 19 are Data Deficient (meaning the IUCN evaluators don't have enough information to make a determination). Some researchers have come to even more dire conclusions. Johnson et al. (2015) rated 72.7 percent of Central American anole species as "high vulnerability," their most grave designation, according to their "Environmental Vulnerability Score." So although mainland anoles may be all right as a group relative to famously endangered species, the continuing existence of many species appears not to be assured.

If mainland anoles are at risk, what are the sources of that risk? As with any other organism, the greatest threat to mainland anoles undoubtedly is habitat loss[28] (IPBES 2019), which these days is overwhelmingly caused by human development.[29] Many species are extremely resilient to traditionally identified threats such as disease, pet trade collectors, hunting, "invasive" species, and even pollution and climate change. This adaptability is accomplished via evolution, behavioral change, dispersal to new areas, cryptic appearance and behavior, or some combination of these factors (e.g., Moritz and Agudo 2013). But if you turn a species' entire habitat into a parking lot, well, extinction is certain to follow.[30] Outside of habitat loss, it is not clear to which of the alternative threats mainland anoles are most subject. My sense from personal experience is that as long as habitat is preserved, threats like human collecting and invasive species are nonissues for anoles.[31] But I see no scenario where mainland anoles would not be vulnerable to habitat loss.

Aside from my own opinions and official designations, are there actual data to bear on the issue of mainland anole population status? It is a conspiratorial secret of conservation biology that only a tiny percentage of species have been evaluated with the detailed, geographically broad, long-term studies needed even to begin to assess their conservation status. For species that are not large mammals or birds, this percentage approaches zero. Species populations vary naturally in geographic size, abundance of individuals, and habitat use and, consequently, human detectability seasonally, yearly, and decadally. Such variability is present, and similarly difficult to evaluate, in mainland anoles. And in the same way that global warming was difficult to establish as attributable to human activity,[32] species declines are difficult to track for blame. Of course, the climate has changed for as long as there has been a climate, and of course species have gone extinct for as long as there have been species. The trick is to decide whether, when, and how humans are responsible.[33]

Pounds et al. (1999) documented declines in two montane anole species (*Anolis altae* [now *A. monteverde*], *A. tropidolepis*) in the 1990s at a site in Costa Rica. Whitfield et al. (2007) noted declines of leaf litter reptiles, including anoles, over 35 years at a lowland site in Costa Rica. Stapley et al. (2015) reported population declines in *A. apletophallus* over 40 years in central Panama. But it is not known if, for example, these species became more prevalent in other areas (e.g., at higher or lower elevations), or if populations have rebounded since these studies. The reviled[34] effects of the human-transported island species *A. sagrei* on mainland *A. carolinensis* appear restricted to slight changes in perch height and display behavior in *A. carolinensis*, and its evolution of larger toepads[35] (Campbell 2000; Edwards and Lailvaux 2011; Stuart et al. 2014). And most exotic anole "invasions" to the mainland result in no new anole species interactions, as these invasions tend to be restricted to human-created habitats like gardens and city parks (see, e.g., Hoogmoed 1981). In sum, there are no compelling data suggesting that any mainland anole species is at risk of extinction, nor that humans are directly accountable for any supposed decline. Note, importantly, that this statement is different from saying that mainland anoles are not at risk. This is a statement of ignorance, not of known status.

I will add two personal observations to the above broader treatment, pertinent to the subjects of this book. First, I have not observed any anole species to require pristine habitat. I have found several supposedly rare or endangered species thought to require undisturbed habitat in patches of trees in highly disturbed habitat (examples: *Anolis brooksi*, *A. fungosus*, *A. cusuco*). *Anolis proboscis*, for example, appears common in pastures in the Mindo area of Ecuador if you can find the right trees. Second, I estimate that every species of mainland anole is best categorized as "Data Deficient." I am not aware of intensive long-term studies of the populations of any mainland anole species throughout its range. Whether these facts should bear on the conservation status of anoles is debatable. We may want to err on the side of caution, for example, and so designate and protect some or all "Data Deficient" species.

I favor the preservation of anole species because I like the idea of a world with more anole species. I think such a world is more attractive, and more interesting, and a more enriching and transformative[36] place for this and future generations. I hold each of these characteristics as valuable nearly axiomatically. But as aesthetic justifications of ethical principles do not carry much weight among most professional ethicists (see Parfit 2011; Maier 2012), I do not pretend to hold any moral high ground in support of this view. It must further be admitted that mainland anoles have little to offer human health–based ethical views. Anoles do not harbor substances that cure disease, or possess unique mechanical or regenerative properties that lead to innovations that benefit humans, or sound an alarm call portending doom (or flourishing) for humanity.[37] There is no political, health, or economic reason to save all or any species of mainland anole. If we are to preserve mainland anole species, it must be for their own sake. I hope that is reason enough.

CHAPTER 2
COLLECTING
HOW TO FIND AND CATCH MAINLAND ANOLES

Anybody can catch a lizard ... It is catching the particular lizard you want that may be difficult. High up on a sodden, windy peak, with a thunderstorm visibly and rapidly approaching to windward, a magnificent specimen of some montane rain forest form, squirreling farther and farther out on the caclin trunks over a great abyss, can lead to intemperance in a man.

—JAMES LAZELL (1972:9), ON CATCHING ANOLES

Museum biologists refer to the process of obtaining specimens from nature as *collecting*. This chapter is about collecting anoles. Although the material in this chapter was written with museum biologists in mind—specifically those engaged in biodiversity research—the practices detailed here should aid any individual who desires to find and secure anoles.[1]

Collecting anoles is a form of *herping*. To herp is to search for amphibians and nonavian reptiles.[2] Herping is performed by professional herpetologists conducting biological research at field sites and by children catching frogs in city parks. The verb *to herp* is an informal term, not yet in standard dictionaries as of this writing, whereas the term herpetology has been in use since 1824 according to merriam-webster.com. These terms derive from the Greek root *herpein*, meaning to creep, a pejorative[3] reference to the supposed creeping nature of reptiles and amphibians.

Herping for individuals of a particular species or group of species may warrant the use of specialized techniques that herpetologists have developed and passed on over academic generations. For example, snake biologists will drive roads slowly at night, searching for sinuous three-dimensional shapes that show up bright against pavement in the glare of headlights. Turtle biologists may lay out baited floating traps or snorkel deep in ponds to find aquatic species. Lizard biologists may use blowguns or fire scattershot to secure individuals that are too fast or too high in the habitat to grab by hand. These techniques sometimes develop in concert with technological advancements (fig. 2.1). For example, extremely bright lights marketed toward security professionals or night mountain bikers have enhanced our ability to find reptiles and amphibians at night, and suspended walkways have rendered previously unreachable treetop canopy faunas accessible.

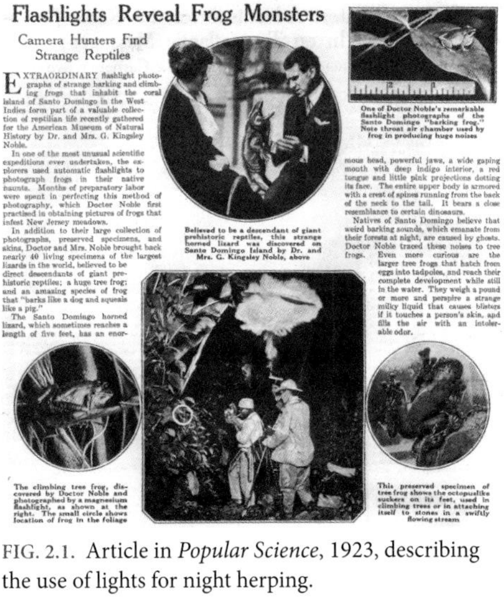

As with other herps, herping for anoles involves some preparations, techniques, and tools that are tailored to the group. Below I describe the practices used by my lab to collect anoles. We have used these approaches to collect 275 species of *Anolis*, including several new to science.

FIG. 2.1. Article in *Popular Science*, 1923, describing the use of lights for night herping.

— PLANNING A FIELD TRIP TO COLLECT ANOLES —

I am a taxonomist and phylogeneticist, which means my research involves describing new species and figuring out how species are related to each other. These research goals necessitate collecting as many species of anole as possible, and my lab group plans its field trips accordingly. Below I describe

27

COLLECTING: HOW TO FIND AND CATCH MAINLAND ANOLES

steps that we take before leaving on a trip to help ensure successful collecting. Some weeks before a trip to collect anoles I will compile a list of anole species present in the area to be visited. The area to be visited may be selected based on a specific research question (e.g., whether Isla Escudo de Veraguas in Panama contains endemic species) or because that area contains a high level of anole species diversity (e.g., the Colombian Andes). I use resources such as VertNet (an online database of museum holdings of vertebrate specimens, including the locations where individuals were found), field guides, iNaturalist, and primary literature to determine the species content of the country or region where fieldwork is planned and to get some idea of the relative rarity of species.[4] In addition, I will use these resources to identify areas (e.g., parks, roads, mountains) where particular rare species have been found, and to search for areas where anole species diversity has been shown or is expected to be high. Google Earth is a wonderful tool for identifying promising roads that extend through potential anole habitat at appropriate elevations. All of this information is processed with reference to the specific goals of the trip, which generally involve collecting as much anole diversity as possible but may reflect targeting particular anole species for collection, or a desire to find species that are new to science. An ecotourist might want to maximize diversity, or to target particular spectacular anoles.

Efficient collection of anoles requires special tools, which will be discussed below, as well as gear that is typical to most tropical biologists (fig. 2.2). The ubiquity of rain in the tropics means that nearly every item brought on a trip must be evaluated with respect to its performance when wet. Rubber shell boots are preferred over hiking boots for this reason and for their protection against plants and insects.[5] Light pants and long-sleeved shirts that dry quickly are desirable, as is a thin, completely waterproof raincoat. A global positioning system (GPS) unit is useful for documenting localities and not getting lost, and a high-resolution digital camera for specimen and dewlap photos is essential. A tablet or small laptop computer may be brought along for convenient internet access, but this should be a cheap model, essentially considered disposable due to the high likelihood of damage and theft. A field notebook and pencils or pens that work in wet conditions are used to document the trip. Many biologists prefer small loose-leaf binders so pages may be rearranged, and Rite in the Rain or similar paper that resists the effects of moisture. In addition to the high-powered herping lights discussed below, a mini headlamp for reading, writing field notes, and otherwise operating in darkness is needed. A small, battery-powered fan can bring relief when sleeping in hot conditions. For temporary storage of anoles, I prefer typical grocery-store gallon-sized ziplock bags with the more secure "slider" closure. These are thin enough that anoles may breathe, and thick enough that tearing is unlikely. Supplies (and techniques) for preserving anoles and other herps are listed in Pisani (1973). A standard approach to field notes was developed by Joseph Grinnell and is described in Remsen (1977).[6]

For museum biologists, the final aspect of planning a trip is to obtain permits to collect and possibly export anoles. Administrative bodies such as the Autoridad Nacional de Ambiente de Panamá (ANAM) and the Instituto Nacional de Recursos Naturales (INRENA) in Peru regulate collecting and distribute permits detailing restrictions on allowable species and specimen counts in response to applications justifying a request to collect. Successful completion of the permitting process ranges from impossible to

FIG. 2.2. Equipment for anole herping: mountain biking light, telescoping pole, GPS unit, waterproof boots, water-resistant pants, plastic ziplock bags.
Photo by Tom Kennedy.

achievable with a lot of effort and several months of planning. The best option is to work closely with local museum or university scientists, which of course brings myriad additional advantages for learning and collaboration. But the permitting process may be difficult even for local scientists well-versed in the administrative guidelines of their pertinent agencies. Permitting agencies frequently are staffed with nonbiologists with little or no training in conservation, and the decisions of such agencies regarding species and specimen number restrictions often are mysterious. Still, although obtaining scientific collecting permits can be a frustrating process, following all legal guidelines is a necessity. Understanding local regulations and obtaining appropriate permits are critical parts of trip planning

— HOW TO FIND AND CATCH ANOLES: SNARING[7] —

There are two main approaches to catching anoles: snaring and spotlighting. These approaches correspond mostly but not completely to daytime and nighttime work, respectively.

Snaring involves placing a loop of string suspended from a pole around the neck of a lizard and pulling tight to snare the individual. The string is structured as a tiny loop (~ 5 mm in diameter) tied tight with a larger, snaring loop extending through the tiny loop such that when a lizard neck is inside the larger loop, pulling tight will secure the larger loop around that item (fig. 2.3). Herpetologists vary in their choice of snaring string. Many use dental floss for its smoothness, flexibility, and relative visibility against forest backgrounds. Others use fishing line, especially kinds that are highly visible and can be molded to stay rigid when extending a snare in the wind.[8] For the pole component of a snare, most herpetologists use fishing poles, but any stick will do.[9] Retractable poles are especially useful for ease of transport, but bring concerns about stiffness. A rigid pole is essential for successful snaring at a distance, especially in inclement weather.

Successful snaring requires dexterity, steadiness, and patience. It usually is most effective to bring the loop of the snare toward a lizard slowly, allowing the anole to notice it and become comfortable with its presence. Sometimes an anole will ignore the invading loop completely. Other times they will attack it, presumably mistaking it for a prey item. While lowering the loop toward the anole, it sometimes helps to look just past the lizard to another point in the substrate as the loop is maneuvered into position. Strangely enough, lizards are often sensitive to eye contact. Once a lizard detects that you are trying to catch it, the process of capture becomes much more difficult. And probably nothing says "I want to catch/eat you" to a small animal more than an extremely large animal moving toward it while staring at it.[10]

Once the anole is comfortable with the presence of the snare or has not detected it, the next step is to ease the loop over the snout of the animal, encircling its head, and move the loop back to the neck region. This move should be done slowly; many lizards have been lost by hasty movements at this stage of capture. After the loop is positioned around the neck of the targeted anole, the pole should be pulled back over the body toward the tail of the animal, thereby securing the loop tight around the lizard's neck. Many collectors jerk back abruptly to ensure a

FIG. 2.3. A snared anole hanging from a snaring pole and (inset) in hand.

COLLECTING: HOW TO FIND AND CATCH MAINLAND ANOLES

tight loop, but this step does not have to be done especially rapidly to be successful.[11] A common mistake at this stage is to pull the snare up (i.e., dorsally) rather than back (i.e., posteriorly). Pulling up may remove the loop from the anole's neck, or may fling the animal upward. Fortunately, unlike skinks and whiptails, anoles have a discrete head well set off from the neck, so once a loop is tightened the anole is seldom lost. Still, it is best to remove the loop from the anole in a timely manner in order to minimize discomfort of the animal.

Snaring is most commonly done to catch active anoles during the day. Its effectiveness depends on the anole-finding ability of the snarer, her snaring proficiency, and the activity levels of the sought anoles. In my experience, most anole species are diurnally inactive during inclement weather such as heavy rain. Normal activity levels in anoles have been studied by several authors and seem to be population- and area-specific.[12] For example, an anole species living at 2500 meters elevation may have a single brief period of sunny time during which it forages and patrols its territory, whereas lowland tropical anole species may have multiple activity peaks during a day. As ectotherms beholden to the environment for their energy levels, anoles can be especially hard to find during cooler periods, and especially wary and difficult to snare during the heat of the day.

Snaring is a satisfying and really fun way to catch anoles. When performed by a competent practitioner, it is an effective way to secure large numbers of common anoles during the day. However, if you want to find rare or cryptic species, and collect common ones most easily, you must go out at night.

— HOW TO FIND AND CATCH ANOLES: SPOTLIGHTING —

When the pioneering hip-hop group Whodini rapped that "Freaks Come Out at Night," they might have been referring to anoles. The great naturalist Ken Miyata, in his 1985 description of *Anolis lynchi*, noted that anoles collected at night "have sometimes been the ones most poorly represented in museum collections; in some cases it seems as if a different anole fauna can be found at night." What Miyata and possibly Whodini realized was that strange, cryptic anoles are only seldom seen during the day, but are visible and potentially even common at night. Additional anecdotal evidence supports this view. Twenty-eight of the first 31 individuals of the tiny and cryptic anole *A. occultus* were collected by Richard Thomas at night.[13] In 2001 I spent approximately equal amounts of time day and night herping over 36 hours at Soroa, Cuba, and found 27 individuals of the fantastic chameleon-like *A. (Chamaeleolis) barbatus* at night—and zero individuals during the day. I have had similar experiences with several mainland forms, including rarely collected species such as *A. brooksi*, *A. calimae*, and *A. fungosus*: Hours of day herping produced few or no specimens, but night herping resulted in several to dozens of individuals of these species.

So how does one find anoles at night? Fortunately for herpetologists, most anoles make it easy on us. Individuals of almost all species (exceptions discussed below and in the species accounts) sleep on leaves or narrow branches at night. The body of an anole contrasts with surrounding vegetation when illuminated at night, so sleeping individuals are easily seen when a light is shined on them (fig. 2.4). This being the case, the best approach to finding maximum anole diversity is to search at night with a bright light, slowly scanning vegetation for the telltale contrast of anole and twig or leaf.

But why do anoles sleep so exposed? The idea, which never has been tested but makes great sense, is that a lizard sleeping on delicate vegetation can detect when a predator is stalking it. A predatory snake climbing a frail plant with an anole sleeping on it may shake that plant and disturb that anole. In response, the anole can drop to the ground and possibly evade the predator.[14] This approach to sleeping is reasonable according to natural selection and exploitable if you are a herpetologist seeking anoles.

Night herping for anoles requires two tools. The most important of these is a bright light. For lights, brighter generally is better, although I seldom use my brightest light on its highest level, as doing so tends to wash out the lizard–vegetation contrast in close quarters. The best night-herping lights strike a balance, emphasizing beam throw (greater distance) in order to see distant lizards, but incorporating enough flood (greater spread) to search large areas quickly. I find a discrete central circle of light to be distracting and prefer even flood across the lit area. Lumens (the absolute light output) are

FIG. 2.4. *Anolis lemurinus* sleeping on a leaf in Belize (top). *Anolis kemptoni* on a twig in Panama (bottom).

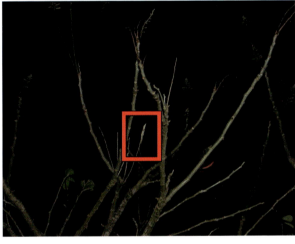

important but of secondary concern to candelas (amount of light in a particular direction), and white (cool) light seems to elicit better contrast than yellow (warm) light.[15] Other desirable qualities include rechargeabilty, long battery life, light weight, compact size, hands-free usability, and reasonable price.

I have settled on high-powered lights designed for night mountain biking as best meeting these criteria. I have been satisfied with offerings by NiteRider, Fenix, and Dinotte, and there undoubtedly are other companies producing anole-worthy lights. As of this writing, several companies offer bikelights of 10,000 lumens or higher, and battery and charging options are myriad. Light brightness, battery life, and charging speed are technologies that develop rapidly, and any specific light I might recommend here could be obsolete within months. Reviews from bicycle-oriented websites are useful, as are tech sheets from outdoor-oriented stores and recommendations from hunters, police, and others who rely on bright light.

The two lighting options that some anologists favor over mountain biking lights are headlamps and focused-beam flashlights. These are excellent choices. My reservations regarding traditional headlamps are that, first, I prefer to use all lights, even headlamps, in hand rather than mounted on the head, so the hands-free advantage of headlamps is negated. Manipulating the light in hand rather than leaving it fixed upon the head seems to allow finer distinction of anole–vegetation contrast because different lighting angles may be examined while maintaining optimal view of a potential lizard. Second, it is difficult to find a headlamp with the brightness of other light options. Flashlights can be extremely bright, and their narrow beam is desirable when long throw is needed. The problems with high-powered flashlights are that they tend to be heavy and unwieldy, such that it is cumbersome to hold one while attempting to secure an anole with both hands, and their limited flood inhibits rapid search of large arboreal areas. Mountain biking lights may be effortlessly transferred from hand to mouth[16] during anole capture such that the light is directed by mouth while both hands are available to make the capture. Headlamps obviously also leave one's hands free, but flashlights require awkward shouldering or uncomfortable mouthing to keep the hands free during anole capture.[17]

Besides a light, the other necessary piece of equipment for night anole herping is a long pole, often fixed with a loop on the end so as to double as one's snaring pole. We have used collapsible fishing poles, retractable golf ball retrievers, and "natural" items such as saplings for this purpose.[18] The function of the pole is to knock the sleeping anole off its perch so the lizard may fall and be secured.[19] Ideally, the anole is caught out of the air after being knocked down, but if missed the anole may be grabbed from the ground. It is important not to pursue anoles too chaotically if they hit the ground after being knocked down. Anoles usually do not flee, but rather tend to freeze after falling.[20] A loop of string on the distal end of the pole may be employed to pull down a sleeping anole in the same way snaring is done during the day (see above). Night snaring often is difficult with small anoles that sleep flush with vegetation, such as *Anolis orcesi*.

There are exceptions to the norm of anoles sleeping on delicate vegetation. A few saxicolous forms such as *Anolis gadovii* may sleep on boulders in addition to vegetation, and individuals of many semiaquatic species sleep on rock surfaces bordering or within streams. In some of these species (e.g., *A. aquaticus*, *A. maculigula*), individuals commonly sleep in rocky splash zones, seemingly oblivious to the splattering cacophony surrounding them (fig. 2.5). Although most diurnally terrestrial anoles sleep in the traditional anole manner on leaves and twigs, some, such as *A. tropidonotus*, commonly sleep in leaf litter.[21] Perhaps most unusually, *A. alvarezdeltoroi* has been observed sleeping deep within caves, sometimes more than 10 meters above ground on the cave ceiling (fig. 2.6; a trait shared with Cuban *A. bartschi*). Notwithstanding these unusual cases, I estimate perhaps 90–95 percent of anole species sleep on twigs, branches, saplings, and/or leaves. Targeting the other 5–10 percent may require unusual approaches, and the species accounts will give some guidance here. But in general, if anole diversity, including especially rare species, is the goal, it is best to focus on leaves and twigs at night.

As a final caveat to the favorability of focusing on night work, I will note that in my estimation perhaps 5 percent of anole species are easier to find (if not catch) while active during the day. In my experience, these species tend to be either terrestrial forms that evade nocturnal detection by sleeping near the ground (e.g., in leaf litter) or species that favor upper tree trunks diurnally and sleep so high on leaves that night observation is difficult. Ground anole *Anolis humilis* is often earmarked as one of the most abundant anoles in diurnal surveys

ABOVE LEFT: FIG. 2.5. *Anolis barkeri* sleeping on a boulder in a splash zone of a stream, Mexico.

LEFT: FIG. 2.6. *Anolis alvarezdeltoroi* sleeping on the ceiling of a cave, Mexico. The individual is approximately 10 meters up from the cave floor and 20 meters from the nearest obvious cave entrance.

(e.g., Perez-Martinez et al. 2021) but seems rare at most localities at night. At All America Park in Miami, Florida, *A. distichus* is present on seemingly every suitable tree trunk during the day but may become eerily absent at night.[22] I focus on nighttime work because this approach remains the best way to procure difficult anoles, but an open mind and species-specific flexibility must be maintained.[23]

— HOW TO FIND AND CATCH ANOLES: OTHER APPROACHES —

Snaring and spotlighting are the commonest means to catch anoles, but there are other techniques. Rodrigues et al. (2002) used sticky trays attached to tree trunks to trap the rarely collected anole *Anolis phyllorhinus*. Several researchers have used blowguns or guns filled with scattershot to procure lizards,[24] and presumably these techniques would work on anoles. Several "invasive" anoles have been observed to be nocturnally active, snareable, around artificial lights. I have very occasionally found anoles under rocks (e.g., *A. armouri*) or bark (e.g., *A. cybotes*), or sleeping on large tree stumps (e.g., *A. notopholis*), or esconced in hanging moss (e.g., *A. quercorum*). But these behaviors do not seem to me to be common for most species in most areas.[25]

In addition to searching in nonstandard areas for anoles, it is sometimes fruitful to engage the help of others. During the heyday of Ernest Williams and Albert Schwartz,[26] "lizard markets" were commonly employed in the Caribbean to accumulate anoles. Researchers would enter a rural village and set up an impromptu anole bazaar for children and adults to receive coins in exchange for specimens. Prices were fluid and set according to demand. At first, one peso would be offered for any lizard, resulting in several individuals of whatever anole was most visible (usually the local "trunk-ground" anole, e.g., *A. cybotes*). The appearance of anything new (say, a green "trunk-crown" anole like *A. callainus*) was paid with two pesos, and so on. If a rare species came in (say, a "twig" anole like *A. sheplani*), a five-peso bounty might be offered. This approach resulted in huge collections of common species and was sometimes but not always effective for securing rare species.[27] My experience with enlisting local people to help find anoles is that you must find the right person—the local naturalist—for this approach to be successful with rare species. I have had local people surprise me with remarkable species of anole I did not expect, and also I have shown local people species they had never seen before even though they lived all their lives in close proximity to them.

The reliable and tested approaches of snaring and spotlighting are commonly used for good reason. Perhaps some new technique will come along to revolutionize anole herping and supplant the usual techniques,[28] as night herping did. But until that time, if you want to find a lot of interesting anoles I recommend getting a bright light and a long pole, training your eyes, and staying out until the sun comes up.

— EXECUTING A FIELD TRIP TO COLLECT ANOLES —

Once an anole fauna is studied, equipment is accumulated, permits are obtained, and techniques are understood, the trip itself proceeds according to principles of *specialization* and *targeting*. Specialization refers to my lab's focus on anoles rather than other species or general herp collecting. As explained above, we herp for anoles almost exclusively at night because that approach yields the highest payoff in terms of rare species and anole diversity (and the approach allows us to collect data on how anoles sleep). Targeting refers to our pursuit of particular, usually rare species, with the assumption that other commoner species will be collected as bycatch during the course of the trip. For example, on our first trip to Ecuador, we targeted the (at that time) rarely collected anole *Anolis proboscis* by visiting Mindo, the only area where that species had been collected. While searching for *A. proboscis*, we also collected multiple individuals of the common Mindo species *A. gemmosus*, *A. aequatorialis*, and *A. fraseri*, which obviated making a special visit somewhere else to obtain these species. As a trip proceeds, the target list may change as some species thought to be common turn out to be difficult, and some rare species may be found in an unexpected area.

COLLECTING: HOW TO FIND AND CATCH MAINLAND ANOLES

General herping

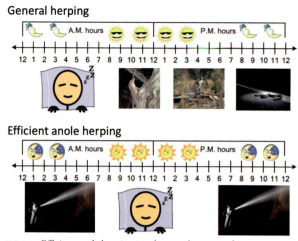

Efficient anole herping

FIG. 2.7. Efficient anole herping involves working at night.

During the course of a trip, we usually will stay at a location only long enough to collect the targeted anoles from that location. We may stay at field stations or "ecolodges" that allow convenient use of trails through good habitat. We also frequently spend our nights driving, making stops and searching at areas along the road that seem likely to harbor anoles. This latter approach can be a highly efficient way to collect significant anole diversity in a single evening, for example by driving an elevational transect from sea level up to 2000 meters and stopping to herp at several elevations along the way.

We often will search for desirable anole habitats during the day as preparation for herping those habitats at night. The best habitats in which to find anoles are not necessarily the habitats considered to be most "natural" (fig. 0.3). Intact old-growth forest with hiking trails is desirable and must be herped when available—after all, anoles obviously evolved in and lived in "undisturbed" areas long before humans evolved. But anoles are often difficult to see in such areas due to habitat complexity, canopy height, and trail overgrowth. Multiple times, I have had the experience of hiking hours through apparently pristine natural habitat and finding few anoles, then herping an adjacent disturbed area such as a roadside and finding several anole species quickly. The success of herping disturbed areas is not limited to common species, either. I have not found any species of anole to require undisturbed forest to survive. In my view, the most productive areas to herp for anoles are disturbed areas near pristine areas. Edge habitats such as roadsides, fields at the margins of forest, or streams through agricultural areas that are bordered by a few meters of adjacent trees are especially productive.[29] Of course, certain species such as semiaquatic and saxicolous forms have very specific habitat needs, and these are discussed in the species accounts.

Habitats and regions vary greatly in anole abundance, and it can be difficult to anticipate the abundance of a particular area before visiting. Anoles are famously common on Caribbean islands, but can seem frustratingly rare at many mainland sites. A seasoned anole collector might catch well over a dozen anoles per hour in El Yunque National Forest in Puerto Rico, but only a few in an entire evening herping Panama's El Copé National Park. Anole abundances in some highly human-affected mainland areas can approach island abundances.[30] Planted hotel grounds, golf courses, and city parks are prime and often convenient locations for observing large numbers of anoles.

A normal day during one of our trips to collect anoles starts with the group waking up during the afternoon, desirably no earlier than 2:00 or 3:00 p.m. We spend the late afternoon preparing for the evening's herping: ensuring the health of previously collected specimens, making final mapping preparations for the evening's localities to be visited, buying provisions, searching for habitat, and discussing plans. As the sun goes down, we will have a large meal—generally our only one of the day—before proceeding out for the night's anole herping. We generally herp for anoles from sunset to sunrise, fueled by snacks and the fun and excitement that come from finding anoles (fig. 2.7).[31] It can be hard to stay up all night, both mentally and physically. The carrot-goal of reaching sunrise often enables hours of labor beyond what may be considered a normal work effort. As I have emphasized in this chapter, efficiency is key for collecting anoles. But also, as in any endeavor worth pursuing, success in anole herping is sometimes a matter of just putting in the time.

CHAPTER 3
HOW TO IDENTIFY MAINLAND ANOLES

A group of small brown lizards that all look the same.
—DAVID HILLIS (1996), IN REFERENCE TO *ANOLIS*

The aim of this chapter is to provide direction for accurate species identification of anoles. Before we continue, I would like to present the best approach to identifying live anoles in the field. This is an approach that my lab and other anole-familiar labs (see Savage 2002: 448) use in practice, and an approach that is embraced throughout this book. Here are the steps:

1. Know where you are, so you can assess which species are likely to be present (see appendix 2; figs. 3.1, 3.2).
2. Catch a male anole and compare its dewlap to the plates in this book.

That's it. This procedure will allow you to identify most species in the "difficult" group *Anolis*. You are welcome to stop reading this chapter now and continue on to the species accounts.

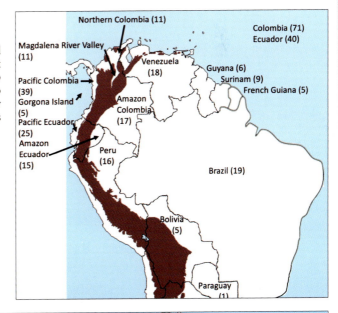

RIGHT: FIG. 3.1. Map of South America showing regions recognized in this book and species counts for countries and regions. Darkened areas show elevations above 2000 meters.

FIG. 3.2. Map of Middle America showing regions recognized in this book and anole species counts for countries and regions.

HOW TO IDENTIFY MAINLAND ANOLES

— PROBLEMS AND APPROACHES —

Cases solvable by the above procedure notwithstanding, identification of anoles is notoriously difficult.[1] Numerous similar species of anole may live in sympatry (e.g., fig. 0.6), and copious variation may be present within species or even within an individual at different times of day (see "Variation within Species" below). Different classes of individuals present distinct challenges for anole identification. Preserved animals are more difficult to identify than live ones, males are easier than females, and juveniles are especially challenging.[2] The alternative difficulties of these classes suggest the need for multiple tools for identifying anoles. Accordingly, I present here several means, including dewlap and body photos, a trait-matching computer program (the *AnoleKey*, described below), species accounts including information on distinguishing similar species, range maps, and geographic species lists. Particular approaches suited to particular classes of anole are discussed below.

Determining the sex of the individual in hand may aid identification. Anoles may be sexed by examination of postcloacal scales, which are noticeably enlarged in males of many (not all) species and undifferentiated in nearly all female anoles (fig. 3.3). Also, in many anole species, males have a large dewlap, whereas the dewlap is small or absent in females. Finally, many species have hemipenial bulges in males but not females. These three traits—postcloacal scales, dewlap, and hemipenial bulges—will suffice to sex nearly all anoles.

LIVE MALE ANOLE This is the easiest identification case. The best approach for identifying a male anole is to know the possible species for the region where the individual was found (appendix 2) and examine the dewlap plates for a match. If no dewlap matches, or if multiple dewlaps seem plausible, the species accounts should provide enough information for a confident identification.

LIVE UNSEXED ANOLE WITH A DEWLAP The species in hand may be female or, more likely, male. Here too it is best to examine the dewlap plates with guidance from geography, and refer to the species accounts for additional information if necessary.

LIVE ANOLE LACKING A DEWLAP Individuals that fit this description are female[3] or possibly hatchling male. The best approach to identification in this case is to examine species accounts and species photos of geographically candidate forms according to appendix 2. The "Similar Species" sections of the species accounts will be helpful in difficult cases.

LIVE ANOLE LACKING LOCALITY INFORMATION In this unusual case, one may scan the dewlap panels for a match or use the *AnoleKey* to narrow down the possibilities before examining species accounts of likely candidates.

PRESERVED ANOLE WITH LOCALITY INFORMATION The species accounts are likely to suffice for accurate identification. The *AnoleKey* may provide corroboration, or may be consulted in order to narrow down the candidates for an especially refractory specimen.

FIG. 3.3. Enlarged postcloacal scales in *Anolis lamari*: absent in females (left); present in males (right).

PRESERVED ANOLE LACKING LOCALITY INFORMATION These cases are likely to be difficult, and may be unsolvable for small female anoles. Candidate species are best identified using the *AnoleKey*. The species accounts may narrow the field further, and the "Similar Species" sections of the species accounts may allow definitive identification.

Note that circumstances may dictate a combination of approaches. You may have only general locality information. There is no dewlap photo available for some species. You may have collected a rare species associated with very little information, or the evaluated specimen may be damaged and display few scoreable characters. By presenting this diversity of approaches and information in the species accounts and *AnoleKey*, I hope to eliminate situations where an individual anole cannot be identified.[4] But of course, no identification system is infallible. If you are faced with a poorly preserved juvenile female small brown anole with no locality information that you would like to identify, well, you probably are screwed. (Your best bet in these situations is to enter as many trait scores as possible into the *AnoleKey*.[5])

1   Lizard lacks expanded toepads or is active nocturnally. Go to 2
      Lizard has expanded toepads and is active diurnally. Go to 3

2   Not an anole.

3   Brown, has 7 or fewer supralabial scales: *Anolis sagrei*
      Brown or green (green when sleeping), has 8 or more supralabial scales: *Anolis carolinensis*

FIG. 3.4. Dichotomous key to the anoles of Texas, USA.

Before continuing on to descriptions of tools for identifying anoles, I would like briefly to discuss how not to identify anoles. The traditional means to identify an unknown animal is the dichotomous key, whereby an individual specimen is scored according to a series of sequential couplets that lead to a choose-your-own-ending species conclusion (fig. 3.4). I have eschewed this venerable approach for several reasons,[6] but summarily because I believe the means presented here are functionally superior. Furthermore, I believe these means better reflect the actions of practicing field taxonomists. Seasoned fieldworkers do not mentally work through an inflexible series of couplets when identifying individual animals. Rather, we first focus on environment (what species are known from the area where I have found the animal to be identified?) and gestalt (what is the overall appearance of the animal?) to narrow down the possibilities, then check scoreable key traits that enable identification to species. The approaches presented in this book were developed with this rough sequence in mind.

— CHARACTERS —

Various external characters have proven useful for identifying anoles. Williams (1995) and Köhler (2014) have attempted standardizations of these characters. This book generally follows Williams (1995) but includes some additional characters.[7] Below I describe the characters used in the species accounts and the *AnoleKey*. Figures 3.5, 3.6, and 3.7 show some of the variation in head scalation in anoles.

Maximum *body size* for males (M) and females (F) is measured from snout to vent (SVL) in millimeters. Large individuals are telling for identification; smaller individuals may be adults of smaller species or juveniles of large species. Body proportions (larger head, smaller dewlap in juveniles) may help to distinguish adults from juveniles.

Keeling refers to the presence or absence of raised ridges on scales. The most standardly used scoring of this trait is on the belly (fig. 3.8). Entire clades of anoles may be characterized by *smooth* or *keeled* ventral scales.[8] Other body surfaces, most commonly dorsal head and hand scales, also may vary interspecifically in presence or absence of keeling, and the keeling may be categorized as single- (unicarinate) or multikeeled (multicarinate). Keeling within body regions usually is consistent,[9] but degree of keeling between regions may differ. That is, for example, all scales of the ventral torso tend strongly to be either keeled or unkeeled, but ventral and dorsal body surfaces may differ in the presence of keeling.[10]

FIG. 3.5. Lateral head scales in *Anolis magnaphallus* (top left), *A. frenatus* (top right), *A. biporcatus* (lower left), *A. kunayalae* (lower right).

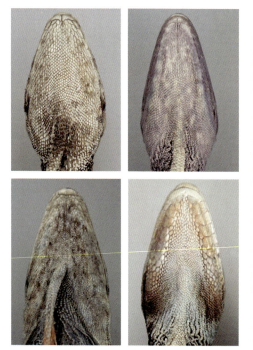

Condition of *middorsal trunk scales* (fig. 3.9) varies little within species, and certain conditions are restricted to one or two species per locality. In the species accounts I recognize four broad categories of middorsals, but note that there is some overlap between conditions within a few species. The most common middorsal condition among anoles is presence of approximately two longitudinally arranged rows of scales that are slightly enlarged relative to more lateral scales. This enlargement often is barely detectable, and some species appear not to have any differentiated middorsals at all, so I recognize a single "middorsal scales enlarged in 0–2 rows" blanket state to encompass these conditions. A second, distinctive condition is the presence of a band of abruptly enlarged, often strongly keeled middorsal scales. In these cases there is a short transition of 1–3 scales from smaller, sometimes granular flank scales to a set of 5 to 20 middorsal scales that are at least twice the size of the flank scales. Within

FIG. 3.6. Ventral head scales in *Anolis pachypus* (top left), *A. soinii* (top right), c) *A. charlesmyersi* (bottom left), *A. transversalis* (bottom right).

FIG. 3.7. Dorsal head scales in *Anolis transversalis* (upper left), *A. heterodermus* (upper right), *A. magnaphallus* (lower left), *A. kunayalae* (lower right).

the species accounts I describe these middorsals as "greatly enlarged." The third recognized middorsal condition is structurally intermediate between the first two. This condition involves a gradual transition across multiple scale rows from enlarged middorsal scales to smaller flank scales. In the species accounts I describe this state as "middorsal scales gradually enlarged in several rows." The key difference between this "gradually enlarged" condition and the "enlarged in 0–2 rows" condition is the greater number of apparently enlarged middorsal scales in the "gradually enlarged" condition. The key difference between the "gradually" and "greatly enlarged" conditions is the abruptness of the transition from large to small scales in the "greatly enlarged" condition. If the number of gradually transitioning scales from enlarged to flank scales is large (> ~8), the lines between these three conditions blur. Examples of these states of dorsal body scalation are shown in figure 3.8. A fourth condition, presence of a middorsal crest of raised scales, is present in a few species (fig. 1.1).

Several scale counts may be employed in identifying anoles. Unless stated otherwise, listed counts are minimums for an individual.

Number of *toe lamellae* refers to the laterally expanded scales on the underside of the fourth (longest) toe of the hind foot—i.e., the "toepad"—as counted by Williams et al. (1995). The count is made from the bend between phalanges IV and V to the tip of the toepad below the proximal attachment of the claw (fig. 3.10). The IV/V bend occurs approximately at the proximal point along the underside of the toe where, moving from the origin of the toe to the tip, the scales begin to become expanded laterally (i.e., wider than long)

ABOVE: FIG. 3.8. Ventral body scales smooth (left), keeled (right).

BELOW: FIG. 3.9. Middorsal body scales abruptly enlarged (left), gradually enlarged in multiple rows (middle), undifferentiated or slightly enlarged in 1 or 2 rows (right).

HOW TO IDENTIFY MAINLAND ANOLES

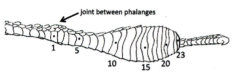

FIG. 3.10. Method for counting the number of toe lamellae in anoles.

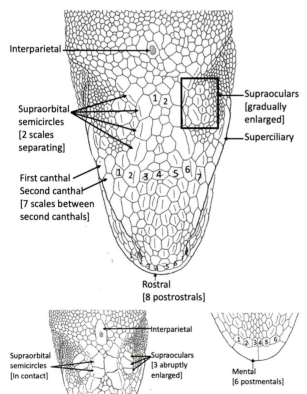

FIG. 3.11. Anole head scales and associated character states. Dorsal (top, lower left), ventral (lower right).

(fig. 3.10). This can be a difficult count, best done with a magnifying glass or from a photo, and some interobserver variation in number is to be expected.

Number of *scales across the snout at the second canthals* is a standard measure of the size of dorsal head scales (fig. 3.11). The canthals are elongate scales along the interface of the dorsal and lateral surfaces of the head anterior to the eye.

The *supraorbital semicircles* may be in contact or separated by one or more rows of scales. The supraorbital semicircles are large squarish scales that follow a medial arc around the smaller scales above the eye (fig. 3.11).

Rows of *subocular* and *supralabial* scales may be in contact or separated by one or two rows of scales (fig. 3.12). Suboculars are enlarged scales below the eye; supralabials are elongate scales along the upper edge of the mouth.

There is variation in the size of the tympanum, or external membrane of the ear, and of the *interparietal* scale among anoles. The interparietal is an enlarged scale on the posterior dorsal surface of the head, usually with a dark central spot (fig. 3.11). Comparison of tympanum size and interparietal size will distinguish some species.

The *supraocular* scales are dorsally placed scales above the eye. These usually include some enlarged scales and some smaller scales presenting in a continuous range. Extreme conditions of approximately equal-sized supraocular scales, or especially of including two to four abruptly and greatly enlarged scales, are telling for anole species identification (fig. 3.11).

The extent of the tip of the *anteriorly adpressed hindlimb* (i.e., the end of the longest toe's claw) is compared to landmarks on the head on live specimens as a crude means to assess relative limb length (fig. 3.13). This measure, traditionally important in mainland anole identification, invites

— CHARACTERS

FIG. 3.12. Lateral head scales in an anole.

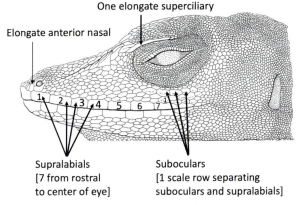

One elongate superciliary

Elongate anterior nasal

Supralabials
[7 from rostral to center of eye]

Suboculars
[1 scale row separating suboculars and supralabials]

some subjectivity but remains useful for gross comparisons.

The overall dorsal *body color* of an anole is generally green, brown, or whitish. Many species are always some shade of brown. Those that are green may change between green and brown. Anoles also may vary in body pattern, which may be solid or composed of transverse bands, blotches, longitudinal stripes, or other adornments. The body plates show several anole color patterns.

The *dewlap color* pattern is the single best character for distinguishing anole species. In some species of mainland anole both males and females display large dewlaps, which may or may not be colored similarly between sexes. In others only the male has a dewlap (see "Variation within Species," below). Dewlap variation among species is shown in the dewlap plates.

Scores for the traits listed above are included for each species in the species accounts (chapter 4). Additional traits are used in the *AnoleKey* and variably in the species accounts. Among these potentially useful traits, a deep axillary pocket (aptly called "tubelike" by Köhler [2008] and "puncturelike" by Savage [2002]) is distinctive and seldom found in more than one species at a given locality. Eye color may be a helpful trait in a few cases, as blue eyes are uncommon in anoles (fig. 3.14). Some anole species possess a narrow toepad that does not appear distinct from the scales at the base of the claw (fig. 3.15).

toe adpressed to level of eye

eye

FIG. 3.13. Method for adpressing hindlimb and marking toe reach to assess anole hindlimb length.

FIG. 3.14. Most anoles have brown, copper, or red eyes as in *Anolis orcesi* (top; photo by Tom Kennedy), but some have blue eyes, as in *A. cristifer* (bottom).

HOW TO IDENTIFY MAINLAND ANOLES

FIG. 3.15. Most anoles possess a distinct expanded toepad (top); some species display a narrow pad continuous with the terminal phalanx (bottom).

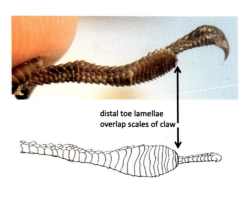

Structure of the scales in the nostril region varies in species-specific ways in anoles. Many species have an elongate teardrop-shaped scale reaching from the nostril to the rostral (called an "anterior nasal"), whereas others lack differentiated scales in the nasal region (fig. 3.12). Counts of scales in the loreal region (i.e., anterior to the eye), above the mouth ("supralabials"), at the dorsal edge of the orbit ("superciliaries") and posterior to and in contact with the mental ("postmentals") or rostral ("postrostrals") scales may be useful (figs. 3.11, 3.12). Most species display a single row of middorsal tail scales which may form a crest, but a double row is characteristic of a few species. Dewlap scalation may include rows of single or multiple scales according to species (compare among dewlap plates). Postcloacal scales are abruptly enlarged in males of some species, undifferentiated or variably enlarged in males of others, and always absent in females (fig. 3.3). Variation in relative tail, toe, head, and limb length will distinguish some groups of anole species.

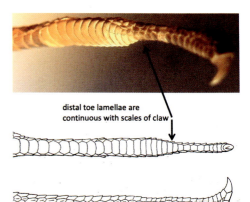

— COMPUTER IDENTIFICATION KEY[11] —

The program *AnoleKey* (Poe, manuscript) operates on the Lucid platform (Lucidcentral.org 2014) to present a set of candidate anole species, given some set of entered character states for an unidentified individual. As data are entered, incompatible species are downgraded and matching species are presented. Species may be ranked according to their degree of matching or completely eliminated as candidates depending on the settings entered by the key user. During this process of species matching, as states are entered any species can be evaluated for its suitability according to the traits it possesses. That is, you can access what trait values caused a species to be eliminated or favored. Photo links offer the opportunity to view images of species and of traits. Ranges of trait values may be examined for any species, and differences between species may be listed in order to focus on decisive traits.

For example, suppose you are in Podocarpus National Park in Ecuador on the eastern slope of the Andes at 1600 meters and you find a large anole. You may enter information on approximate body size, approximate elevation, and location into the *AnoleKey* to see what candidate species are suggested. Limited and easily scored information such as this, coupled with a dewlap photo or data on some other key character, often is enough for a secure identification. In the above case, entering just location, elevation, and body size reduces the number of candidate species to two, *Anolis fitchi* and *A. podocarpus* (fig. 3.16). The user can then compare photos of these two species or assess distinguishing traits recommended by the key. In this case, dewlap photos are available for both species, and the "differences" function decrees that these species may be distinguished by dewlap scalation (rows of single scales in *A. fitchi*, rows of multiple scales in *A. podocarpus*). With just a handful of traits and the *AnoleKey*, a potentially difficult identification has been rendered straightforward.

— SPECIES ACCOUNTS

FIG. 3.16. Panel from AnoleKey showing identification of a large anole found at high elevation on Ecuador's eastern Andean slope. Based on provided information, (1600 meters elevation, 90 mm SVL, eastern Andes), *Anolis fitchi* and *A. podocarpus* are likely identifications. Knowledge of the arrangement of scales on the dewlap ("Different features") would allow distinction between these two species.

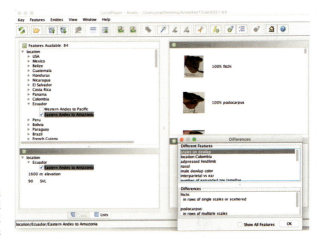

The *AnoleKey* is described in detail in Poe (manuscript), and that paper should be consulted as an introduction to its use. This paper also describes caveats on the reliability of the key; most of these concern difficulties with poorly known or inadequately sampled species.

— SPECIES ACCOUNTS —

The species accounts of chapter 4 of this book are intended to serve as a resource for anole identification in the lab and (especially) the field. In particular, the "Morphology," "Similar Species," and "Range" sections should be informative. Figure 3.17 gives an example procedure for anole identification using the resources of chapter 4, described in more detail below.

A pillar of the identification aspect of the species accounts is the categorization of anoles by morphological and in some cases ecological gestalt. Herpetologists in the field have some sense of the general semblance of lizard they have grabbed. We tend first to determine what appearance-category of lizard (anole) we have in hand—small and brown, aquatic, etc.—and then move on to subtleties of scales and proportion, all the while focusing on geographically likely species (see figs. 3.1, 3.2 for geographic divisions assumed here). In the spirit of these tenets, I have categorized each anole species according to its gestalt so that candidate species may be narrowed. The categories:

Small brown: This is the commonest version of anole and the version, in my experience, that causes the most difficulty for identification.[12] The anole fauna of some localities is composed entirely of species of small brown anole (e.g., most of Pacific Mexico). These anoles have maximum body length less than 60 mm and a constant brown or grayish-brown dorsum. The shade of brown may vary based on mood or temperature, but these forms do not have the capacity to turn green. Many species of small brown anole may be alternatively solid brown or strongly patterned with bars, blotches, or stripes, depending on mood or temperature.

Large brown: These species range in maximum body size from 60 mm up to approximately 90 mm. As in small brown anoles, large brown species may be patterned or solidly colored and may turn many shades of brown to white but will never appear completely green.

Small green: A few species of mainland anole are both small (body < 55 mm) and predominantly green. The green dorsum is the baseline appearance for these species, but it must be noted that most anoles that usually are green may change to brown. For all species categorized as *small green* or *large green* (see below), individuals are green during sleep. It is only when agitated or otherwise influenced (e.g., when turning brown to better absorb heat) that these species change from green.

Large green: These species range from about 65 to 90 mm. All the color comments for *small green*

HOW TO IDENTIFY MAINLAND ANOLES

FIG. 3.17. Example procedure for identifying anoles in the field using the species accounts and dewlap plates of this book.

1	Note your geographic region. See Figures 3.1, 3.2.
	e.g.: Northwest Honduras, 1200 meters

2	Determine what gestalt category and dewlap of anole you have in hand
	i.e.: *Small brown*, Large brown, Small green, Large green, Giant, Semiaquatic, Twig
	Dewlap orange

3	Consult Appendix 2 to determine which species are likely candidates according to geography and gestalt.

Honduran Small brown species: *Anolis amplisquamosus, A. beckeri, A. caceresae, A. cupreus, A. cusuco, A. heteropholidotus, A. humilis, A. kreutzi, A. laeviventris, A. morazani, A. muralla, A. ocelloscapularis, A. pijolense, A. purpurgularis, A. rodriguezii, A. rubriboaris, A. sagrei, A. sericeus, A. sminthus, A. tropidonotus, A. uniformis, A. wermuthi, A. yoroensis, A. zeus*

4	Examine dewlap plates of candidate species.

amplisquamosus — rodriguezii — yoroensis — ocelloscapularis

caceresae — sminthus — sagrei — tropidonotus

5	If no obvious dewlap match (or if a female), compare SIMILAR SPECIES in the species accounts of Chapter 4 of candidate species. Examine pertinent traits as warranted.

Of small brown species with an orange male dewlap in Honduras, only **A. rodriguezii** lacks enlarged middorsal scale rows and has smooth ventral scales.

small brown anole in Honduras + orange dewlap + smooth ventrals + uniform middorsals =

Anolis rodriguezii

species apply to *large green*. That is, all large green species may turn brown under some circumstances, but green is the usual color.

Giant: Traditionally for Caribbean anoles (e.g., Williams 1983), the moniker "giant" is applied to species larger than 100 mm body length regardless of color. This category is useful on the mainland as well. Large species of mainland anole often are limited in range overlap—unlike "small brown" anoles, there generally are few "giant" anoles per locality—and several giant species incorporate both greens and browns in dorsal coloration (examples: *A. ventrimaculatus, A. brooksi, A. fitchi*), so it seems reasonable to treat very large anoles as their own category.

Twig: This category is reserved for species that share a particular complement of unusual traits and usually are not easily categorized as "brown" or "green." Twig anoles have a long head and short limbs, a whitish to brownish lichenous dorsal coloration, slender body, usually a short tail, and ecology wherein they tend to occupy twigs, at least at night. Most South American species that fit this category are highland forms, whereas Central American twig species may be high- or lowland inhabitants.

Semiaquatic: The ecologically most distinctive anoles are the semiaquatic forms. These species generally are some shade of brown, of intermediate to large size (60–100 mm SVL), usually with long hindlimbs and a light lateral stripe extending along the body. They are most distinctive in behavior and habitat use. They swim, frequently diving into the water to escape capture; may sleep on boulders rather than vegetation; and are almost never found farther than 2 meters from the edge of a stream or river. The ecology of semiaquatic anoles makes their elimination as a potential identification simple in most cases—that is, if the anole you have found is not by a stream or river, it is highly unlikely that you have found a semiaquatic anole.

I would like to be clear that *small brown*, *large brown*, *small green*, *large green*, *giant*, *twig*, and *semiaquatic* are informal categories, my attempt at communicating how we in my lab preliminarily identify anoles. As will be clear in the species accounts, certain species cross over categories. But there are many more species that are easily categorized as one of the above. These categories are a convenient means to narrow down potential identifications and determine which species are appropriate for comparison.

Once the gestalt of a species is hypothesized, the next step is to examine candidate species according to geography in appendix 2. If an adult male is evaluated, dewlap photos of the candidate species of appendix 2 may be surveyed for a match. Supplementally and in the case of females and juveniles, the "Similar Species" section of candidate species may be examined, as well as the body photos. This procedure (gestalt + geography → dewlaps, body, similar species distinctions; fig. 3.17) should suffice to identify almost any mainland anole in the field.

— VARIATION WITHIN SPECIES —

Apparently the herpetology gods deemed the degree of interspecific variation in anoles insufficiently challenging for species identification and so interjected additional inconstancies into the group. Intraspecies variation between sexes, across ontogeny, between individuals of the same sex, and within a single individual at different times of day further complicate anole identification. Moreover, degree of intraspecific variability varies between traits and between species. Here I review these additional sources of variation.

Variation between sexes is common in body size, dewlap presence and size, and color pattern. In 89 mainland anole species, the size of the largest known male is at least 5 percent larger than the size of the largest known female. The pattern is reversed, with females larger than males, in 32 mainland species.[13] Sexual size differences tend to be more extreme in species where the male is larger. In well-sampled species *Anolis carolinensis*, females are about 75 percent the length of males. Among extreme and well-sampled species with the reverse pattern, *A. vittigerus* males are about 85 percent the length of females. Body size may be a useful identifying trait if an extremely large individual is evaluated.

Dewlap size and color tend to be uniform among comparably sized males of most species[14] (exceptions: *Anolis nebulosus*, *A. fuscoauratus*; fig. 3.18), especially within populations, but variation in each of these characteristics is not uncommon in females of some species. The dewlap may be absent in females (e.g., *A. chloris*, *A. auratus*), present but smaller than the male's (e.g., *A. pentaprion*, *A. vittigerus*), or, most infrequently, of equal size to the male's (e.g., *A. brooksi*, *A. aequatorialis*). Female dewlap color often duplicates the male's but may be white or otherwise different than the male's

FIG. 3.18. Intraspecific variation in male dewlap color pattern in *Anolis fuscoauratus*: from Pilcopata, Peru (top left); Tingo María, Peru (top right); Estación Tunquini, Bolivia (middle left); Trinidad, Bolivia (middle right); Iquitos, Peru (bottom left); Camiaco, Peru (bottom right). The Camiaco specimen is from the type locality for *A. fuscoaurautus*; the Trinidad specimen is from near the type locality for *A. fuscoauratus*. Genetic divergences between these populations are well below levels between distinct species (data not shown).

FIG. 3.19. Sexual dimorphism in dewlap color pattern in anoles. Males, left; females, right. *Anolis johnmeyeri* (top), *A. brooksi* (second from top), *A. alvarezdeltoroi* (second from bottom), *A. menta* (bottom).

(fig. 3.19). In some species, the female dewlap is patterned but not colored like the male's (e.g., *A. johnmeyeri*, *A. vittigerus*), whereas in others the sexes possess differently patterned and colored dewlaps (e.g., *A. biporcatus*, *A. brooksi*). A wide range of female dewlap color patterns has been documented in *A. lemurinus* (White et al. 2019) and *A. salvini* (Bientreau et al. 2013).

Body color and pattern may vary between and within sexes (figs. 3.20, 3.21), across ontogeny, and by mood within an individual (fig. 3.22). In some species, relatively distinct male and/or female morphs exist (e.g., *Anolis proboscis*), whereas in others variation appears continuous (e.g., *A. ventrimaculatus*). A common female morph of many species is the addition of a broad dark or light middorsal stripe, sometimes bordered by contrasting narrow parallel stripes, layered upon solid green or brown or upon a male-like pattern (fig. 3.20). This morph is present as a variant among one or more pattern alternatives, often with one of these alternatives similar to the male's pattern (fig. 3.21); I know of no species for which all females possess the middorsal stripe pattern. Males and females also may differ in the amount of green, brown, or blue. For example, males but not females of *A. kunayalae* often display a blue wash anteriorly. Most species show little variation in color pattern across ontogeny, but some (e.g., *A. brooksi*, *A. parilis*) are more strongly patterned as juveniles; I know of no species with the reverse condition of stronger pattern in adults. As discussed above, green species may change to brown, but brown species are always some shade of brown or gray. Some green species become more starkly patterned when brown (e.g., *A. soinii*, *A. biporcatus*; fig. 3.22). Individuals of several brown species may appear either solid brown or gray or patterned with blotching, banding, or striping of different shades of brown.

Scale counts such as the number of postmentals or of scales across the snout vary among individuals of all species, and this variation standardly is not correlated with sex or ontogeny. In a few species (e.g., *Anolis punctatus*) ventral scales may be keeled or smooth, but degree of keeling usually is constant within species for a given anatomical region. The specific number of enlarged middorsal scale rows and of enlarged supraocular scales varies within species, but the condition itself does not. That is, for example, all individuals of *A. tropidonotus* display rows of abruptly enlarged middorsal scales, but there is variation in number of rows, from approximately 10 to 15. Cresting on the tail and/or middorsum may be more pronounced in large individuals or males or may vary without regard for sex (see, e.g., Lazell 1969: fig. 1). Males of two species possess scaly rostral extensions that are absent in females (*A. proboscis*, *A. phyllorhinus*; fig. 1.1).[15]

The bewildering degree of intraspecific variation in anoles complicates species identification. The species accounts of chapter 4 discuss intraspecific variation and should be invaluable for adjudicating

— VARIATION WITHIN SPECIES

FIG. 3.20. Sexual dimorphism in body color pattern in anoles. Males, left; females, right. *Anolis apollinaris* (top), *A. cuscoensis* (second from top), *A. vittigerus* (middle), *A. anchicayae* (second from bottom; left photo by Guido F. Medina-Rangel), *A. proboscis* (bottom; photos by Tom Kennedy).

HOW TO IDENTIFY MAINLAND ANOLES

FIG. 3.21. Sexual dimorphism in body color pattern in *Anolis solitarius*: female (top), female (middle), male (bottom). The three lizards are from the same locality.

many difficult cases, as should the figures in this chapter taken as examples. Catching and examining multiple individuals for comparison, if possible, also is likely to help. Finally, awareness of the above issues will go a long way toward avoiding identification errors and difficulties related to intraspecific variation. Expect that females may have a middorsal stripe and that complexly patterned species are likely to come in different pattern flavors. Understand that males, females, and possibly juveniles may be patterned differently.[16] If these precepts are kept in mind, the hurdle of intraspecific variation for anole identification will be lowered considerably.

— CONCLUSION —

As implied by the epigraph for this chapter, identifying anoles can be difficult. There is no definitive trait or foolproof system that can guarantee accurate identification of all individual anoles. In this chapter I have presented a diversity of methods and information for anole species identification, and the species accounts should be helpful as complements for each species. Male dewlap color pattern coupled with geography is a supremely valuable combination but the species accounts, body photos and *AnoleKey* can supplement dewlap information and apply to females or preserved specimens for which male dewlap details are unavailable. L. C. Stuart (1955) stated of his key to Guatemalan anoles, "The worker who knows what species he has before him should experience few difficulties in its use." I hope that the combination of tools presented here will be useful for workers, even in those unfortunate cases where species identity is a priori unknown.

FIG. 3.22. Intraindividual color pattern variation in *Anolis punctatus* (left, top and bottom), *A. soinii* (middle), *A. biporcatus* (right). In each case, the same individual is shown in both photos.

CHAPTER 4

SPECIES ACCOUNTS

PLATE SET 1 - *DEWLAPS*

1.1 *A. aenus* p.66 RS

1.2 *A. aequatorialis* p.67

1.3 *A. agassizi* p.68
Small, apparently white.
(PERSONAL OBSERVATION OF PHOTOS)

1.4 *A. allisoni* p.69 CA

1.5 *A. altae* p.70

1.6 *A. alvarezdeltoroi* p.71

1.7 *A. amplisquamosus* p.72 TK

1.8 *A. anatoloros* p.73
"Turquoise-green (bluer on center and anteriorly, pale green elsewhere) with pale or dark greenish scales."
(UGUETO ET AL. 2007)

1.9 *A. anchicayae* p.74 GM

1.10 *A. anisolepis* p.75

1.11 *A. annectens* p.76 CBA

1.12 *A. anoriensis* p.77 JD

1.13 *A. antioquiae* (female) p.78 EA

1.14 *A. antonii* p.80

1.15 *A. apletophallus* p.81

DEWLAPS – PLATE SET 1

1.16 *A. apollinaris* p.82

1.17 *A. aquaticus* p.83

1.18 *A. arenal* p.85

1.19 *A. auratus* p.86

1.20 *A. baleatus* p.87

1.21 *A. barkeri* p.88

1.22 *A. beckeri* p.89

"Pale yellow … with rows of light brown scales that show through strongly translucent skin from one side to the other."
(MYERS AND DONNELLY 1996)

1.23 *A. bellipeniculus* p.90

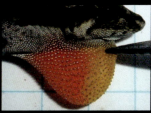
1.24 *A. benedikti* p.91

1.25 *A. bicaorum* p.92

1.26 *A. binotatus* p.93

1.27 *A. biporcatus* p.95

1.28 *A. biscutiger* p.96

1.29 *A. boettgeri* p.97

1.30 *A. bombiceps* p.98

PLATE SET 1 – *DEWLAPS*

1.31 *A. boulengerianus* p.99

1.32 *A. brasiliensis* p.100 MT

1.33 *A. brianjuliani* p.101 GK

1.34 *A. brooksi* p.102

1.35 *A. caceresae* p.103 DJ

1.36 *A. calimae* p.104

1.37 *A. callainus* p.105 DM

"Reddish-pink."
KÖHLER AND SMITH 2008

1.38 *A. campbelli* p.106

1.39 *A. capito* p.107

Dewlap color unknown.

1.40 *A. caquetae* p.109

"Bluish gray color on the gular fan when extended."
("FROM GORZULA'S FIELD NOTES" IN WILLIAMS ET AL. 1996)

1.41 *A. carlostoddi* p.110

1.42 *A. carolinensis* p.111

1.43 *A. carpenteri* p.112

1.44 *A. casildae* p.113

1.45 *A. charlesmyersi* p.114 KP

DEWLAPS – PLATE SET 1

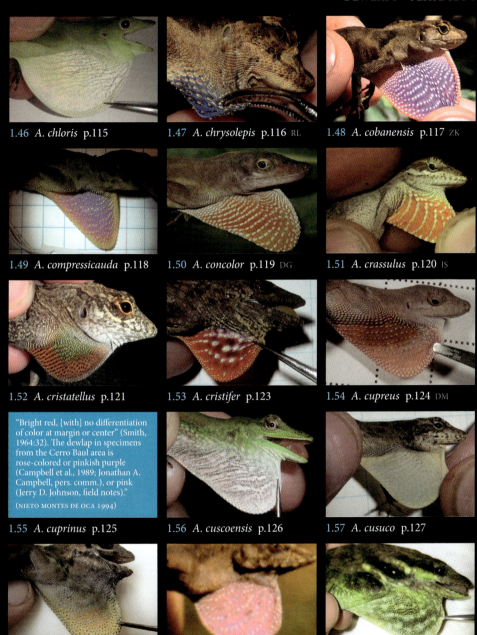

1.46 *A. chloris* p.115

1.47 *A. chrysolepis* p.116 RL

1.48 *A. cobanensis* p.117 ZK

1.49 *A. compressicauda* p.118

1.50 *A. concolor* p.119 DG

1.51 *A. crassulus* p.120 JS

1.52 *A. cristatellus* p.121

1.53 *A. cristifer* p.123

1.54 *A. cupreus* p.124 DM

"Bright red, [with] no differentiation of color at margin or center" (Smith, 1964:32). The dewlap in specimens from the Cerro Baul area is rose-colored or pinkish purple (Campbell et al., 1989; Jonathan A. Campbell, pers. comm.), or pink (Jerry D. Johnson, field notes)."

(NIETO MONTES DE OCA 1994)

1.55 *A. cuprinus* p.125

1.56 *A. cuscoensis* p.126

1.57 *A. cusuco* p.127

1.58 *A. cybotes* p.128 RG

1.59 *A. cymbops* p.129 ANi

1.60 *A. danieli* p.130

PLATE SET 1 – *DEWLAPS*

1.61 *A. datzorum* p.131

1.62 *A. dissimilis* p.132

1.63 *A. distichus* p.133

1.64 *A. dollfusianus* p.134

1.65 *A. dracula* p.136

1.66 *A. duellmani* p.137

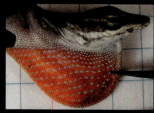

1.67 *A. dunni* p.138

1.68 *A. eleopeensis* p.139

1.69 *A. equestris* p.140

1.70 *A. eulaemus* p.141

1.71 *A. euskalerriari* p.142

1.72 *A. extremus* p.143

1.73 *A. fasciatus* p.144

1.74 *A. festae* p.146

1.75 *A. fitchi* p.147

DEWLAPS – PLATE SET 1

1.76 *A. fortunensis* p.148

1.77 *A. fraseri* p.149 TK

1.78 *A. frenatus* p.150

1.79 *A. fungosus* (female) p.152

1.80 *A. fuscoauratus* p.153

1.81 *A. gadovii* p.154

1.82 *A. gaigei* p.155

1.83 *A. garmani* p.157 IL

1.84 *A. gemmosus* p.158 TK

1.85 *A. ginaelisae* p.159

1.86 *A. gorgonae* p.160 DM

1.87 *A. gracilipes* p.161 TK

1.88 *A. granuliceps* p.162

1.89 *A. gruuo* p.163

1.90 *A. heterodermus* p.164

PLATE SET 1 – *DEWLAPS*

1.91 *A. heteropholidotus* p.166
HV

1.92 *A. hobartsmithi* p.167

1.93 *A. huilae* p.168 DB

1.94 *A. humilis* p.169

1.95 *A. hyacinthogularis* p.170
OT

1.96 *A. ibanezi* p.171

1.97 *A. immaculogularis* p.172
RA

> Distal portion of the dewlap reddish-brown, rows of pale green scales, basally yellow-orange.
> (TRANSLATED FROM RUEDA AND HERNANDEZ-CAMACHO 1988)

1.98 *A. inderenae* p.173

1.99 *A. insignis* p.174

1.100 *A. jacare* p.175 EA

1.101 *A. johnmeyeri* p.176

1.102 *A. kathydayae* p.176

1.103 *A. kemptoni* p.178

> "Dewlap pale yellow with purple gorgetal scales."
> (MCCRANIE ET AL. 2000)

1.104 *A. kreutzi* p.179

1.105 *A. kunayalae* p.180

DEWLAPS – PLATE SET 1

1.106 *A. laevis* p.181 PV

1.107 *A. laeviventris* p.182

1.108 *A. lamari* p.183

1.109 *A. latifrons* p.185

1.110 *A. lemurinus* p.186

1.111 *A. limifrons* p.187

1.112 *A. limon* p.188 JH

1.113 *A. lineatus* p.189 AO

1.114 *A. liogaster* p.190

1.115 *A. lionotus* p.191

1.116 *A. lososi* p.192 SR

1.117 *A. loveridgei* p.193 JG

1.118 *A. lynchi* p.194 OT

1.119 *A. macrinii* p.195 MK

1.120 *A. macrolepis* p.196

PLATE SET 1 – *DEWLAPS*

Dewlap pale pink with dull orange blotch proximally.
(PERSONAL OBSERVATION OF LIVE SPECIMEN)

1.121 *A. macrophallus* p.197

1.122 *A. maculigula* p.199

1.123 *A. maculiventris* p.200

1.124 *A. magnaphallus* p.201

1.125 *A. maia* p.202

1.126 *A. mariarum* p.204

1.127 *A. marmoratus* p.205

1.128 *A. marsupialis* p.206 MR

1.129 *A. matudai* p.207 ANa

1.130 *A. medemi* p.208 DM

1.131 *A. megalopithecus* p.209

1.132 *A. megapholidotus* p.210

1.133 *A. menta* p.211

1.134 *A. meridionalis* p.212 RL

1.135 *A. microlepidotus* p.214

DEWLAPS – PLATE SET 1

1.136 *A. microtus* (female) p.215
MR

1.137 *A. milleri* p.216

1.138 *A. mirus* p.217

"Red with cream gorgetal scales."
(TOWNSEND AND WILSON 1999)

Red with brown scales.
(PERSONAL OBSERVATION OF PHOTOS)

1.139 *A. monteverde* p.218

1.140 *A. morazani* p.219

1.141 *A. muralla* p.220

1.142 *A. nasofrontalis* (female) p.221 PS

1.143 *A. naufragus* p.223

1.144 *A. neblininus* p.224 IP

1.145 *A. nebuloides* p.225

1.146 *A. nebulosus* p.226

1.147 *A. neglectus* p.227 PS

Dull yellow-orange with white scales.
(PERSONAL OBSERVATION OF ILLUSTRATION BY GABRIEL UGUETO)

1.148 *A. nemonteae* p.228 AN

1.149 *A. nicefori* p.229

1.150 *A. notopholis* p.230

PLATE SET 1 – DEWLAPS

1.151 A. ocelloscapularis p.231 TK

1.152 A. omiltemanus p.232 See dewlap of A. peucephilus.

1.153 A. onca p.233 AM

1.154 A. orcesi p.234 TK

1.155 A. ortonii p.235

1.156 A. otongae p.237 OT

1.157 A. oxylophus p.238 CA

1.158 A. pachypus p.239

1.159 A. parilis p.240 DQ

1.160 A. parvauritus p.241

1.161 A. parvicirculatus p.242

1.162 A. pentaprion p.243

1.163 A. peraccae p.244

1.164 A. peruensis p.245

1.165 A. petersii p.247

DEWLAPS – PLATE SET 1

1.166 A. *peucephilus* p.248

1.167 A. *phyllorhinus* p.249 MT

Pink, purple-pink centrally with light scales.
(PERSONAL OBSERVATION OF PHOTOS)

1.168 A. *pijolense* p.250

1.169 A. *pinchoti* p.251 DG

1.170 A. *planiceps* p.252 FOm

1.171 A. *podocarpus* p.253 SR

1.172 A. *poecilopus* p.254

1.173 A. *poei* p.255 OT

1.174 A. *polylepis* p.256

1.175 A. *princeps* p.257

1.176 A. *proboscis* p.258 TK

Unknown.

1.177 A. *propinquus* p.259

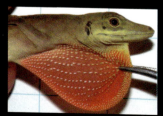
1.178 A. *pseudokemptoni* p.260

"Dewlap uniform Orange Yellow (18) with Dusky Brown (19) gorgetals."
(KÖHLER ET AL. 2007)

1.179 A. *pseudopachypus* p.261

1.180 A. *pseudotigrinus* (female) p.262 MT

61

PLATE SET 1 – *DEWLAPS*

1.181 *A. punctatus* p.263

1.182 *A. purpurescens* p.265

"Dewlap Purple (color 6) with dirty white scales."
(MCCRANIE ET AL 1993)

1.183 *A. purpurgularis* p.266

1.184 *A. purpuronectes* p.267

1.185 *A. pygmaeus* (female) p.268 AH

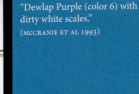

1.186 *A. quercorum* p.269

1.187 *A. quimbaya* p.270

"Pink or white-cream."
MORENO-ARIAS ET AL. 2023

1.188 *A. richteri* p.272

1.189 *A. riparius* p.273 DL

1.190 *A. rivalis* p.274

"Pink red … with suffusion of black pigment centrally, gorgetal scales white."
(KÖHLER AND MCCRANIE 2001)

1.191 *A. roatanensis* p.275

1.192 *A. robinsoni* p.276

Pale red with light scales.
(PERSONAL OBSERVATION OF PHOTOS)

1.193 *A. rodriguezii* p.277 TK

1.194 *A. rubiginosus* p.278

1.195 *A. rubribarbaris* p.280

DEWLAPS – PLATE SET 1

Pale yellow with white scales.
(TRANSLATED FROM RUEDA AND WILLIAMS 1986)

1.196 *A. ruizii* p.281

1.197 *A. sagrei* p.282

1.198 *A. salvini* p.283

1.199 *A. santamartae* p.284 JPG

1.200 *A. savagei* p.285

1.201 *A. schiedii* p.286

1.202 *A. scypheus* p.287

1.203 *A. sericeus* p.289 TK

1.204 *A. serranoi* p.290

1.205 *A. sminthus* p.291

1.206 *A. soinii* p.292

1.207 *A. solitarius* p.293

Yellow posteriorly, red-orange anteriorly, with light-greenish scales.
(PERSONAL OBSERVATION OF ILLUSTRATION BY GABRIEL UGUETO)

1.208 *A. squamulatus* p.294

1.209 *A. subocularis* p.295

1.210 *A. sulcifrons* p.296

PLATE SET 1 – *DEWLAPS*

1.211 *A. tandai* p.297 IP

1.212 *A. taylori* p.298

1.213 *A. tenorioensis* p.299

1.214 *A. tequendama* p.301

1.215 *A. tetarii* p.302 AMCZ

1.216 *A. tigrinus* p.303 RC

1.217 *A. tolimensis* p.304

1.218 *A. townsendi* p.305 OK

1.219 *A. trachyderma* p.306

1.220 *A. transversalis* p.307 OT

1.221 *A. trinitatis* p.308 RP

1.222 *A. triumphalis* p.309 RM

1.223 *A. tropidogaster* p.310

1.224 *A. tropidolepis* p.311

1.225 *A. tropidonotus* p.313

DEWLAPS – PLATE SET 1

1.226 *A. uniformis* p.314

1.227 *A. urraoi* p.315 FG

1.228 *A. ustus* p.316 JL

1.229 *A. utilensis* p.317 TK

1.230 *A. vanzolinii* p.318

"Dewlap skin black; scales of dewlap grey."
(F. PYBURN [COLLECTOR OF HOLOTYPE] IN WILLIAMS 1982)

"Dewlap 'morado'—violet, mulberry-colored."
(F. MEDEM IN WILLIAMS 1982)

1.231 *A. vaupesianus* p.319

1.232 *A. ventrimaculatus* p.321

1.233 *A. vicarius* p.322 JV

"Dark, chocolate-colored."
(FITCH AND HENDERSON 1976)

1.234 *A. villai* p.323

1.235 *A. vittigerus* p.324

"Flame Scarlet (Color 15) with white scales."
(KÖHLER AND OBERMEIER 1998)

1.236 *A. wermuthi* p.325

1.237 *A. williamsmittermeierorum* p.326

1.238 *A. woodi* p.327

1.239 *A. yoroensis* p.328 TK

1.240 *A. zeus* p.330 TK

SPECIES ACCOUNTS

— *ANOLIS AENEUS* — PLATE 1.1, 2.156

DESCRIPTION — Gray, J. E. 1840. Catalogue of the species of reptiles collected in Cuba by W. S. MacLeay, esq.; with some notes on their habits extracted from his MS. *Annals and Magazine of Natural History* 5 (1):108–115.

TYPE SPECIMEN — BMNH 1946.8.28.7.

TYPE LOCALITY — None given. Restricted by Lazell (1972) to "Point Saline, Grenada." Point Salines is at the southwest tip of Grenada, at approximately lat 12.01 lon −61.78, 20 m.

MORPHOLOGY — Body length to 80 M, 55 F; ventral scales smooth; middorsal scales enlarged in 0–2 rows; interparietal usually larger than tympanum; scales across the snout between the second canthals 5–11; toe lamellae 19–28; suboculars and supralabials in contact; supraorbital semicircles in contact; supraocular scales gradually enlarged; hindlimbs short, adpressed toe to ear or just beyond; body color gray to brown to bright green with blue and/or yellow, sometimes with speckling, a lateral stripe, and/or spots arranged in transverse bars (regarding the color pattern of *Anolis aeneus*, Lazell [1972: 80] stated "There is so much variation that verbal description is wasteful"); iris brown; male dewlap gray-green to yellow; female dewlap absent.

SIMILAR SPECIES — *Anolis aeneus* is a *large brown* or *large green* anole that can be distinguished from potentially sympatric congeners on the mainland by its large posterior head scales, wherein the supraorbital semicircles strongly abut (usually several pairs in contact) and are in contact with the interparietal. *Anolis ortonii* and *A. punctatus* may have the supraorbital semicircles in contact but *A. punctatus* is solid green (variable in *A. aeneus*) and *A. ortonii* is lichenous white to brown and smaller than *A. aeneus* (SVL to 59 mm; to 80 mm in *A. aeneus*). Also, on the mainland *A. aeneus* is most likely to be found in human-altered environments, whereas the other forms are more likely to occur in less disturbed areas.

RANGE — From 0 to 300 meters in northern and central Guyana; also throughout Grenada, the Grenadines, and associated cays, and Trinidad and Tobago.

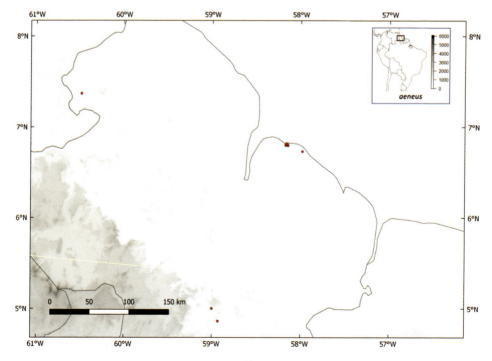

NATURAL HISTORY — Very common on vegetation and the ground. Caribbean populations of this species are well-studied by Judy Stamps (e.g., 1983) and others. Also see, e.g., Schwartz and Henderson (1991) and Lazell (1972) for information on the Caribbean populations of this species.
COMMENT — I consider *Anolis deltae* to be a synonym of one of the *roquet* species that has "invaded" Venezuela and the Guyana shield (*A. extremus, A. aeneus,* possibly *A. roquet*).

— *ANOLIS AEQUATORIALIS* — PLATE 1.2, 2.113

DESCRIPTION — Werner, F. 1894. Über einige Novitäten der herpetologischen Sammlung des Wiener zoolog. vergl. anatom. Instituts. *Zoologischer Anzeiger* 17:155–157.
TYPE SPECIMEN — NMW 16233.
TYPE LOCALITY — "Ecuador." Cisneros-Heredia's (2017) detailed study restricted the type locality to "The mountain pass between the south-western slopes of the Pichincha volcano and the northern slopes of the Atacazo volcano, at 2300 m elevation, province of Pichincha, Ecuador. Coordinates focus point: −0.278660° −78.709697°, radius: 1 km."
MORPHOLOGY — Body length to 92 M, 83 F; ventral scales smooth; middorsal scales enlarged in 0–2 rows; interparietal smaller than tympanum; scales across the snout between the second canthals 10–16; toe lamellae 19–26; suboculars and supralabials usually separated by a row of scales; supraorbital semicircles separated by at least three scales; supraocular scales subequal; hindlimbs long, adpressed toe reaches anterior to eye, often beyond tip of snout; overall body color may appear brown or green, usually with complex pattern of dorsal bands and lateral ocelli; male and female dewlap both large, black and green with blue, sometimes with orange; iris usually blue, may appear brown or green.
SIMILAR SPECIES — *Anolis aequatorialis* is a *large brown* or *large green* anole that is unlikely to be confused with sympatric forms due to its large distinctive dewlaps in males and females (dewlap black

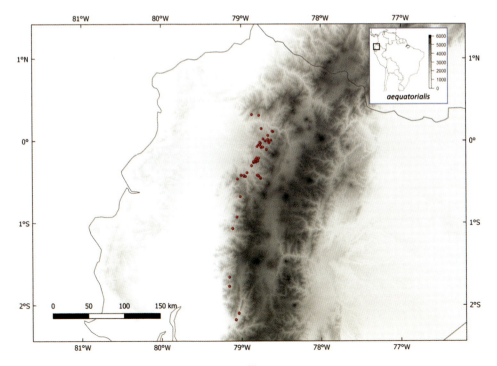

SPECIES ACCOUNTS

and green with blue, sometimes with orange). *Anolis dracula* shares similar gestalt and dewlap color with *A. aequatorialis* and geographically replaces it in far northern Ecuador (these species appear to be molecularly distinct [Yañez-Muñoz et al. 2018]).
RANGE — From 1100 to 2400 meters on the western Andean slope of central Ecuador.
NATURAL HISTORY — Common at several mid-elevation localities on the Andean slope of central Ecuador, in forest and edge habitat, often sympatrically with *Anolis gemmosus*, abundant but less abundant than *A. gemmosus*. Diurnally active on vegetation or the ground. See Fitch et al. (1976); Savit (2006); Miyata (2013); Arteaga et al. (2013); Narváez (2017); Ramirez-Jaramillo (2018); Torres-Carvajal et al. (2019); Arteaga et al. (2023).

— *ANOLIS AGASSIZI* — PLATE 1.3

DESCRIPTION — Stejneger, L. 1900. Description of two new lizards of the genus *Anolis* from Cocos and Malpelo Islands. *Bulletin of the Museum of Comparative Zoology* 36:161–4.
TYPE SPECIMEN — USNM 22101.
TYPE LOCALITY — "Malpelo Island, Pacific Ocean, off Columbia, South America." Malpelo Island is at lat 4.00 lon −81.61, 0–360 m.
MORPHOLOGY — Body length to 127 M, 97 F; ventral scales keeled; middorsal scales enlarged in 0–2 rows; interparietal equal to or larger than tympanum; scales across the snout between the second canthals 4–9; toe lamellae 29–39; suboculars and supralabials in contact; scales separating supraorbital semicircles 0–2; supraocular scales gradually enlarged; hindlimbs long, appressed toe to between eye and tip of snout; body color brown with tiny light dots; male dewlap small, white; female dewlap absent.
SIMILAR SPECIES — *Anolis agassizi* is a *giant* or *large brown* anole; the only anole found on Malpelo Island.

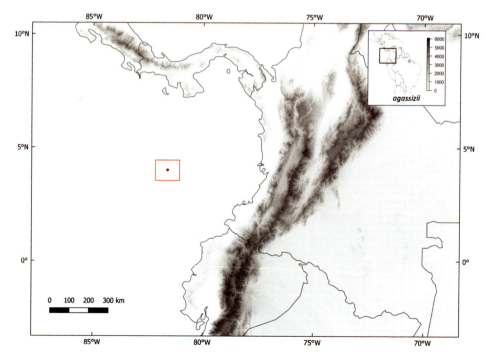

RANGE — Throughout Malpelo Island off the western coast of Colombia.
NATURAL HISTORY — Highly abundant and nonterritorial on most surfaces of Malpelo Island; sleeps on "large rocks, surfaces on hills or on man-made structures" (Lopez-Victoria et al. 2011). See Rand et al. (1975); Wolda (1975); Lopez-Victoria (2006); Lopez-Victoria et al. (2011); Phillips et al. (2019).

— *ANOLIS ALLISONI* — PLATE 1.4, 2.170

DESCRIPTION — Barbour, T. 1928. Reptiles from the Bay Islands. *Proceedings of the New England Zoölogical Club* 10:55–61.
TYPE SPECIMEN — MCZ 26725.
TYPE LOCALITY — "Coxen Hole, Ruatan, Bay Islands of Honduras." Coxen Hole, the largest town on the island of Roatán, is at lat 16.32 lon −86.54, 10 m.
MORPHOLOGY — Body length to 91 M, 66 F; ventral scales keeled; middorsal scales enlarged in 0–2 rows; interparietal larger or smaller than tympanum; scales across the snout between the second canthals 5–8; toe lamellae 23–34; suboculars and supralabials in contact; scales separating supraorbital semicircles 0–2; supraocular scales gradually enlarged; hindlimbs short, adpressed toe to ear or posterior to ear; body color green, blue anteriorly in large males, changeable to brown; male dewlap pink; female dewlap absent.
SIMILAR SPECIES — *Anolis allisoni* is a *large* or *small green* anole. Sympatric anoles on islands off Honduras are brown. In Florida, this species is unlikely to be confused with brown species *A. sagrei*, *A. cristatellus*, *A. cybotes*, *A. trinitatis*, or giant species *A. equestris* and *A. garmani*. *Anolis allisoni* differs from green species *A. callainus* in its keeled ventral scales (smooth in *A. callainus*). *Anolis allisoni* and *A. carolinensis* are very similar species but differ in external ear shape (anteroposteriorly elongate in *A. allisoni*, small and round in *A. carolinensis*).

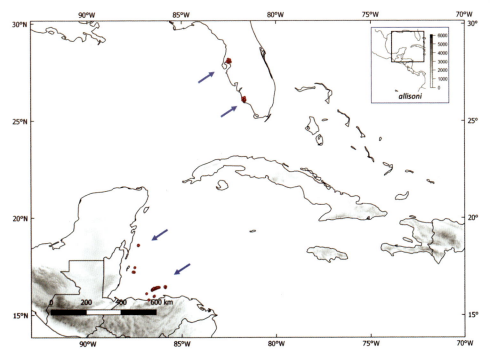

SPECIES ACCOUNTS

RANGE — From 0 to 50 meters, on Caribbean islands off Belize and Honduras, and southwestern Florida, USA. Native to Cuba (and the Bay Islands).
NATURAL HISTORY — Common; a "trunk-crown" anole (Williams 1983), active diurnally on tree trunks and branches, usually above 2 meters above ground; sleeps on leaves. See Medina et al. (2016). See Schwartz and Henderson (1991) and references therein for ecology on Cuba.

— *ANOLIS ALTAE* — PLATE 1.5, 2.33

DESCRIPTION — Dunn, E. R. 1930. Notes on Central American *Anolis*. *Proceedings of the New England Zoological Club* 12:15–24.
TYPE SPECIMEN — MCZ 29385.
TYPE LOCALITY — "Near the Acosta Farm, at about 7,000 feet elevation on the Volcan Barba in Costa Rica." Savage (1974) stated that the Acosta farm is "On the southwestern slope of Volcan Barba, at 2133 m, on the main road from Alajuela to Varablanca, Heredia Province." Given the elevation and Savage's description, I estimate the locality to be at approximately lat 10.1 lon −84.1.
MORPHOLOGY — Body length to 49 M, 51 F; ventral scales weakly keeled; middorsal scales enlarged in 0–2 rows; interparietal usually equal to or smaller than tympanum; scales across the snout between the second canthals 5–12; toe lamellae 13–17; suboculars and supralabials in contact or separated by a row of scales; scales separating supraorbital semicircles 0–3; supraocular scales gradually enlarged; hindlimbs short, appressed toe to ear or to between ear and eye; body color brown; male dewlap orange.
SIMILAR SPECIES — *Anolis altae* is a *small brown* highland anole that may be distinguished from geographically proximal similar forms as follows: *A. humilis* has greatly enlarged middorsal scale rows and a puncturelike axillary pocket (middorsals undifferentiated, puncturelike axillary pocket absent in *A. altae*); *A. pachypus*, *A. tropidolepis*, *A. cupreus*, *A. limifrons*, *A. biscutiger*, and *A. polylepis* have

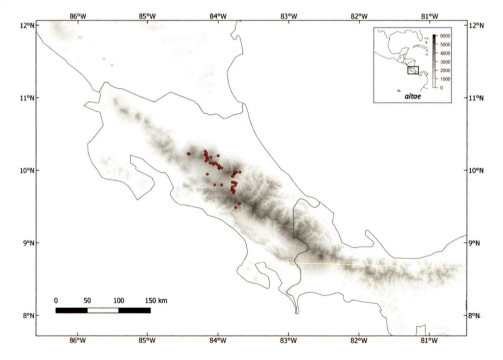

longer hindlimbs (adpressed toe reaches to beyond eye; to ear in *A. altae*); *A. kemptoni* has smooth ventral scales (weakly keeled in *A. altae*); *A. laeviventris* and *A. sericeus* have strongly keeled ventral scales. *Anolis arenal* is a lowland form (known only from 600 m; 1200+ m in *A. altae*) and differs from *A. altae* in male dewlap color (dark centrally, orange-red distally in *A. arenal*; orange in *A. altae*). *Anolis tenorioensis* and *A. monteverde* are very similar to *A. altae* but have been found only around Volcán Tenorio and Monteverde, respectively, allopatric to *A. altae*. *Anolis tenorioensis* differs from *A. altae* in its darkly spotted male dewlap; *A. monteverde* is reported to differ from *A. altae* in its unilobed hemipenes (bilobed in *A. altae*) and slightly shorter tail (Köhler 2009).

RANGE — From 1200 to 2100 meters, from central Costa Rica to the northwest limit of the Cordillera Central.

NATURAL HISTORY — Common to rare, depending on locality, in forest and forest edge; active diurnally on low vegetation, tree trunks, and ground; sleeps on leaves, grasses, or twigs from 1 to 3+ meters above ground. See Savage (2002).

— *ANOLIS ALVAREZDELTOROI* — PLATE 1.6, 2.6

DESCRIPTION — Nieto-Montes de Oca, A. 1996. A new species of *Anolis* (Squamata: Polychrotidae) from Chiapas, México. *Journal of Herpetology* 30:19–27.
TYPE SPECIMEN — ENCB 12940.
TYPE LOCALITY — "19.5 km N, 8.1 km W Ocozocoautla, Chiapas, Mexico (16 56'N, 93 27'W), 940 m."
MORPHOLOGY — Body length to 74 M, 67 F; ventral scales smooth; middorsal scales enlarged in 0–2 rows; interparietal smaller than tympanum; scales across the snout between the second canthals 7–11; toe lamellae 11–18; suboculars and supralabials in contact or occasionally separated by a row of scales; scale separating supraorbital semicircles 0–3; supraocular scales gradually enlarged; hindlimbs long

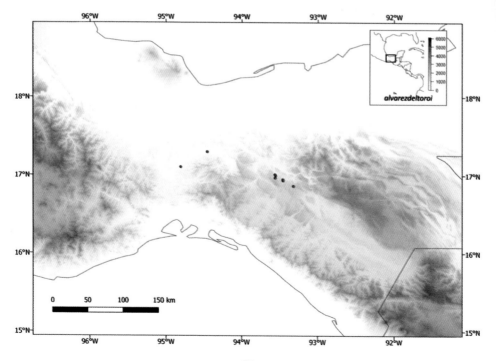

SPECIES ACCOUNTS

and body short, adpressed toe to anterior to tip of snout; body color mostly brown, with broken banding and spots; tail is sharply banded; male dewlap red; female dewlap black; iris blue.
SIMILAR SPECIES — This distinctive *large brown* anole combines a short body with long hindlmbs (adpressed toe reaches anterior to snout). These traits, its cave/rock ecology, and its dewlap colors (solid red in males; black in females) should render identification straightforward.
RANGE — From 50 to 1200 meters near the intersection of Veracruz, Oaxaca, and Chiapas States in southern Mexico.
NATURAL HISTORY — Common on boulders, rock walls, and caves in open areas; sleeps on rock surfaces including cave interiors up to at least 10 meters above ground, and (less frequently) on twigs, leaves, or vines. See Scarpetta et al. (2015).

— *ANOLIS AMPLISQUAMOSUS* — PLATE 1.7, 2.43

DESCRIPTION — McCranie, J. R., L. D. Wilson, K. L. Williams. 1992. A new species of anole of the *Norops crassulus* group (Sauria: Polychridae) from northwestern Honduras. *Caribbean Journal of Science* 28 (3–4):208–215.
TYPE SPECIMEN — KU 219924.
TYPE LOCALITY — "El Cusuco (15° 31' N, 88° 12' W), a finca located 5.6 km WSW Buenos Aires, 1550 m elevation, Sierra de Omoa, Departamento de Cortes, Honduras."
MORPHOLOGY — Body length to 46 M, 49 F; ventral scales weakly keeled; middorsal scales greatly enlarged in about 10 rows; interparietal equal to or larger than tympanum; scales across snout between the second canthals 4–8; toe lamellae 14–16; suboculars and supralabials in contact; scales between supraorbital semicircles 0–2; supraocular scales abruptly or gradually enlarged; adpressed toe to between ear and eye or to eye; body color tan; male and female dewlaps orange with light scales.

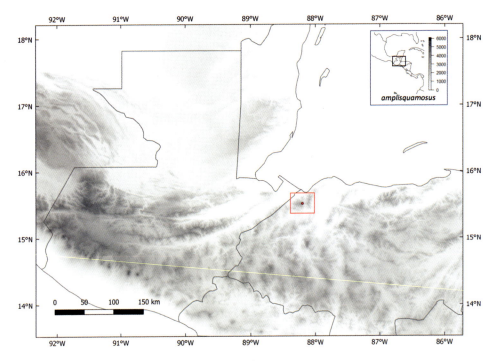

SIMILAR SPECIES — *Anolis amplisquamosus* is a *small brown* anole known only from high elevations around Cusuco National Park in Honduras. Its band of abruptly enlarged rows of keeled middorsal scales and orange dewlap in males and females serve to distinguish it from other Cusuco anoles. *Anolis yoroensis* has an orangish male dewlap and somewhat enlarged middorsal scale rows, but smaller head scales than *A. amplisquamosus* (usually 9–12 across the snout; 4–8 in *A. amplisquamosus*) and a distinctive lateral body pattern of two parallel light lines. Highland forms *A. crassulus* (found west and south of the range of *A. amplisquamosus*) and *A. caceresae* (found east of the range of *A. amplisquamosus*) differ from *A. amplisquamosus* in their strongly keeled ventral scales (keeling barely noticeable in *A. amplisquamosus*).

RANGE — From 1500 to 2000 meters in Cusuco National Park area, northwest Honduras.

NATURAL HISTORY — Common in forest; active diurnally low on vegetation or on ground; sleeps on twigs or leaves usually below 1.5 meters above ground. See Townsend and Wilson (2008); McCranie and Köhler (2015).

— *ANOLIS ANATOLOROS* — PLATE 1.8

DESCRIPTION — Ugueto, G. N., G. R. Fuenmayor, T. Barros, S. J. Sánchez-Pacheco, J. E, García-Pérez. 2007. A revision of the Venezuelan Anoles I: A new *Anolis* species from the Andes of Venezuela with the redescription of *Anolis jacare* Boulenger 1903 (Reptilia: Polychrotidae) and the clarification of the status of *Anolis nigropunctatus* Williams 1974. Zootaxa 1501:17–28.
TYPE SPECIMEN — MHNLS 17872.
TYPE LOCALITY — "San Isidro, Barinas (8°50'83'N; 70°34'23"W), 1480 m, Venezuela."
MORPHOLOGY — Body length to 68 M, 66 F; ventral scales smooth; middorsal scales enlarged in 0–2 rows; interparietal larger or smaller than tympanum; scales across the snout between the second

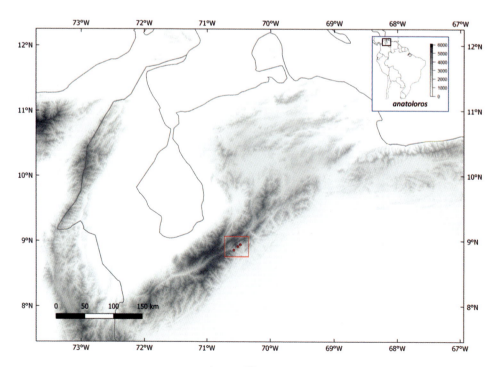

SPECIES ACCOUNTS

canthals 7–10; toe lamellae 20–24; subocular and supralabial scales in contact; scales separating supraorbital semicircles 0–1; supraocular scales gradually enlarged; body color green or brown, possibly with banding and/or speckling; tail banded; male dewlap green; female dewlap black.

SIMILAR SPECIES — *Anolis anatoloros* is a *twig* or *small green* highland anole that has not been found sympatrically with other anoles (Ugueto et al. 2007). It is at least parapatric with *A. jacare*, which differs in its tan male and tan with black blotches female dewlap (male dewlap green, female dewlap black in *A. anatoloros*).

RANGE — From 1400 to 2000 meters on the eastern slope of the Venezuelan Andes.

NATURAL HISTORY — Only information is from the type description: collected in wet forest and pasture; most individuals were found sleeping on leaves at 1.5 meters above ground.

— *ANOLIS ANCHICAYAE* — FIGURE 3.20; PLATE 1.9

DESCRIPTION — Poe, S., J. Velasco, K. Miyata, E. E. Williams. 2009. Descriptions of two nomen nudum species of *Anolis* lizard from northwestern South America. *Breviora* 516:1–16.

TYPE SPECIMEN — MCZ 160234.

TYPE LOCALITY — "Colombia, Valle, San Isidro." There are multiple San Isidros in Valle, Colombia. Given collector Helen Chin's temporally proximal travels (ascertained from her collecting localities), the known elevational distribution of this species, and information in the MCZ catalog, the type locality is likely to be in San Isidro along Rio Calima near Bahia de Malaga, lat 4.050 lon −77.064, 10 m.

MORPHOLOGY — Body length to 63 M, 56 F; ventral scales smooth; middorsal scales gradually enlarged in several rows; interparietal larger or smaller than tympanum; scales across the snout between the second canthals 7–14; toe lamellae 14–20; suboculars and supralabials in contact; scales separating supraorbital semicircles 1–3; supraocular scales gradually enlarged; appressed toe usually to between

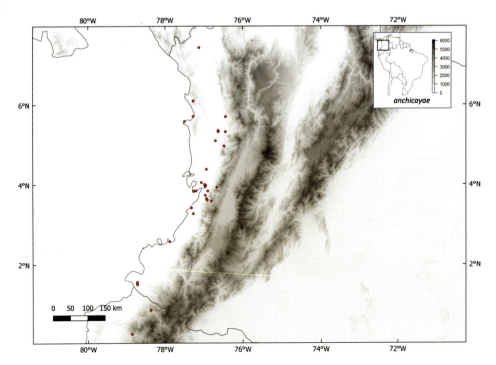

ear and eye; body color green and brown, strongly patterned with dorsal bands, blotches and/or lateral stripes; male dewlap dull greenish-yellow to yellow-orange with brown spots; female dewlap absent; iris brown to blue.

SIMILAR SPECIES — *Anolis anchicayae* is a *small brown* or *green* lowland anole patterned with stripes and/ or bars. Among potentially sympatric similar species, *A. chloris* is solid green and has keeled ventral scales (smooth in *A. anchicayae*); *A. fasciatus* is larger than *A. anchicayae* (SVL to 72 mm; to 63 mm in *A. anchicayae*) and nearly completely green (patterned with brown in *A. anchicayae*); *A. festae* and *A. peraccae* are smaller than *A. anchicayae* (SVL to 55 mm, 52 mm, respectively) and differ slightly in hindlimb length (addressed toe usually reaches to ear in *A. festae*, to eye in *A. peraccae*, to between ear and eye in *A. anchicayae*); all these species differ from *A. anchicayae* in male dewlap color (white in *A. chloris*, *A. fasciatus*, *A. peraccae*; white with dark basal blotch in *A. festae*; greenish-yellow in *A. anchicayae*).

RANGE — From 0 to 1100 meters in Pacific Colombia and extreme northern Ecuador. The map symbols are reversed in the species description for this form (i.e., the *Anolis lyra* points correspond to localities for *A. anchicayae*, and vice versa; Poe et al. 2009: figure 7).

NATURAL HISTORY — Common in forest and disturbed habitats; observed diurnally on tree trunks or branches at all visible heights; also on bushes; may appear rare at night, probably due to sleeping high on leaves. See Castro-Herrera (1988); Moreno et al. (2007); Rengifo et al. (2014, 2015, 2019, 2021); Pinto-Erazo et al. (2020); Moreno-Arias et al. (2020); Torres-Carvajal et al. (2019); Arteaga et al. (2023).

— *ANOLIS ANISOLEPIS* — PLATE 1.10, 2.24

DESCRIPTION — Smith, H. M., F. W. Burley, T. H. Fritts. 1968. A new anisolepid *Anolis* (Reptilia: Lacertilia) from Mexico. *Journal of Herpetology* 2 (3/4):147–151.
TYPE SPECIMEN — UIMNH 73899.

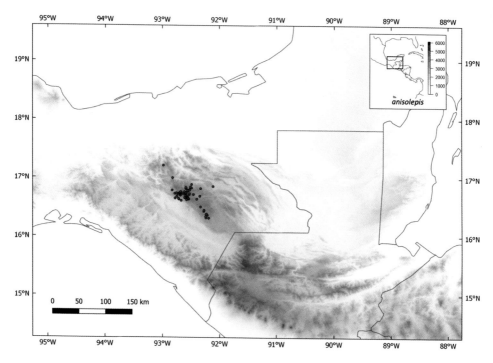

SPECIES ACCOUNTS

TYPE LOCALITY — "Ten miles SE San Cristobal Las Casas, Chiapas." This locality is at approximately lat 16.632 lon −92.533, 2200 m.

MORPHOLOGY — Body length to 47 M, 50 F; ventral scales keeled; middorsal scales greatly enlarged in several rows; interparietal usually equal to or smaller than tympanum; scales across the snout between the second canthals 5–7; toe lamellae 14–16; suboculars and supralabials in contact; scales separating supraorbital semicircles 0–2; supraocular scales gradually enlarged; adpressed toe to between ear and eye or to eye; body color brown, sometimes with middorsal dark markings and a light lateral line extending back from mouth; male and female dewlaps orange with yellow along edge and along rows of large scales.

SIMILAR SPECIES — *Anolis anisolepis* is a *small brown* highland anole. Among geographically proximal highland species, only *A. crassulus*, to the east of the range of *A. anisolepis*, shares its combination of abruptly enlarged keeled middorsal scales, strongly keeled ventral scales, and lack of a puncturelike axillary pocket. *Anolis crassulus* is not straightforwardly differentiable from *A. anisolepis* but these forms appear to be molecularly distinct (Hofmann and Townsend 2018) and should be tentatively differentiated by geography. *Anolis laeviventris* frequently is sympatric with *A. anisolepis* and differs from it in its white male dewlap (orange in *A. anisolepis*), shorter hindlimbs (adpressed toe reaches to ear; usually to eye in *A. anisolepis*), and elongate anterior nasal scale (scale undifferentiated in *A. anisolepis*).

NATURAL HISTORY — Common, active diurnally and sleeps nocturnally on low vegetation.

COMMENT — *Anolis anisolepis* sometimes is differentiated from *A. crassulus* by Smith et al.'s (1968) apparently erroneous assertion that the dewlap of *A. anisolepis* is "bright pink" (see, e.g., Hofmann and Townsend [2018: 100]). The male dewlap of *A. anisolepis* is nearly identical to that of *A. crassulus*, and among anole species found near San Cristobal, only *A. hobartsmithi*, a species that is very different from *A. crassulus*, has a dewlap that could be interpreted as "bright pink." Work remains to be done on the species boundaries of *A. crassulus*, *A. anisolepis*, and *A. caceresae* (see discussion in Hofmann and Townsend 2018).

— *ANOLIS ANNECTENS* — PLATE 1.11, 2.158

DESCRIPTION — Williams, E. E. 1974a. A case history in retrograde evolution: the *onca* lineage in anoline lizards. I. *Anolis annectens* new species, intermediate between the genera *Anolis* and *Tropidodactylus*. Breviora 421:3–10.

TYPE SPECIMEN — FMNH 5679.

TYPE LOCALITY — "Lago de Maracaibo." Lake Maracaibo (lat 9.8 lon −71.5) is a large brackish inlet in western Venezuela bordered mainly by the state of Zulia. Based on collector Osgood's (1912) itinerary and considerations of habitat, Barros et al. (2007:51) suggested El Panorama (approximately lat 10.40 lon −71.25 fide Paynter [1982]) or "the Empalado savannas 30 miles further east [of Panorama]" as potential type localities.

MORPHOLOGY — Body length to 78 M, 63 F; ventral scales keeled; middorsal scales greatly enlarged in approximately 10–20 rows; interparietal equal to or smaller than ear; scales across the snout between the second canthals 8–12; toe narrow, no laterally enlarged lamellae; suboculars and supralabials separated by a row of scales; scales separating supraorbital semicircles 1–2; supraocular scales gradually enlarged; hindlimbs long, adpressed toe to between eye and tip of snout; body color tan with white and dark brown markings; male dewlap red with black blotching; female dewlap yellow.

SIMILAR SPECIES — *Anolis annectens* is a *large brown* lowland anole. The lack of expanded toepads and unusual (for an anole) ecology of xeric ground dweller will separate *A. annectens* from all geographically proximal species except *A. onca*. *Anolis onca* lacks the enlarged middorsal scale rows of *A. annectens* and differs in male dewlap color (red-orange with vermiculations; red with black blotches in *A. annectens*). *Anolis auratus* is a ground anole that is smaller than *A. annectens* (SVL to 54 mm; to 78 mm in *A. annectens*), has a (narrow) toepad, and differs in body pattern (light lateral stripe in *A. auratus*; stripe absent or only faint and at shoulder in *A. annectens*) and male dewlap color

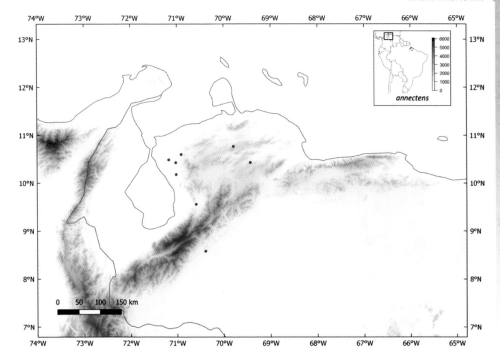

(blackish-blue). *Anolis planiceps* has a (narrow) toepad, very long hindlimbs (adpressed toe reaches past tip of snout; to between eye and tip of snout in *A. annectens*), and a different male dewlap color than *A. annectens* (pale orange-red).
RANGE — From 0 to 200 meters in northern Venezuela.
NATURAL HISTORY — In open arid areas; active diurnally on ground or vegetation up to 1.5 meters; sleeps on leaves. See Barros et al. (2007).

— *ANOLIS ANORIENSIS* — PLATE 1.12, 2.118

DESCRIPTION — Velasco, J. A., P.D.Λ. Gutiérrez-Cárdenas, A. Quintero-Angel. 2010. A new species of *Anolis* of the *aequatorialis* group (Squamata: Iguania) from the central Andes of Colombia. *The Herpetological Journal* 20:231–6.
TYPE SPECIMEN — MHUA 11719.
TYPE LOCALITY — "Colombia: Vereda El Retiro, Anori municipality, Antioquia department, 6 59 00 N, 75 8 05 W, 1374 m."
MORPHOLOGY — Body length to 94 M, 78 F; ventral scales smooth; middorsal scales enlarged in 0–2 rows; interparietal smaller than tympanum; scales across the snout between the second canthals 11–15; toe lamellae 21–25; suboculars and supralabials separated by rows of scales; scales separating supraorbital semicircles 4–5; supraocular scales subequal to gradually enlarged; hindlimbs long, adpressed toe to between eye and tip of snout; body color with complex pattern of greens and browns, appearing banded with spots and/or middorsal chevrons; tail banded; male dewlap dirty white, darker brown anteriorly; female dewlap small, brownish-white.
SIMILAR SPECIES — *Anolis anoriensis* is a *large brown* highland anole with some green dorsally. Similarly sized geographically proximal anoles may be distinguished by dewlap (*A. vittigerus*: red with

SPECIES ACCOUNTS

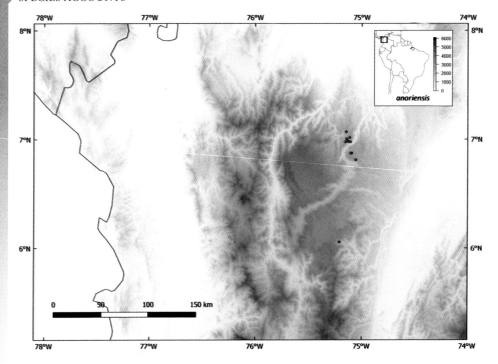

dark central spot in males, blue with dark central spot in females; *A. sulcifrons*: red with black markings in males, absent in females; *A. anoriensis*: dirty white, darker anteriorly in males, brownish-white in females) or scalation (*A. poecilopus* has rows of greatly enlarged middorsal scales and keeled ventral scales; *A. quimbaya* has heterogeneous flank scales and a serrated middorsal crest; middorsals and flank scales uniform, ventrals smooth in *A. anoriensis*). Other geographically proximal large anoles are completely or nearly completely green and possess greatly expanded toepads (*A. frenatus, A. limon, A. biporcatus, A. danieli*; toepads narrow in *A. anoriensis*). *Anolis antioquiae* and *A. megalopithecus* differ from *A. anoriensis* in their lack of a differentiated interparietal scale (present in *A. anoriensis*) and in dewlap color (male dewlap red with black in *A. megalopithecus*, female dewlap white or red-orange with black spots in *A. antioquiae*). *Anolis eulaemus*, found southwest of the range of *A. anoriensis*, differs from *A. anoriensis* in lacking a dark anterior blotch on the brown male dewlap and in tending to possess larger posterior head scales (1–3 scales between supraorbital semicircles; 4–5 in *A. anoriensis*).
RANGE — From 1350 to 1900 meters in the northern central Andes of Colombia.
NATURAL HISTORY — Common in forest; sleeps on twigs, leaves, or vines. See Molina Zuluaga and Gutiérrez-Cárdenas (2007).

— *ANOLIS ANTIOQUIAE* — PLATE 1.13, 2.124

DESCRIPTION — Williams, E. E. 1985. New or problematic *Anolis* from Colombia. IV. *Anolis antioquiae*, new species of the *eulaemus* subgroup from western Colombia. *Breviora* 482:1–9.
TYPE SPECIMEN — INDERENA 0277.
TYPE LOCALITY — "Along a road paralleling Quebrada Chaparral, Rio San Juan drainage, 10 km E of Andes (town), western Antioquia, Colombia, 2200–2300 m." A paratype (LACM 72763) that is stated

Anolis antioquiae

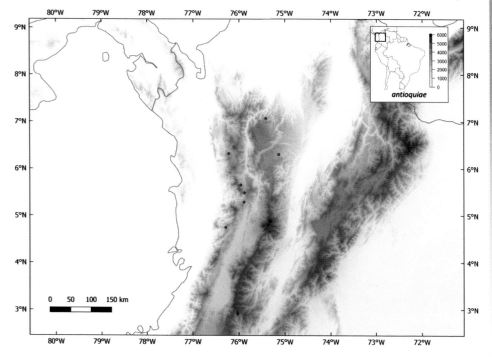

to be from "same locality as type" in the species description is listed as from "10 km W of Andes (town)" in LACM sources (my italics). Given collector P. A. Silverstone's temporally proximal collections (several from 5 km west of Andes, none from 5 km east) and the presence of Rio Chaparral on the west side of the Rio San Juan drainage (e.g., Cisneros 1880; CIA 1953), the correct type locality appears to be west rather than east of the town of Andes. I estimate this locality to be at approximately lat 5.65 lon −75.98, 2100 m; other collections from the "10 km W" site have listed elevations of 2100 m.
MORPHOLOGY — Body length to 79 F (males unknown); ventral scales smooth; middorsal scales enlarged in 0–2 rows; interparietal absent, if present, smaller than tympanum; scales across the snout between the second canthals 16–19; toe lamellae 21–22; suboculars and supralabials in contact; scales separating supraorbital semicircles 4–5; supraocular scales gradually enlarged; hindlimbs long, adpressed toe to between eye and tip of snout; body color green with black and/or white middorsal markings, sometimes with red on flanks; male dewlap unknown (but see below); female dewlap red-orange or possibly white, with black blotches.
SIMILAR SPECIES — *Anolis antioquiae* is a *large green* highland anole that differs from potentially sympatric large green anoles in body and dewlap color pattern (body green with black and/or white middorsal markings, female dewlap red-orange or possibly white with black blotches in *A. antioquiae*). *Anolis danieli* is usually solid green or faintly banded with a white or yellow stripe extending back from the mouth and a cream dewlap in female and male (stripe absent in *A. antioquiae*) and has larger head scales than *A. antioquiae* (8–12 scales across the snout between the second canthals in *A. danieli*; 16–19 in *A. antioquiae*); *A. ventrimaculatus* lacks a female dewlap (male dewlap is dirty orange); *A. anoriensis* and *A. eulaemus* have dark brown dewlaps in females and tend to appear brown or gray (green in *A. antioquiae*); *A. megalopithecus* is dark brown and red with a red and black male dewlap (female dewlap possibly similar; but see Comment below). *Anolis frenatus*, *A. latifrons*, *A. purpurescens*, *A. limon*, *A. biporcatus*, and *A. parvauritus* are large green anoles found at lower elevations than *A. antioquiae* (see species accounts for these forms).

SPECIES ACCOUNTS

RANGE — from 2100 to 2650 meters on the Pacific Andean slope of Colombia, Antioquia, Valle del Cauca, and presumably Risaralda Departments.

NATURAL HISTORY — The type series was collected from ferns next to pasture. I have found one individual of this form sleeping on a fern at the edge of a road through good habitat.

COMMENT — My group has collected females of apparent *Anolis antioquiae* from Serraniagua, near Cairo, Valle del Cauca, that possess a white (rather than red-orange) female dewlap with black blotches. This population may represent intraspecific variation within *A. antioquiae* or an undescribed species. *Anolis antioquiae* may represent the female version of *A. megalopithecus* (D. L. Mahler, personal communication 2021), which would render the name *megalopithecus* a junior synonym of *antioquiae* and substantially expand the known color variation in *A. antioquiae* (see Comments under *A. megalopithecus*).

— *ANOLIS ANTONII* — PLATE 1.14, 2.90

DESCRIPTION — Boulenger, G. A. 1908. Descriptions of new Batrachians and Reptiles discovered by Mr. M. G. Palmer in southwestern Colombia. *Annals and Magazine of Natural History* 8:515–22.

TYPE SPECIMEN — BMNH 1946.8.12.98.

TYPE LOCALITY — "San Antonio." M. G. Palmer collected extensively in Valle del Cauca, including the area on the eastern slope of the western Andes around Cauca River valley, 10 km NW of Cali, referred to as San Antonio (Paynter 1997). This locality is at roughly lat 3.51 lon −76.62, 1970 m.

MORPHOLOGY — Body length to 55 M, 44 F; ventral scales smooth; middorsal scales uniform or gradually enlarged in several rows; interparietal larger or smaller than tympanum; scales across the snout between the second canthals 6–12; toe lamellae 14–19; suboculars and supralabials in contact; scales separating supraorbital semicircles 0–2; supraocular scales gradually enlarged; adpressed toe to eye

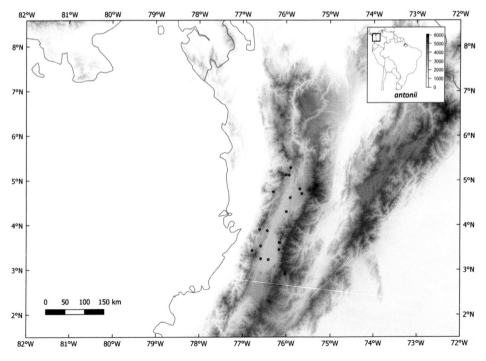

or just anterior to eye; body color brown; male dewlap pink posteriorly, orange-red anteriorly; female dewlap absent.

SIMILAR SPECIES — *Anolis antonii* is a *small brown* highland anole. Similar Colombian Andean anoles are best distinguished by male dewlap and locality: The male dewlap of *A. tolimensis* is solid pinkish-red; *A. tolimensis* is found in the Cordillera Central in upper regions of the southern Magdalena River valley. The male dewlap of *A. fuscoauratus* usually is pale pink in Colombia; *A. fuscoauratus* is found on the Amazonian slope of the Cordillera. The male dewlap of *A. mariarum* is red-orange anteriorly, yellow posteriorly; *A. mariarum* is found in the northern Cordillera Central. The male dewlap of *A. urraoi* is pale pink posteriorly, orange anteriorly; *A. urraoi* is found in the northern Cordillera Occidental. The male dewlap of *A. antonii* is red-orange to orange-red anteriorly, pale pink posteriorly; *A. antonii* is found in the central to southern Cordillera Occidental and Cordillera Central.

RANGE — From 1550 to 2300 meters, on the western Andes of southern Colombia.

NATURAL HISTORY — Common in forest and disturbed areas; active diurnally on tree trunks or other arboreal perches; sleeps on a variety of perches but common on twigs, at all visible perch heights. See Castro-Herrera (1978); Hoyos-Hoyos (2012); Gallego et al. (2012); Gallego-Carmona (2016); Vanegas-Guerrero et al. (2016); Lopez-Herrera (2016); Arango (2017).

COMMENT — See Espitia Sanabria (2023) for issues of species boundary and geographic range in *Anolis antonii*. The results of that work suggest that the taxonomy of northern and western Andean *fuscoauratus*-like anoles is a mess that encourages skepticism of many current and previous species identifications of *fuscoauratus*-like anoles from that region. Ergo, the range maps and species boundaries described in this book for Colombian Andean *fuscoauratus*-like anoles should be treated with appropriate caution.

— *ANOLIS APLETOPHALLUS* — PLATE 1.15, 2.28

DESCRIPTION — Köhler, G., J. Sunyer. 2008. Two new species of anoles formerly referred to as *Anolis limifrons* (Squamata: Polychrotidae). *Herpetologica* 64:98–101.

TYPE SPECIMEN — SMF 85307.

TYPE LOCALITY — Panama City, Metropolitan National Park (8°58'60"N, 79°32'46"W), 45 m elevation, Panama Province, Panama.

MORPHOLOGY — Body length to 47 M, 47 F; ventral scales smooth; middorsal scales enlarged in 0–2 rows; interparietal usually smaller than tympanum; scales across the snout between the second canthals 9–16; toe lamellae 15–18; suboculars and supralabials in contact; scales separating supraorbital semicircles 1–3; supraocular scales gradually enlarged; hindlimbs long, adpressed toe to anterior to eye; body color mostly brown, sometimes with patterning including middorsal rectangles; tail banded; male dewlap yellow; female dewlap usually absent.

SIMILAR SPECIES — *Anolis apletophallus* is a *small brown* anole, the common small lowland anole found from the Panama Canal east to northwest Colombia. It is generally identifiable by its smooth ventral scales (keeled in *A. gaigei*, *A. tropidogaster*, *A. humilis*, *A. auratus*, *A. vittigerus*, *A. poecilopus*), long hindlimbs (adpressed toe reaches to ear in *A. elcopeensis*, *A. pentaprion*, *A. triumphalis*; to anterior to eye in *A. apletophallus*), and yellow male dewlap. *Anolis limifrons* geographically replaces *A. apletophallus* to the northwest around the Panama Canal. It differs from *A. apletophallus* in its predominantly white male dewlap.

RANGE — From 0 to 500 meters, possibly higher in Colombia, from the Panama Canal east to northwest Colombia.

NATURAL HISTORY — Very common in forest, edge, and disturbed areas; sleeps on twigs, vines, or leaves at all visible vertical levels. See Sexton et al. (1963); Sexton (1967); Ballinger et al. (1969); Sexton et al. (1971); Sexton et al. (1972); Andrews and Rand (1974); Hover and Jenssen (1976); Jenssen and Hover (1976); Andrews (1979b); Wright (1979); Andrews and Sexton (1981); Andrews (1982); Andrews and Rand (1983); Rand et al. (1983); Wright et al. (1984); Andrews (1989); Andrews et al. (1989); Andrews and Nichols (1990); Andrews (1991); Andrews and Stamps (1994); Stapley et al. (2015); Batista et al.

SPECIES ACCOUNTS

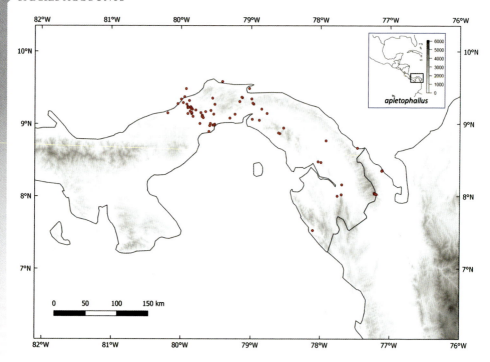

(2020); Cox et al. (2020); Neel et al. (2020); Logan et al. (2021). All references before 2008, including those listed above, considered populations of *Anolis apletophallus* to be *A. limifrons*.
COMMENT — Köhler and Sunyer (2008) split *Anolis limifrons* into three species, including *A. limifrons* sensu strictu, with a predominantly white dewlap, west of the Panama Canal and *A. apletophallus*, with a yellow dewlap, from the Panama Canal east to Darién. Stapley et al. (2011) found evidence of gene flow between these putative species, suggesting the need for additional work on species boundaries in this complex.

— *ANOLIS APOLLINARIS* — PLATE 1.16, FIG. 3.20

DESCRIPTION — Boulenger, G. A. 1919. Descriptions of two new lizards and a new frog from the Andes of Colombia. Proceedings of the Zoological Society of London 1919:79–80.
TYPE SPECIMEN — BMNH 1946.8.13.22.
TYPE LOCALITY — "Near Bogotá." Bogotá, Colombia, is located at approximately lat 4.7 lon −74.1, 2560 m.
MORPHOLOGY — Body length to 106 M, 94 F; ventral scales weakly keeled; middorsal scales enlarged in 0–2 rows; interparietal larger or smaller than tympanum; scales across the snout between the second canthals 8–13; toe lamellae 23–29; suboculars and supralabials usually separated by a row of scales; scales separating supraorbital semicircles 1–5; supraocular scales gradually enlarged; hindlimbs long, adpressed toe to eye or just beyond; body color solid green or with middorsal diamonds, frequently blue anteriorly; male and female dewlaps green.
SIMILAR SPECIES — *Anolis apollinaris* is a *giant* or *large green* highland anole. It differs from similar large to giant green highland Colombian species in its usually unpatterned green flanks and weakly keeled ventral scales. Lowland forms *Anolis frenatus*, *A. latifrons*, and *A. princeps* have oblique lateral rows of ocelli; *A. huilae* has dense yellow spotting laterally; *Anolis dracula*, *A. fitchi*, *A. eulaemus*,

Anolis aquaticus

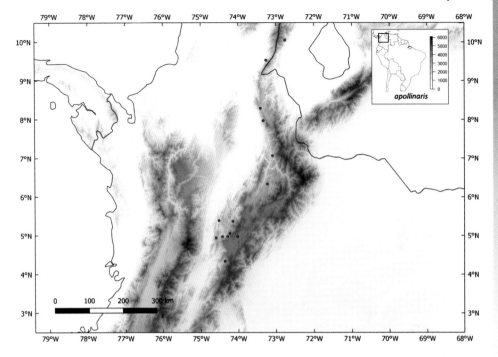

A. *anoriensis*, A. *maculigula* are strongly patterned and brown or brown and green, usually with lateral bands and/or ocelli, and possess narrow toepads (broad, usual anoline toepad in *A. apollinaris*). All the above-compared species possess smooth ventral scales (weakly keeled in *A. apollinaris*). *Anolis limon* is smaller than *A. apollinaris* (SVL to 82 mm; to 106 mm in *A. apollinaris*) and has smooth ventral scales. *Anolis danieli* is very similar to *A. apollinaris* but found north and west of it, and differs in lacking a differentiated scale anterior to the naris (i.e., *A. danieli* possesses a "circumnasal" sensu Williams et al. [1995]; differentiated elongate nasal scale in *A. apollinaris*). *Anolis propinquus*, known only from a juvenile but likely to be similar to *A. apollinaris*, is allopatric to *A. apollinaris* (to date found only near Lago Calima, well south and west of all known *A. apollinaris* localities).

RANGE — From 950 to 1750 meters along the Magdalena River valley north to César Department, Colombia, and in neighboring Venezuela in the western Serranía del Perijá.

NATURAL HISTORY — Common in forest, edge, and disturbed areas; sleeps above 1 meter above ground, on twigs or (more frequently) leaves. See Hernández-Ruiz et al. (2001); Moreno-Arias et al. (2009).

— *ANOLIS AQUATICUS* — PLATE 1.17, 2.85

DESCRIPTION — Taylor, E. H. 1956. A review of the lizards of Costa Rica. *University of Kansas Science Bulletin* 38:141–145.

TYPE SPECIMEN — KU 34276.

TYPE LOCALITY — Palmar, Puntarenas Province, Costa Rica. Savage (1974) states that "Palmar" of early authors refers to Palmar Norte, Canton de Osa, Puntarenas Province. Palmar Norte is at lat 8.962 lon −83.459, 25 m.

MORPHOLOGY — Body length to 73 M, 65 F; ventral scales keeled; middorsal scales enlarged in 0–2 rows; interparietal equal to or smaller than tympanum; scales across the snout between the second

SPECIES ACCOUNTS

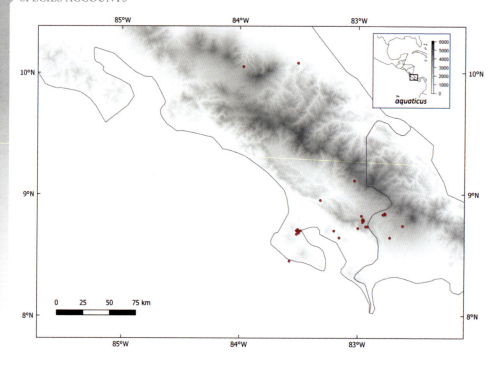

canthals 8–16; toe lamellae 12–19; suboculars and supralabials separated by rows of scales; scales separating supraorbital semicircles 2–4; supraocular scales gradually enlarged; hindlimbs long, adpressed toe to eye or anterior to eye; body color mostly brown with a light lateral line extending back from shoulder and a light line just above mouth, usually with dorsal and lateral banding and spotting and some greenish tint; male dewlap red with broad yellow streaks; female dewlap absent; iris blue.

SIMILAR SPECIES — *Anolis aquaticus* is a *semiaquatic* anole. Semiaquatic ecology and large brown body will distinguish *A. aquaticus* from all potentially sympatric congeners except *A. lionotus*, *A. oxylophus*, and *A. woodi*. *Anolis woodi* possesses a very different male dewlap (orange varying to olive in *A. woodi*; red with broad yellow streaks in *A. aquaticus*) and tends to sleep on vegetation (on boulders in splash zones of streams in *A. aquaticus*). Fellow semiaquatic anoles *A. lionotus* and *A. oxylophus* have 10–15 rows of enlarged middorsal scales (middorsals uniform in *A. aquaticus*). *Anolis capito* and *A. lemurinus* are large brown anoles that differ from *A. aquaticus* most obviously in dewlap size and color (small and red with no yellow in male, blue or white in female *A. lemurinus*; small and tan in male and female *A. capito*). *Anolis aquaticus* is extremely similar in ecology and morphology to *A. robinsoni* and *A. riparius*, but these species are not found sympatrically with *A. aquaticus* and differ in either male dewlap color (dark brown in *A. robinsoni*) or size of head scales (tiny in *A. riparius*: 15–19 scales across the snout between the second canthals; 8–16 in *A. aquaticus*).

RANGE — From 0 to 1200 meters in southern Pacific Costa Rica and adjacent western Panama. The Caribbean slope localities for this species (see map) may represent human introductions (Chaves et al. 2023).

NATURAL HISTORY — Common in forest and open areas; generally found within one meter of a stream; active diurnally on rocks, logs, or low vegetation; often abundant at boulder-filled splash zones, where it sleeps on boulders or (less frequently) low vegetation; stays underwater by "rebreathing" via an air bubble (Bocci et al. 2021). See Fitch (1975); Fitch et al. (1976); Márquez (1994); Márquez et al. (2005); Márquez et al. (2009); Muñoz et al. (2009); Eifler and Eifler (2010); Muñoz et al. (2015); Frank and Flanders (2016); Herrmann (2017); Putnam et al. (2019); Talavera et al. (2021).

— ANOLIS ARENAL — PLATE 1.18; PLATE 2.38

DESCRIPTION — Köhler, G., J Vargas. 2019. A new species of anole from Parque Nacional Volcán Arenal, Costa Rica (Reptilia, Squamata, Dactyloidae: Norops). *Zootaxa* 4608:261–278.
TYPE SPECIMEN — SMF 103506.
TYPE LOCALITY — "Costa Rica, Province Alajuela, Parque Nacional Volcán Arenal (10.46048°; –84.75434°, 585 m elevation above sea level)."
MORPHOLOGY — Body length to 42 M, 39 F; ventral scales smooth; middorsal scales enlarged in 0–2 rows; interparietal equal to or smaller than tympanum; scales across the snout between the second canthals 7–8; toe lamellae about 15; suboculars and supralabials in contact; scales separating supraorbital semicircles 1; supraocular scales gradually enlarged; hindlimbs short, adpressed toe reaches posterior to ear; body color brown; male dewlap centrally dark brown, peripherally orange-red; female dewlap small, red-orange.
SIMILAR SPECIES — *Anolis arenal* is a *small brown* lowland anole. Many potentially confusing nearby forms have strongly keeled ventral scales (*A. cupreus, A. humilis, A. laeviventris, A. lemurinus, A. oxylophus, A. sagrei, A. sericeus*; ventrals smooth in *A. arenal*). *Anolis limifrons, A. cristatellus, A. pachypus*, and *A. tropidolepis* have longer hindlimbs than *A. arenal* (adpressed toe reaches at least to eye; to posterior to ear in *A. arenal*). *Anolis carpenteri* is olive green (*A. arenal* is brown). *Anolis pentaprion* and *A. charlesmyersi* differ from *A. arenal* in lichenous white dorsal coloration and larger body size (SVL to 77 mm; to 42 mm in *A. arenal*). *Anolis altae, A. monteverde*, and *A. tenorioensis* are very similar to *A. arenal* but lack the dark central shading in the male dewlap of *A. arenal*, and none of these has been found below 1100 meters (*A. arenal* is known from 585 meters).
RANGE — Known only from the area of the type locality.
NATURAL HISTORY — Information from the type description states individuals were captured at 4 and 8 meters above ground sleeping on twigs. My observations of this species are congruent with these.

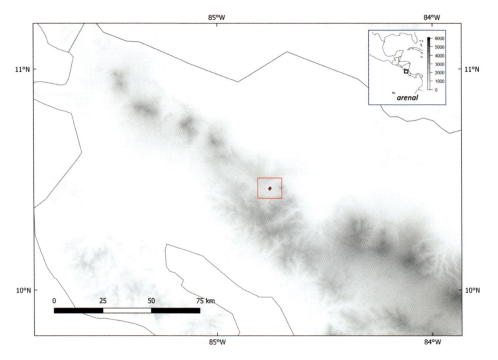

SPECIES ACCOUNTS

— *ANOLIS AURATUS* — PLATE 1.19, 2.107

DESCRIPTION — Daudin, F. M. 1802. *Histoire Naturelle, Générale et Particulière des Reptiles*, vol. 4. F. Dufart, Paris. p. 89.
TYPE SPECIMEN — Presumably MNHN. Workers have been unable to locate a holotype specimen for *Anolis auratus* (e.g., Ávila-Pires 1995: 48).
TYPE LOCALITY — "Surinam."
MORPHOLOGY — Body length to 55 M, 49 F; ventral scales keeled; middorsal scales greatly enlarged in several rows; interparietal larger than tympanum; scales across the snout between the second canthals 6–11; toe lamellae 11–18; toepad narrow, not distinct from claw; suboculars and supralabials in contact or separated by rows of scales; scales separating supraorbital semicircles 0–2; supraocular scales gradually enlarged; hindlimbs long, appressed toe to anterior to eye; body color mostly brown with light white to greenish-yellow lateral line extending back from shoulder and light line just above mouth, often with dorsal and lateral banding and spotting and some greenish tint; male dewlap bluish-black with white to greenish scales; female dewlap absent.
SIMILAR SPECIES — *Anolis auratus* is a widespread *small brown* lowland anole. This species is distinguishable from all similar sympatric anoles by its bluish-black male dewlap. In addition, its combination of abruptly enlarged keeled middorsal scale rows, strongly keeled ventral scales, inhabitation of low vegetation in open (often disturbed) habitats, lack of a puncturelike axillary pocket, and light lateral stripe are distinctive. Amazonian forest anoles (*A. bombiceps, A. brasiliensis, A. chrysolepis, A. tandai, A. scypheus*)—which may have a blue dewlap, but the blue is lighter than in *A. auratus*—are larger than *A. auratus* (maximum SVL > 70 mm; to 54 mm in *A. auratus*) and possess a dewlap in females (dewlap absent in female *A. auratus*), and tend not to inhabit open areas. *Anolis meridionalis* is similar to *A. auratus* but found south of its range and differing in possessing a greater number of supralabial scales counted from the rostral to the center of the eye (*A. meridionalis*: 6–9; *A. auratus*: 4–6).

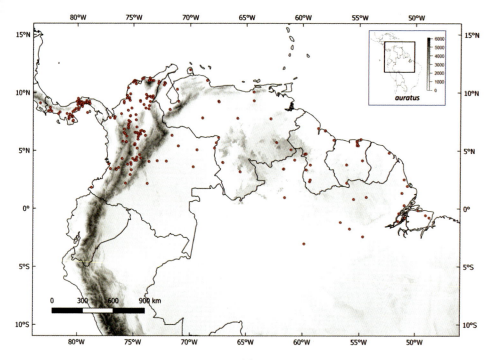

Anolis baleatus

RANGE — From 0 to 1800 meters from extreme southern Costa Rica to Pacific Colombia and northern Amazonia, including Brazil, French Guiana, Suriname, Guyana, and Venezuela.

NATURAL HISTORY — Very common in open and disturbed areas; usually observed in tall grasses or low shrubs, both active diurnally and sleeping nocturnally. See Ruthven (1922); Kastle (1963); Valdivieso and Tamsitt (1963); Sexton and Heatwole (1968); Ballinger et al. (1970); Sexton et al. (1971); Vanzolini (1972); Hoogmoed (1973); Ayala and Spain (1975); Kiester et al. (1975); Kiester (1979); Cunha (1981); Magnusson et al. (1985); Fleishman (1988a, 1988b, 1988c); Fleishman (1992); Magnusson (1993); Magnusson and Vieira da Silva (1993); Vitt and Carvalho (1995); Ávila-Pires (1995); Hernández-Ruiz et al. (2001); Magnusson et al. (2001); Faria et al. (2004); Mesquita et al. (2006); Medina-Rangel (2011, 2013); Carvajal et al. (2013); de Oliveira et al. (2014); Calderón-Espinosa and Barragán-Contreras (2014); Carvajal-Cogollo and Urbina-Cordona (2015); Medina-Rangel and Cárdenas-Árevalo (2015); Moreno-Arias et al. (2020); Pinto-Erazo et al. (2020).

— *ANOLIS BALEATUS* — PLATE 1.20, 2.171

DESCRIPTION — Cope, E. D. 1864. Contributions to the herpetology of tropical America. *Proceedings of the Academy of Natural Sciences of Philadelphia* 16:166–181.
TYPE SPECIMEN — BMNH 1946.8.29.22.
TYPE LOCALITY — "St. Domingo." Presumably Santo Domingo, Dominican Republic (lat 18.49 lon −69.87, 0 m).
MORPHOLOGY — Body length to 180 M, 151 F; ventral scales smooth to weakly keeled; middorsal scales form a crest of triangular plates; interparietal usually smaller than tympanum; scales across the snout between the second canthals 2–7; toe lamellae 27–37; suboculars and supralabials in contact or separated by a row of scales; scales separating supraorbital semicircles 1–4; supraocular scales

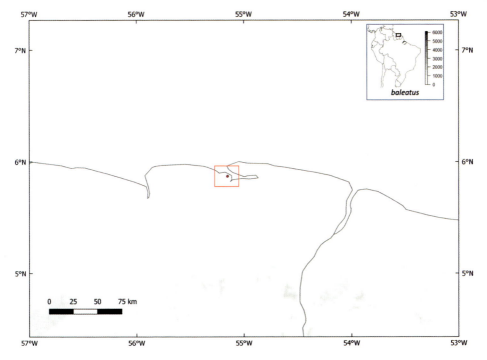

gradually enlarged; adpressed toe to between ear and eye; body color variable, usually green with brown; male and female dewlaps variable, pink to yellow-orange (see Schwartz 1974).
SIMILAR SPECIES — *Anolis baleatus* is a *giant* anole, unlikely to be confused with any anole species in Suriname due to its large size and serrated dorsal crest, and presence in disturbed areas.
RANGE — Known from "a garden at the northern edge of Paramaribo," Suriname (Hoogmoed 1981: 280). *Anolis baleatus* is native to the Dominican Republic.
NATURAL HISTORY — A "crown-giant" anole (Williams 1983) inhabiting higher aspects of trees. Ecologically unstudied in Suriname; see Schwartz and Henderson (1991) and references therein for Caribbean studies.
COMMENT — It was not clear whether the population of *Anolis baleatus* resulting from intentional release of specimens in Paramaribo, Suriname, had persisted even in 1981 (Hoogmoed 1981), and I know of no recent sightings. Obviously, more survey work is needed on this form.

— *ANOLIS BARKERI* — PLATE 1.21, 2.86

DESCRIPTION — Schmidt, K. P. 1939. A new lizard from Mexico with a note on the genus *Norops*. *Zoological Series of the Field Museum of Natural History* 24:7–10.
TYPE SPECIMEN — BMNH 1946.9.4.28.
TYPE LOCALITY — "Cascajal, Upper Uzpanapa River, Vera Cruz, Mexico." The collector, R. W. Barker, informed Meyer (1968) that "the holotype was probably taken about 3 km from Cascajal in the Las Cuevas hills, which rise to about 150 m." Of the 12 Cascajals located in Veracruz (http://mexico.places-in-the-world.com/), only one is along the Uxpanapa River (lat 17.61 lon −94.15, 15 m). This site appears to be the hypothesized type locality mapped by Meyer (1968) and Gray et al. (2016). The closest elevations to Cascajal above 100 meters are in some forested hills just over 5 km due south

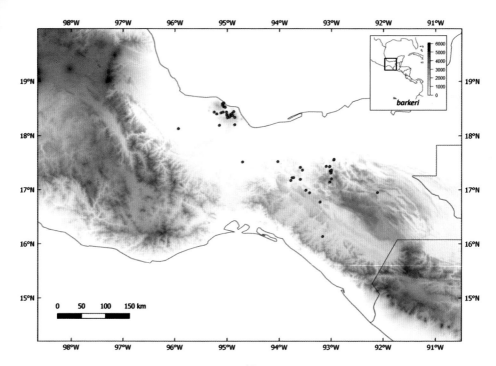

Anolis beckeri

of town; these rise to nearly 500 m. Perhaps these hills (~lat 17.55 lon −94.15) are the appropriate type locality.
MORPHOLOGY — Body length to 101 M, 78 F; ventral scales keeled; middorsal scales enlarged in 0–2 rows; interparietal usually smaller than tympanum; scales across the snout between the second canthals 7–11; toe lamellae 14–19; suboculars and supralabials usually separated by a row of scales; scales separating supraorbital semicircles 0–2; supraocular scales gradually enlarged; hindlimbs long and body very long, adpressed toe usually to between ear and eye; body color mostly brown, usually with light lateral stripe, usually with some dorsal and lateral spotting, and often a reddish tint especially anteroventrally; male dewlap red with orange; female dewlap usually absent.
SIMILAR SPECIES — *Anolis barkeri* is a *semiaquatic* anole. Similar species *A. capito* has a small dark dewlap (extending approximately to forelimbs) in males and females (large red and orange dewlap extending to chest in males, dewlap absent in females of *A. barkeri*) and a short body (adpressed toe reaches anterior to snout; to between ear and eye in *A. barkeri*). *Anolis lemurinus* has a distinctively colored, comparatively small dewlap in males and females (red in males; reddish, blue-gray, or white in females; dewlap reaches approximately to forelimbs) and long hindlimbs (adpressed toe reaches anterior to eye). *Anolis alvarezdeltoroi* has distinctive male and female dewlaps (male: solid red; female: black) and long hindlimbs (adpressed toe reaches beyond snout) and tends to be found in caves or rocky areas. *Anolis petersii* has a distinctively colored, comparatively small dewlap in males and females (pink with black blotches in males; pale with black blotches in females), and tends to be found at higher elevations than *A. barkeri* (over 1100 m; to 550 m in *A. barkeri*). *Anolis sagrei* is smaller than *A. barkeri* (SVL to 70 mm; to 101 mm in *A. barkeri*), found in disturbed areas rather than streams, and has fewer supralabial scales counted from the rostral to center of eye (4–7 in *A. sagrei*; 9–11 in *A. barkeri*). Semiaquatic *A. purpuronectes* is very similar to *A. barkeri* but found south and west of its range and differs from *A. barkeri* in its purple male dewlap.
RANGE — From 0 to 550 meters in the states of Veracruz, Tabasco, and Chiapas in southern Mexico.
NATURAL HISTORY — Common in forest along streams; active diurnally on streamside rocks, logs, and roots; sleeps on boulders or vegetation; stays underwater by "rebreathing" via an air bubble (Bocci et al. 2021). See Robinson (1962); Kennedy (1965); Brandon et al. (1966); Meyer (1968a, b); Benítez (1997); Birt et al. (2001); Powell and Birt (2001); Urbina-Cardona et al. (2006); Cardona and Reynoso (2017); Muñoz-Nolasco et al. (2019); Flores Villela (2019).

— *ANOLIS BECKERI* — PLATE 1.22, 2.77

DESCRIPTION — Boulenger, G. A. 1881. Description of a new species of *Anolis* from Yucatan. *Proceedings of the Zoological Society of London* 1881:921–922
TYPE SPECIMEN — Syntypes IRSNB 2010 (two specimens; Köhler 2010).
TYPE LOCALITY — "Yucatan." Restricted to Chichen Itza, Yucatan, Mexico (lat 20.68 lon −88.57, 40 m) by Smith and Taylor (1950). Lee (1996:234) disagreed, stating of the only recorded instance of *A. beckeri* in Yucatan: "On both distributional and ecological grounds, I consider the Chichen Itza record highly doubtful."
MORPHOLOGY — Body length to 61 M, 55 F; ventral scales smooth; middorsal scales enlarged in 0–2 rows; interparietal larger than tympanum; scales across the snout between the second canthals 6–14; toe lamellae 19–21; suboculars and supralabials in contact; scales separating supraorbital semicircles 0–1; supraocular scales gradually enlarged; hindlimbs short, adpressed toe to ear or posterior to ear; body color mostly gray with some brown, white, and black; male and female dewlaps reddish-pink.
SIMILAR SPECIES — *Anolis beckeri* is a *twig* or *small brown* lowland anole. Its combination of smooth ventral scales, lichenous dorsal color, and short hindlimbs (adpressed toe reaches to ear) will distinguish it from geographically proximal congeners. *Anolis pentaprion* is found south and east of the range of *A. beckeri* and is similar in color, limb proportion, and scalation but larger (SVL to at least 77 mm) and with a different male dewlap color (purple-pink; reddish-pink in *A. beckeri*).

SPECIES ACCOUNTS

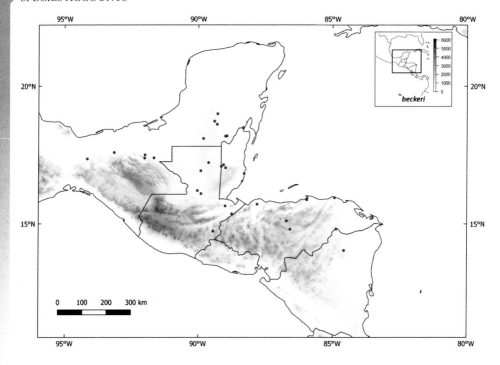

RANGE — From 0 to 1400 meters, from Caribbean Mexico south to northern Nicaragua.
NATURAL HISTORY — Common to rare in forest and disturbed areas; active diurnally on tree trunks or branches; usually sleeps on large leaves or (less frequently) twigs at least 2 meters above ground. See Smith and Kerster (1955; as *Anolis pentaprion*); Lee (1996; as *A. pentaprion*); Villareal-Benítez (1997; as *A. pentaprion*); McCranie and Köhler (2015; as *Norops beckeri*); Suarez-Varon et al. (2016); Brown et al. (2017); Urbina-Cardona and Reynoso (2017; as *A. pentaprion*).

— *ANOLIS BELLIPENICULUS* — PLATE 1.23

DESCRIPTION — Myers, C. W., M. A. Donnelly. 1996. A new herpetofauna from Cerro Yaví, Venezuela: first results of the Robert G. Goelet American Museum-Terramar expedition to the northwestern tepuis. *American Museum Novitates* 3172:31–41.
TYPE SPECIMEN — EBRG 3120.
TYPE LOCALITY — "The summit of Cerro Yavi, 2150 m elevation, Amazonas, Venezuela. lat 5.715 lon −65.906."
MORPHOLOGY — Body length to 70 M, 59 F; ventral scales smooth; middorsal scales form a crest; interparietal larger than tympanum; scales across the snout between the second canthals 4–6; toe lamellae 21; suboculars and supralabials in contact; supraorbital semicircles in contact; supraocular scales abruptly enlarged; hindlimbs short, adpressed toe to posterior to ear; body color greenish-brown; male dewlap pale yellow; female dewlap blue proximally, pale brown distally.
SIMILAR SPECIES — *Anolis bellipeniculus* is a highland *twig* anole. With its large smooth platelike head scales and short limbs and tail, *A. bellipeniculus* is unlikely to be confused with its potentially sympatric forms (*A. planiceps*, *A. fuscoauratus*), which have much smaller head scales (8–14 scales across the snout between the second canthals in *A. fuscoauratus*, *A. planiceps*; 4–6 in *A. bellipeniculus*).

Anolis benedikti

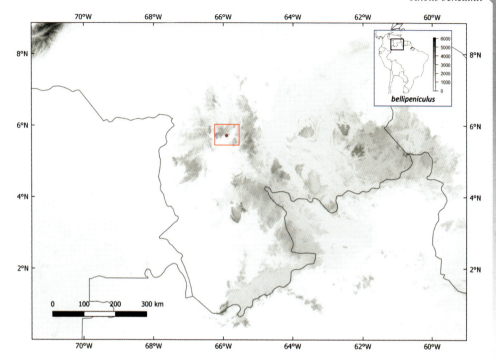

RANGE — Known only from the type locality.
NATURAL HISTORY — Only information is from the type description: One individual was found diurnally in a bromeliad; others were found sleeping at night in edge habitats 1 to 2.5 meters above ground on vegetation.

— *ANOLIS BENEDIKTI* — PLATE 1.24, 2.48

DESCRIPTION — Lotzkat, S., J. F. Bienentreu, A. Hertz, G. Köhler. 2011b. A new species of *Anolis* (Squamata: Iguania: Dactyloidae) formerly referred to as *A. pachypus* from the Cordillera de Talamanca of western Panama and adjacent Costa Rica. *Zootaxa* 3125:1–21.
TYPE SPECIMEN — SMF 90149.
TYPE LOCALITY — "The north slope of Cerro Pando, leaving the cattle trail to the right after following it to about 1000 m airline north of the large border monument (Fig. 8D), 8.9333°N, 82.7131°W, 2310 m a.s.l., Parque Internacional La Amistad (PILA), Bocas del Toro Province, Panama, close to the border with Costa Rica."
MORPHOLOGY — Body length to 49 M, 48 F; ventral scales weakly keeled; middorsal scales enlarged in 0–2 rows; interparietal usually smaller than tympanum; scales across the snout between the second canthals 11–18; toe lamellae approximately 14; suboculars and supralabials separated by rows of scales; scales separating supraorbital semicircles 4–5; supraocular scales gradually enlarged; hindlimbs long, appressed toe to anterior to eye; body color mostly brown, sometimes with tan chevrons or diamonds or a middorsal stripe; a dark band extends back from the eye, bordered dorsally and ventrally by lighter bands (lines appear to radiate from eye); male dewlap with orange posteriorly and yellow anteriorly; female dewlap small, dirty white or colored as in male.
SIMILAR SPECIES — *Anolis benedikti* is a *small brown* highland anole distinguishable from many

SPECIES ACCOUNTS

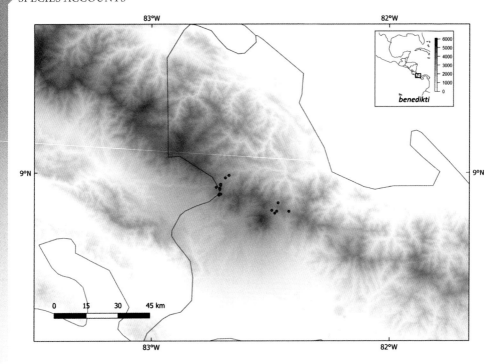

common sympatric small brown *Anolis* by its long hindlimbs (adpressed toe reaches anterior to eye in *A. benedikti*). *Anolis datzorum* is distinguishable from *A. benedikti* by male and female dewlap color (solid orange in both sexes; orange posteriorly, yellow anteriorly in male *A. benedikti*). Long-limbed *Anolis polylepis*, *A. biscutiger*, and *A. limifrons* are lowland forms with smooth ventral scales (keeled in *A. benedikti*). *Anolis benedikti* has much smaller head scales than most small potentially sympatric species (11–18 scales across the snout, 4–5 scales between supraorbital semicircles). Geographically proximal highland forms *Anolis magnaphallus*—which may be syntopic with *A. benedikti*—*A. pachypus*, and *A. pseudopachypus* are very similar to *A. benedikti* and most straightforwardly distinguished by male dewlap color (*A. magnaphallus*: red; *A. pseudopachypus*: dull yellow; *A. pachypus*: red-orange with yellow centrally).
RANGE — From 1600 to 2400 meters in the Cordillera de Talamanca near the border between Chiriquí and Bocas del Toro Provinces in Panama to extreme eastern Costa Rica.
NATURAL HISTORY — Common in forest or among bushes in open pastures and other disturbed areas; active diurnally on ground or low vegetation; sleeps on twigs or (more frequently) leaves, especially of shrubs, usually near ground but up to 2.5 meters above ground.

— *ANOLIS BICAORUM* — PLATE 1.25, 2.59

DESCRIPTION — Köhler, G. 1996b. Additions to the known herpetofauna of Isla de Utila (Islas de la Bahia, Honduras) with the description of a new species of the genus *Norops* (Reptilia: Sauria: Iguanidae). *Senckenbergiana Biologica* 76:19–28.
TYPE SPECIMEN — SMF 77100.
TYPE LOCALITY — "Honduras, Islas de la Bahia, Isla de Utila, on the trail to Rock Harbour, about 3 km north of the town of Utila (16°6.34' N; 86° 53.94' W)."

Anolis binotatus

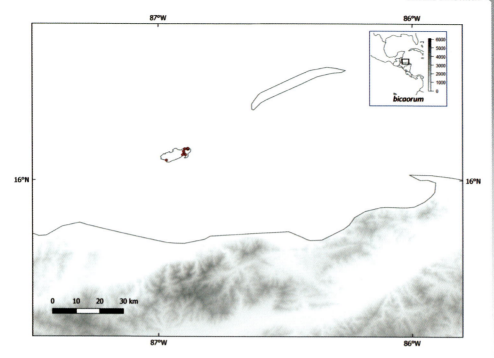

MORPHOLOGY — Body length to 73 M, 86 F; ventral scales keeled; middorsal scales enlarged in 0–2 rows; interparietal usually smaller than tympanum; scales across the snout between the second canthals 8–18; toe lamellae 17–18; suboculars and supralabials in contact or separated by rows of scales; scales separating supraorbital semicircles 1–3; supraocular scales gradually enlarged; hindlimbs long, adpressed toe to between eye and tip of snout, or to eye; body color tan, often with two light parallel lateral stripes; tail usually banded; male dewlap dull red with white scales; female dewlap white or pale blue, sometimes suffused with red.

SIMILAR SPECIES — *Anolis bicaorum* is a *large brown* anole that differs from each of the other four anole species on Utila by male and female dewlap color (dull red in males, white or pale blue in females of *A. bicaorum*). *Anolis bicaorum* further differs from Utila species *A. sagrei*, *A. sericeus*, and *A. utilensis* in its larger size (maximum SVL of these species is 70 mm; to 86 mm in *A. bicaorum*) and from *A. allisoni* and *A. utilensis* in body color (green in *A. allisoni*; lichenous white, gray, brown, and green in *A. utilensis*). Similar mainland species *A. lemurinus* has a much smaller dewlap than *A. bicaorum* (approximately to axillae; well on to chest in *A. bicaorum*).

RANGE — Throughout Utila Island off the northern coast of Honduras.

NATURAL HISTORY — Common in forest, also in edge and disturbed areas; active diurnally on tree trunks or ground; sleeps on twigs, leaves, or vines at all observable heights above 0.5 meters. See Logan et al. (2013); McCranie and Köhler (2015; as *Norops bicaorum*); Brown et al. (2017; as *Norops bicaorum*).

— *ANOLIS BINOTATUS* — PLATE 1.26, 2.101

DESCRIPTION — Peters, W.C.H. 1863. Derselbe machte eine Mittheilung über einige neue Arten der Saurier-Gattung *Anolis*. *Monatsberichte der Königlich Preussischen Akademie der Wissenschaften zu Berlin*.

SPECIES ACCOUNTS

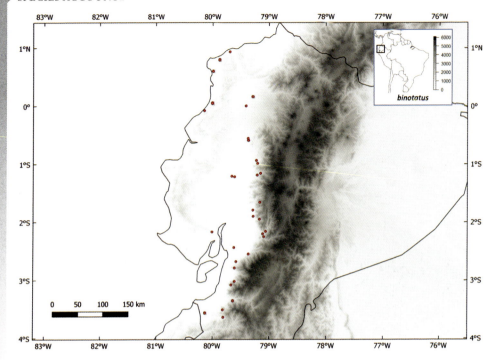

TYPE SPECIMEN — ZMB 4685.
TYPE LOCALITY — "Guayaquil." That is, Guayaquil, Guayas, Ecuador (lat −2.2 lon −79.9, 50 m).
MORPHOLOGY — SVL to 51 M, 51 F; ventral scales keeled; middorsal scales greatly enlarged in several rows; interparietal larger than tympanum; scales across the snout between the second canthals 8–13; toe lamellae 12–17; suboculars and supralabials in contact or separated by a row of scales; scales separating supraorbital semicircles 1–2; supraocular scales gradually enlarged; adpressed toe to eye or just beyond; body color brown, with lateral longitudinal stripe; male dewlap orange.
SIMILAR SPECIES — *Anolis binotatus* is a *small brown* lowland anole with keeled ventral scales. Among potentially sympatric brown anoles with keeled ventrals, *A. granuliceps* has faintly keeled, or possibly smooth ventral scales and a small male dewlap (extends to arms in *A. granuliceps*; past arms well on to chest in *A. binotatus*); *A. lynchi* is semiaquatic, exclusively associated with waterways, and has smaller head scales (16–29 scales across the snout; 8–13 in *A. binotatus*); *A. vittigerus* differs in dewlap (male: red with dark central blotch; female: usually blue or white with dark central blotch; male dewlap orange, large in *A. binotatus*). *Anolis gracilipes* is very similar to *A. binotatus* but differs in its weakly enlarged middorsal scales (middorsals abruptly enlarged and strongly keeled in *A. binotatus*) and strongly keeled head scales (some head scales smooth in *A. binotatus*).
RANGE — From 0 to 1200 meters in Pacific Ecuador. Probably also in Colombia (Tailor-Rengifo and Renteria-Moreno 2011), but I cannot find a confirmable specimen record.
NATURAL HISTORY — Common to rare in forest and disturbed areas with tree cover, usually within 1 meter of the ground actively and sleeping, often terrestrial diurnally. Miyata (2013; as *A. bitectus*); Narváez (2017); Torres-Carvajal et al. (2019); Arteaga et al. (2023).
COMMENT — I consider *Anolis bitectus* (type locality: west Ecuador) to be a junior synonym of *A. binotatus* (see appendix 1 for justification).

— *ANOLIS BIPORCATUS* — FIGURE 3.22; PLATE 1.27

DESCRIPTION — Wiegmann, A.F.A. 1834. Herpetologica Mexicana, seu Descriptio amphibiorum Novae Hispaniae, quae itineribus Comitis de Sack, F. Deppe et Chr. Guil. Schiede in Museum Zoologicum Berolinense pervenerunt. Pars prima, Saurorum species amplectens, Adiecto Systematis Saurorum Prodromo, additisque multis in hunc amphibiorum ordinem observationibus. Lüderitz, Berlin. 54 pp.

TYPE SPECIMEN — Neotype MNHN 2426 (= holotype of *Anolis copei* Bocourt 1873) designated by Köhler and Bauer (2001) in order to maintain prevailing usage of the names *A. biporcatus* and *A. petersii*. The original *biporcatus* holotype ZMB 524 is conspecific with *A. petersii* (Köhler and Bauer 2001).

TYPE LOCALITY — The neotype of *A. biporcatus* (i.e., the holotype of *Anolis copei*) is from "Santa Rosa de Pansos (Guatemala)." Panzos in Santa Rosa Department, Guatemala (shown on collector Bocourt's 1873 itinerary map) is at lat 15.40 lon −89.64, 20 m.

MORPHOLOGY — Body length to 103 M, 108 F; ventral scales keeled; middorsal scales enlarged in 0–2 rows; interparietal larger or smaller than tympanum; scales across the snout between the second canthals 8–12; toe lamellae 20–25; suboculars and supralabials in contact or separated by rows of scales; scales separating supraorbital semicircles 0–5; supraocular scales gradually enlarged; hindlimbs short, adpressed toe to ear or just beyond; body color usually solid green, changeable to mottled brown under duress; male dewlap red anteriorly, yellow posteriorly, blue proximally (sometimes nearly or completely lacking blue or yellow); female dewlap sky blue or white, with black flecks.

SIMILAR SPECIES — *Anolis biporcatus* is a distinctive widespread *large green* lowland anole with keeled ventral scales that typically is patternless green unless stressed. The small (barely to forelimbs), distinctively colored male and female dewlaps of *A. biporcatus* (blue, red, and yellow in male, sky blue or white in female) serve to distinguish this species from all potentially sympatric forms. *Anolis chloris* is solid green with keeled ventral scales, but differs from *A. biporcatus* in its gracile body (chunky

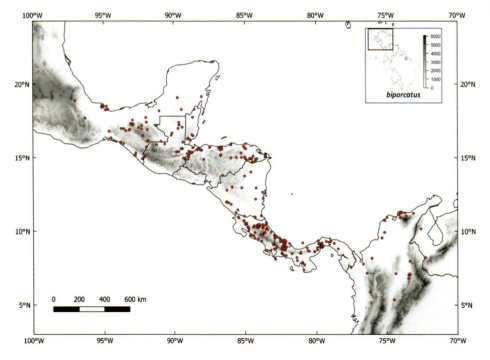

SPECIES ACCOUNTS

in *A. biporcatus*), large white male and absent female dewlaps. *Anolis parvauritus* is very similar to *A. biporcatus* and replaces it geographically to the south. *Anolis parvauritus* differs from *A. biporcatus* in possessing black scales on an otherwise typical *biporcatus*-like male dewlap.
RANGE — From 0 to 1250 meters from southern Mexico to northern Pacific and southwestern Caribbean Colombia east at least to Venezuela. The species may reach Brazil, but Williams (1966) was skeptical of the locality of the Brazil specimen, from the Vienna Museum.
NATURAL HISTORY — Common in edge and disturbed habitat; also in forest but difficult to see; often high on tree trunks and branches diurnally; sleeps high (>2 meters) on the upper surface of low canopy vegetation (e.g., at roadsides or along trails) among green leafy foliage. See Slevin (1942); Beebe (1944); Taylor (1956); Smith et al. (1972); Fitch (1975); Corn (1981); Rand and Myers (1990); Villareal-Benítez (1997); Rengifo et al. (2014, 2015); Carvajal-Cogollo and Urbina-Cardona (2015); Armstead et al. (2017); Rengifo et al. (2019); Batista et al. (2020); Hilje et al. (2020); Perez-Martinez et al. (2021).

— *ANOLIS BISCUTIGER* — PLATE 1.28, 2.29

DESCRIPTION — Taylor, E. H. 1956. A review of the lizards of Costa Rica. *University of Kansas Science Bulletin* 38:81–84.
TYPE SPECIMEN — KU 40771.
TYPE LOCALITY — "Golfito, Puntarenas Province, Costa Rica." Golfito is at lat 8.629 lon –83.156, 5 m.
MORPHOLOGY — Body length to 38 M, 39 F; ventral scales smooth; middorsal scales enlarged in 0–2 rows; interparietal larger or smaller than tympanum; scales across the snout between the second canthals 9–16; toe lamellae 14–17; suboculars and supralabials usually in contact, occasionally separated by a row of scales; scales separating supraorbital semicircles 1–2; supraocular scales gradually enlarged; appressed toe to eye; body color brown; male dewlap dirty white to gray with yellow proximal blotch.
SIMILAR SPECIES — *Anolis biscutiger* is a *small brown* lowland anole that essentially is the version of

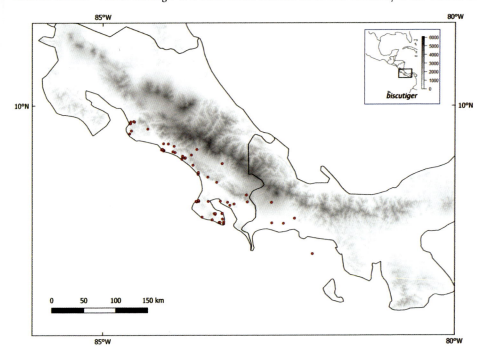

Anolis boettgeri

A. *limifrons* that inhabits southern Pacific Costa Rica and western Pacific Panama. Among small brown potentially sympatric anoles that share its long hindlimbs (adpressed hindlimb to eye) and smooth ventral scales, *A. polylepis* has a solid orange male dewlap (male dewlap mostly ashy, with orange basal blotch, in *A. biscutiger*) and smaller posterior head scales than *A. biscutiger* (2–4 scales separating supraorbital semicircles; 1–2 in *A. biscutiger*). *Anolis limifrons* has longer hindlimbs than *A. biscutiger* (adpressed hindlimb past eye; to eye in *A. biscutiger*) and a white male dewlap (male dewlap ashy in *A. biscutiger*).
RANGE — From 0 to 1150 meters in Pacific southwestern Costa Rica and adjacent Panama. I am not confident of the accuracy of my presented range map. The Panama range of this species has not been investigated. I hypothesize that all specimens recorded as *Anolis limifrons* west of the Azuero peninsula in Pacific Panama may be assignable to *A. biscutiger*.
NATURAL HISTORY — Common in open areas and forest; sleeps on twigs or leaves up to 2 meters above ground. See Echelle et al. (1971a); Fitch et al. (1976); Barquero and Bolaños (2018; as *A. limifrons*).
COMMENT — This species frequently has been treated as a synonym of *A. limifrons*, following Savage (1973, 2002). I tentatively follow Fitch et al. (1976) in recognizing this species based on their arguments and my own field observations.

— *ANOLIS BOETTGERI* — PLATE 1.29, 2.136

DESCRIPTION — Boulenger, G. A. 1911. Descriptions of new reptiles from the Andes of South America, preserved in the British Museum. *The Annals and Magazine of Natural History* 8:19–20.
TYPE SPECIMEN — Syntypes BMNH 1946.8.8.27-30.
TYPE LOCALITY — "Huancabamba, E. Peru, above 3000 feet." Barbour (1934) suggested Oxapampa (lat −10.58 lon −75.40, 1810 m), which is about 20 km south of Huancabamba in the Department of Pasco, to be a more likely type locality.

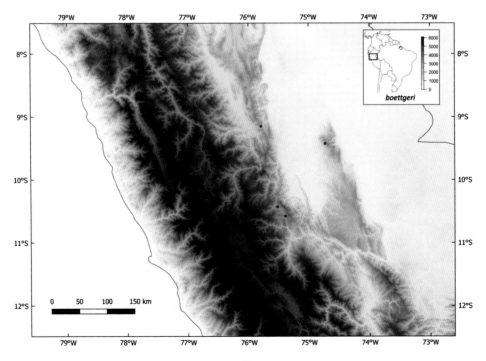

97

SPECIES ACCOUNTS

MORPHOLOGY — Body length to 69 M, 68 F; ventral scales smooth; middorsal scales enlarged in 0–2 rows; interparietal equal to or larger than tympanum; scales across the snout between the second canthals 7–11; toe lamellae 16–21; suboculars and supralabials in contact; scales separating supraorbital semicircles 0–1; supraocular scales gradually enlarged; hindlimbs long, adpressed toe to between eye and tip of snout; body color mostly green, usually with light lateral spots and a dark shoulder spot with light central flecks; male dewlap light yellow with white scales; female dewlap absent.

SIMILAR SPECIES — *Anolis boettgeri* is a *large green* highland anole. The only potentially sympatric congener likely to be confused with *A. boettgeri* is *A. punctatus*, which has a very different male dewlap (solid yellow-orange with rows of single scales; pale yellow with rows of multiple scales in *A. boettgeri*) and shorter limbs (adpressed toe usually reaches to ear; to beyond eye in *A. boettgeri*). *Anolis cuscoensis* is very similar to *A. boettgeri* but geographically separate, apparently replacing *A. boettgeri* to the south. *Anolis cuscoensis* differs from *A. boettgeri* in its shorter hindlimbs (adpressed toe reaches to between ear and eye) and white male dewlap.

RANGE — From 1600 to 2000 meters on the Amazonian Andean slope of central Peru.

NATURAL HISTORY — Unknown; probably similar to its close relative *Anolis cuscoensis* (see account for *A. cuscoensis*).

— *ANOLIS BOMBICEPS* — PLATE 1.30, 2.94

DESCRIPTION — Cope, E. D. 1876. Report on the reptiles brought by Professor James Orton from the middle and upper Amazon, and western Peru. *Journal of the Academy of Natural Sciences of Philadelphia* 8:168–169.

TYPE SPECIMEN — Lost, according to Vanzolini and Williams (1970). Unpublished notes of Ernest Williams state that syntypes of *Anolis bombiceps* were at ANSP but are "lost, fide Barbour."

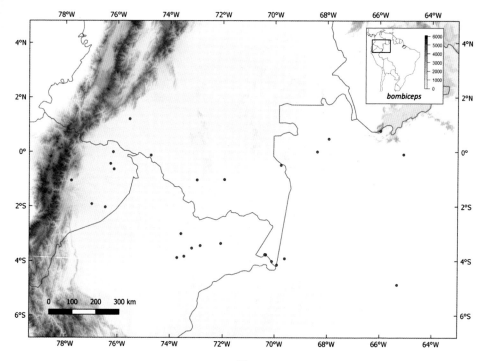

Anolis boulengerianus

TYPE LOCALITY — "Nauta." That is, Nauta, Loreto Province, Peru (lat −4.507 lon −73.585, 100 m).
MORPHOLOGY — Body length to 67 M, 71 F; ventral scales keeled; middorsal scales enlarged in 0–2 rows; interparietal usually larger than tympanum; scales across the snout between the second canthals 9–12; toe lamellae 12–18; suboculars and supralabials in contact or separated by a row of scales; scales separating supraorbital semicircles 1–4; supraocular scales subequal to gradually enlarged; hindlimbs long, adpressed toe to anterior to tip of snout; body color brown; male and female dewlaps blue.
SIMILAR SPECIES — *Anolis bombiceps* is a *small* to *large brown* lowland anole with keeled ventral scales, frequently sympatric with one of three other similar Amazonian anoles that inhabit low parts of vegetation: *A. scypheus*, *A. planiceps*, and *A. tandai*. *Anolis bombiceps* is distinguishable from each of these by female dewlap color (red distally, blue proximally in *A. scypheus*; blue with cream distally in *A. tandai*; red in *A. planiceps*; blue in *A. bombiceps*). *Anolis auratus* differs from *A. bombiceps* in its several abruptly enlarged rows of heavily keeled middorsal scales (middorsals barely enlarged in *A. bombiceps*).
RANGE — From 0 to 500 meters in Amazonian Colombia, Ecuador, Peru, and Brazil.
NATURAL HISTORY — Common in forest; active diurnally near or on ground or on low vegetation, often in shade; sleeps on leaves or twigs below 1 meter above ground. See Dixon and Soini (1975, 1986); Ávila-Pires (1995); Landauro and Morales (2007); Arteaga et al. (2023).

— *ANOLIS BOULENGERIANUS* — PLATE 1.31, 2.17

DESCRIPTION — Thominot, A. 1887. Description de trois espèces nouvelles d'*Anolis* et d'un amphisbaenien. *Bulletin de la Société Philomathique de Paris* 1887:182–183.
TYPE SPECIMEN — Syntypes MNHN 6554, 1994.1670, 1994.1671.
TYPE LOCALITY — "Téhuantépec (Mexique)." Presumably the Isthmus of Téhuantépec (i.e., approximately lat 16.6 lon −94.9).

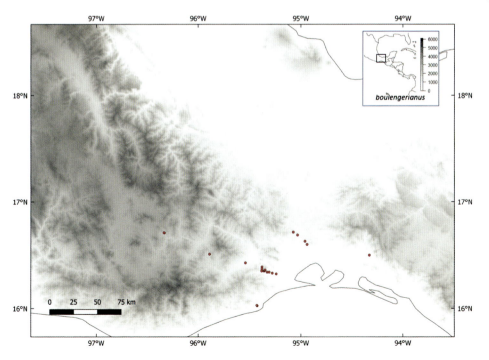

SPECIES ACCOUNTS

MORPHOLOGY — Body length to 63 M, 58 F; ventral scales keeled; middorsal scales greatly enlarged in several rows; interparietal larger than tympanum; scales across the snout between the second canthals 7–8; toe lamellae 13–14; suboculars and supralabials in contact; scales separating supraorbital semicircles 0–1; supraocular scales abruptly or gradually enlarged; adpressed toe to eye; body color brown; male dewlap orange-red with yellow along scale rows.

SIMILAR SPECIES — *Anolis boulengerianus* is a *small brown* lowland anole commonly found in coastal dry forest at the Isthmus of Téhuantépec in sympatry with *A. sericeus*. *Anolis sericeus* differs from *A. boulengerianus* in its shorter hindlimbs (adpressed toe reaches to ear; to eye in *A. boulengerianus*). *Anolis immaculogularis* is very similar to *A. boulengerianus* but differs in male dewlap color (pinkish-red without yellow along scale rows; orange-red with yellow along scale rows in *A. boulengerianus*) and geographically, replacing *A. boulengerianus* west of the Isthmus of Téhuantépec. *Anolis nebulosus* has shorter hindlimbs than *A. boulengerianus* (adpressed toe reaches to ear; to eye in *A. boulengerianus*) and larger head scales (strong contact between supraorbital semicircles, 5–7 scales across the snout; often one row separating supraorbital semicircles, 7–8 scales across the snout in *A. boulengerianus*).

RANGE — From 0 to 1350 meters on the Pacific slope of the Isthmus of Téhuantépec.

NATURAL HISTORY — Common in dry forest; active diurnally on rocks or low on trees in rocky areas; sleeps on twigs or leaves up to 1.5 meters above ground. See Fitch (1978; as *Anolis isthmicus*); Köhler et al. (2014a).

— *ANOLIS BRASILIENSIS* — PLATE 1.32

DESCRIPTION — Vanzolini, P. E., E. E. Williams. 1970. South American anoles: the geographic differentiation and evolution of the *Anolis chrysolepis* species group (Sauria, Iguanidae). *Arquivos de Zoologia* 19:85–86.

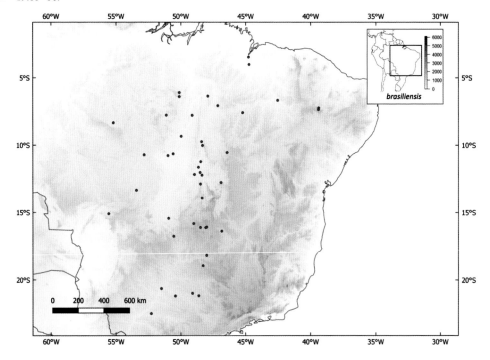

Anolis brianjuliani

TYPE SPECIMEN — MZUSP 10319.
TYPE LOCALITY — "Barra do Tapirapés, Mato Grosso, Brasil." (Vanzolini and Williams 1970:85). Ribeiro-Junior (2015) locates this site at lat −10.65 lon −50.60, 180 m.
MORPHOLOGY — Body length to 66 M, 65 F; ventral scales keeled; middorsal scales enlarged in 0–2 rows; interparietal usually larger than tympanum; scales across the snout between the second canthals 8–14; toe lamellae 15–19; subocular and supralabial scales separated by a row of scales; scales separating supraorbital semicircles 0–3; supraocular scales gradually enlarged; hindlimbs long, adpressed toe to anterior to tip of snout; body color brown; male and female dewlap blue or blue-green.
SIMILAR SPECIES — *Anolis brasiliensis* is a *small brown* lowland Amazonian ground anole that may be sympatric with other small brown anoles *A. meridionalis*, *A. fuscoauratus*, *A. ortonii*, and possibly *A. trachyderma*. *Anolis trachyderma*, *A. ortonii*, and *A. fuscoauratus* have smooth ventral scales (keeled in *A. brasiliensis*). *Anolis meridionalis* has shorter hindlimbs than *A. brasiliensis* (adpressed toe to between ear and eye; past snout in *A. brasiliensis*). *Anolis tandai* is very similar to *A. brasiliensis*, geographically replaces it to the north, and differs in female dewlap color (blue with cream distally; blue with no cream in *A. brasiliensis*).
RANGE — Sea level to 600 meters in Amazonia, central Brazil.
NATURAL HISTORY — Common in cerrado, including open areas and forest; active diurnally on low vegetation, especially tree trunks, or ground. See Ávila-Pires (1995); Costa et al. (2008; as *Anolis chrysolepis*); Vitt et al. (2008); Gainsbury and Colli (2014); Mesquita et al. (2015); Amorim and Ávila (2019).

ANOLIS BRIANJULIANI — PLATE 1.33

DESCRIPTION — Köhler, G., C.B.P. Petersen, F. R. Méndez de la Cruz. 2019. A new species of anole from the Sierra Madre del Sur in Guerrero, Mexico (Reptilia, Squamata, Dactyloidae: *Norops*). *Vertebrate Zoology* 69:145–160.

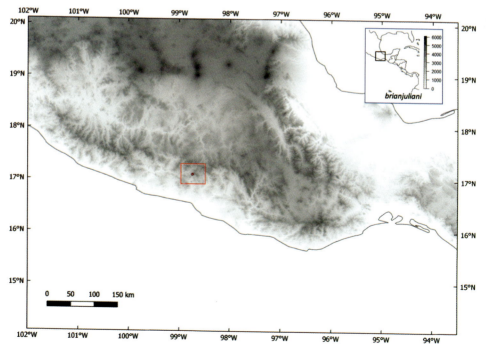

SPECIES ACCOUNTS

TYPE SPECIMEN — SMF 96360.
TYPE LOCALITY — "Near Espino Blanco, along road from Santa Cruz El Rincón to Tlapa (17.10336°N, 98.73065°W, WGS84), 2055 m, Estado de Guerrero, Mexico."
MORPHOLOGY — Body length to 50 M, adult females unknown; ventral scales smooth; middorsal scales greatly enlarged in about 10 rows; interparietal larger or smaller than tympanum; scales across the snout between the second canthals 5–6; toe lamellae 13–16; suboculars and supralabials in contact; scales separating supraorbital semicircles 0–1; supraocular scales abruptly enlarged; adpressed toe to eye; body color brown, with light lateral stripe; male dewlap pink.
SIMILAR SPECIES — *Anolis brianjuliani* is a *small brown* anole with smooth ventral scales. *Anolis brianjuliani* is distinguished from all other northern Pacific Mexico anoles with smooth ventral scales except *A. liogaster* by its solid pink dewlap and band of at least 10 rows of abruptly enlarged middorsal scales. *Anolis brianjuliani* essentially is an eastern version of *A. liogaster* with slightly larger middorsal scales than *A. liogaster*.
RANGE — Known only from the type locality.
NATURAL HISTORY — Only information is from the type description: "Collected at night while the lizards were sleeping on low vegetation along the road, 0.5–1.5 m above the ground. The habitat in the vicinity of the type locality is montane pine-oak forest."

— *ANOLIS BROOKSI* — PLATE 1.34, 2.72

DESCRIPTION — Barbour, T. 1923. Notes on reptiles and amphibians from Panama. *Occasional Papers of the Museum of Zoology*, University of Michigan 129:7–9.
TYPE SPECIMEN — MCZ 16297.
TYPE LOCALITY — "Mt. Sapo, eastern Panama, 2,500 feet elevation." Cerro Sapo is at lat 7.9753 lon −78.3619, 890 m.

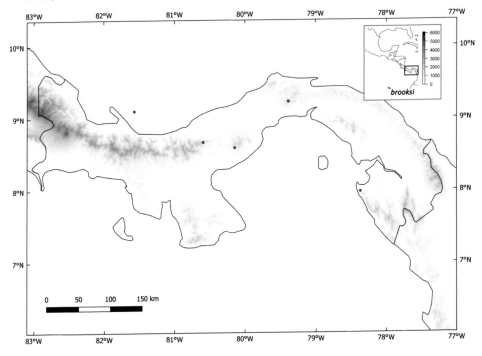

102

Anolis caceresae

MORPHOLOGY — Body length to 176 M, 134 F; ventral scales smooth; middorsal scales enlarged in 0–2 rows; interparietal smaller than tympanum; scales across the snout between the second canthals 10–11; toe lamellae 25–28; suboculars and supralabials in contact; scales separating supraorbital semicircles 3–4; supraocular scales subequal to gradually enlarged; hindlimbs short, adpressed toe to ear or posterior to ear; body color gray, brown, and/or green, usually with some banding; tail banded; male dewlap tan-peach; female dewlap white with dark lines.

SIMILAR SPECIES — *Anolis brooksi* is a *giant* anole with smooth ventral scales differentiated from other potentially sympatric large anoles with smooth ventrals by its short hindlimbs (adpressed toe past eye in *A. frenatus*, *A. latifrons*, *A. casildae*; to ear in *A. brooksi*) and dewlap colors (white in males and females of *A. frenatus*, *A. latifrons*, *A. casildae*; white with yellow edge in males, greenish with yellow edge in females in *A. kunayalae*; pink in males and females of *A. microtus*, *A. ginaelisae*; tan-peach in males, white with dark lines in females of *A. brooksi*). *Anolis brooksi* is replaced geographically to the west by similar anole *A. kathydayae*, which differs in dewlap (greenish-white in males and females).

RANGE — From 0 to 1000 meters from Darién west at least to Santa Fe, Veragua, Panama.

NATURAL HISTORY — Rare in wet forest (I have found one individual in a disturbed area); sleeps laterally on thick twigs or branches (narrower than the body) above 2 meters above ground.

COMMENT — The population of apparent *Anolis brooksi* on Isla Escudo de Veragua may represent an undescribed species.

— *ANOLIS CACERESAE* — PLATE 1.35

DESCRIPTION — Hofmann, E. P., J. H. Townsend. 2018. A cryptic new species of anole (Squamata: Dactyloidae) from the Lenca Highlands of Honduras, previously referred to as *Norops crassulus* (Cope, 1864). *Annals of the Carnegie Museum* 85:91–111.

TYPE SPECIMEN — CM 161315.

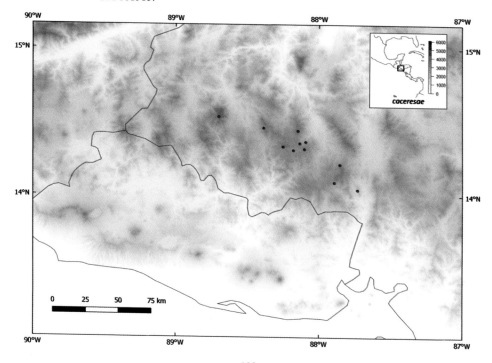

SPECIES ACCOUNTS

TYPE LOCALITY — "Near Río Agua Negra, 14.459°N, 88.385°W, 1,940 m above sea level, Reserva Biológica Opalaca, Departamento de Intibucá, Honduras."
MORPHOLOGY — Body length to 47 M, 49 F; ventral scales keeled; middorsal scales greatly enlarged in several rows; interparietal larger or smaller than tympanum; scales across the snout between the second canthals 4–7; toe lamellae 14–16; suboculars and supralabials in contact; scales separating supraorbital semicircles 0–2; supraocular scales gradually enlarged; adpressed toe to eye; body color brown with lateral stripe; male and female dewlaps orange with yellow edge.
SIMILAR SPECIES — *Anolis caceresae* is a *small brown* highland anole. Hofmann and Townsend (2018) observed *A. caceresae* sympatric with *A. laeviventris* and *A. heteropholidotus*. *Anolis laeviventris* differs from *A. caceresae* in its middorsal scales (some gradually enlarged middorsal scale rows in *A. laeviventris*; apruptly enlarged rows of keeled middorsal scales in *A. caceresae*) and short hindlimbs (adpressed toe reaches approximately to ear; to eye in *A. caceresae*). *Anolis heteropholidotus* differs from *A. caceresae* in its smooth ventral scales (keeled in *A. caceresae*). A set of other highland Honduran species share most of the "*crassulus*-group" characteristics of greatly enlarged, strongly keeled overlapping ventral scales, abruptly enlarged keeled middorsal scales, heterogeneous lateral scalation, and dewlap some shade of orange or red with *A. caceresae*. These species are more or less allo- or parapatric to *A. caceresae* and endemic to particular peaks in Honduras: Montaña de Yoro National Park (*Anolis morazani*), La Muralla National Park (*A. muralla*), Santa Bárbara Mountain (*A. rubribarbaris*), Cusuco National Park (*A. amplisquamosus*). Differences between these species and *A. caceresae* range from obvious to subtle; geography is the simplest diagnostic trait. *Anolis sminthus* differs from *A. caceresae* in its weakly keeled ventral scales, as does predominantly Nicaraguan species *A. wermuthi* (strongly keeled in *A. caceresae*). *Anolis crassulus* replaces *A. caceresae* to the west in Guatemala and differs from *A. caceresae* subtly in proportion and in some modal scale counts (these species appear to be genetically distinct [Hoffman and Townsend 2018]); as with the Honduran *crassulus* group forms, geography is your best distinguishing trait in this case.
RANGE — From 1200 to 2300 meters in southwestern Honduras, Comayagua, Intibucá, La Paz, and Lempira Departments.
NATURAL HISTORY — Common in forest and edge habitats, often in areas with forest patches mixed with agriculture; active diurnally low on vegetation or ground; sleeps on twigs or leaves below 1.5 meters above ground. See McCranie and Köhler (2015; as *Anolis crassulus*).

— *ANOLIS CALIMAE* — PLATE 1.36, 2.125

DESCRIPTION — Ayala, S., D. Harris, E. E. Williams. 1983. New or problematic *Anolis* from Colombia. I. *Anolis calimae*, new species, from the cloud forest of western Colombia. *Breviora* 475:1–11.
TYPE SPECIMEN — MCZ 158392.
TYPE LOCALITY — "San Antonio, Television Tower Mountain, Depto. Valle del Cauca, Colombia. (3 28 N 76 40 W) 1,800 m elevation."
MORPHOLOGY — Body length to 59 M, 58 F; ventral scales smooth; middorsal scales enlarged in 0–2 rows; interparietal smaller than tympanum; scales across the snout between the second canthals 7–10; toe lamellae 15–18; suboculars and supralabials in contact; scales separating supraorbital semicircles 0–2; supraocular scales gradually enlarged, occasionally subequal; hindlimbs short, adpressed toe to ear or posterior to ear; body color lichenous green with brown and cream, sometimes with lateral spots and/or middorsal bars; tail banded; male and female dewlaps blue-greenish cream.
SIMILAR SPECIES — *Anolis calimae* is a distinctive *twig* or *small green* or *small brown* highland anole, unlikely to be confused with other species due to its combination of intermediate size, complex lichenous (green and brown) body pattern, short hindlimbs (adpressed toe reaches to ear), and distinctive dewlap (cream with blue-greenish tint in both sexes).
RANGE — From 1300 to 2150 meters, on the Pacific Andean slope of Colombia.
NATURAL HISTORY — Rare in cloud forest or disturbed areas near forest; sleeps from 0.5 to 3.5 meters above ground on leaves (especially ferns) or twigs.

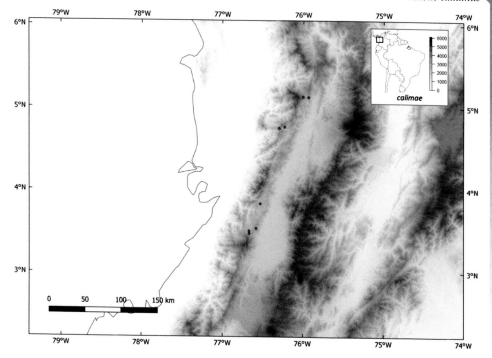

— ANOLIS CALLAINUS — PLATE 1.37, 2.162

DESCRIPTION — Köhler, G., S. B. Hedges. 2020. A replacement name for the Hispaniolan anole formerly referred to as *Anolis chlorocyanus* Duméril & Bibron, 1837. *Caribbean Herpetology* 70:1–3.
TYPE SPECIMEN — SMF 97845.
TYPE LOCALITY — "El Limón, Samaná Peninsula (19.28929, −69.43118), 30 m."
MORPHOLOGY — Body length to 80 M, 61 F; ventral scales smooth; middorsal scales enlarged in 0–2 rows; interparietal larger or smaller than tympanum; scales across the snout between the second canthals 5–9; toe lamellae 25–33; suboculars and supralabials in contact; scales separating supraorbital semicircles 0–2; supraocular scales gradually enlarged; adpressed toe approximately to ear; body color grayish-green, changeable to brown; male dewlap pale bluish-gray anteriorly, black posteriorly.
SIMILAR SPECIES — *Anolis callainus* is a *small* to *large green* anole distinguishable from potentially sympatric Florida green anoles *A. carolinensis* and *A. allisoni* by its smooth ventral scales (keeled in *A. carolinensis*, *A. allisoni*). *Anolis trinitatis* is green and has smooth ventral scales like *A. callainus*, but displays vivid dorsal coloration with blue and yellow, and further differs in its yellow male dewlap (bluish-gray and black in *A. callainus*). *Anolis garmani* (SVL to 131 mm) and *A. equestris* (SVL to 188 mm) are larger than *A. callainus* and have a serrated middorsal crest (middorsals undifferentiated in *A. callainus*). In Suriname, most anoles are brown (*A. chrysolepis*, *A. cybotes*, *A. fuscoauratus*, *A. lineatus*, *A. ortonii*). Green species *A. punctatus* differs from *A. callainus* in its smaller head scales (8–14 scales across the snout; 5–9 in *A. callainus*) and yellow-orange male dewlap. *Anolis baleatus* is much larger than *A. callainus* (SVL to 180 mm) and has a serrated dorsal crest.
RANGE — Krysko et al. (2011) reported established populations of *A. callainus* (as *A. chlorocyanus*) in Broward and Palm Beach Counties in southern Florida. Hoogmoed (1979) reported a breeding population in Paramaribo, Suriname. *Anolis callainus* is native to Hispaniola.
NATURAL HISTORY — A "trunk-crown" anole (Williams 1983), clearly adaptable to disturbed habitats.

SPECIES ACCOUNTS

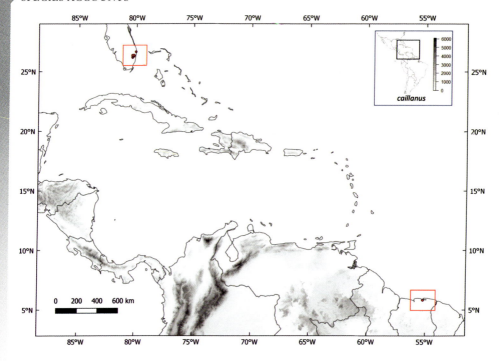

In Hispaniola, commonly observed diurnally high (above 1.5 meters above ground) on tree trunks and branches; sleeps on twigs or (more frequently) leaves above 2 meters above ground. See, e.g., Schwartz and Henderson (1991; as *A. chlorocyanus*) for information on the Caribbean populations of this species.

COMMENT — *Anolis callainus* is the Hispaniolan anole that is well-known as *A. chlorocyanus*. Köhler and Hedges (2016) demonstrated that the syntypes of *A. chlorocyanus* are conspecific with *A. coelestinus*, and appropriately attempted to retain the name *A chlorocyanus* for this form. Unfortunately, the ICZN failed to appreciate the entrenchment of the name *A. chlorocyanus*, and decreed that some other name must be applied to this population. Hence, Köhler and Hedges recognized this species as *A. callainus*.

— *ANOLIS CAMPBELLI* — PLATE 1.38

DESCRIPTION — Köhler, G., E. N. Smith. 2008. A new species of anole of the *Norops schiedei* group from western Guatemala (Squamata: Polychrotidae). *Herpetologica* 64:216–223.
TYPE SPECIMEN — UTA R-46038.
TYPE LOCALITY — "Trail to Laguna Yolnabaj, Aldea Yalambojoch, Municipalidad Nentón, Departamento de Huehuetenango, Guatemala, 1540 m elevation, 16° 00' 17" N, 91° 34' 05" W."
MORPHOLOGY — Body length to 51 M, 53 F; ventral scales keeled; middorsal scales enlarged in 0–2 rows; interparietal smaller than tympanum; scales across the snout between the second canthals 7–8; toe lamellae 15–17; suboculars and supralabials usually in contact, occasionally separated by a row of scales; scales between supraorbital semicircles 2; supraocular scales gradually enlarged; hindlimbs long, appressed toe to between eye and tip of snout; body color brown; male dewlap pink.
SIMILAR SPECIES — *Anolis campbelli* is a *small brown* highland anole that differs from geographically

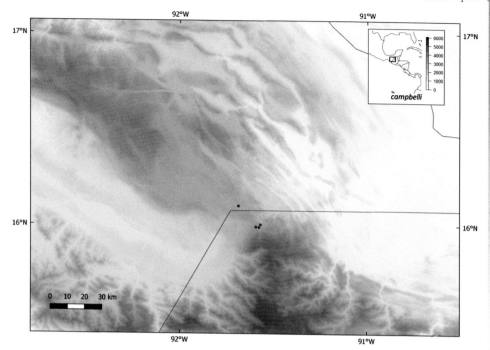

proximal small brown forms in hindlimb length (adpressed toe reaches to ear in *A. laeviventris*, *A. sericeus*; to between eye and snout in *A. campbelli*) and ventral keeling (strongly, continuously keeled scales in *A. laeviventris*, *A. sericeus*, *A. crassulus*, *A. caceresae*, *A. dollfusianus*, *A. serranoi*, *A. lemurinus*; weak keeling in *A campbelli*). *Anolis cobanensis* to the east, *A. hobartsmithi* to the west, and *A. matudai* to the south are similar highland species near *A. campbelli*; they all differ from *A. campbelli* in their predominantly smooth ventral scales (distinctly keeled in *A. campbelli*).

RANGE — From 1500 to 1700 meters in the Sierra de los Cuchumatanes, extreme western Guatemala and eastern Mexico.

NATURAL HISTORY — Only information is from the type description (Köhler and Smith 2008: 221): specimens were "active during the day along trails in secondary vegetation."

— *ANOLIS CAPITO* — PLATE 1.39, 2.62

DESCRIPTION — Peters, W.C.H. 1863. Derselbe machte eine Mittheilung über einige neue Arten der Saurier-Gattung *Anolis*. *Monatsberichte der Königlich Preussischen Akademie der Wissenschaften zu Berlin* 1:142–143.

TYPE SPECIMEN — Lectotype ZMB 4684 designated by Stuart (1955).

TYPE LOCALITY — "Costa Rica." Restricted to Palmar by Smith and Taylor (1950). Savage (1974), who lists "Costa Rica" as the type locality of *A. capito*, states that "Palmar" of early authors refers to Palmar Norte, Canton de Osa, Puntarenas Province. Palmar Norte is at lat 8.962 lon –83.459, 25 m.

MORPHOLOGY — Body length to 95 M, 102 F; ventral scales keeled; middorsal scales enlarged in 0–2 rows; interparietal equal to or smaller than tympanum; scales across the snout between the second canthals 8–13; toe lamellae 12–21; suboculars and supralabials usually in contact, occasionally separated by rows of scales; scales separating supraorbital semicircles 0–3; supraocular scales subequal to

SPECIES ACCOUNTS

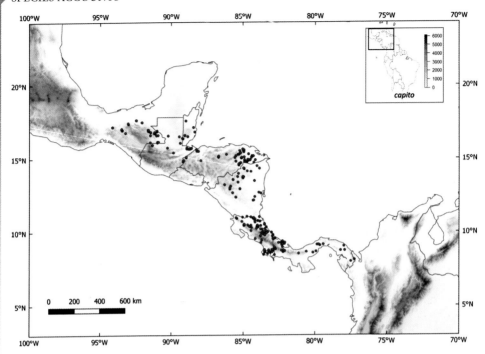

gradually enlarged; hindlimbs long, adpressed toe to anterior to tip of snout; body color lichenous brown and/or green; male and female dewlaps small, barely reaching forelimbs, yellow-brown.

SIMILAR SPECIES — *Anolis capito* is a (usually) *large brown* or (sometimes) *large green* anole that is unusual in combining large body size, brown body color, very long hindlimbs (adpressed toe reaches beyond snout), and a very short snout. Its distinctive male and female dewlaps (small, to axillae, and yellow-brown) will separate *A. capito* from all potentially sympatric species. *Anolis capito* is broadly sympatric with and occasionally confused with fellow large brown anoles *A. vittigerus* (South America to Panama) and *A. lemurinus* (Panama to Mexico). *Anolis capito* possesses a greater number of supralabial scales from the rostral to the center of the eye (8–11 in *A. capito*, usually 6–8 in *A. vittigerus*, *A. lemurinus*) and longer hindlimbs than these species (adpressed toe usually to eye or to between eye and snout in *A. lemurinus*, *A. vittigerus*).

RANGE — From 0 to 1550 meters from southeastern Mexico through Panama.

NATURAL HISTORY — Common to rare in forest; active diurnally low on tree trunks; associated with shade; sleeps on twigs, leaves, vines, or trunks of saplings up to 2 meters above ground; McCranie and Köhler (2015) reported this species to sleep on tree trunks. See Fitch (1975); Corn (1981); Lee (1996); Savage (2002); Vitt and Zani (2005); D'Cruze (2005); D'Cruze and Stafford (2006); Townsend and Wilson (2008); McCranie and Köhler (2015); Hilje et al. (2020); Perez-Martinez et al. (2021).

— ANOLIS CAQUETAE — PLATE 1.40

DESCRIPTION — Williams, E. E. 1974b. South American *Anolis*: three new species related to *Anolis nigrolineatus* and *A. dissimilis*. *Breviora* 422:8–11.
TYPE SPECIMEN — MCZ 136176 (erroneously listed as 131176 in Williams [1974b]).
TYPE LOCALITY — "Camp Soratama, Upper Apaporis, Caqueta, Colombia." Paynter (1997) describes Soratama as "camp on right bank of Rio Apaporis [0123S/6925 (USBGN)], ca. 20 km below mouth of Rio Pacoa" in department Amazonas. This location is at approximately lat 0.06 lon −71.24, 160 m, near the intersection of Caqueta, Vaupes, and Amazonas Departments.
MORPHOLOGY — Body length to 58 M (females unknown); ventral scales keeled; middorsal scales enlarged in 0–2 rows; interparietal larger than tympanum; scales across the snout between the second canthals 10; toe lamellae 22; suboculars and supralabials in contact; scales separating supraorbital semicircles 1; supraocular scales gradually enlarged; hindlimbs appear short (the single known specimen was not assessed for this trait).
SIMILAR SPECIES — *Anolis caquetae* is a lowland Amazonian form known from a single specimen (but see comment under *A. vaupesianus*). Its body and dewlap colors are unknown. The large number of toe lamellae for *A. caquetae* separates it from all potentially sympatric species except *A. punctatus* and *A. transversalis* (maximum of 20 lamellae among *A. bombiceps*, *A. scypheus*, *A. trachyderma*, *A. ortonii*, *A. fuscoauratus*; 22 in *A. caquetae*). *Anolis transversalis*, *A. fuscoauratus*, and *A. ortonii* differ from *A. caquetae* in their smooth ventral scales (keeled in *A. caquetae*). *Anolis punctatus* has smooth to weakly keeled ventral scales, likely is larger than *A. caquetae* (SVL to 89 mm) and has an enlarged snout in mature males.
RANGE — Known only from the type locality.
NATURAL HISTORY — Unknown.

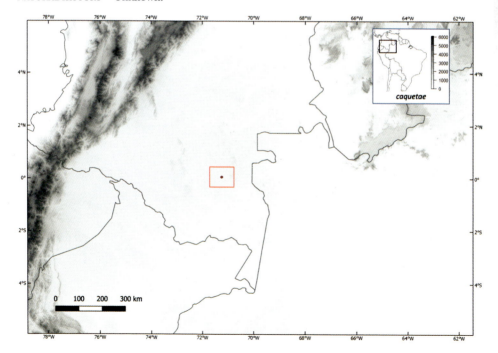

SPECIES ACCOUNTS

— *ANOLIS CARLOSTODDI* — PLATE 1.41

DESCRIPTION — Williams, E. E., M. J. Praderio, S. Gorzula. 1996. A phenacosaur from Chimanta Tepui, Venezuela. *Breviora* 506:1–15.
TYPE SPECIMEN — SCN 10351.
TYPE LOCALITY — "The southern high plateau of Abacapa-tepui (05 12' N, 62 19' W) (Chimanta V.), Estado Bolivar, Venezuela, 2,200 m."
MORPHOLOGY — Body length to 55 F (males unknown); middorsal scales form a low uneven crest of triangular plates; interparietal smaller than tympanum; ventral scales smooth; scales across the snout between the second canthals 6; toe lamellae 18; suboculars and supralabials in contact; supraorbital semicircles in contact; supraocular scales abruptly enlarged; hindlimbs apparently short; dewlap "bluish gray" in female (Williams et al. 1996).
SIMILAR SPECIES — *Anolis carlostoddi* is a highland *twig* anole potentially sympatric with *A. fuscoauratus* and/or *A. planiceps* and amply distinct from both in its large smooth head and body scales (8–14 scales across the snout in *A. fuscoauratus* and *A. planiceps*; 6 in *A. carlostoddi*) and crested middorsum (no crest in *A. fuscoauratus*, *A. planiceps*).
RANGE — Known only from the type locality.
NATURAL HISTORY — The type specimen was collected "at about 11:00 A.M. in a small crack of sandstone, near the top of a deep crevasse on a very exposed rock escarpment" with only some scrub vegetation nearby (Williams et al. 1996: 10).
COMMENT — To my knowledge this species is known from a single collected specimen but has been photographed at least once since its description (Barrio-Amorós and Fuentes-Ramos 2013).

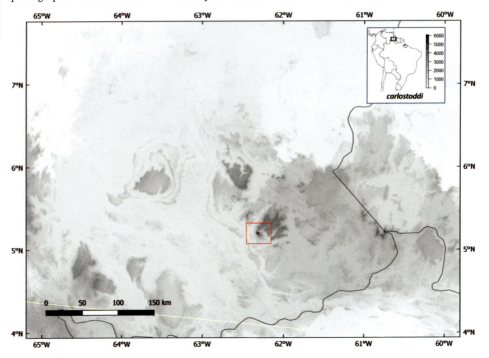

— ANOLIS CAROLINENSIS — PLATE 1.42, 2.161

DESCRIPTION — Voigt, F. S. 1832: *Das Thierreich, geordnet nach seiner Organisation: als Grundlage der Naturgeschichte der Thiere und Einleitung in die vergleichende Anatomie*, vol. 2:71.
NEOTYPE DESCRIPTION — Vance, T. 1991. Morphological variation and systematics of the green anole, *Anolis carolinensis* (Reptilia: Iguanidae). *Bulletin of the Maryland Herpetological Society* 27 (2):60–74.
TYPE SPECIMEN — Holotype undesignated, and/or lost (Vance 1991). Neotype NCSM 93545 (originally CR 862). See Vance (1991) for nomenclatural history.
TYPE LOCALITY — Neotype: "Charleston, South Carolina." Charleston, South Carolina, USA, is at lat 32.78 lon 79.93, 0 m.
MORPHOLOGY — Body length to 71 M, 54 F; ventral scales keeled; middorsal scales enlarged in 0–2 rows; interparietal larger or smaller than tympanum; scales across the snout between the second canthals 6–9; toe lamellae 19–29; suboculars and supralabials in contact; scales separating supraorbital semicircles 1–2; supraocular scales gradually enlarged; hindlimbs short, adpressed toe approximately to ear; body color green, changeable to dark brown; male dewlap pink; female dewlap usually absent.
SIMILAR SPECIES — *Anolis carolinensis* is a *small* to *large green* anole frequently sympatric with *A. sagrei*, which is a *small brown* anole easily distinguishable by body color. Giant green species *A. equestris* and *A. garmani* possess serrated dorsal crests (middorsals uniform in *A. carolinensis*). Potentially sympatric green anole *A. callainus* has smooth ventral scales (keeled in *A. carolinensis*). *Anolis allisoni* is very similar to *A. carolinensis* but differs in its anteroposteriorly elongate tympanum (round or oval in *A. carolinensis*). In Brazil, *A. carolinensis* may be distinguished from other green anoles by its keeled ventral scales (smooth in *A. dissimilis*, *A. phyllorhinus*, *A. nasofrontalis*, *A. pseudotigrinus*, *A. neglectus*). Green species *A. punctatus* is larger than *A. carolinensis* (SVL to 89 mm; to 71 mm in *A. carolinensis*) and further differs in its smaller head scales (8–14 scales across the snout; 6–9 in *A. carolinensis*).

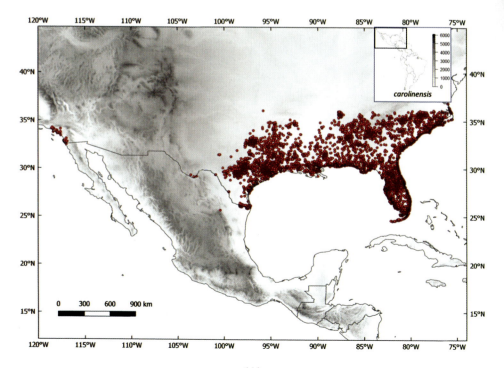

SPECIES ACCOUNTS

RANGE — Within the geographic purview of this book, *Anolis carolinensis* is known from southeastern mainland USA, Mexico, and Brazil (usually recognized there as *A. porcatus*). This species also inhabits Hawaii, Guam, Japan, Saipan, Palau, and several Caribbean islands.

NATURAL HISTORY — Very common in areas with trees, including forest, edge, and disturbed habitats; a "trunk-crown" anole (Williams 1983); active diurnally on tree branches and trunks above 2 meters above ground; sleeps on twigs or leaves above 1.5 meters above ground. See Hamlett (1952); Gordon (1956); Ruby (1984); Meshaka (1999a, b; as *A. porcatus*); Jenssen et al. (1995a, b; 1996); Nunez et al. (1997); Jenssen and Nunez (1998); Jenssen et al. (1998); Campbell (2000); Cosgrove et al. (2002); Lailvaux et al. (2004); Irschick et al. (2005a, b); Stehle et al. (2017); Tiatragul et al. (2019); Culbertson and Herrmann (2019).

COMMENT — This species is the "native" anole of the United States. I include Floridian *Anolis* "*porcatus*" in *A. carolinensis* based on Wegener et al. (2019).

— ANOLIS CARPENTERI — PLATE 1.43, 2.64

DESCRIPTION — Echelle, A. A., A. F. Echelle, H. S. Fitch. 1971b. A new anole from Costa Rica. *Herpetologica* 27:354–362.

TYPE SPECIMEN — KU 132506.

TYPE LOCALITY — "East bank of Rio Reventazon, 500+ m elev, about 7 km ESE Turrialba, where Highway 10 crosses the river, Cartago Province, Costa Rica." This location is at lat 9.881 lon −83.649, 560 m.

MORPHOLOGY — Body length to 45 M, 47 F; middorsal scales enlarged in 0–2 rows; interparietal larger or smaller than tympanum; ventral scales smooth; scales across the snout between the second canthals 9–17; toe lamellae 13–18; suboculars and supralabials in contact or separated by rows of scales; scales separating supraorbital semicircles 1–3; supraocular scales gradually enlarged; adpressed toe to between the ear and eye; body color olive green, frequently with some mottling, changeable to dark

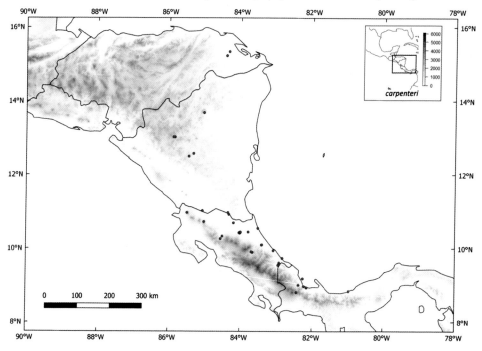

brown; tail usually banded; male dewlap orange; female dewlap absent.
SIMILAR SPECIES — *Anolis carpenteri* is a *small green* anole. All comparably sized potentially sympatric anoles are brown or whitish. *Anolis charlesmyersi*, *A. salvini*, *A. pentaprion*, and *A. fungosus* may appear green but differ from *A. carpenteri* in their red or purple male dewlaps (orange in *A. carpenteri*) and slightly shorter hindlimbs (adpressed toe reaches to shoulder or ear; to between ear and eye in *A. carpenteri*).
RANGE — From 0 to 1100 meters from Honduras to Caribbean slope Costa Rica and western Panama.
NATURAL HISTORY — Common in forest; occupies diverse arboreal perch substrates and heights diurnally and nocturnally. See Echelle et al. (1971a); Fitch (1975); Fitch et al. (1976); Corn (1981); Guyer and Donnelly (2005); Savage (2002); McCranie and Köhler (2015); Hilje et al. (2020); Perez-Martinez et al. (2021).

— *ANOLIS CASILDAE* — PLATE 1.44

DESCRIPTION — Arosemena, F. A., R. Ibáñez, F. de Sousa. 1991. Una especie nueva de *Anolis* (Squamata: Iguanidae) del grupo *latifrons* de Fortuna, Panamá. *Revista de Biologia Tropical* 39:255–262.
TYPE SPECIMEN — MVUP 755.
TYPE LOCALITY — "En las márgenes de Quebrada Frank (8° 44' N, 82° 13' O), a 1100 m sobre el nivel del mar, en el lado oeste de la carretera Chiriquí-Bocas del Toro, en un área de bosque primario dentro de la Reserva Forestal de Fortuna, en la Cordillera Central, Provincia de Chiriquí, República de Panamá."
MORPHOLOGY — Body length to 114 M, 99 F; ventral scales smooth; middorsal scales enlarged in 0–2 rows; interparietal smaller than tympanum; scales across the snout between the second canthals 11–17; toe lamellae 21–26; suboculars and supralabials in contact or separated by rows of scales; scales separating supraorbital semicircles 2–5; supraocular scales subequal to gradually enlarged; hindlimbs long, adpressed toe extends to between eye and snout or occasionally anterior to snout; body patterned green, sometimes with banding; male dewlap white with greenish scales; female dewlap dark.

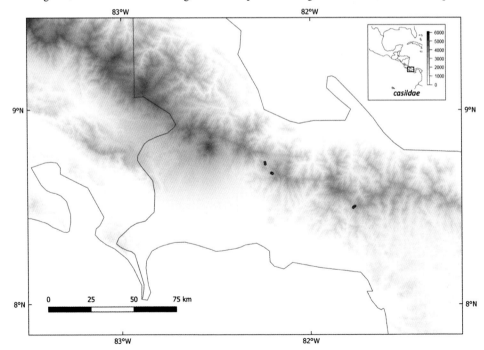

SPECIES ACCOUNTS

SIMILAR SPECIES — *Anolis casildae* is a *giant anole* distinguishable from fellow nearby giant anoles *A. insignis*, *A. brooksi*, *A. ginaelisae*, *A. savagei*, and *A. kathydayae* by its long hindlimbs (adpressed toe reaches to ear or to between ear and eye in these species; to between eye and snout in *A. casildae*). *Anolis kunayalae* differs from *A. casildae* in toepad structure (narrow, not distinct from claw, with 11–15 lamellae in *A. kunayalae*; broad, distinct, with 21–26 lamellae in *A. casildae*). *Anolis biporcatus* differs from *A. casildae* in ventral scalation (keeled in *A. biporcatus*; smooth in *A. casildae*). *Anolis casildae* is indistinguishable in scalation from *A. frenatus* but differs in some details of color pattern: *Anolis casildae* has a strongly patterned female dewlap (white in *A. frenatus*), may appear banded or solid green (*A. frenatus* displays distinct regular oblique rows of dark ocelli), and has dark green or yellow scales on the white male dewlap (scales pale, colored similarly to dewlap skin in *A. frenatus*).
RANGE — From 950 to 1750 meters in the Fortuna area of Panama.
NATURAL HISTORY — Common in forest; active diurnally on tree trunks below 2 meters above ground; sleeps on branches or leaves above 1 meter above ground. See Nicholson et al. (2001).

— *ANOLIS CHARLESMYERSI* — PLATE 1.45

DESCRIPTION — Köhler, G. 2010. A revision of the Central American species related to *Anolis pentaprion* with the resurrection of *A. beckeri* and the description of a new species (Squamata: Polychrotidae). Zootaxa 2354:6–9.
TYPE SPECIMEN — SMF 89688.
TYPE LOCALITY — "From trail to Río Majagua, Los Algarrobos, 8.48927°N, 82.43333°W, ca. 130 m a.s.l., Province Chiriquí, Panama."
MORPHOLOGY — Body length to 78 M, 67 F; middorsal scales enlarged in 0–2 rows; interparietal larger than tympanum; ventral scales smooth; scales across the snout between the second canthals 6–9; toe lamellae 22–24; suboculars and supralabials in contact; scales separating supraorbital semicircles 0–2;

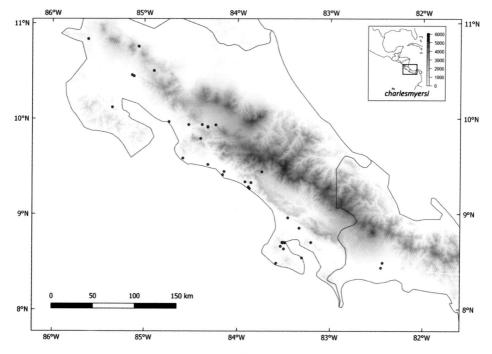

supraocular scales gradually enlarged; hindlimbs short, adpressed toe to ear or short of ear; body color usually whitish with dark flecks, lichenous; male and female dewlaps red.

SIMILAR SPECIES — *Anolis charlesmyersi* is a *twig*, or *small* to *large brown* lowland anole. Its combination of short hindlimbs (adpressed toe reaches to ear) and smooth ventral scales will distinguish it from most potentially sympatric similar species. Among geographically proximal short-legged forms with smooth ventrals, its white lichenous dorsal color will distinguish it most easily from *A. gruuo*, *A. kemptoni*, *A. pseudokemptoni*, *A. arenal*, *A. altae*, *A. tenorioensis*, and *A. monteverde*, which are brown and often patternless (also see male dewlap descriptions for these species) and live at high elevations (except *A. arenal*). *Anolis carpenteri* is green and smaller than *A. charlesmyersi* (SVL to 48 mm; to 78 mm in *A. charlesmyersi*). *Anolis fungosus* is similar to *A. charlesmyersi* but smaller (SVL to 48 mm) and with fewer toe lamellae (13–16), and lives at high elevations (1000+ m; to 800 m in *A. charlesmyersi*). *Anolis pentaprion* is similar to *A. charlesmyersi* and parapatric to it, and differs from *A. charlesmyersi* in its purple-pink dewlap and close-set dewlap scales (17–25 close-set scales in *A. pentaprion*; dewlap red, 4–9 scales per dewlap scale row in *A. charlesmyersi*).

RANGE — From 0 to 800 meters on the Pacific versant of Costa Rica and adjacent Panama.

NATURAL HISTORY — Rare in forest and edge habitat, and on large trees in disturbed areas; diurnally arboreal up to at least 10 meters; sleeps at all visible heights, on twigs or leaves. See Echelle et al. (1971; as *A. pentaprion*); Fitch (1975; as *A. pentaprion*).

— *ANOLIS CHLORIS* — PLATE 1.46, 2.126

DESCRIPTION — Boulenger, G. A. 1898. An account of the reptiles and batrachians collected by Mr. W. F. H. Rosenberg in western Ecuador. *Proceedings of the Zoological Society of London* 1898:110.

TYPE SPECIMEN — BMNH 1946.8.13.20.

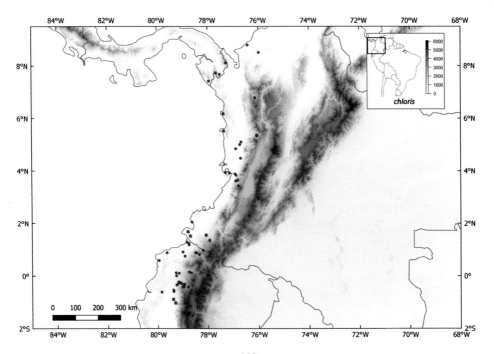

SPECIES ACCOUNTS

TYPE LOCALITY — "Paramba." Rosenberg collected at Hacienda Paramba, in northern Imbabura on the south bank of Rio Mira (Paynter 1993), lat 0.817 lon −78.350, 700 m.
MORPHOLOGY — Body length to 62 M, 58 F; ventral scales keeled; middorsal scales enlarged in 0–2 rows; interparietal larger or smaller than tympanum; scales across the snout between the second canthals 10–16; toe lamellae 16–20; suboculars and supralabials in contact; scales separating supraorbital semicircles 1–3; supraocular scales subequal to gradually enlarged; adpressed toe to eye; body color solid green; very large male dewlap white with yellowish-green scales; female dewlap absent.
SIMILAR SPECIES — *Anolis chloris* is a *small green* lowland anole distinguishable from most potentially confusing sympatric green anoles by its combination of keeled ventral scales and unpatterned green body (smooth ventrals and usually patterned body in *A. festae*, *A. fasciatus*, *A. gemmosus*, *A. poei*, *A. purpurescens*, *A. maia*). *Anolis biporcatus* and *A. parvauritus* differ from *A. chloris* in body proportion (stocky; gracile in *A. chloris*) and dewlap size and color (small and red with blue in male, small and blue or white in female; large and white in male, absent in female *A. chloris*).
RANGE — From 0 to 1800 meters in Darién Panama and Pacific Colombia to northern Ecuador.
NATURAL HISTORY — Common in forest and edge and open habitats; usually observed diurnally high in trees, often on trunk; sleeps on large leaves, usually at least 3 meters above ground. See Fitch et al. (1976); Miyata (2013); Castro-Herrera (1988); Rios et al. (2011); Viteri (2015); Narváez (2017); Rengifo et al. (2014, 2015, 2019, 2021); Pinto-Erazo et al. (2020); Moreno-Arias et al. (2020); Arteaga et al. (2023).

— *ANOLIS CHRYSOLEPIS* — PLATE 1.47

DESCRIPTION — Duméril, A.M.C., G. Bibron. 1837. *Erpétologie Générale ou Histoire Naturelle Complète des Reptiles*, vol. 4. Librairie Encyclopédique de Roret, Paris. 94–95.
TYPE SPECIMEN — Lectotype MNHN 2436.

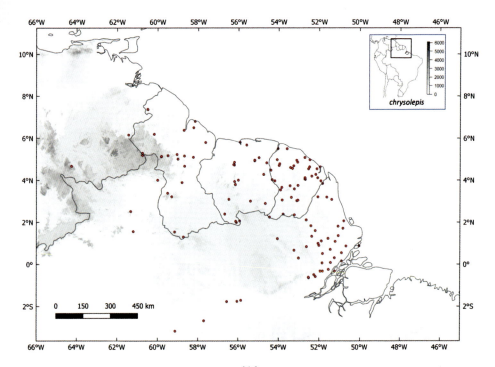

TYPE LOCALITY — "Mana, French Guiana" for Lectotype. Mana is at lat 5.67 lon −53.78, 0 m.
MORPHOLOGY — Body length to 74 M, 59 F; ventral scales keeled; middorsal scales greatly enlarged in several rows; interparietal larger or smaller than tympanum; scales across the snout between the second canthals 9–15; toe lamellae 12–15; subocular and supralabials separated by a row of scales; scales separating supraorbital semicircles 1–3; supraocular scales subequal to gradually enlarged; hindlimbs long, adpressed toe to anterior to tip of snout; body color brown; male dewlap blue; female dewlap cream.
SIMILAR SPECIES — *Anolis chrysolepis* is a *small* to *large brown* anole that is found in the Amazon basin along with similar species *A. bombiceps*, *A. scypheus*, *A. brasiliensis*, *A. planiceps*, and *A. tandai*. Among these species only *A. bombiceps* may be sympatric with *A. chrysolepis*, so consultation of range maps will be helpful for proper identification. All these species may be distinguished by female dewlap (red distally, blue proximally in *A. scypheus*; blue with cream distally in *A. tandai*; blue in *A. brasiliensis*, *A. bombiceps*; apparently red in *A. planiceps*; cream in *A. chrysolepis*). The other small brown anoles frequently sympatric with *A. chrysolepis* (*A. fuscoauratus*, *A. ortonii*, *A. trachyderma*) are smaller than *A. chrysolepis* (SVL < 60 mm; to 74 mm in *A. chrysolepis*) and have smooth or weakly keeled ventral scales (strongly keeled in *A. chrysolepis*). *Anolis auratus* is found in open areas, is smaller than *A. chrysolepis* (SVL to 54 mm), and has shorter hindlimbs (adpressed toe reaches to between eye and snout; past snout in *A. chrysolepis*). *Anolis transversalis* is larger than *A. chrysolepis* (SVL to 84 mm) and has blue eyes (brown in *A. chrysolepis*).
RANGE — From 0 to 500 meters in southern Guyana, Suriname, French Guiana, and northern Brazil (Amapa, Para).
NATURAL HISTORY — Common in forest; active diurnally low on vegetation or ground; sleeps on twigs or leaves below 1.5 meters above ground. See Hoogmoed (1973); Martins (1991); de Oliveira et al. (2014); Faria et al. (2019).

— *ANOLIS COBANENSIS* — PLATE 1.48

DESCRIPTION — Stuart, L. C. 1942. Comments on several species of *Anolis* from Guatemala, with descriptions of three new forms. *Occasional Papers of the Museum of Zoology, University of Michigan* 464:6–8.
TYPE SPECIMEN — UMMZ 90232.
TYPE LOCALITY — "Wet pine and broadleaf forest 3 km. south of Finca Samac (6 km. [straight line] due west of Cobán), Alta Verapaz, Guatemala. Altitude, about 1350 m." This locality is at approximately lat 15.5 lon −90.5.
MORPHOLOGY — Body length to 50 M, 50 F; ventral scales smooth to faintly keeled; middorsal scales enlarged in 0–2 rows; interparietal equal to or smaller than tympanum; scales across the snout between the second canthals 8–11; toe lamellae 14–19; subocculars and supralabials usually in contact; scales separating supraorbital semicircles 2–3; supraocular scales gradually enlarged; hindlimbs long, adpressed toe to anterior to tip of snout; body color brown; male dewlap pinkish purple.
SIMILAR SPECIES — *Anolis cobanensis* is a *small brown* highland anole distinguishable from potentially sympatric small brown highland anoles by its long hindlimbs (adpressed toe reaches posterior to eye in *A. laeviventris*, *A. crassulus*; to anterior to tip of snout in *A. cobanensis*) and middorsal scale rows (rows of abruptly enlarged middorsal rows in *A. crassulus*; middorsals undifferentiated in *A. cobanensis*). *Anolis campbelli* and *A. matudai* are very similar species to the west of the range of *A. cobanensis*. *Anolis campbelli* differs from *A. cobanensis* in its distinctly keeled ventral scales (ventrals usually smooth in *A. cobanensis*). *Anolis matudai* differs from *A. cobanensis* in having undifferentiated scales in the nasal region (elongate nasal scale present in *A. cobanensis*) and longitudinally keeled dorsal head scales anterior to the orbits (keels transversely oriented in *A. cobanensis*).
RANGE — From 1000 to 2100 meters in central Guatemala, southwestern Alta Verapaz and northwestern Baja Verapaz.

SPECIES ACCOUNTS

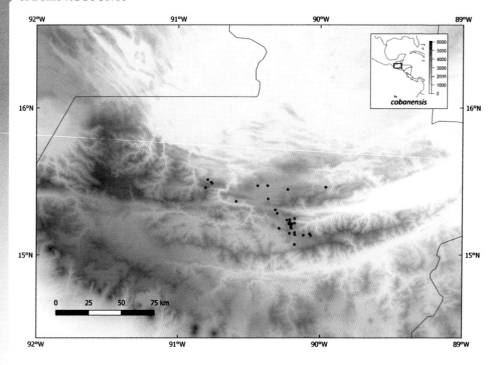

NATURAL HISTORY — Common in cloud forest; active diurnally low on tree trunks, other low vegetation, and ground; sleeps on twigs or leaves up to 2 meters above ground; also found under loose bark on trees. See Nieto Montes de Oca (1994a).

— *ANOLIS COMPRESSICAUDA* — PLATE 1.49, 2.21

DESCRIPTION — Smith, H. M., H. W. Kerster. 1955. New and noteworthy Mexican lizards of the genus *Anolis*. *Herpetologica* 11:193–198.
TYPE SPECIMEN — UIMNH 35625.
TYPE LOCALITY — "Near La Gloria, Oaxaca, Mexico." Multiple La Glorias exist in Oaxaca. Wylie and Grunwald (2016) placed collector MacDougall's La Gloria at lat 16.803972 lon −94.609012, 500 m.
MORPHOLOGY — Body length to 55 M, 52 F; ventral scales keeled; middorsal scales greatly enlarged in several rows; interparietal smaller than tympanum; scales across the snout between the second canthals 6–9; toe lamellae 12–17; usually one row of scales separating suboculars and supralabials; scales separating supraorbital semicircles 2–3; supraocular scales gradually enlarged; adpressed toe to between eye and snout; body patterned copper brown, with light lateral banding and striping; male dewlap pinkish purple with yellow edge; female dewlap absent; iris blue; puncturelike axillary pocket.
SIMILAR SPECIES — *Anolis compressicauda* is a *small brown* anole. Its combination of blue eyes, abruptly enlarged middorsal scales, and a puncturelike axillary pocket will distinguish *A. compressicauda* from all similar sympatric species.
RANGE — From 50 to 1200 meters, north and east of the Isthmus of Téhuantépec, in the states of Oaxaca, Veracruz, Chiapas, and Tabasco, Mexico.

Anolis concolor

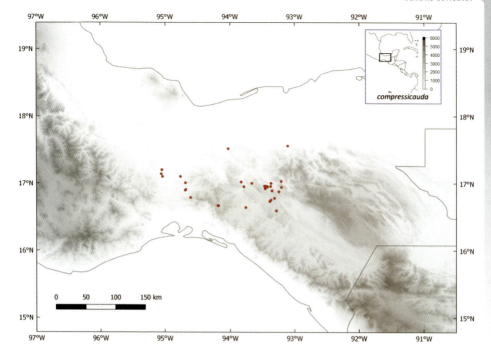

NATURAL HISTORY — In forest; active diurnally on rocks or ground, or low on vegetation; sleeps on twigs or leaves below 1 meter above ground, possibly in leaf litter. See Rios Rodas et al. (2017); Muñoz-Alonso et al. (2017).

— *ANOLIS CONCOLOR* — PLATE 1.50; MAP overleaf

DESCRIPTION — Cope, E. D. 1862. Contributions to Neotropical saurology. *Proceedings of the Academy of Natural Sciences of Philadelphia* 14:180–181.
TYPE SPECIMEN — Syntypes USNM 6055 (n=3), MCZ 22341. The MCZ website lists specimen 22341 as "missing" and the MCZ catalog includes a note: "on loan since 1968 (Colombia)—probably lost."
TYPE LOCALITY — "Nicaragua." The species is endemic to San Andrés Island, Colombia, which was part of Nicaragua at the time of the species' description. Corn and Dalby (1973) restricted the type locality to San Andrés (lat 12.58 lon −81.70, 0–55 m).
MORPHOLOGY — Body length to 80 M, 60 F; ventral scales keeled; middorsal scales gradually enlarged in several rows; interparietal usually equal to or smaller than tympanum; scales across the snout between the second canthals 7–11; toe lamellae 18–24; suboculars and supralabials in contact; scales separating supraorbital semicircles usually 1–2; supraocular scales gradually enlarged; adpressed toe to between eye and tip of snout; body color brown; male dewlap orange; female dewlap absent.
SIMILAR SPECIES — *Anolis concolor* is a *small brown* anole, the only anole species on San Andrés.
RANGE — Throughout San Andrés island, east of Nicaragua in the Caribbean Sea.
NATURAL HISTORY — Common on tree trunks. See Dunn and Saxe (1950); Tamsitt and Valdivieso (1963); Corn and Dalby (1973); Calderón-Espinosa and Barragán-Forero (2011).

SPECIES ACCOUNTS

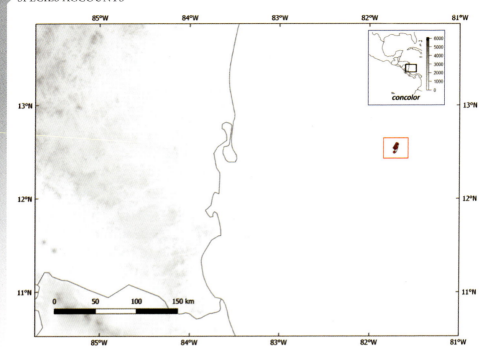

— *ANOLIS CRASSULUS* — PLATE 1.51

DESCRIPTION — Cope, E. D. 1864. Contributions to the herpetology of tropical America. *Proceedings of the Academy of Natural Sciences of Philadelphia* 16:173.
TYPE SPECIMEN — Lectosyntypes ANSP 8023-27 designated by Stuart (1942).
TYPE LOCALITY — "Coban, Vera Paz." Coban, Alta Verapaz, Guatemala, is at lat 15.48 lon −90.37, 1320 m.
MORPHOLOGY — Body length to 53 M, 56 F; ventral scales keeled; middorsal scales greatly enlarged in several rows; interparietal usually equal to or smaller than tympanum; scales across the snout between the second canthals 5–8; toe lamellae 13–18; suboculars and supralabials in contact; scales separating supraorbital semicircles 1–2; supraocular scales gradually enlarged; adpressed toe to between ear and eye; body color brown, sometimes with middorsal dark markings and a light lateral line extending back from the mouth; male and female dewlaps orange with yellow along edge and along rows of large scales.
SIMILAR SPECIES — *Anolis crassulus* is a *small brown* highland anole. It frequently is sympatric with *A. laeviventris*, from which it differs in dewlap (white in *A. laeviventris*; orange in *A. crassulus*) and its abruptly enlarged keeled middorsal scale rows (sometimes a few slightly enlarged middorsal rows in *A. laeviventris*). Other geographically proximal highland species also differ from *A. crassulus* in male dewlap color (*A. dollfusianus*: yellow; *A. cobanensis*: pinkish-purple; *A. matudai*: pink with light edge) and condition of middorsal scales (0–4 slightly enlarged middorsal rows in *A. cobanensis, A. dollfusianus, A. matudai*). *Anolis caceresae* is very similar to *A. crassulus*, differing only modally in body proportions, and geographically replaces it to the east in Honduras (these species appear to be genetically distinct [Hofmann and Townsend 2018]). See *A. caceresae* account for comparison with similar high-elevation Honduran forms.
RANGE — From 1300 to 3000 meters in El Salvador, central Guatemala, and Pacific Chiapas, Mexico. My map of *Anolis crassulus* is fairly speculative for Mexico. I am unable to discern the geographic

120

Anolis cristatellus

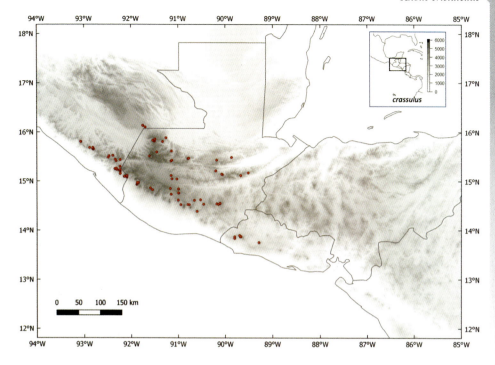

limits of *A. anisolepis* relative to *A. crassulus*, especially east of the Central Depression of Chiapas. I am assuming all specimens deemed *A. crassulus* in the Central Highlands of Guatemala are true *A. crassulus* based mostly on Hofmann and Townsend (2017), and that the Central Depression of Chiapas acts as a geographic barrier between *A. anisolepis* in the northern Central Highlands and *A. crassulus* in the southern Sierra Madre in Chiapas. This inference ignores the inconvenient complication of the phylogenetic placement of MZFC 6458, purportedly an *A. crassulus* specimen from the Sierra Madre (locality "Mpio. Union Juarez, Est. Chiapas"), with *A. anisolepis* (Hofmann and Townsend 2017).

NATURAL HISTORY — In forest and edge habitats; active diurnally on ground or low on vegetation; sleeps on low vegetation, usually below 1.5 meters above ground. See Fitch et al. (1976).

COMMENT — I follow McCranie et al. (1992) and McCranie and Köhler (2015) in considering *Anolis haguei* to be a synonym of *A. crassulus*.

— *ANOLIS CRISTATELLUS* — PLATE 1.52, 2.168

DESCRIPTION — Duméril, A.M.C., G. Bibron. 1837. *Erpétologie Générale ou Histoire Naturelle Complete des Reptiles*, vol. 4. Librairie Encyclopédique de Roret, Paris. 143–146.

TYPE SPECIMEN — Syntypes MNHN 2353, 2447, 2451.

TYPE LOCALITY — Duméril and Bibron (1837) state that specimens of this species were sent from "Martinique," which is standardly given as the (certainly erroneous) type locality. Syntype 2451 was received from "Guyane," also a doubtful locality (as realized by Duméril and Bibron [1837:146]: "Cette origine nous parait douteuse.").

MORPHOLOGY — Body length to 80 M, 73 F; ventral scales smooth; middorsal scales enlarged in 0–2 rows; interparietal larger or smaller than tympanum; scales across the snout between the second

SPECIES ACCOUNTS

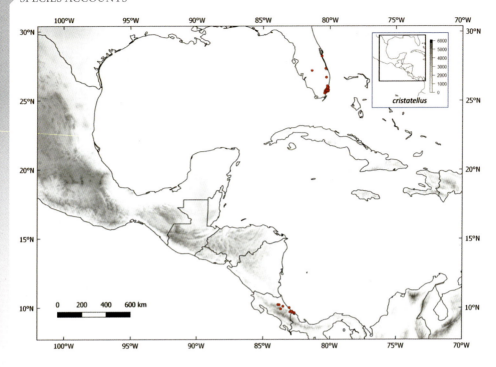

canthals 3–9; toe lamellae 16–24; suboculars and supralabials in contact or separated by a row of scales; scales separating supraorbital semicircles 0–1; supraocular scales gradually enlarged; addressed toe approximately to eye; body color brown, sometimes with light lateral stripe; male dewlap red distally, green proximally, varying between pale blending of colors and distinct red and green depending on locality.

SIMILAR SPECIES — *Anolis cristatellus* is a lowland *small* or *large brown* anole that is found in south Florida, USA, and Costa Rica outside of its native range in Caribbean islands. There are three other similar but distantly related small brown anoles in south Florida, *A. sagrei*, *A. distichus*, and *A. cybotes*. These four species differ most obviously in male dewlap (red with yellow edge in *A. sagrei*; small [reaching to forelimbs], pale yellow sometimes with red proximally in *A. distichus*; pale yellow in *A. cybotes*; red distally and yellow-green proximally in *A. cristatellus*). Females of these forms are not easy to distinguish. *Anolis sagrei* differs from *A. cristatellus* in its keeled ventral scales (smooth in *A. cristatellus*). *Anolis distichus* is smaller than *A. cristatellus* (SVL to 58 mm; to 80 mm in *A. cristatellus*), and may appear mottled pale gray or greenish (*A. cristatellus* is always brown). *Anolis cybotes* has a robust head (less so in *A. cristatellus*), and lacks a pronounced caudal crest (crest present in adult males of *A. cristatellus*). In Caribbean Costa Rica, other small brown anoles with smooth ventrals differ from *A. cristatellus* in limb length and male dewlap (*A. limifrons*: addressed toe reaches past eye, male dewlap white usually with yellow basal blotch; *A. pentaprion*: addressed toe to ear, male dewlap reddish-purple; *A. cristatellus*: addressed toe to eye).

RANGE — From 0 to 300 meters. Native to Puerto Rico and surrounding islands; now present in USA (Florida) and Costa Rica (Caribbean slope; Limón, Cartago). A single questionable specimen is recorded from the Yucatan peninsula, Mexico (see Lee 1996).

NATURAL HISTORY — Very common in edge and disturbed habitats; a "trunk-ground" anole (Williams 1983) that is commonly diurnally observed at the bases of tree trunks, facing down, less than 1 meter above ground; sleeps on twigs or leaves from 0.5 to 3 meters above ground. For natural history information on mainland populations, see Salzburg (1984); Rogowitz (1996); Kolbe et al. (2012); Kolbe et al. (2016); Tiatragul et al. (2017); Hall and Warner (2017); Kahrl and Cox (2017); Battles et al. (2018); Stroud (2018); Battles and Kolbe (2019); Battles et al. (2019); Thawley et al. (2019); Tiatragul and Warner (2019); Avilés-Rodríguez and Kolbe (2019); Tiatragul et al. (2019); Kolbe et al. (2021). For natural history in its native range, see, e.g., Schwartz and Henderson (1991).

— ANOLIS CRISTIFER — PLATE 1.53, 2.79

DESCRIPTION — Smith, H. M. 1968a. A new pentaprionid anole (Reptilia: Lacertilia) from Pacific slopes of Mexico. *Transactions of the Kansas Academy of Science* 71(2):195–200.
TYPE SPECIMEN — UIMNH 37066.
TYPE LOCALITY — "The vicinity of a small lake near Acacoyagua, Chiapas." There are numerous bodies of water around Acacoyagua. The town itself is at lat 15.34 lon –92.67, 90 m.
MORPHOLOGY — Body length to 88 M, 75 F; ventral scales smooth; middorsal scales enlarged in 0–2 rows; interparietal larger than tympanum; scales across the snout between the second canthals 6–10; toe lamellae 20–27; suboculars and supralabials in contact; scales separating supraorbital semicircles 1–2; supraocular scales gradually enlarged; hindlimbs short, adpressed toe to ear or short of ear; body color gray with white, black, and brown; tail banded; male and female dewlaps dark red; iris blue.
SIMILAR SPECIES — *Anolis cristifer* is a *large brown* or *twig* anole unlikely to be confused with sympatric forms due to its dewlap (large and red in males and females), smooth body scales, short hindlimbs (adpressed toe reaches to ear), and white to gray, often lichenous, body color.
RANGE — From 0 to 800 meters on Pacific slope in Guatemala and extreme southern Mexico.
NATURAL HISTORY — Rare in disturbed areas with tall trees; presumably historically inhabited forests; sleeps on vines or large leaves above 2 meters above ground.

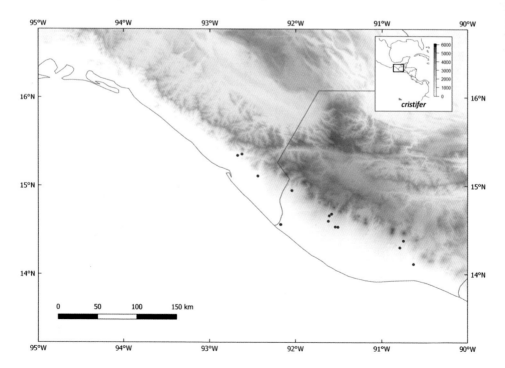

SPECIES ACCOUNTS

— *ANOLIS CUPREUS* — PLATE 1.54, 2.55

DESCRIPTION — Hallowell, E. 1861. Report upon the Reptilia of the North Pacific exploring expedition, under command of Capt. John Rogers, U.S.N. *Proceedings of the Academy of Natural Sciences of Philadelphia* 12:481–482.
TYPE SPECIMEN — 14 syntypes located by McCranie and Köhler (2014): MCZ R-17631–32, UIMNH 40733, USNM 12211 (11).
TYPE LOCALITY — "Nicaragua."
MORPHOLOGY — Body length to 55 M, 52 F; ventral scales keeled; middorsal scales enlarged in 0–2 rows; interparietal larger or smaller than tympanum; scales across the snout between the second canthals 7–13; toe lamellae 13–16; suboculars and supralabials in contact or separated by a row of scales; scales separating supraorbital semicircles 1–3; supraocular scales gradually enlarged; appressed toe to eye or just beyond; body color brown, often with light lateral stripe, often patterned dorsally with blotches or bars; male dewlap orange-brown proximally, pale pink-peach distally.
SIMILAR SPECIES — *Anolis cupreus* is a *small brown* anole distinguishable from all potentially sympatric anoles by its orange-brown proximally, pale pink-peach distally male dewlap. *Anolis cupreus* is further distinguished from other potentially sympatric small brown anoles as follows: *A. laeviventris*, *A. sericeus*, *A. altae*, *A. tenorioensis*, *A. monteverde*, and *A. arenal* have short limbs (adpressed toe reaches to ear; to eye or just beyond in *A. cupreus*); *A. humilis*, *A. tropidonotus*, and *A. uniformis* have abruptly enlarged rows of keeled middorsal scales (middorsals not or slightly enlarged in *A. cupreus*) and a puncturelike axillary pocket (absent in *A. cupreus*); *A. polylepis*, *A. limifrons*, *A. zeus*, *A. rodriguezii*, *A. cristatellus*, *A. pentaprion*, and *A. charlesmyersi* have smooth ventral scales (keeled in *A. cupreus*). *Anolis capito* and *A. lemurinus* are larger than *A. cupreus* (SVL to 80 mm in *A. lemurinus*; to 103 mm in *A. capito*; to 55 mm in *A. cupreus*) and have proportionately smaller dewlaps (posterior edge of dewlap approximately at forelimbs in *A. capito*, *A. lemurinus*; well into chest in *A. cupreus*). *Anolis*

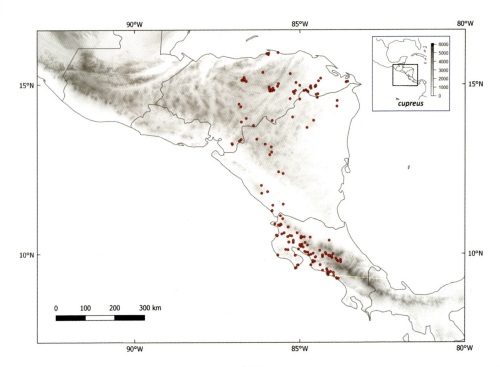

Anolis cuprinus

ocelloscapularis, *A. yoroensis*, and *A. sagrei* differ from *A. cupreus* in possessing an elongate anterior nasal scale (scales of the nasal region undifferentiated in *A. cupreus*).

RANGE — From 0 to 1450 meters from Honduras to Costa Rica.

NATURAL HISTORY — Very common in forest, edge, and disturbed areas; often highly abundant in dry forest; active diurnally on low vegetation or ground, to much higher on vegetation during wet season; sleeps on twigs, leaves, or vines up to at least 5 meters above ground. See Echelle et al. (1971a); Clark (1973); Fitch (1973a, b, c; 1975); Fitch et al. (1972); Fleming and Hooker (1975); Berkum (1986); Savage (2002); McCranie and Köhler (2015); Serrano (2018).

— *ANOLIS CUPRINUS* — PLATE 1.55

DESCRIPTION — Smith, H. M. 1964. A new *Anolis* from Oaxaca, Mexico. *Herpetologica* 20:31–33.
TYPE SPECIMEN — UIMNH 52959.
TYPE LOCALITY — "Zanatepec, Oaxaca, Mexico." Zanatepec, which is a coastal city, is almost certainly in error as a type locality, as *A. cuprinus* is known only from high elevations (Fitch et al. 1976). Fitch et al. (1976) and Nieto Montes de Oca (1994a) considered the Sierra Madre north of Zanatepec to be the most likely type locality (approximately lat 16.9 lon −94.4).
MORPHOLOGY — Body length to 69 M, 55 F; ventral scales keeled; middorsal scales gradually enlarged in several rows; interparietal smaller than tympanum; scales across the snout between the second canthals 7–11; toe lamellae 14–18; suboculars and supralabials usually separated by one row of scales; scales separating supraorbital semicircles 2–3; supraocular scales gradually enlarged; hindlimbs long, addressed toe to between eye and tip of snout; body color greenish brown; male dewlap reddish-orange with yellow edge.
SIMILAR SPECIES — *Anolis cuprinus* is a *small brown* highland anole that may be sympatric with *A. laeviventris* and *A. petersii* (Nieto Montes de Oca 1994a). *Anolis cuprinus* is differentiated from these species

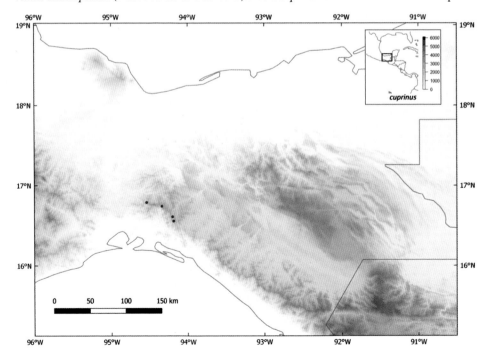

SPECIES ACCOUNTS

by male dewlap color (white in *A. laeviventris*, pink with dark blotches in *A. petersii*; reddish-orange with yellow edge in *A. cuprinus*), body size (SVL to 118 mm in *A. petersii*; to 69 mm in *A. cuprinus*), and long hindlimbs (adpressed toe usually to ear in *A. laeviventris*; to between eye and tip of snout in *A. cuprinus*). *Anolis parvicirculatus* and *A. matudai*, similar highland anoles found north and east of the range of *A. cuprinus*, respectively, differ from *A. cuprinus* in their smooth ventral scales (keeled in *A. cuprinus*).

RANGE — From 1200 to 1900 meters in the southern Sierra Madre of Mexico east of the Isthmus of Téhuantépec.

NATURAL HISTORY — In cloud forest; active diurnally on ground or vegetation. See Nieto Montes de Oca (1994).

COMMENT — A body of research mainly by Fitch and collaborators (e.g., Henderson and Fitch [1975]) describes the ecology of supposed *Anolis "cuprinus"* from lowland areas, i.e., xeric forest around Zanatepec. These individuals are misidentified, apparently *A. boulengerianus* (Nieto Montes de Oca 1994).

— *ANOLIS CUSCOENSIS* — FIGURE 3.20; PLATE 1.56

DESCRIPTION — Poe, S., C. Yañez-Miranda, E. Lehr. 2008. Notes on variation in *Anolis boettgeri* Boulenger 1911, assessment of the status of *Anolis albimaculatus* Henle and Ehrl 1991, and description of a new species of *Anolis* (Squamata: Iguania) similar to *Anolis boettgeri*. Journal of Herpetology 42:251–259.

TYPE SPECIMEN — MZUNAP 02.000191

TYPE LOCALITY — "Peru, Department of Cusco, 72 km north of Paucartambo on Paucartambo-Itahuania Road, 13 03 30 S 71 33 54 [W], 1624 m elevation." The listed point is off by about a kilometer. A more accurate type locality is lat −13.070 lon −71.569, 1624 m.

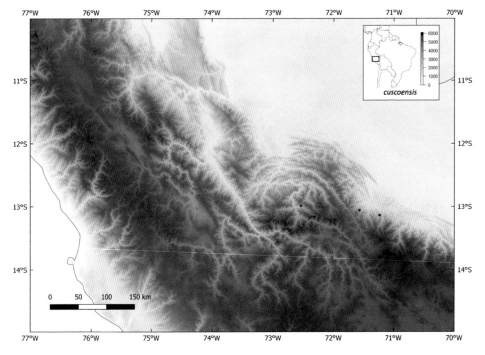

Anolis cusuco

MORPHOLOGY — Body length to 58 M, 63 F; ventral scales smooth; middorsal scales enlarged in 0–2 rows; interparietal larger or smaller than tympanum; scales across the snout between the second canthals 9–11; toe lamellae 22–28; suboculars and supralabials in contact; supraorbital semicircles in contact or occasionally separated by one scale; supraocular scales gradually enlarged; adpressed toe to between ear and eye; body color mostly green, with lateral yellow spots; male dewlap skin cream with dark blotching, with blue and green scales; female dewlap absent.
SIMILAR SPECIES — *Anolis cuscoensis* is a *large green* highland anole likely to be sympatric only with *A. fuscoauratus*, a highly dissimilar *small brown* anole.
RANGE — From 1600 to 1700 meters on the Amazonian slope of the southern Andes of Peru.
NATURAL HISTORY — Common in wet forest and forest edge; sleeps on leaves (especially ferns) from 1 to 5 meters above ground.

— *ANOLIS CUSUCO* — PLATE 1.57, 2.45

DESCRIPTION — McCranie, J. R., G. Köhler, L. D. Wilson. 2000. Two new species of anoles from northwestern Honduras related to *Norops laeviventris* (Wiegmann 1834) (Reptilia, Squamata, Polychrotidae). *Senckenbergiana Biologica* 80:214–218.
TYPE SPECIMEN — SMF 78842.
TYPE LOCALITY — "Parque Nacional El Cusuco Centro de Visitantes (15° 29.92' N, 88° 12.88' W), 1550 m elevation, Departamento de Cortés, Honduras."
MORPHOLOGY — Body length to 46 M, 43 F; ventral scales keeled; middorsal scales appear uniform or gradually enlarged in several rows; interparietal larger or smaller than tympanum; scales across the snout between the second canthals 7–10; toe lamellae 13–17; suboculars and supralabials usually in contact; scales separating supraorbital semicircles 0–1; supraocular scales gradually enlarged; hindlimbs short, adpressed toe to ear or posterior to ear; body color grayish brown; male dewlap white.

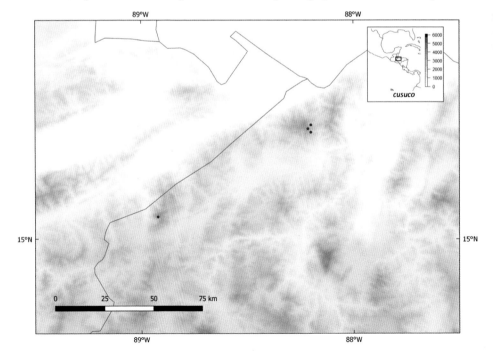

SPECIES ACCOUNTS

SIMILAR SPECIES — This *small brown* (or gray to white) anole is essentially a form of *Anolis laeviventris* that occurs at high elevations in northwestern Honduras and has a large dewlap in males (male dewlap extends to the chest; approximately to the forelimbs in *A. laeviventris*). It is distinguishable from all other area anoles except *A. laeviventris* by its white male dewlap. Other small-bodied potentially highland geographically proximal forms are distinguishable by dorsal scales (abruptly enlarged in several rows in *A. amplisquamosus, A. tropidonotus, A. uniformis*; slightly enlarged in few rows in *A. cusuco*), ventral scales (smooth in *A. rodriguezii*; keeled in *A. cusuco*), or long hindlimbs (adpressed toe reaches at least to eye in *A. yoroensis, A. ocelloscapuris*; to ear in *A. cusuco*). See the account of *A. laeviventris* for additional comparisons.

RANGE — From 1350 to 2000 meters in Cusuco National Park and Cerro Azul National Park in northwestern Honduras.

NATURAL HISTORY — Common in edge and disturbed areas, also in forest; active diurnally on vegetation up to 2.5 meters; sleeps on vegetation from 0.5 to 3.5 meters above ground, often on twigs. See Townsend and Wilson (2008); McCranie and Köhler (2015); Clause and Brown (2017).

— *ANOLIS CYBOTES* — PLATE 1.58, 2.166

DESCRIPTION — Cope, E. D. 1862. Contributions to Neotropical saurology. *Proceedings of the Academy of Natural Sciences of Philadelphia* 14:176–188.

TYPE SPECIMEN — Lectotype MCZ 14346 designated by Köhler et al. (2019).

TYPE LOCALITY — "Western Hayti; from near Jeremie." Jeremie, Haiti, is at lat 18.64 lon –74.12, 50 m.

MORPHOLOGY — Body length to 81 M, 66 F; ventral scales smooth; middorsal scales enlarged in 0–2 rows; interparietal larger or smaller than tympanum; scales across the snout between the second canthals 5–9; toe lamellae 14–23; suboculars and supralabials in contact or separated by row of scales; scales separating supraorbital semicircles 0–1; supraocular scales gradually enlarged; adpressed toe

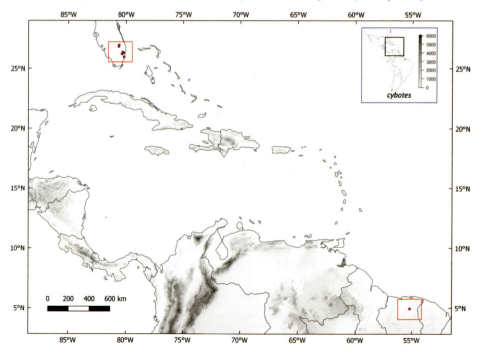

128

Anolis cymbops

to between eye and tip of snout; body color grayish brown, sometimes with light lateral stripe; male dewlap pale yellow to gray.

SIMILAR SPECIES — *Anolis cybotes* is a *small brown* anole that is currently found only in south Florida outside of Caribbean islands. There are three other similar but distantly related small brown anoles in south Florida, *A. sagrei*, *A. distichus*, and *A. cristatellus*. These four species differ most obviously in male dewlap color (red with yellow edge in *A. sagrei*; pale yellow sometimes with red proximally in *A. distichus*; red distally and greenish-yellow proximally in *A. cristatellus*; pale yellow in *A. cybotes*). *Anolis sagrei* further differs from *A. cybotes* in its keeled ventral scales (smooth in *A. cybotes*). *Anolis distichus* is smaller than *A. cybotes* (SVL to 58 mm; to 81 mm in *A. cybotes*), with a smaller male dewlap (reaching to forelimbs; male dewlap extends well onto chest in *A. cybotes*), and may appear mottled pale gray or greenish (*A. cybotes* is always brown). *Anolis cristatellus* has a pronounced caudal crest in adult males (crest absent in *A. cybotes*).

RANGE — On the mainland, from 0 to 300 meters in southern Florida and Suriname. Native to Hispaniola.

NATURAL HISTORY — Very common in edge and disturbed habitats; a "trunk-ground" anole (Williams 1983), commonly observed facing down on tree trunks below 1 meter above ground; sleeps on twigs or leaves from 0.5 to 3 meters above ground. See, e.g., Schwartz and Henderson (1991) for natural history information on Hispaniolan populations.

COMMENT — The species identity of the Florida "cybotoid" may properly be referred to as *A. hispaniolae* or one of the other flavors of cybotoid indicated by Köhler et al. (2019).

— *ANOLIS CYMBOPS* — PLATE 1.59, 2.9

DESCRIPTION — Cope, E. D. 1864. Contributions to the herpetology of tropical America. *Proceedings of the Academy of Natural Sciences of Philadelphia* 16:173–174.

TYPE SPECIMEN — BMNH 1946.8.5.84.

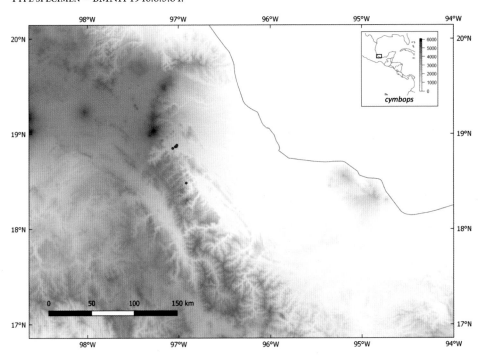

SPECIES ACCOUNTS

TYPE LOCALITY — "Vera Cruz." Workers have interpreted this locality as being somewhere in the state of Veracruz, Mexico.
MORPHOLOGY — Body length to 47 M, 47 F; ventral scales keeled; middorsal scales gradually enlarged in several rows; interparietal approximately equal to tympanum; scales across the snout between the second canthals 8–9; toe lamellae 15; suboculars and supralabials separated by one row of scales; scales separating supraorbital semicircles 1; supraocular scales gradually enlarged; hindlimbs long, adpressed toe to between eye and tip of snout; body color brown; male dewlap pink.
SIMILAR SPECIES — *Anolis cymbops* is a *small brown* highland anole. Nieto Montes de Oca (1994a) noted potential sympatric small brown congeners *A. laeviventris* and *A. sericeus*. These species differ from *A. cymbops* in having shorter hindlimbs (adpressed toe reaches to ear in *A. sericeus*, *A. laeviventris*; to between eye and tip of snout in *A. cymbops*). *Anolis tropidonotus* has approximately 10 rows of abruptly enlarged keeled middorsal scales (approximately 5 gradually enlarged middorsals in *A. cymbops*) and a puncturelike axillary pocket (absent in *A. cymbops*). *Anolis rodriguezii* has undifferentiated middorsal scales and broad contact between suboculars and supralabials (suboculars and supralabials usually separated by a row of scales in *A. cymbops*). Highland species *A. schiedii* and *A. naufragus* are very similar to *A. cymbops* and geographically proximal but differ in lacking enlarged postcloacal scales in males (present in males of *A. cymbops*) and in male dewlap color (orange-red in *A. schiedii*, *A. naufragus*; pink in *A. cymbops*).
RANGE — From 1000 to 1250 meters on the eastern slope of the Sierra Madre Oriental in Veracruz, Mexico.
NATURAL HISTORY — Sleeps on low (<1 meter) twigs or leaves in disturbed areas that formerly were cloud forest.

— *ANOLIS DANIELI* — PLATE 1.60, 2.134

DESCRIPTION — Williams, E. E. 1988. New or problematic *Anolis* from Colombia. V. *Anolis danieli*, a new species of the *latifrons* species group and a reassessment of *Anolis apollinaris* Boulenger, 1919. Breviora 489:1–25.
TYPE SPECIMEN — CSJ 3405 (formerly CSJ 111, ICN 5997).
TYPE LOCALITY — "Urrao, Antioquia, Colombia." Urrao is at lat 6.3156 lon −76.1342, 1820 m.
MORPHOLOGY — Body length to 117 M, 104 F; ventral scales smooth to weakly keeled; middorsal scales enlarged in 0–2 rows; interparietal smaller than tympanum, occasionally absent; scales across the snout between the second canthals 8–12; toe lamellae 22–26; suboculars and supralabials in contact or separated by a row of scales; scales separating supraorbital semicircles 3–4; supraocular scales gradually enlarged; hindlimbs long, adpressed toe to between eye and tip of snout; body color olive green with (possibly faint) dark crossbands, a thick dark line extending posteriorly from eye and from shoulder, these separated by a thick light line extending back from mouth; male dewlap cream to dark green with dark elongate streaks (but see below).
SIMILAR SPECIES — *Anolis danieli* is a *giant* highland anole. Geographically proximal large to giant anoles may be distinguished as follows: *Anolis eulaemus*, *A. ventrimaculatus*, *A. megalopithecus*, *A. anoriensis*, *A. maculigula*, and *A. antioquiae* have narrow toepads (broad, typical anole toepad in *A. danieli*) and much smaller head scales than *A. danieli* (11–21 scales across the snout between the second canthals; 8–12 in *A. danieli*). *Anolis frenatus*, *A. latifrons*, and *A. princeps* usually are found at lower elevations than *A. danieli* and differ in color pattern (oblique lateral rows of ocelli on an olive green background in these species; approximately solid olive green or banded green in *A. danieli*). *Anolis biporcatus* has strongly keeled ventral scales (smooth in *A. danieli*). *Anolis mirus* has fewer toe lamellae than *A. danieli* (12–15, 22–26 in *A. danieli*).
RANGE — From 1250 to 2300 meters on the Pacific Andean slope of central Colombia.
NATURAL HISTORY — In forest, edge, and disturbed habitats. Sleeps 1.1 to 4 meters above ground, most commonly on leaves.

Anolis datzorum

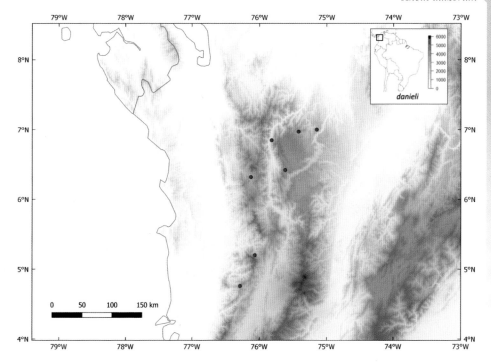

COMMENT — Williams (1988) discussed a "Parque las Orquideas sibling" species that may or may not be considered proper *A. danieli*. The photo and description of the male dewlap in this book are of the Parque Orquideas form. Notably, the type description of *A. danieli* includes a secondhand color description from Marco Antonio Serna stating that the male dewlap is "yellow with whitish scales," i.e., different from the male dewlap color pattern presented here. Clearly, additional taxonomic work is needed to clarify the species boundaries of *A. danieli*.

— *ANOLIS DATZORUM* — PLATE 1.61, 2.46

DESCRIPTION — Köhler, G., M. Ponce, J. Sunyer, A. Batista. 2007. Four new species of anoles (genus *Anolis*) from the Serranía de Tabasará, West-Central Panama (Squamata: Polychrotidae). *Herpetologica* 63:385–388.
TYPE SPECIMEN — SMF 85093.
TYPE LOCALITY — "La Nevera, 8° 29' 45" N, 81° 46' 35" W, 1600 m elevation, Serranía de Tabasará, Comarca Ngöbe Bugle, Distrito de Nole Düima, Corregimiento de Jadeberi, Panama." Corrected by Lotzkat et al. (2010) to 8° 30' N, 81° 46' 20" W.
MORPHOLOGY — Body length to 47 M, 48 F; ventral scales keeled; middorsal scales gradually enlarged in several rows; interparietal approximately equal to tympanum; scales across the snout between the second canthals 7–9; toe lamellae 16–17; suboculars and supralabials in contact; scales separating supraorbital semicircles 0–2; supraocular scales gradually enlarged; hindlimbs long, adpressed toe to between ear and eye; body color yellowish-greenish-brown, with middorsal chevrons; tail banded; male and female dewlaps orange.
SIMILAR SPECIES — *Anolis datzorum* is a *small brown* or *green* highland anole with keeled ventral scales and well developed orange dewlaps in males and females. Geographically proximal highland anoles

SPECIES ACCOUNTS

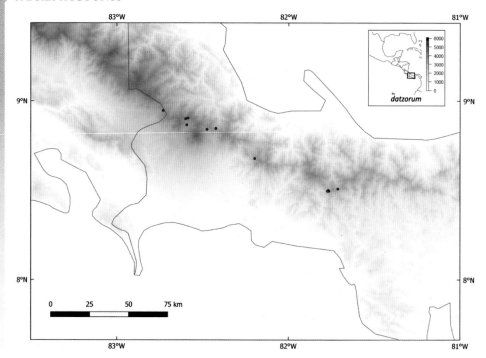

with keeled ventrals have small or absent female dewlaps and very different male dewlap colors than *A. datzorum* (*A. magnaphallus*: red; *A. benedikti*: orange and yellow; *A. pachypus*: yellow with reddish-orange; *A. tropidolepis*: red; *A. humilis*: red with yellow edge; *A. salvini*: red; *A. laeviventris*: white; *A. pseudopachypus*: dull yellow). *Anolis woodi* is larger than *A. datzorum* (SVL to 92 mm; to 48 mm in *A. datzorum*) and has blue eyes (brown in *A. datzorum*).
RANGE — From 1650 to 2400 meters from western Panama to extreme eastern Costa Rica.
NATURAL HISTORY — Rare in forest and forest patches in disturbed areas; sleeps on twigs, leaves, vines and epiphytes above 1.5 meters above ground.

— *ANOLIS DISSIMILIS* — PLATE 1.62, 2.127

DESCRIPTION — Williams, E. E. 1965. South American *Anolis* (Sauria, Iguanidae): Two new species of the *punctatus* group. *Breviora* 233:2–4.
TYPE SPECIMEN — CNHM 81369.
TYPE LOCALITY — "Itahuania, upper Rio Madre de Dios, Madre de Dios Province, Peru." Itahuania (lat −12.64 lon −71.22, 400 m) is the town that currently is at the end of the road that extends east from the Andes through Pilocapata and Shintuya towards Manu national park.
MORPHOLOGY — Body length to 56 M, 59 F; ventral scales weakly keeled; middorsal scales enlarged in 0–2 rows; interparietal larger than tympanum; scales across the snout between the second canthals 6–7; toe lamellae 16–17; suboculars and supralabials in contact; supraorbital semicircles in contact; supraocular scales gradually enlarged; hindlimbs short, adpressed toe to ear or posterior to ear; body color green; male dewlap white.
SIMILAR SPECIES — *Anolis dissimilis* is a distinctive *small green* lowland Amazonian anole with a very long head that may be sympatric with one other predominantly green anole, *A. punctatus*. *Anolis*

Anolis distichus

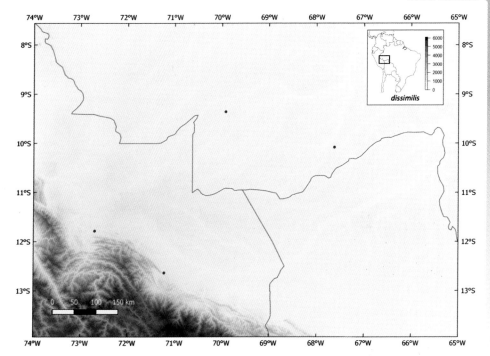

punctatus is larger (to 89 mm SVL; to 59 mm in *A. dissimilis*) and has smaller head scales than *A. dissimilis* (8-14 scales across the snout between the second canthals; 6-7 in *A. dissimilis*).
RANGE — From 200 to 500 meters in southeastern Peru and adjacent Brazil.
NATURAL HISTORY — Rare. Only information is for the specimen reported by de Freitas et al. (2013): in a guava tree grove 200 meters from primary forest "on the trunk (10-20 cm) of a guava tree at a height of about 4 meters."

— *ANOLIS DISTICHUS* — PLATE 1.63, 2.169

DESCRIPTION — Cope, E. D. 1861. Notes and descriptions of anoles. *Proceedings of the Academy of Natural Sciences of Philadelphia* 13:208-209.
TYPE SPECIMEN — Syntypes ANSP 7780-87.
TYPE LOCALITY — "New Providence Island. Bahamas." New Providence Island is at lat 25.03 lon −77.41, 0 m.
MORPHOLOGY — Body length to 58 M, 50 F; ventral scales smooth; middorsal scales enlarged in 0-2 rows; interparietal equal to or larger than tympanum; scales across the snout between the second canthals 4-8; toe lamellae 15-21; suboculars and supralabials in contact or separated by a row of scales; supraorbital semicircles in contact; supraocular scales gradually enlarged; adpressed toe to eye; body color brownish gray; male dewlap pale yellow sometimes with red proximally; female dewlap absent.
SIMILAR SPECIES — *Anolis distichus* is a *small brown* to white, occasionally greenish anole that is found only in south Florida outside of Caribbean islands. There are three other similar but distantly related small brown anoles in south Florida, *A. sagrei*, *A. cybotes*, and *A. cristatellus*. These four species differ most obviously in male dewlap color (red with yellow edge in *A. sagrei*, red distally

SPECIES ACCOUNTS

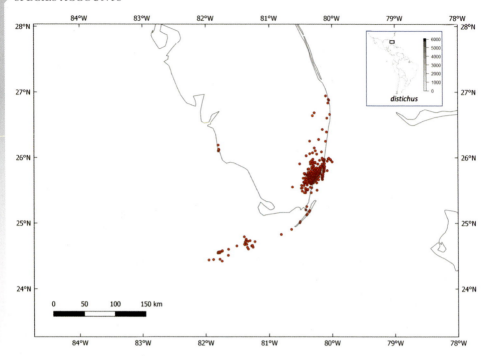

and greenish-yellow proximally in *A. cristatellus*, pale yellow to gray in *A. cybotes*, pale yellow sometimes with red proximally in *A. distichus*). *Anolis sagrei* further differs from *A. distichus* in its keeled ventral scales (smooth in *A. distichus*). *Anolis cybotes* and *A. cristatellus* are larger than *A. distichus* (SVL to 80 mm; to 58 mm in *A. distichus*), with larger male dewlaps (reaching well on to chest; to forelimbs in *A. distichus*), and always appear some shade of brown (*A. distichus* may be mottled pale gray or greenish).
RANGE — From 0 to 50 meters in south Florida. Native to the Bahamas.
NATURAL HISTORY — Very common in edge and disturbed habitats, especially diurnally; a "trunk" anole (Williams 1983) that is often observed diurnally at least one meter above ground on tree trunks; sleeps on twigs or leaves, usually above 2 meters above ground. For natural history information on Florida populations see Corn (1971); Lee (1980); Doan (1996); Paterson (1999); Meshaka (1999a, b); Giery et al. (2013); Mothes et al. (2019); Tiatragul et al. (2019). See, e.g., Schwartz and Henderson (1991) for natural history information on Bahamian and Hispaniolan populations.

— *ANOLIS DOLLFUSIANUS* — PLATE 1.64, 2.56

DESCRIPTION — Bocourt, M. F. 1873. Études sur les reptiles. In: Duméril, M. A., M. F. Bocourt, M. Mocquard, eds., *Mission Scientifique au Mexique et dans l'Amérique Centrale*, livr. 3, 84–85. Paris: Imprimerie impériale.
TYPE SPECIMEN — Syntypes MNHN 2435, 1994.1361-3.
TYPE LOCALITY — Volcán Atitlán at 1200 m is standardly listed as the type locality for this species (e.g., Stuart 1955). In the description, Bocourt stated that the species is found in the forests of "Saint-Augustin (Guatemala)," and that one of the five syntypes was collected (by Dollfus) at the base of

Anolis dollfusianus

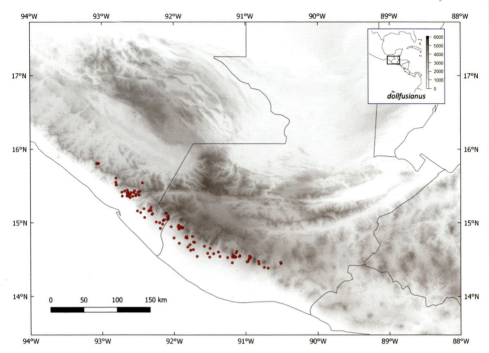

Atitlán at 1200 m. The other syntypes may be assumed to be from the Saint Augustin area, which was in the region of the Pacific slope just east of Atitlán (see, e.g., Godman and Salvin 1901: xxxviii). Volcán Atitlán is at lat 14.58 lon −91.19.

MORPHOLOGY — Body length to 43 M, 40 F; ventral scales keeled; middorsal scales enlarged in 0–2 rows; interparietal usually equal to or smaller than tympanum; scales across the snout between the second canthals 8–13; toe lamellae 12–15; suboculars and supralabials in contact or separated by a row of scales; scales separating supraorbital semicircles 1–2; supraocular scales gradually enlarged; hindlimbs long, adpressed toe to between eye and tip of snout; body color brown with light lateral stripe; tail usually banded; male dewlap yellow.

SIMILAR SPECIES — *Anolis dollfusianus* is a (very) *small brown* anole. Its combination of keeled ventral scales, long hindlimbs (adpressed toe reaches to between eye and tip of snout), yellow male dewlap, and lack of enlarged postcloacal scales in males distinguishes this form from potentially sympatric small brown anoles. In particular for potentially confusing species: *Anolis rodriguezii* has smooth ventral scales; *A. cobanensis*, *A. matudai*, and *A. campbelli* have pinkish male dewlaps and a greater number of scales separating supraorbital semicircles (usually 2–3; 1–2 in *A. dollfusianus*); *A. laeviventris* and *A. sericeus* have shorter hindlimbs (adpressed toe reaches to ear); *A. humilis*, *A. tropidonotus*, and *A. uniformis* have greatly enlarged middorsal scale rows (middorsals uniform in *A. dollfusianus*) and a puncturelike axillary pocket (absent in *A. dollfusianus*).

RANGE — From 200 to 1850 meters on the Pacific slope of eastern Chiapas, Mexico, and western Guatemala. Köhler and Acevedo (2004) note that the species occurs "exceptionally" below 800 meters.

NATURAL HISTORY — Common in forest, edge habitat, and coffee fields; active diurnally on vegetation below 1 meter above ground; sleeps on twigs or leaves below 1.5 meters above ground. See Muñoz et al. (2002); Köhler and Acevedo (2004); Macip-Ríos and Muñoz (2008).

SPECIES ACCOUNTS

ANOLIS DRACULA — PLATE 1.65, 2.114

DESCRIPTION — Yañez-Muñoz, M. H., C. Reyes-Puig, J. P. Reyes-Puig, J. A. Velasco, F. Ayala-Varela, O. Torres-Carvajal. 2018. A new cryptic species of *Anolis* lizard from northwestern South America (Iguanidae, Dactyloinae). *Zookeys* 794:135–163.
TYPE SPECIMEN — DHMECN 12579.
TYPE LOCALITY — "Km 18 road Gualpi-Chical, 0°51'8.26"N, 78°13'52.59"W, 2200 m, near Reserva Dracula, Parroquia El Chical, Cantón Tulcán, Provincia Carchi, Ecuador."
MORPHOLOGY — Body length to 91 (sexual dimorphism unknown); ventral scales smooth; middorsal scales enlarged in 0–2 rows; interparietal smaller than tympanum; scales across the snout between the second canthals 13–17; toe lamellae 18–23; suboculars and supralabials separated by rows of scales; scales separating supraorbital semicircles 2–5; supraocular scales gradually enlarged; hindimbs long, addressed toe to between eye and tip of snout or anterior to snout; body color may appear predominantly green or brown, complexly patterned usually with some banding and lateral rows of ocelli; male dewlap brown with orange-red blotches and cream scales; female dewlap dark brown to black with light (orange-red, pale greenish-blue) blotches and scales.
SIMILAR SPECIES — *Anolis dracula* is a *large brown* or *green* highland anole that is unlikely to be confused with sympatric forms due to its large distinctive dewlaps in males and females (male dewlap brown with orange-red blotches and cream scales; female dewlap dark brown to black with orange-red or pale greenish-blue blotches and scales). *Anolis aequatorialis* has similar male and female dewlaps and geographically replaces *A. dracula* to the south in most of Ecuador (these species appear to be molecularly distinct [Yañez-Muñoz et al. 2018]).
RANGE — From 1150 to 2400 meters on the Pacific Andean slope of northern Ecuador and southern Colombia.

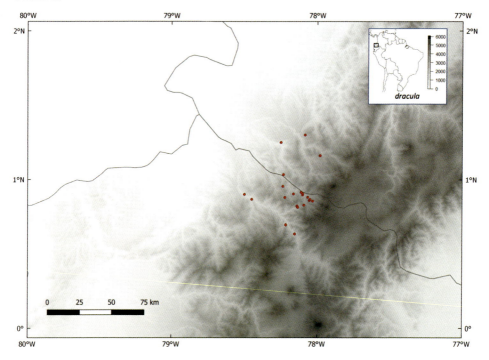

NATURAL HISTORY — The type description states this form is found in forest and disturbed areas; active diurnally arboreally, may be terrestrial or near ground; sleeps on leaves or ferns from 0.6 to 2.3 meters. See Torres-Carvajal et al. (2019); Arteaga et al. (2023).

ANOLIS DUELLMANI — PLATE 1.66, 2.22

DESCRIPTION — Fitch, H. S., R. W. Henderson. 1973. A new anole (Reptilia: Iguanidae) from southern Veracruz, México. *Journal of Herpetology* 7:125–128.
TYPE SPECIMEN — KU 59532.
TYPE LOCALITY — "The south slope of Volcan San Martin, Tuxtla, Veracruz, Mexico, between 800 and 1150m." Volcán San Martin is at lat 18.56 lon −95.20.
MORPHOLOGY — Body length to 41 M, 39 F; ventral scales keeled; middorsal scales greatly enlarged in several rows; interparietal smaller than tympanum; scales across the snout between the second canthals 7; toe lamellae 12–13; suboculars and supralabials in contact or separated by a row of scales; scales separating supraorbital semicircles 1–3; supraocular scales gradually enlarged; appressed toe to eye; body color brown; male dewlap solid pink.
SIMILAR SPECIES — *Anolis duellmani* is a *small brown* anole commonly sympatric with *A. sericeus* (Pavón-Vázquez et al. 2014). *Anolis sericeus* lacks the abruptly enlarged middorsal scales of *A. duellmani* and has strongly keeled ventral scales (weakly keeled in *A. duellmani*). *Anolis compressicauda*, *A. tropidonotus*, and *A. uniformis* have a puncturelike axillary pocket (absent in *A. duellmani*). Other potentially sympatric small brown anoles lack the enlarged keeled middorsal scales of *A. duellmani*. *Anolis pygmaeus*, found south and east of the range of *A. duellmani*, is not clearly distinguishable from *A. duellmani* (see below).

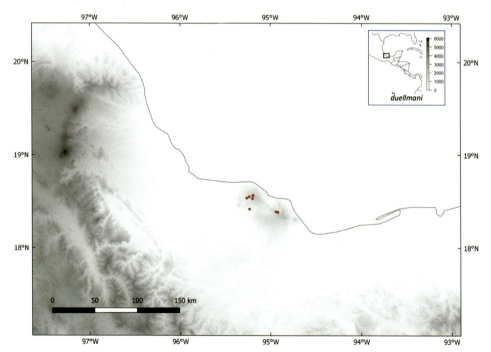

SPECIES ACCOUNTS

RANGE — From 800 to 1450 meters in Sierra de los Tuxtlas, southeastern Veracruz, Mexico.
NATURAL HISTORY — Common in remnant forest patches; active diurnally on logs and ground; sleeps on shrubs below 2 meters above ground. See Pavón-Vázquez et al. (2014).
COMMENT — The status of this form should be evaluated with reference to *Anolis pygmaeus* (Pavón-Vaázquez et al. 2014).

— *ANOLIS DUNNI* — PLATE 1.67, 2.2

DESCRIPTION — Smith, H. M. 1936. A new *Anolis* from Mexico. *Copeia* 1936:9.
TYPE SPECIMEN — FMNH 100109 ("Taylor and Smith Collection No. 1506" in Smith [1936]).
TYPE LOCALITY — "A boulder in the high mountains, within the evergreen zone, between Rincon and Cajones, Guerrero." (approximately lat 17.32 lon −99.47, 950 m)
MORPHOLOGY — Body length to 58 M, 48 F; ventral scales smooth; middorsal scales enlarged in 0–2 rows; interparietal larger than tympanum; scales across the snout between the second canthals 4–8; toe lamellae 15–18; suboculars and supralabials in contact; supraorbital semicircles in contact; supraocular scales abruptly enlarged; adpressed toe to eye or just short of eye; body color brown with light lateral stripe; male dewlap orange-red with yellow.
SIMILAR SPECIES — *Anolis dunni* is a *small brown* anole with smooth ventral scales. Locally proximal species with smooth ventrals are *A. taylori, A. gadovii, A. liogaster, A. brianjuliani, A. peucephilus*, and *A. omiltemanus*. *Anolis omiltemanus* and *A. peucephilus* are each small and short-legged (adpressed toe reaches ear; to eye in *A. dunni*) with a solid orange dewlap (orange-red with yellow in *A. dunni*), and so unlikely to be confused with *A. dunni*. *Anolis liogaster* and *A. brianjuliani* have a band of enlarged middorsal scales (middorsals uniform in *A. dunni*) and a solid pink dewlap. *Anolis gadovii* and *A. taylori* are very similar to *A. dunni*. *Anolis taylori* and *A. gadovii* are boulder specialists with

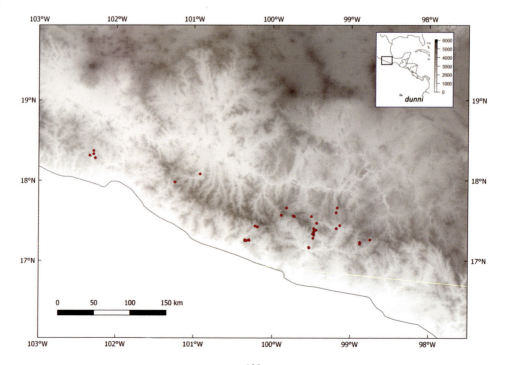

pinkish to purpleish dewlaps, whereas *A. dunni* is more catholic in habitat use and possesses an orange-red dewlap.

RANGE — From 500 to 1750 meters on the Pacific slope of Guerrero and eastern Michoacán, Mexico.
NATURAL HISTORY — In forest, edge, and disturbed areas; sleeps on twigs or leaves from 0.2 to 3 meters above ground. See Fitch et al. (1976); Köhler et al. (2014a).
COMMENT — There may be multiple evolutionary lineages, i.e., species, within *Anolis dunni* (see Köhler et al. 2014a). The taxonomic status of *A. dunni* relative to *A. gadovii* and *A. taylori* should be investigated.

— *ANOLIS ELCOPEENSIS* — PLATE 1.68, 2.31

DESCRIPTION — Poe, S., S. Scarpetta, E. W. Schaad. 2015b. A new species of *Anolis* from Panama. *Amphibian and Reptile Conservation* 9 (general section):1–13.
TYPE SPECIMEN — MSB 95571.
TYPE LOCALITY — "Parque Nacional G.D. Omar Torrijos H., Coclé Province, Panama (lat 8.66815, lon −80.59267, 801 m)."
MORPHOLOGY — Body length to 45 M, 44 F; ventral scales smooth; middorsal scales enlarged in 0–2 rows; interparietal larger or smaller than tympanum; scales across the snout between the second canthals 10–14; toe lamellae 13–16; suboculars and supralabials in contact; scales separating supraorbital semicircles 1–3; supraocular scales gradually enlarged; hindlimbs short, adpressed toe to ear or posterior to ear; body color usually grayish-brown or brown, sometimes appearing whitish; male dewlap solid orange, or (in Panama Canal area and some areas east) with orange anteriorly and yellow posteriorly; female lacks dewlap or with small orangeish-red dewlap.
SIMILAR SPECIES — *Anolis elcopeensis* is a *small brown* lowland anole distinguishable from most potentially sympatric small *Anolis* by its short hindlimbs (adpressed toe reaches to ear; anterior to eye in

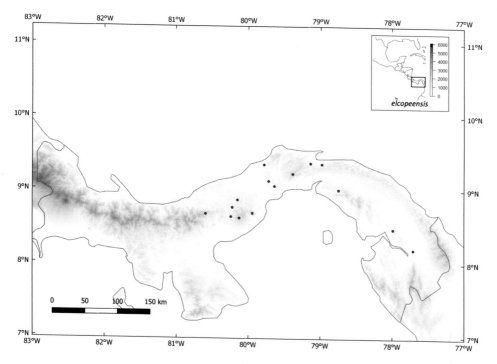

SPECIES ACCOUNTS

A. limifrons, A. humilis, A. auratus, A. apletophallus, A. gaigei) and smooth ventral scales (keeled in *A. humilis, A. gaigei, A. auratus*). *Anolis pentaprion* and *A. fungosus* usually appear lichenous and whitish dorsally (*A. elcopeensis* is gray-brown), and have larger head scales than *A. elcopeensis* (5–9 scales across the snout; 10–14 in *A. elcopeensis*). *Anolis carpenteri* is green, with slightly longer hindlimbs than *A. elcopeensis* (adpressed toe reaches to between ear and eye). *Anolis gruuo* is very similar to *A. elcopeensis*, geographically replacing it at higher elevations to the west. *Anolis gruuo* has slightly larger head scales (usually 8–10 scales across the snout between the second canthals; 10–14 in *A. elcopeensis*) and a noticeably bulging hemipenial area in larger males (area not expanded in *A. elcopeensis*).

RANGE — From 0 to 900 meters in Panama from Coclé Province east to Darién.

NATURAL HISTORY — Common in forest interior and edge habitat, and in disturbed areas; sleeps on twigs or (less frequently) leaves across a wide vertical range, from 0.5 meters up to the limits of observable individuals.

COMMENT — The populations from the Panama Canal area and farther east that we called *Anolis* cf. *elcopeensis* in our description (Poe et al. 2015b) may represent one or more undescribed species.

— *ANOLIS EQUESTRIS* — PLATE 1.69, 2.165

DESCRIPTION — Merrem, B. 1820. Versuch eines Systems der Amphibien. *Tentamen Systematis Amphibiorum*. J. C. Krieger, Marburg, 191 pp. p. 45.

TYPE SPECIMEN — None. Unpublished notes of Ernest Williams state of the holotype of *A. equestris*: "Based on CUVIER, 1817, Reg. Anim. Pl. 5, Fig. 1," indicating Williams's view that Merrem's description is based on Cuvier's (1817) plate.

TYPE LOCALITY — Unknown. Restricted by Schwartz and Garrido (1972) to Havana, La Habana Province, Cuba (lat 23.1 lon −82.4, 0 m).

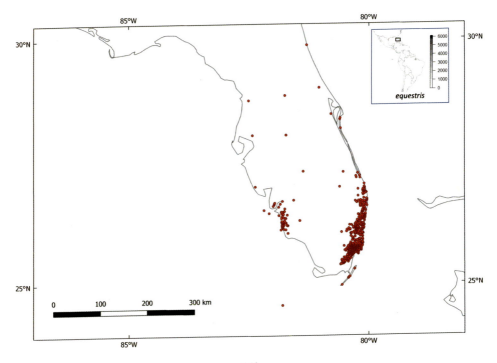

SIMILAR SPECIES — *Anolis equestris* is a *giant* green anole. Its mainland localities currently are restricted to Florida, USA, where it could only be confused with *A. garmani* among anoles. *Anolis garmani* is much less widespread than *A. equestris*, and differs in its longer hindlimbs (adpressed toe to between ear and eye; to ear in *A. equestris*) and dewlap (yellow-orange in males, usually absent in females; large, pink, in males and females of *A. equestris*). *Anolis equestris* may be confused with the green iguana, *Iguana iguana*, in Florida. Iguanas lack expanded toepads and an extensible dewlap. Other green anoles in Florida (*A. carolinensis*, *A. callainus*, *A. allisoni*) are smaller than *A. equestris* (SVL < 80 m; to 188 mm in *A. equestris*) and lack its serrated dorsal crest.

MORPHOLOGY — Body length to 188 M, 170 F; ventral scales smooth; middorsal scales form a low crest of triangular plates; interparietal larger or smaller than tympanum; scales across the snout between the second canthals 3–8; toe lamellae 24–39; suboculars and supralabials in contact; scales separating supraorbital semicircles 1–4; supraocular scales gradually enlarged; adpressed toe to ear or posterior to ear; body color green, changeable to brown, usually with a short broad light stripe extending back from the shoulder; male and female dewlaps pink.

RANGE — On the mainland, from 0 to 50 meters in peninsular Florida; originally evolved on Cuba.

NATURAL HISTORY — Very common in edge and disturbed habitats; a "crown giant" anole (Williams 1983), commonly observed diurnally above 2 meters above ground on tree trunks and branches; sleeps on leaves and (more frequently) narrow branches usually above 2 meters above ground; has been observed foraging nocturnally (Stroud and Giery 2013). For natural history information on Florida populations, see Brach (1976); Dalrymple (1980); Meshaka (1993; 1999a, b; 2010); Meshaka and Rice (2005); Meshaka et al. (2008); Camposano et al. (2008); Nicholson and Richards (2011); Giery et al. (2017). See, e.g., Schwartz and Henderson (1991) for natural history information on Cuban populations.

— *ANOLIS EULAEMUS* — PLATE 1.70, 2.119

DESCRIPTION — Boulenger, G. A. 1908. Descriptions of new batrachians and reptiles discovered by Mr. M. G. Palmer in south-western Colombia. *Annals and Magazine of Natural History* 8:516–517.
TYPE SPECIMEN — BMNH 1946.8.13.31.
TYPE LOCALITY — "Near Pavas." Presumably Pavas, La Cumbre, Valle del Cauca, Colombia (lat 3.68 lon −76.58, 1400 m).
MORPHOLOGY — Body length to 101 M, 87 F; ventral scales smooth; middorsal scales enlarged in 0–2 rows; interparietal smaller than tympanum; scales across the snout between the second canthals 11–19; toe lamellae 21–23; suboculars and supralabials usually separated by a row of scales; scales separating supraorbital semicircles 1–4; supraocular scales subequal; hindlimbs long, adpressed toe to between eye and tip of snout; body color brown with black and cream banding and spotting; male and female dewlaps brown with black flecks especially proximally.
SIMILAR SPECIES — *Anolis eulaemus* is a *giant* or *large brown* to olive green highland anole. It is distinctive among giant Colombian anoles in being predominantly shades of pale brown and gray, often with little to no green. *Anolis eulaemus* differs from similar and geographically proximal forms as follows: *A. anoriensis* has a greater contrast between dark anterior and much lighter posterior shading in the dewlap (dewlap pale brown with diffuse black flecks in *A. eulaemus*, anterior aspect only slightly darker than posterior), is greener, and has smaller posterior head scales (4–5 scales between supraorbital semicircles; 1–3, rarely 4 in *A. eulamus*); *A. antioquiae* is bright green (but see discussion under that species and *A. megalopithecus*), usually lacks a differentiated interparietal scale (interparietal present in *A. eulaemus*), and has a different female dewlap (red-orange or white with black spots; brown with black flecks in *A. eulaemus*); *A. danieli* and lowland species *A. biporcatus*, *A. purpurescens*, and *A. frenatus* are green and have broad, distinct toepads (i.e., the usual *Anolis* condition; toepads narrow in *A. eulaemus*); *Anolis ventrimaculatus* is smaller (maximum SVL: 80 mm; to 101 mm in *A. eulaemus*), lacks a dewlap in females (present in *A. eulaemus*), and often appears green; *Anolis megalopithecus* is distinctively colored in body (dark brown with red,

SPECIES ACCOUNTS

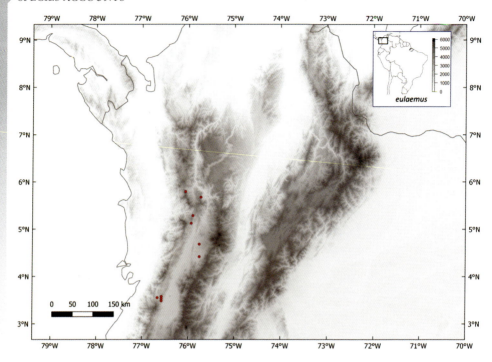

tan, and green) and male dewlap (red with black markings) and usually lacks a differentiated interparietal scale.
RANGE — From 1600 to 2000 meters on the Pacific Andean slope of Colombia.
NATURAL HISTORY — Common to rare in forest, depending on locality; sleeps low on twigs and (especially) leaves and ferns. See López-Herrera et al. (2016).

— *ANOLIS EUSKALERRIARI* — PLATE 1.71, 2.142

DESCRIPTION — Barros, T. R., E. E. Williams, A. Viloria. 1996. The genus *Phenacosaurus* (Squamata: Iguania) in western Venezuela: *Phenacosaurus tetarii*, new species, *Phenacosaurus euskalerriari*, new species, and *Phenacosaurus nicefori* Dunn, 1944. Breviora 504:16–23.
TYPE SPECIMEN — MBLUZ 308.
TYPE LOCALITY — "The canyons of Mesa Turik, Sierra de Perijá, Estado Zulia, Venezuela (72 44' 27" W, 10 22' 23" N), 1,600 m elevation."
MORPHOLOGY — Body length to 53 M, 53 F; ventral scales smooth; middorsal scales enlarged in 0–2 rows; interparietal larger than tympanum; scales across the snout between the second canthals 2–3; toe lamellae 21–25; suboculars and supralabials in contact; supraorbital semicircles in contact; supraocular scales abruptly or gradually enlarged; body color gray to green, with black, brown and white marks, with broad banding and lichenous pattern; male dewlap pink.
SIMILAR SPECIES — *Anolis euskalerriari* is a highland *twig* anole that may be sympatric with *A. fuscoauratus* and/or *A. planiceps*, and is amply distinct from both in its large smooth head scales (scales across the snout at second canthals 8–14 in *A. fuscoauratus* and *A. planiceps*; 2–3 in *A. euskalerriari*). *Anolis frenatus* and *A. apollinaris* are much larger than *A. euskalerriari* (maximum SVL > 100 mm; to 53 mm in *A. euskalerriari*), possess smaller head scales (8–15 scales across the snout), and are not

142

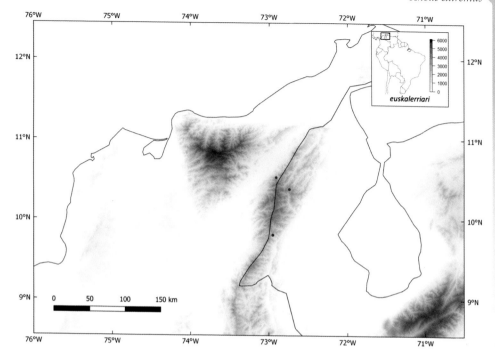

likely to be found at the high elevations inhabited by *A. euskalerriari*. *Anolis tetarii* is very similar to *A. euskalerriari* and found west of its range in the Sierra de Perijá. These species differ in lateral body scalation (heterogeneous in *A. tetarii*, homogeneous in *A. euskalerriari*) and presence of a serrated middorsal crest in *A. tetarii* (absent in *A. euskalerriari*).

RANGE — From 1600 to 1700 meters in the southern Perijá mountains, Venezuela and Colombia.

NATURAL HISTORY — Only information is from the type description: "Individuals perch on small shrubs and bushes or on moss."

— *ANOLIS EXTREMUS* — PLATE 1.72, 2.159

DESCRIPTION — Garman, S. 1887. On West Indian reptiles (Iguanidae). *Bulletin of the Essex Institute* 19:35–36.

TYPE SPECIMEN — MCZ 6183, lectotype selected by Lazell (1972).

TYPE LOCALITY — "Barbadoes." That is, Barbados, southern Lesser Antilles. Lazell (1972) listed Bridgetown, Barbados (lat 13.11 lon −59.61, 0 m) as the likely type locality.

MORPHOLOGY — Body length to 83 M, 60 F; ventral scales smooth; middorsal scales enlarged in 0–2 rows; interparietal equal to or larger than tympanum; scales across the snout between the second canthals 6–11; toe lamellae 22–28; suboculars and supralabials in contact; supraorbital semicircles in contact; supraocular scales gradually enlarged; addpressed toe to between ear and eye; body color green with light and dark flecks, often with blue-gray on the head; dewlap orange with yellow-green scales in males, absent in females.

SIMILAR SPECIES — *Anolis extremus* is a *large green* anole. Identification of this form should be straightforward due to its large size, vivid green body coloration with blue-gray on the head, presence in

143

SPECIES ACCOUNTS

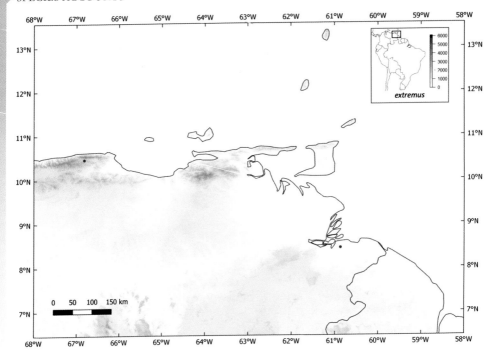

human-altered environments, and distinctive dorsal head scalation of broad contact between the supraorbital semicircles and between these scales and the interparietal.

RANGE — On the mainland, established in Caracas, Venezuela. Rivas et al. (2012) reported observing an individual at a nursery in Macaracuay, eastern Caracas. Native to Barbados, southern Lesser Antilles. Also established in Saint Lucia and Bermuda and has been observed but apparently is not currently breeding in Florida, USA, and Trinidad.

NATURAL HISTORY — Not studied in Venezuela. On Barbados, "utterly ubiquitous" on most terrestrial and arboreal surfaces (Lazell 1972). Lazell (1972:85) described Barbour's (1930:112) statement that the species is "almost if not quite extinct on Barbados" as "one of the great verbal monuments of all time, but whether to a lizard's incredible fecundity or a man's incredible myopia, I cannot be sure." See, e.g., Lazell (1972), Daltry (2009) for natural history on Barbados.

COMMENT — I consider *Anolis deltae* to be a junior synonym of *A. extremus* or *A. aeneus* (see appendix 1).

— *ANOLIS FASCIATUS* — PLATE 1.73, 2.103

DESCRIPTION — Boulenger, G. A. 1885. Catalogue of the lizards in the British Museum (Natural History), 2nd ed., vol. 2, 59.

TYPE SPECIMEN — BMNH 1946.8.20.13.

TYPE LOCALITY — "Ecuador ... Guayaquil." Guayaquil, Guayas, Ecuador is at lat −2.14 lon −79.89, 10 m.

MORPHOLOGY — Body length to 72 M, 71 F; ventral scales smooth; middorsal scales enlarged in 0–2 rows; interparietal larger or smaller than tympanum; scales across the snout between the second canthals 8–14; toe lamellae 18–23; subocular and supralabials in contact; scales separating

Anolis fasciatus

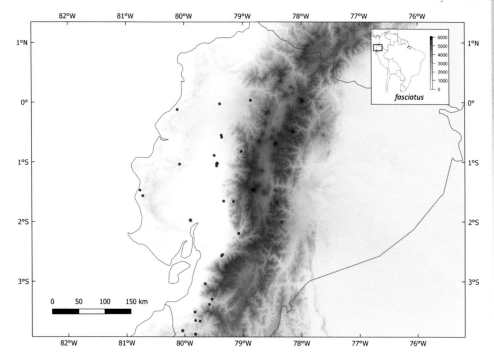

supraorbital semicircles 0–2; supraocular scales subequal to gradually enlarged; hindlimbs long, adpressed toe to between eye and tip of snout; body, limbs, and tail banded green and brown; male dewlap white, possibly with blue tint, with dark spots along scale rows; female dewlap absent; iris blue.

SIMILAR SPECIES — *Anolis fasciatus* is a *large green* anole that is separable from most similar geographically proximal anoles by its blue eyes. The iris of *A. chloris* may appear blue, but this species has keeled ventral scales (smooth in *A. fasciatus*) and is generally patternless green, whereas *A. fasciatus* is complexly patterned with bands and rows of ocelli. *Anolis festae* may have blue eyes but is smaller than *A. fasciatus* (maximum SVL 55 mm; to 72 mm in *A. fasciatus*) and has shorter hindlimbs (adpressed toe reaches to ear; to between eye and tip of snout in *A. fasciatus*). *Anolis peraccae* is smaller (SVL to 52 mm) than *A. fasciatus* and has fewer toe lamellae (13–19; 18–23 in *A. fasciatus*). Other potentially confusing green-bodied blue-eyed species (*A. gemmosus*, *A. otongae*, *A. poei*) live at higher elevations and have multicolored male dewlaps (solid white in *A. fasciatus*).

RANGE — From 0 to 1600 meters on the Pacific Andean slope of Ecuador.

NATURAL HISTORY — Patchily distributed in forest and disturbed areas: very common at some localities, rare in others; active diurnally on low vegetation or leaf litter; sleeps on twigs or (more frequently) leaves above 1 meter above ground. See Miyata (2013); Narváez (2017); Torres-Carvajal et al. (2019); Arteaga et al. (2023).

COMMENT — The reports of Ecuadorian species *Anolis fasciatus* from Gorgona Island, Colombia (Parker 1926; Castro-Herrera et al. 2012) apparently refer to BMNH 1926.1.20.106, a juvenile *A. purpurescens* originally identified as *A. fasciatus* (Ayala and Williams 1988).

SPECIES ACCOUNTS

— ANOLIS FESTAE — PLATE 1.74, 2.102

DESCRIPTION — Peracca, M. G. 1904. Viaggio del Dr. Enrico Festa nell'Ecuador e regioni vicine. Bollettino dei Musei di Zoologia ed Anatomia comparata della R. Universita di Torino 19:4–5.
TYPE SPECIMEN — MRSN 2895.
TYPE LOCALITY — "Balzar." That is, Balzar, Guayas, Ecuador (lat −1.36 lon −79.90, 40 m).
MORPHOLOGY — Body length to 55 M, 51 F; ventral scales smooth; middorsal scales enlarged in 0–2 rows; interparietal larger or smaller than tympanum; scales across the snout between the second canthals 8–12; toe lamellae 15–21; suboculars and supralabials in contact; scales separating supraorbital semicircles 1–2; supraocular scales gradually enlarged; hindlimbs short, adpressed toe to ear; body brown, gray, and green, with light spots and light lateral bands; male dewlap white distally, black near throat; female dewlap absent; iris usually dark blue.
SIMILAR SPECIES — *Anolis festae* is a *small green* or *brown* lowland anole. The combination of small body size, smooth ventral scales, short hindlimbs (adpressed toe reaches to ear), and an unusually shaped and colored male dewlap (relatively shallow for its length, white distally with black proximally) will serve to distinguish *A. festae* from other lowland anoles of western Ecuador. *Anolis peraccae* is similar to *A. festae* but has slightly longer hindlimbs (adpressed reaches to eye), a dark spot posterior to the eye (absent in *A. festae*), and a normally shaped anole dewlap (i.e., not shallow). *Anolis anchicayae* is similar to *A. festae* but slightly larger (SVL to 63 mm; to 55 mm in *A. festae*) and with a normally shaped, yellow-green male dewlap. *Anolis fasciatus* is larger than *A. festae* (SVL to 72 mm) and has longer hindlimbs (adpressed toe reaches to between eye and tip of snout).
RANGE — From 0 to 450 meters in Pacific Ecuador and neighboring Colombia and Peru.
NATURAL HISTORY — Common in forest and especially in disturbed areas. Active diurnally on low vegetation and tree trunks; sleeps on twigs, leaves, or vines below 1.5 meters above ground. See Miyata (2013); Pinto-Erazo et al. (2020); Torres-Carvajal et al. (2019); Arteaga et al. (2023).

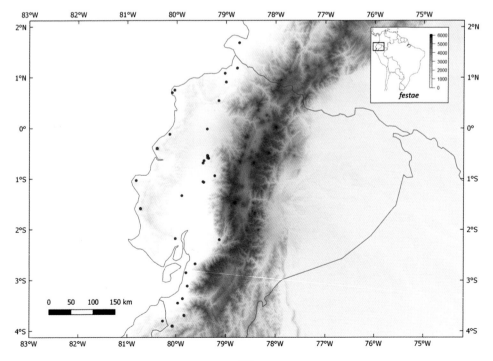

Anolis fitchi

COMMENT — I consider *Anolis nigrolineatus* (type locality: "Machala, El Oro, Province, Ecuador") to be a junior synonym of *A. festae* (see appendix 1 for justification). For those who wish to recognize *A. nigrolineatus*, it was considered to be the version of *A. festae* found in southern Ecuador.

— *ANOLIS FITCHI* — PLATE 1.75, 2.120

DESCRIPTION — Williams, E. E., W. E. Duellman. 1984. *Anolis fitchi*, a new species of the *Anolis aequatorialis* group from Ecuador and Colombia. In Seigel, R. A., L. E. Hunt, J. L. Knight, L. Malaret, N. L. Zuschlag. *Vertebrate Ecology and Systematics. A Tribute to Henry S. Fitch*, 257–266. The University of Kansas Museum of Natural History.
TYPE SPECIMEN — KU 142865.
TYPE LOCALITY — "16.5 km (by road) north-northeast of Santa Rosa, Provincia Napo, Ecuador, 1700 m elevation." This locality is at lat −0.204 lon −77.711.
MORPHOLOGY — Body length to 91 M, 78 F; ventral scales smooth; middorsal scales enlarged in 0–2 rows; interparietal larger or smaller than tympanum; scales across the snout between the second canthals 11–23; toe lamellae 19–24; suboculars and supralabials in contact or separated by a row of scales; scales separating supraorbital semicircles 1–4; supraocular scales subequal; hindlimbs long, appressed toe to anterior to snout; body color complex mix of browns and greens, banded, often appearing brown with yellow or greenish bands; male and female dewlap dark gray with greenish scales; iris dark green to brown or blue.
SIMILAR SPECIES — *Anolis fitchi* is a *large brown* highland anole that tends to be sympatric with *A. orcesi* and *A. fuscoauratus*, which are unlikely to be confused with it as both are much smaller than *A. fitchi* and differ in numerous other traits. *Anolis fitchi* is geographically replaced to the south by similar anole *A. podocarpus*, with which it differs mainly in dewlap (dark gray to brown dewlap skin in males

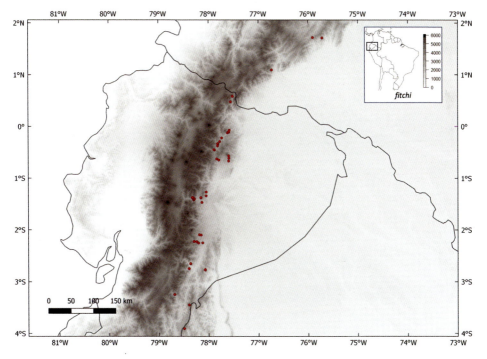

SPECIES ACCOUNTS

with scale rows composed of multiple scales in males and females; lighter, yellowish-gray dewlap skin with scales in close-set rows of single scales in males and females of *A. fitchi*). Geographically proximal large brown Amazonian species have keeled ventral scales (*A. scypheus*, *A. bombiceps*; ventrals smooth in *A. fitchi*) or large head scales (*A. transversalis*: 4–9 scales across the snout; *A. fitchi*: 11–23) and tend to inhabit lower elevations than *A. fitchi*.

RANGE — From 1250 to 2200 meters on the north and central Amazonian Andean slope of Ecuador and southern Colombia.

NATURAL HISTORY — Common in forest and open areas, diurnally on leaf litter, tree trunks, and low vegetation; sleeps on twigs or (more frequently) leaves, often ferns, from 0.5 to 2.5 meters above ground. See Yañez-Muñoz (2001); Narváez (2017); Torres-Carvajal et al. (2019); Arteaga et al. (2023).

— *ANOLIS FORTUNENSIS* — PLATE 1.76, 2.36

DESCRIPTION — Arosemena, F. A., R. Ibañez. 1993. Una nueva especie de *Anolis* (Squamata: Iguanidae) del grupo *fuscoauratus* de Fortuna, Panamá. *Revista de Biología Tropical* 41:267–272.

TYPE SPECIMEN — MVUP 756.

TYPE LOCALITY — "Along Rio Chiriquí (8 45 N, 82 10 W), in front of the meteorological station of the Instituto de Recursos Hidráulicos y Electrificación in Bijau, between 1050–1075 m, in Fortuna Reserve, Chiriquí, Panamá." (Translated from Spanish; the coordinates point is about 9 km east of the hydroelectric dam).

MORPHOLOGY — Body length to 49 M, 49 F; ventral scales smooth; middorsal scales enlarged in 0–2 rows; interparietal larger than tympanum; scales across the snout between the second canthals 9–14; toe lamellae 14–15; suboculars and supralabials in contact; scales separating supraorbital semicircles 1–3; supraocular scales gradually enlarged; hindlimbs short, adpressed toe to ear; body color

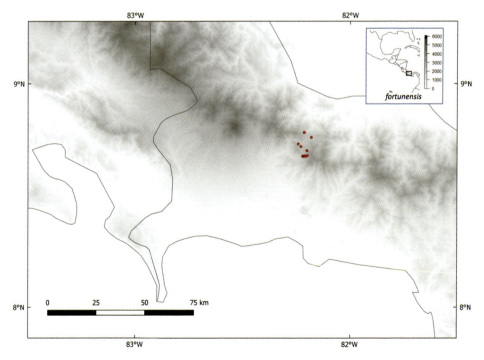

Anolis fraseri

grayish-brown; male dewlap orange or red anteriorly, yellow to orange posteriorly; female dewlap white.

SIMILAR SPECIES — *Anolis fortunensis* is a *small brown* anole with smooth ventral scales currently known only from the Fortuna pass area in Panama. *Anolis fortunensis* differs from similar geographically proximal anoles as follows: *A. limifrons* and *A. polylepis* have longer hindlimbs (adpressed toe reaches at least to eye; to ear in *A. fortunensis*); *A. carpenteri* is olive green; *A. fungosus*, *A. pentaprion*, and *A. salvini* have a lichenous body pattern, appearing whitish or greenish-yellow, and possess solid red (*A. salvini*, *A. fungosus*) or purple-pink (*A. pentaprion*) male and female dewlaps (male dewlap red to orangish-red anteriorly, orange or yellow posteriorly in *A. fortunensis*). *Anolis fortunensis* differs from all the above species, and *A. gruuo*, *A. elcopeensis*, *A. kemptoni*, and *A. pseudokemptoni*, in male dewlap color (orange in *A. gruuo*, *A. elcopeensis*; pink posteriorly, reddish anteriorly in *A. kemptoni*, *A. pseudokemptoni*).
RANGE — From 900 to 1200 meters along the Fortuna pass in Panama.
NATURAL HISTORY — Common in forest and open areas. Sleeps on twigs or leaves from 0.5 to 2 meters above ground.
COMMENT — The dewlap variation (see above) may signal different species within *Anolis fortunensis*. *Anolis exsul* from the Fortuna pass region was described by Arosemena and Ibañez (1994) and synonymized with *A. fortunensis* by Ponce and Köhler (2008).

— *ANOLIS FRASERI* — FIGURE 0.1; PLATE 1.77

DESCRIPTION — Günther, A. 1859. Second list of cold-blooded vertebrata collected by Mr. Fraser in the Andes of Western Ecuador. *Proceedings of the Zoological Society of London* 27:407–408.
TYPE SPECIMEN — Lectotype BMNH 1946.8.8.47.
TYPE LOCALITY — "The Andes of western Ecuador." Louis Fraser, the collector of the type specimen,

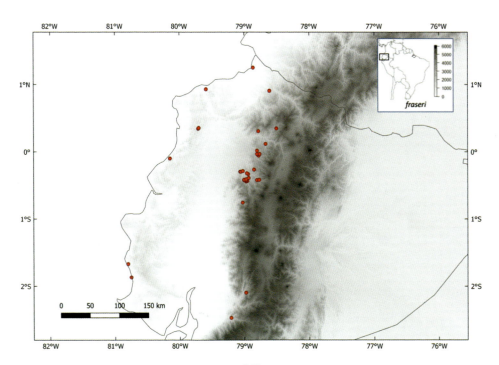

SPECIES ACCOUNTS

traveled extensively in Ecuador. Some of the species in the lot including the *Anolis fraseri* type donated to Günther and retained for the British Museum are restricted in range to northern Ecuador, so perhaps the Pacific slope of the northern Andes is the more likely type region.

MORPHOLOGY — Body length to 109 M, 116 F; ventral scales smooth to weakly keeled; middorsal scales enlarged in 0–2 rows; interparietal larger or smaller than tympanum; scales across the snout between the second canthals 5–10; toe lamellae 19–24; suboculars and supralabials in contact; scales separating supraorbital semicircles 1–4; supraocular scales subequal to gradually enlarged; hindlimbs short, adpressed toe to ear or posterior to ear; body with colorful array of greens, browns, and grays, with banding on dorsum and tail; sometimes with red on the head; male dewlap greenish-white with orange-yellow anteriorly; female dewlap similar to male's but smaller and duller; iris red to reddish-brown.

SIMILAR SPECIES — *Anolis fraseri* is a *giant* anole with short hindlimbs (adpressed toe reaches to ear) that differs from similarly proportioned, potentially sympatric giant anoles as follows: *A. parilis* has narrow, indistinct toepads continuous with a large claw (broad, distinct, typical anoline toepad in *A. fraseri*) and few toe lamellae (13–16; 19–24 in *A. fraseri*); *A. biporcatus* and *A. parvauritus* are solid green (*A. fraseri* usually is strongly patterned); *A. princeps* and *A. aequatorialis* are giant anoles with long hindlimbs (adpressed toe reaches beyond eye).

RANGE — From 100 to 1800 meters on the Pacific Andean slope of Ecuador.

NATURAL HISTORY — Common in edge and disturbed habitat, also in forest; active high on tree trunks, branches, and leaves; usually sleeps high (4–8 meters) on twigs, narrow branches, leaves, or vines. See Arteaga et al. (2013); Narváez (2017); Viteri (2015); Torres-Carvajal et al. (2019); Arteaga et al. (2023).

COMMENT — The Colombian populations referred to as *Anolis fraseri* near Lago Calima and Yotoco Reserve represent an undescribed species. The geographic boundaries between this undescribed form and true *A. fraseri* from Ecuador are not clear to me. I have excluded all Colombian populations of "*A. fraseri*" from the range map, as there appears to be a geographic gap between northern Ecuadorean and Colombian *fraseri*-like anoles.

— *ANOLIS FRENATUS* — PLATE 1.78, 2.131

DESCRIPTION — Cope, E. D. 1899. Contributions to the herpetology of New Granada and Argentina, with descriptions of new forms. *The Philadelphia Museums Scientific Bulletin* 1:6–7.

TYPE SPECIMEN — Lost, according to Lotzkat et al. (2013), Barbour (1934).

TYPE LOCALITY — "Colombia." In the introduction to the paper describing the species, Cope stated, "I have not been able to ascertain the exact localities at which the specimens were obtained, but most of them, it is believed, were found in the neighborhood of Bogota." Bogota itself is too high for *Anolis frenatus*, but the species is present in the Magdalena River valley, west of Bogota. Barbour (1934) states, without explanation, that collection of the specimen was "apparently made near Barranquilla." Pending more detailed elucidation of the collection of the type specimen, I favor restricting the type locality to the Magdalena River valley west of Bogota.

MORPHOLOGY — Body length to 143 M, 118 F; ventral scales smooth; middorsal scales enlarged in 0–2 rows; interparietal smaller than tympanum; scales across the snout between the second canthals 9–15; toe lamellae 21–28; suboculars and supralabials in contact or separated by a row of scales; scales separating supraorbital semicircles 1–5; supraocular scales subequal; hindlimbs long, adpressed toe to between eye and tip of snout; body color olive green with lateral rows of ocelli; tail banded; male and female dewlaps white.

SIMILAR SPECIES — *Anolis frenatus* is a widespread *giant* green anole potentially sympatric with several large species across its range. *Anolis insignis*, *A. brooksi*, *A. kathydayae*, *A. savagei*, *A. ginaelisae*, *A. microtus*, *A. fraseri*, and *A. biporcatus* differ from *A. frenatus* in their shorter hindlimbs (adpressed toe reaches approximately to ear; to between eye and tip of snout in *A. frenatus*). *Anolis anoriensis*, *A. ventrimaculatus*, *A. aequatorialis*, *A. dracula*, *A. antioquiae*, and *A. eulaemus* are usually found at

Anolis frenatus

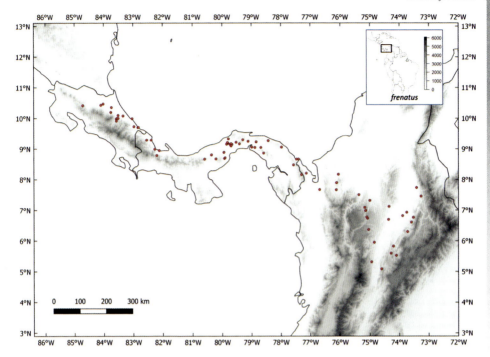

higher elevations than *A. frenatus*, possess narrow toepads (toepads broad, distinct in *A. frenatus*), and lack the distinctive dorsal body pattern of *A. frenatus* of diagonal lateral rows of ocelli on an olive green background. *Anolis maculigula* is bluish in body color and has narrow toepads and fewer toe lamellae than *A. frenatus* (16–22; 21–28 in *A. frenatus*). *Anolis huilae*, *A. danieli*, and *A. apollinaris* usually are found at higher elevations than *A. frenatus* and differ in body pattern (*A. danieli*, *A. apollinaris*: approximately solid bright green or banded green; *A. huilae*: green with densely packed lateral yellow spots) and male dewlap color (green in *A. apollinaris*, cream to dark green in *A. danieli*, yellow in *A. huilae*, white in *A. frenatus*). *Anolis kunayalae*, *A. mirus*, and *A. parilis* have narrow, indistinct toepads, often some blue dorsally in males, and few toe lamellae (11–15; 21–28 in *A. frenatus*). *Anolis purpurescens*, *A. maia*, *A. ibanezi*, and *A. limon* are smaller (SVL < 85 mm) than *A. frenatus* with fewer toe lamellae (<22) and suboculars and supralabials in contact (suboculars and supralabials often separated by a row of scales in *A. frenatus*). *Anolis capito* has a very short snout, keeled ventral scales (smooth in *A. frenatus*), and small (barely to axillae; well on to chest in *A. frenatus*) dark dewlaps in males and females. *Anolis casildae* is indistinguishable in scalation from *A. frenatus*, but thus far has been found only in the Fortuna, Panama area (*A. frenatus* is widespread) and differs in some details of color pattern: *Anolis casildae* has a strongly patterned female dewlap (white in *A. frenatus*), may appear banded or solid green, and has dark green or yellow scales on the male dewlap (scales pale, usually colored similarly to dewlap skin in *A. frenatus*). *Anolis latifrons* is very similar to *A. frenatus* but differs in its squarish scales in the superciliary area (*A. frenatus* has a single elongate superciliary scale). I know of no consistent trait to separate *A. frenatus* from *A. princeps*, which generally is considered to replace *A. frenatus* to the south along the Pacific coast lowlands of northern South America (see Comment).

RANGE — From 0 to 1500 meters, from northern Costa Rica south to Darién Panama, northern Colombia, and the Magdalena River valley.

NATURAL HISTORY — Common in forest; active diurnally on tree trunks; sleeps on twigs, vines, or (less frequently) leaves, usually above 1.5 meters above ground. See Ballinger et al. (1970); Scott et al.

SPECIES ACCOUNTS

(1976); Andrews (1976); Scott et al. (1976); Rand and Myers (1990); Losos et al. (1991); Moreno-Arias et al. (2020); Batista et al. (2020).

COMMENT — *Anolis frenatus*, *A. latifrons*, and *A. princeps* are not clearly distinguishable by morphology, and molecular studies (e.g., Castañeda and de Queiroz 2011) have found minimal mitochondrial divergence between individuals identified as these species (see also Savage and Talbot 1978). Molecular studies incorporating broad geographic sampling of these forms would be taxonomically informative.

— *ANOLIS FUNGOSUS* — PLATE 1.79, 2.76

DESCRIPTION — Myers, C. W. 1971. Central American lizards related to *Anolis pentaprion*: two new species from the Cordillera de Talamanca. *American Museum Novitates* 2471:3–10.
TYPE SPECIMEN — KU 113451.
TYPE LOCALITY — " 'Campo Mojica,' a clearing on a trail at 1450 meters elevation, on the north slopes of Cerro Pando, upper watershed of Rio Changena, in the Cordillera de Talamanca, Bocas del Toro Province, Republic of Panama." Fairchild and Handley (1966) describe and place Rancho Mojica "Near 9° 02' N–82°41' W," which fits with Myers's description of Campo Mojica.
MORPHOLOGY — Body length to 48 M, 48 F; ventral scales smooth; middorsal scales enlarged in 0–2 rows; interparietal usually larger than tympanum; scales across the snout between the second canthals 7; toe lamellae 13–16; suboculars and supralabials in contact; scales separating supraorbital semicircles 1–2; supraocular scales gradually enlarged; hindlimbs short, adpressed toe to ear or posterior to ear; body color grayish-white, lichenous; male dewlap red; female dewlap yellow.
SIMILAR SPECIES — *Anolis fungosus* is a small highland *twig* anole that is distinctive in its white coloration, elongate body, and short limbs (adpressed toe reaches to posterior to ear) and tail. Similarly sized species that share smooth ventral scales and short hindlimbs differ in having brown body color and

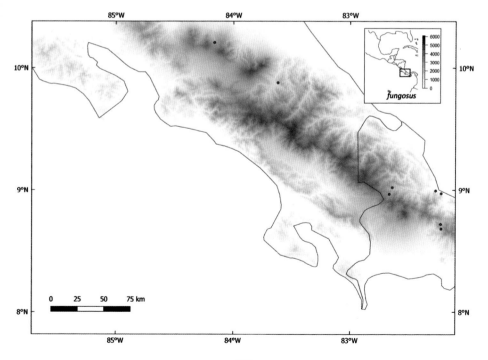

in male dewlap color (*A. kemptoni*, *A. pseudokemptoni*: pink posteriorly, red anteriorly; *A. fortunensis*: orange-red anteriorly and yellow or orange posteriorly; *A. tenorioensis*: reddish-orange with dark spots; *A. arenal*: red with dark red to black centrally; *A. monteverde*, *A. elcopeensis*, *A. altae*, *A. gruuo*: orange; *A. fungosus*: solid red). *Anolis carpenteri* is olive green with an orange male dewlap. *Anolis pentaprion*, *A. charlesmyersi*, and *A. salvini* are similar to *A. fungosus* in body color and limb proportion but are larger (maximum SVL 66–76 mm; to 48 mm in *A. fungosus*) and possess greater numbers of toe lamellae (at least 16, frequently more than 20; 13–16 in *A. fungosus*).
RANGE — From 1000 to 1450 meters from western Panama (Fortuna pass area) to south-central Costa Rica (Volcán Poás area).
NATURAL HISTORY — Rare in forest and edge habitat; sleeps on twigs above 2 meters above ground.

— *ANOLIS FUSCOAURATUS* — PLATE 1.80, 2.88

DESCRIPTION — d'Orbigny in Duméril, A.M.C., G. Bibron. 1837. *Erpétologie Générale ou Histoire Naturelle Complete des Reptiles*, vol. 4, 110–111.
TYPE SPECIMEN — MNHN 2420.
TYPE LOCALITY — "Chili," corrected by d'Orbigny (1847) to Río Mamoré, between Loreto and Río Sara, Bolivia, and by Bocourt (1873) to Moxos Province, Bolivia. Poe et al. (manuscript) identified the town of Camiaco (lat −15.33 lon −64.87, 160 m) as a reasonable type locality.
MORPHOLOGY — Body length to 46 M, 53 F; ventral scales smooth; middorsal scales enlarged in 0–2 rows; interparietal larger or smaller than tympanum; scales across the snout between the second canthals 8–14; toe lamellae 10–17; suboculars and supralabials in contact; scales separating supraorbital semicircles 1–3; supraocular scales gradually enlarged; hindlimbs short, adpressed toe approximately to ear; body color brown; tail banded; male dewlap usually pink or white, may have some orange or

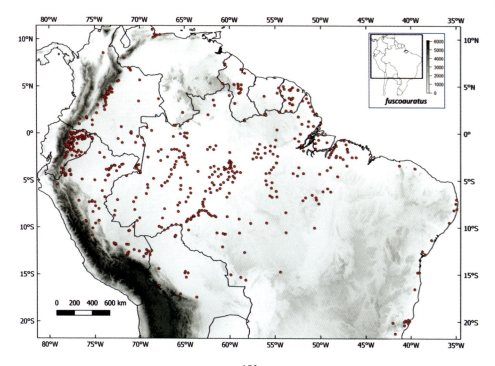

SPECIES ACCOUNTS

yellow, may be bicolor pink and orange at high elevations on the eastern Andean slope; female dewlap usually absent.
SIMILAR SPECIES — *Anolis fuscoauratus* is a common, widespread *small brown* anole that is a member of the complement of anoles usually found at lowland Amazonian sites. Among these species, *A. fuscoauratus* differs from *A. transversalis* and *A. punctatus* in its smaller body size (adults generally over 70 mm SVL in *A. punctatus*, *A. transversalis*; to 53 mm in *A. fuscoauratus*), brown body (green in *A. punctatus*), and brown eyes (blue in *A. transversalis*); from *A. caquetae*, *A. chrysolepis*, *A. bombiceps*, *A. tandai*, *A. planiceps*, and *A. scypheus* in its smooth ventral scales (keeled in these species); from *A. ortonii* in male and female dewlaps (both bright red with yellow along scale rows in *A. ortonii*; pink to white, sometimes with orange in male, usually absent in female *A. fuscoauratus*) and condition of supraorbital semicircles (usually in contact in *A. ortonii*; separated by scales in *A. fuscoauratus*); and from *A. trachyderma* in its shorter hindlimbs (adpressed toe reaches past eye in *A. trachyderma*; to ear in *A. fuscoauratus*). Brown body color and/or smaller body size will separate *A. fuscoauratus* from rare, larger Amazonian green species *A. dissimilis* and *A. phyllorhinus*. *Anolis fuscoauratus* may be sympatric with several anole species at relatively high elevations on the eastern Andean slope. Most comparably sized potentially sympatric eastern Andean slope species are *twig* anoles (*A. lamari*, *A. ruizii*, *A. lososi*, *A. orcesi*, *A. laevis*, *A. peruensis*, *A. williamsmittermeierorum*); each of these differs from *A. fuscoauratus* in possessing large smooth head scales (supraorbital semicircles in contact in these species; separated by rows of scales in *A. fuscoauratus*). *Anolis auratus* and *A. gaigei* differ from *A. fuscoauratus* in their keeled ventral scales (smooth in *A. fuscoauratus*). Several Andean and Pacific lowland *small brown* "fuscoauratid" anoles are very similar to *A. fuscoauratus*. All are allopatric to *A. fuscoauratus* and differ in male dewlap color (see species accounts of *A. maculiventris*, *A. mariarum*, *A. urraoi*, *A. antonii*, *A. tolimensis*).
RANGE — From 0 to 1650 meters in Amazonia; possibly west to the northern Magdalena River valley (Grisales et al. 2017).
NATURAL HISTORY — Common in forest and forest edge; frequently the most abundant anole where it is found; occupies a wide range of arboreal perches and perch heights diurnally and nocturnally; often observed sleeping on twigs at night. See Beebe (1944); Rand and Humphrey (1968); Crump (1971); Vanzolini (1972); Hoogmoed (1973); Duellman (1978); Martins (1991); Ávila-Pires (1995); Vitt and Zani (1996b); Vitt et al. (1999); Vitt (2000); Vitt et al. (2003b); Landauro and Morales (2007); Macedo et al. (2008); Silva et al. (2011); de Oliveira et al. (2014); Campos (2016); Narváez (2017); Araújo-Nieto (2017); Pinto-Aguirre (2014); Diele-Viegas et al. (2019); Faria et al. (2019); Torres-Carvajal et al. (2019); Moreno-Arias et al. (2020); Campos et al. (2022); Arteaga et al. (2023); Pinto and Torres-Carvajal (2023).
COMMENT — I consider *Anolis bocourtii* (type locality: Nauta, Peru) to be a junior synonym of *A. fuscoauratus* (see appendix 1 for justification).

— *ANOLIS GADOVII* — PLATE 1.81, 2.4

DESCRIPTION — Boulenger, G. A. 1905. Descriptions of new reptiles discovered in Mexico by Dr. H. Gadow, F.R.S. *Proceedings of the Zoological Society of London* 1905 (1):245.
TYPE SPECIMEN — BMNH 1946.8.13.1.
TYPE LOCALITY — "Tierra Colorada, South Guerrero." Tierra Colorada, Guerrero, Mexico is located at lat 17.167 lon −99.533, 280 m.
MORPHOLOGY — Body length to 80 M, 69 F; ventral scales smooth; middorsal scales enlarged in 0–2 rows; interparietal larger or smaller than tympanum; scales across the snout between the second canthals 9–12; toe lamellae 17–22; suboculars and supralabials in contact or separated by rows of scales; supraorbital semicircles usually in contact; supraocular scales abruptly or gradually enlarged; hindlimbs long, adpressed toe to between eye and tip of snout; body color light brown with dark brown to black reticulations; male dewlap pink with dark red lines along scale rows; iris blue.
SIMILAR SPECIES — *Anolis gadovii* is a *large brown* lowland anole with smooth ventral scales. Other northern Pacific Mexican anoles with smooth ventrals may be differentiated from *A. gadovii* as follows:

Anolis gaigei

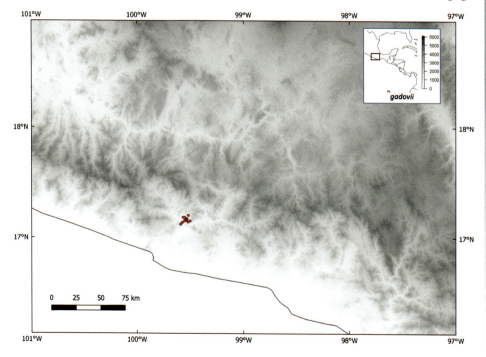

Anolis omiltemanus and *A. peucephilus* are smaller (SVL to 48 mm; to 80 mm in *A. gadovii*) and shorter-limbed than *A. gadovii* (adpressed toe reaches to ear; to between eye and tip of snout in *A. gadovii*). *Anolis liogaster* and *A. brianjuliani* have a band of enlarged middorsal scales (middorsals undifferentiated in *A. gadovii*). *Anolis dunni* differs from *A. gadovii* in its shorter hindlimbs (adpressed toe reaches to eye or just short of eye) and in dewlap color (orange-red; pink with dark red lines in *A. gadovii*). *Anolis gadovii* differs from *A. taylori* mainly in color (contrasting reticulating dorsal pattern in *A. gadovii*; pattern absent in *A. taylori*; male dewlap with dark crescents on a pinkish background in *A. taylori*).
RANGE — From 150 to 600 meters on the Pacific slope of central Guerrero, Mexico.
NATURAL HISTORY — Common in forest and edge habitat; active diurnally on boulders and surrounding vegetation; sleeps on boulders. See Mosauer (1936); Fitch and Henderson (1976a); Fitch et al. (1976); Köhler et al. (2014a); Muñoz-Nolasco et al. (2019).

— *ANOLIS GAIGEI* — PLATE 1.82, 2.105

DESCRIPTION — Ruthven, A. G. 1916. Three new species of *Anolis* from the Santa Marta mountains, Colombia. *Occasional Papers of the Museum of Zoology, University of Michigan* 32:6–8.
TYPE SPECIMEN — UMMZ 48304.
TYPE LOCALITY — "San Lorenzo, Santa Marta Mountains, Colombia, elevation of 2700 ft." Paynter (1997) notes that collector A. G. Ruthven's "San Lorenzo" is Cuchilla San Lorenzo, which is a ridge approximately 25 km SE of Santa Marta. This ridge (lat 11.111 lon −74.054 at Estación Experimental San Lorenzo) reaches over 2800 meters elevation; presumably the specimen was collected on a lower slope of this ridge.
MORPHOLOGY — Body length to 54 M, 52 F; ventral scales keeled; middorsal scales gradually enlarged in several rows; interparietal approximately equal to tympanum; scales across the snout between the

SPECIES ACCOUNTS

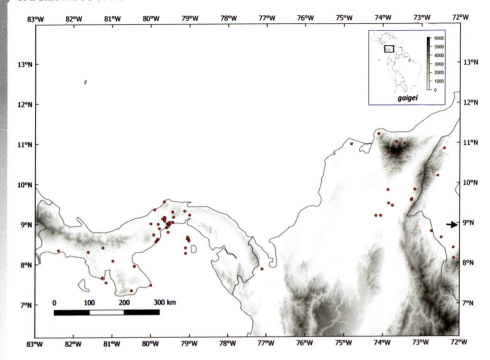

second canthals 8–14; toe lamellae 13–17; suboculars and supralabials separated by rows of scales; scales separating supraorbital semicircles 2–4; supraocular scales gradually enlarged; hindlimbs long, adpressed toe to anterior to eye; body color brown; male and female dewlaps red-orange and yellow.
SIMILAR SPECIES — *Anolis gaigei* is a widespread and common *small brown* lowland anole with keeled ventral scales. *Anolis gaigei* differs from potentially sympatric small brown anoles with keeled ventrals as follows: *Anolis humilis*, *A. auratus*, and *A. notopholis* have abruptly enlarged rows of keeled middorsal scales (middorsals only slightly enlarged in *A. gaigei*); *A. granuliceps* has smaller head scales (12–23 scales across the snout at second canthals; 8–14 in *A. gaigei*), a smaller male dewlap (reaching only to arms; male dewlap reaches well on to chest in *A. gaigei*), and smooth to weakly keeled ventral scales (strongly keeled in *A. gaigei*); *Anolis vittigerus*, *A. lemurinus*, *A. scypheus*, *A. planiceps*, and *A. bombiceps* are larger (SVL > 65 mm in adults; to 53 mm in *A. gaigei*), with blue or red female dewlaps (red-orange and yellow in *A. gaigei*). *Anolis tropidogaster* is externally indistinguishable from *A. gaigei* but is usually allopatric and possesses very different hemipenes (elongated and unilobed in *A. tropidogaster*, bilobed in *A. gaigei*; see Köhler et al. 2012).
RANGE — From 0 to 900 meters. Disjunct distribution, in Panama except most of the Darién and in northern Colombia around Magdalena Department, and western Venezuela to just east of Lake Maracaibo (see Rojas-Runjaic et al. 2023).
NATURAL HISTORY — Common in forest and edge habitat; active diurnally on low vegetation; sleeps on twigs or leaves up to 1.5 meters above ground. See Ballinger et al. (1970; as *Anolis tropidogaster*); Sexton et al. (1971; as *A. tropidogaster*); Kiester (1979; as *A. tropidogaster*); Rand and Myers (1990; as *A. tropidogaster*); Medina-Rangel and Cardenas-Arevalo (2015); Cox et al. (2020).
COMMENT — The status of this form should be evaluated relative to *Anolis tropidogaster*. These putative species are reported to be sympatric in Darién Panama (Köhler et al. 2012a: fig.2) and differ only in hemipenial structure.

ANOLIS GARMANI — PLATE 1.83, 2.164

DESCRIPTION — Stejneger, L. 1899. A new name for the great crested *Anolis* of Jamaica. *American Naturalist* 33:601-602.
TYPE SPECIMEN — None designated.
TYPE LOCALITY — "Jamaica."
MORPHOLOGY — Body length to 131 M, 90 F; ventral scales smooth; middorsal scales form a crest of triangular plates; interparietal equal to or larger than tympanum; scales across the snout between the second canthals 7-11; toe lamellae 23-24; suboculars and supralabials in contact; scales separating supraorbital semicircles 1-3; supraocular scales gradually enlarged; adpressed toe to between ear and eye; body color green; male dewlap yellow-orange; female dewlap absent.
SIMILAR SPECIES — *Anolis garmani* is a *giant* anole, likely to be confused only with *A. equestris* at its mainland (Florida) localities. *Anolis equestris* differs from *A. garmani* in its pink male and female dewlaps (male dewlap yellow-orange in *A. garmani*) and broad white stripe over the shoulder (stripe absent in *A. garmani*). *Anolis garmani* could be confused with juvenile *Iguana iguana*, with which it differs in possessing an extensible dewlap in males and expanded toepads in both sexes (these traits absent in *I. iguana*). The smaller green anoles of Florida (*A. carolinensis, A. allisoni, A. callainus*) lack the serrated dorsal crest of *A. garmani*.
RANGE — On the mainland, from 0 to 50 meters in southern Florida. Native to Jamaica.
NATURAL HISTORY — Rare in Florida, common in Jamaica in edge and disturbed habitats; a "crown giant" anole (Williams 1983), commonly observed above 2 meters above ground on tree trunks or branches; sleeps on leaves or (more frequently) narrow branches. See, e.g., Schwartz and Henderson (1991) for natural history information on Jamaican populations.

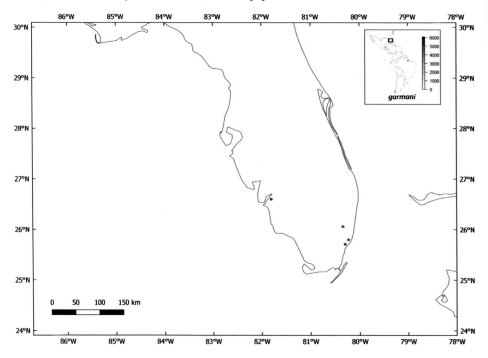

SPECIES ACCOUNTS

— *ANOLIS GEMMOSUS* — PLATE 1.84, 2.110

DESCRIPTION — O'Shaughnessy, A.W.E. 1875. List and revision of the species of Anolidae in the British Museum collection, with descriptions of new species. *Annals and Magazine of Natural History* 15:280.
TYPE SPECIMEN — BMNH 1946.8.13.32.
TYPE LOCALITY — Unknown ("One specimen in the British Museum, the habitat of which is not indicated"; O'Shaughnessy 1875: 280).
MORPHOLOGY — Body length to 66 M, 63 F; ventral scales smooth; middorsal scales enlarged in 0–2 rows; interparietal smaller than tympanum; scales across the snout between the second canthals 12–18; toe lamellae 16–23; suboculars and supralabials in contact or separated by a row of scales; scales separating supraorbital semicircles 1–5; supraocular scales subequal; hindlimbs long, appressed toe to between eye and tip of snout; body color green, usually with lateral rows of white ocelli, sometimes with black spots or dark middorsal crossbars ("Colours prettily variegated" [O'Shaughnessy 1875: 280]); male dewlap somewhat variable, usually yellow with green anteriorly; female dewlap absent; iris brown or bluish.
SIMILAR SPECIES — *Anolis gemmosus* is a *small green* highland anole. *Anolis gemmosus* is frequently sympatric with *A. aequatorialis* and *A. dracula*, from which it differs in smaller body size (maximum SVL to 92 mm in these species; 66 mm in *A. gemmosus*) and male and female dewlap (large and dark, black and green with complexly patterned blue, brown, and/or orange in *A. aequatorialis*, *A. dracula*; two-toned, often yellow with green in male, absent in female *A. gemmosus*). *Anolis proboscis* differs from *A. gemmosus* in its shorter hindlimbs (appressed toe reaches to ear in *A. proboscis*; to between eye and tip of snout in *A. gemmosus*). *Anolis ventrimaculatus* usually appears brown rather than green, is larger than *A. gemmosus* (SVL to 80 mm), and has longer hindlimbs (appressed hindlimb to anterior to snout). Most green anoles that could be confused with *A. gemmosus* are found at lower elevations (e.g., *A. festae*, *A. peraccae*, *A. parilis*, *A. fasciatus*, *A. biporcatus*, *A. parvauritus*,

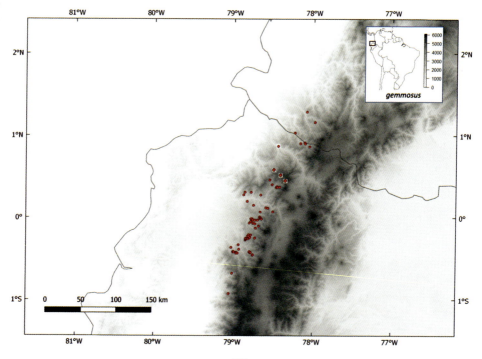

158

Anolis ginaelisae

A. purpurescens, A. chloris; see accounts for these). *Anolis otongae* and *A. poei* are very similar to *A. gemmosus* but parapatric to it. *Anolis otongae* is found at higher elevations than *A. gemmosus*, has a distinctive female dorsal pattern of dark transverse saddles, and has a greater number of scales per dewlap scale row (3–6 in *A. otongae*, 2–3 in *A. gemmosus*). *Anolis poei* overlaps with *A. gemmosus* in scale counts and proportion (it is diagnosed mainly molecularly) and geographically replaces *A. gemmosus* to the south.

RANGE — From 1300 to 2500 meters on the Pacific Andean slope of Ecuador and southern Colombia.

NATURAL HISTORY — Very common in forest and edge habitats; active diurnally in understory vegetation, usually observed within 1.5 meters of ground; sleeps on twigs or (more frequently) leaves, especially ferns, below 2 meters above ground. *Anolis gemmosus* is often the most abundant anole at any cloud forest locality where it is found (e.g., Miyata 2013). See Fitch et al. (1976); Miyata (2013); Arteaga et al. (2013); Narváez (2017); Ramirez-Jaramillo (2018); Torres-Carvajal et al. (2019); Arteaga et al. (2023).

— *ANOLIS GINAELISAE* — PLATE 1.85, 2.74

DESCRIPTION — Lotzkat, S., A. Hertz, J. F. Bienentreu, G. Köhler. 2013. Distribution and variation of the giant alpha anoles (Squamata: Dactyloidae) of the genus *Dactyloa* in the highlands of western Panama, with the description of a new species formerly referred to as *D. microtus*. Zootaxa 3626:14–20.

TYPE SPECIMEN — SMF 19504.

TYPE LOCALITY — "Banks of Quebrada Juglí … on the southeastern slope of Cerro Saguí (also known as Cerro Ratón …) at Finca Alto Cedro, about 2 km north-northeast of the village Ratón, 8.5576°N, 81.8262°W, 1710 m asl, Corregimiento de Piedra Roja, Distrito de Kankintú, Comarca Ngöbe-Buglé, Panama."

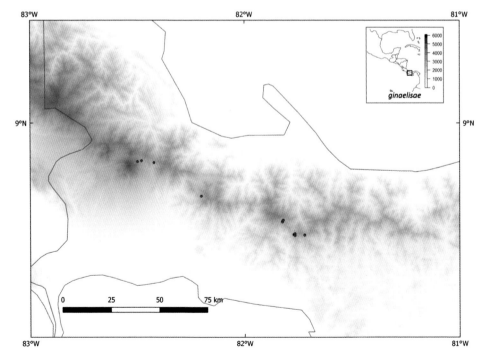

SPECIES ACCOUNTS

MORPHOLOGY — Body length to 112 M, 108 F; ventral scales smooth; middorsal scales enlarged in 0–2 rows; interparietal larger or smaller than tympanum; scales across the snout between the second canthals 5–9; toe lamellae 22–24; suboculars and supralabials in contact; scales separating supraorbital semicircles 1–3; supraocular scales gradually enlarged; hindlimbs short, adpressed toe to eye or posterior to eye; body color variable, often green with broad dark bands; tail banded; male and female dewlaps pink to very pale pink; iris blue.

SIMILAR SPECIES — *Anolis ginaelisae* is a highland *giant* anole that differs from other large geographically proximal species by its short hindlimbs (adpressed toe reaches to between ear and eye; to beyond eye in *A. frenatus, A. casildae, A. capito*) and pink dewlap in males and females (white in males and females of *A. frenatus, A. casildae*; small, yellow-brown in males and females of *A. capito*; white with yellow edge in males, greenish with yellow edge in females in *A. kunayalae*). *Anolis biporcatus* and *A. woodi* have keeled ventral scales (smooth in *A. ginaelisae*). *Anolis brooksi* and *A. kathydayae* have smaller head scales than *A. ginaelisae* (usually at least 10 scales across the snout; 5–9 in *A. ginaelisae*). *Anolis ginaelisae* is replaced geographically to the west by similar anole *A. microtus*, which is reported to differ from *A. ginaelisae* in its slightly shorter hindlimbs.

RANGE — From 1350 to 2150 meters in western Panama.

NATURAL HISTORY — Common in forest and disturbed areas; sleeps high, usually above 2 meters above ground, on narrow branches or leaves.

— *ANOLIS GORGONAE* — PLATE 1.86

DESCRIPTION — Barbour, T. 1905. The Vertebrata of Gorgona Island, Colombia. V. Reptilia and Amphibia. *Bulletin of the Museum of Comparative Zoology* 46:99–100.

TYPE SPECIMEN — MCZ 6984.

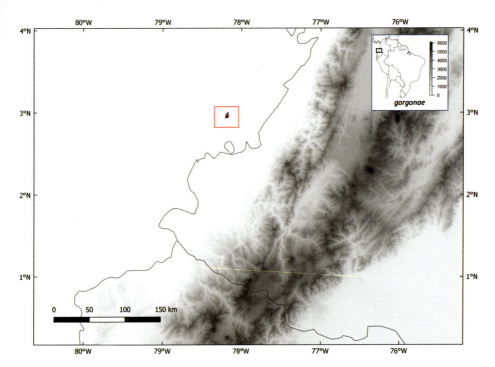

TYPE LOCALITY — "Gorgona Island." Gorgona Island, Cauca, Colombia, is located at lat 2.97, lon −78.18, 0–330 m.
MORPHOLOGY — Body length to 62 M, 58 F; ventral scales weakly keeled; middorsal scales enlarged in 0–2 rows; interparietal smaller than tympanum; scales across the snout between the second canthals 11–13; toe lamellae 17–19; suboculars and supralabials in contact; scales separating supraorbital semicircles 2; supraocular scales gradually enlarged; addpressed toe to between eye and tip of snout; body color blue; male dewlap white; female dewlap absent; iris blue.
SIMILAR SPECIES — *Anolis gorgonae* is easily identifiable on Gorgona due to its blue body color.
RANGE — Throughout Gorgona Island, Pacific Colombia.
NATURAL HISTORY — Observable diurnally on trees at 6 to 10 meters above ground. See Castro-Herrera et al. (2012); Phillips et al. (2019).

— ANOLIS GRACILIPES — PLATE 1.87, 2.99

DESCRIPTION — Boulenger, G. A. 1898. An account of the reptiles and batrachians collected by Mr. W.F.H. Rosenberg in western Ecuador. *Proceedings of the Zoological Society of London* 1898:112–113.
TYPE LOCALITY — "Paramba." Presumably Hacienda Paramba, Imbabura, Ecuador (Paynter 1993), at lat 0.82 lon −78.36, 700 m.
TYPE SPECIMEN — Syntypes: BMNH 1946.8.13.9-12, MCZ 70223 (formerly BMNH 1946.8.13.13), NMW 12816.
MORPHOLOGY — Body length to 58 M, 61 F; ventral scales keeled; middorsal scales gradually enlarged in several rows; interparietal equal to or smaller than tympanum; scales across the snout between the second canthals 8–17; toe lamellae 12–16; suboculars and supralabials in contact or separated by a row of scales; scales separating supraorbital semicircles 1–4; supraocular scales gradually enlarged;

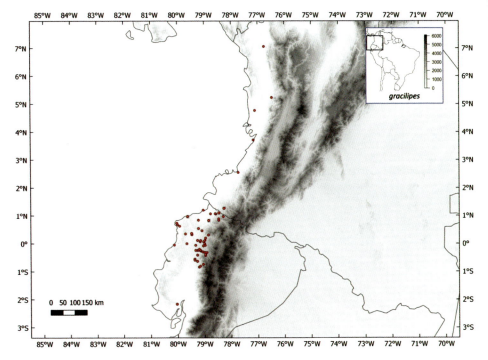

hindlimbs long, adpressed toe to between eye and tip of snout; body color mostly brown, with greenish-yellow ventral tint most noticeable anteriorly; male dewlap large, yellow-orange; female dewlap absent or like male's but smaller; iris may appear brown or blue.

SIMILAR SPECIES — Anolis gracilipes is a *small brown* lowland anole with keeled ventral scales. Among potentially sympatric species that fit this description, *A. granuliceps* is smaller than *A. gracilipes* (SVL to 49 mm; to 61 mm in *A. gracilipes*) and has smooth to very faintly keeled ventral scales and a small male dewlap (extends to arms in *A. granuliceps*; past arms well on to chest in *A. gracilipes*); *A. lynchi* is semiaquatic, exclusively associated with waterways, and has smaller head scales (16–29 scales across the snout; 8–17 in *A. gracilipes*); *A. vittigerus* differs in dewlap color (male: red with dark central blotch; female: usually blue or white with dark central blotch; yellow-orange in *A. gracilipes*) and in possessing greater numbers of toe lamellae (16–20; 12–16 in *A. gracilipes*); *A. rivalis* and *A. macrolepis* are semiaquatic and have abruptly enlarged rows of middorsal scales (middorsals gradually enlarged in *A. gracilipes*). *Anolis binotatus* is very similar to *A. gracilipes* but differs in possessing rows of abruptly enlarged keeled squarish middorsal scales (middorsals only slightly enlarged in *A. gracilipes*) and some smooth head scales (all head scales strongly keeled or rugose in *A. gracilipes*).

RANGE — From 0 to 1200 meters in Pacific Colombia and Ecuador.

NATURAL HISTORY — Rare to common in forest; diurnally active in understory vegetation within 1.5 meters of ground; sleeps on vegetation. See Narváez (2017); Torres-Carvajal et al. (2019); Arteaga et al. (2023).

— *ANOLIS GRANULICEPS* — PLATE 1.88, 2.93

DESCRIPTION — Boulenger, G. A. 1898. An account of the reptiles and batrachians collected by Mr. W.F.H. Rosenberg in western Ecuador. *Proceedings of the Zoological Society of London* 1898: 111–112.

TYPE LOCALITY — "Paramba." Presumably Hacienda Paramba, Imbabura, Ecuador (Paynter 1993), at lat 0.82 lon −78.36, 700 m.

TYPE SPECIMEN — Syntypes: BMNH 1946.8.25-30, UMMZ 59002, MRSN 2357.

MORPHOLOGY — Body length to 47 M, 49 F; ventral scales smooth to weakly keeled; middorsal scales enlarged in 0–2 rows; interparietal larger or smaller than tympanum; scales across the snout between the second canthals 12–23; toe lamellae 12–17; suboculars and supralabials in contact or separated by a row of scales; scales separating supraorbital semicircles 2–5; supraocular scales gradually enlarged; hindlimbs long, adpressed toe to between eye and tip of snout; body color brown, frequently with light lateral stripe; male dewlap small, extending to arms, orangish-yellow; female dewlap usually absent.

SIMILAR SPECIES — *Anolis granuliceps* is a *small brown* lowland anole. Among potentially sympatric species that fit this description, *A. festae* has shorter limbs (adpressed toe approximately to ear; to between eye and tip of snout in *A. granuliceps*) and an unusually shaped male dewlap (long but shallow; normal anoline dewlap shape in *A. granuliceps*). *Anolis maculiventris* has a large pink and orange male dewlap (small orangish-yellow male dewlap in *A. granuliceps*) and usually lacks the light lateral longitudinal stripe of *A. granuliceps* (*A. maculiventris* frequently is banded with lines, bars, and/or rows of ocelli). *Anolis sulcifrons* and *A. fuscoauratus* have larger head scales than *A. granuliceps* (usually 7–13 scales across the snout; 12–23 in *A. granuliceps*). *Anolis antonii*, *A. urraoi*, *A. mariarum*, and *A. tolimensis* have larger head scales than *A. granuliceps* (usually 7–13 scales across the snout) and are found at higher elevations. *Anolis gracilipes*, *A. binotatus*, *A. vittigerus*, *A. poecilopus*, *A. rivalis*, *A. macrolepis*, and *A. lynchi* are larger than *A. granuliceps* (SVL to >60 mm in these species; to 49 mm in *A. granuliceps*) and have strongly keeled ventral scales (smooth to weakly keeled in *A. granuliceps*). *Anolis vicarius* is small and brown and has comparably small head scales to *A. granuliceps* (17 scales across the snout) but differs from *A. granuliceps* in its larger male dewlap (extending past axillae; to axillae in *A. granuliceps*) and completely smooth ventrals (ventrals faintly keeled in *A. granuliceps*).

RANGE — From 0 to 1100 meters in Pacific Colombia and Ecuador, and in the Magdalena River valley of Colombia.

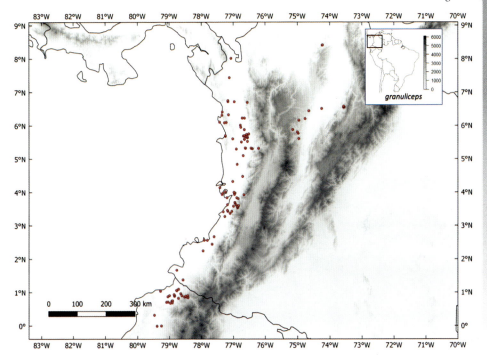

NATURAL HISTORY — Common in forest; often in leaf litter or low understory vegetation; sleeps on twigs, leaves, or vines. See Castro-Herrera (1988); Ríos et al. (2011); Rengifo et al. (2014, 2015, 2019, 2021); Pinto-Erazo et al. (2020); Moreno-Arias et al. (2020); Arteaga et al. (2023).

COMMENT — The population of anoles assigned to *Anolis granuliceps* in the Magdalena River valley may represent an undescribed species.

— *ANOLIS GRUUO* — PLATE 1.89, 2.32

DESCRIPTION — Köhler, G., M. Ponce, J. Sunyer, A. Batista. 2007. Four new species of anoles (genus *Anolis*) from the Serranía de Tabasará, West-Central Panama (Squamata: Polychrotidae). *Herpetologica* 63:376–380.

TYPE SPECIMEN — SMF 85416.

TYPE LOCALITY — "Near the headwaters of Río San Félix, ca. 2 km N Escopeta Camp, ca. 08°32'N, 81°50'W, Serranía de Tabasará, 900 m elevation, Comarca Ngöbe Bugle, Distrito de Nole Düima, Corregimiento de Jadeberi, Panama."

MORPHOLOGY — Body length to 52 M, 51 F; ventral scales smooth; middorsal scales enlarged in 0–2 rows; interparietal larger or smaller than tympanum; scales across the snout between the second canthals 7–13; toe lamellae 13–16; suboculars and supralabials in contact; scales separating supraorbital semicircles 0–3; supraocular scales gradually enlarged; hindlimbs short, adpressed toe approximately to ear; body color grayish-brown; male dewlap orange.

SIMILAR SPECIES — *Anolis gruuo* is a *small brown* anole with short hindlimbs (adpressed toe reaches to ear) and smooth ventral scales. Among geographically proximal species that share these traits, *A. pentaprion*, *A. salvini*, and *A. fungosus* differ from *A. gruuo* in their whitish, lichenous-patterned dorsums and larger head scales (usually 4–8 scales across the snout; usually 8–10 in *A. gruuo*);

SPECIES ACCOUNTS

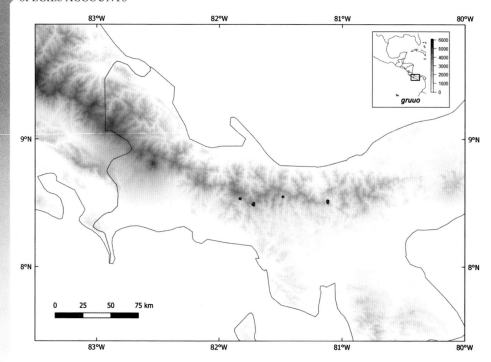

A. *carpenteri* is olive green and has slightly longer hindlimbs (adpressed toe to between ear and eye); A. *kemptoni*, A. *pseudokemptoni*, and A. *fortunensis* differ in male dewlap color (pink posteriorly, red anteriorly in A. *kemptoni*, A. *pseudokemptoni*; orange anteriorly, yellow posteriorly or red anteriorly, orange posteriorly in A. *fortunensis*; orange in A. *gruuo*). Anolis gruuo is very similar to A. *elcopeensis*, which geographically replaces A. *gruuo* at lower elevations to the east. Anolis elcopeensis differs only modally from A. *gruuo* in scale counts (these species appear molecularly distinct; Poe et al. [2015b]) but tends to have smaller head scales (usually 10–14 scales across the snout); another potential difference is the noticeably bulging hemipenial area in larger males of A. *gruuo* (area not expanded in A. *elcopeensis*).

RANGE — From 850 to 1550 meters in Comaraca Ngöbe-Bugle and western Veraguas, Panama.

NATURAL HISTORY — Common in forest, edge, and disturbed areas; sleeps on twigs, leaves, or vines at all visible vertical levels. See Lotzkat et al. (2012).

— *ANOLIS HETERODERMUS* — PLATE 1.90, 2.137

DESCRIPTION — Duméril, M. C., M. A. Duméril. 1851. *Catalogue methodique de la collection des reptiles*. Museum d'Histoire Naturelle de Paris, 59.

TYPE SPECIMEN — Lectotype MNHN 1664 apparently designated by Lazell (1969).

TYPE LOCALITY — "Nouvelle Grenade." Lazell (1969) restricted this locality to the vicinity of Bogota, Colombia (lat 4.7 lon −74.1, 2560 m).

MORPHOLOGY — Body length to 76 M, 73 F; ventral scales smooth; middorsal scales form a crest of triangular plates (see Lazell 1969: fig. 1); interparietal larger than tympanum; scales across the snout between the second canthals 3–4; toe lamellae 16–21; suboculars and supralabials in contact; scales separating supraorbital semicircles 0–1; supraocular scales abruptly enlarged; hindlimbs short,

Anolis heterodermus

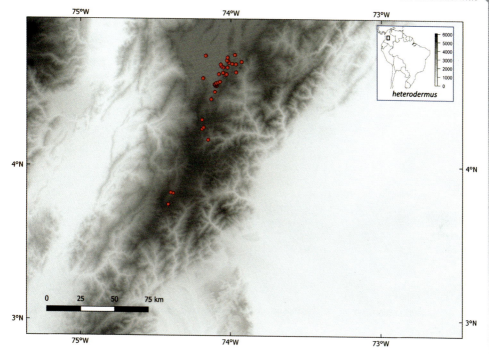

adpressed toe to ear or posterior to ear; body color green or brown, with light stripe from supralabial area back to shoulder; male and female dewlap purple, red, or black, sometimes with yellowish distal edge; heterogeneous lateral scutellation.

SIMILAR SPECIES — *Anolis heterodermus* is a distinctive highland *twig* anole. The heterogeneous lateral scalation, wherein large smooth oval scales are separated by granular interstitial scales, is differentiating in *A. heterodermus*, especially in combination with its short hindlimbs and tail and large smooth head scales. Among potential eastern Andean forms that share these traits, *A. vanzolinii* and *A. inderenae* are larger than *A. heterodermus* (to >100 mm SVL; to 76 mm in *A. heterodermus*); *A. richteri*, *A. quimbaya*, and *A. tequendama* differ from *A. heterodermus* in male and female dewlap color (*A. richteri*: cream to yellow in males and females, sometimes with red distally; *A. quimbaya*: orange in male, green in females; *A. tequendama*: brown with or without spotting in male, yellow with dark stripes in female; *A. heterodermus*: purple, red, or black in males and females).

RANGE — From 2500 to 3600 meters in Cundimarca, Colombia. "Northern populations of *A. heterodermus* have been found in sympatry with southern populations of *A. richteri*." (Moreno-Arias et al. 2023). Despite my attempts to follow localities and maps in Moreno-Arias et al. (2023), I am not confident of the species boundaries among *Anolis heterodermus*, *A. richteri*, and *A. tequendama* in the Bogota area, nor of the southern limit of *A. heterodermus* relative to *A. quimbaya*, as presented in range maps for these species in this book.

NATURAL HISTORY — Common on páramo vegetation; sleeps on twigs or leaves. See Dunn (1944); Osorno and Osorno (1946); Ramírez-Pinilla et al. (1989); Ramírez-Perilla et al. (1991); Moreno-Arias and Urbina-Cardona (2013); Méndez-Galeano and Calderón-Espinosa (2020); Beltrán and Barragán-Contreras (2019).

SPECIES ACCOUNTS

— ANOLIS HETEROPHOLIDOTUS — PLATE 1.91

DESCRIPTION — Mertens, R. 1952. Die Amphibien und Reptilien von El Salvador. *Abhandlung der senckenbergischen naturforschenden Gesellschaft* (Frankfurt) 487.

TYPE SPECIMEN — SMF 43041.

TYPE LOCALITY — "Oberhalb Hacienda Los Planes am Miramundo, 2000 m H., Dept. Santa Ana, El Salvador." Cerro Miramundo in Santa Ana, El Salvador, is at lat 14.417 lon −89.367.

MORPHOLOGY — Body length to 51 M, 59 F; ventral scales smooth; middorsal scales greatly enlarged in several rows; interparietal equal to or larger than tympanum; scales across the snout between the second canthals 5–8; toe lamellae 15–17; suboculars and supralabials in contact; scales separating supraorbital semicircles 0–2; supraocular scales gradually enlarged; adpressed toe to eye; body color brown; male dewlap red-orange; female dewlap small, orange; heterogeneous lateral scutellation.

SIMILAR SPECIES — *Anolis heteropholidotus* is a *small brown* highland anole. Most geographically proximal highland species differ from *A. heteropholidotus* in possessing keeled ventral scales (*A. caceresae, A. crassulus, A. laeviventris, A. rubribarbaris, A. yoroensis*), although in some cases the keeling is weak and may be difficult to see (*A. amplisquamosus, A. sminthus, A. wermuthi, A. ocelloscapularis*; ventral scales smooth in *A. heteropholidotus*).

RANGE — From 1850 to 2200 meters in northwest El Salvador, southeast Guatemala, and western Honduras.

NATURAL HISTORY — In forest and edge habitats; active diurnally on low vegetation or ground. See McCranie and Köhler (2015).

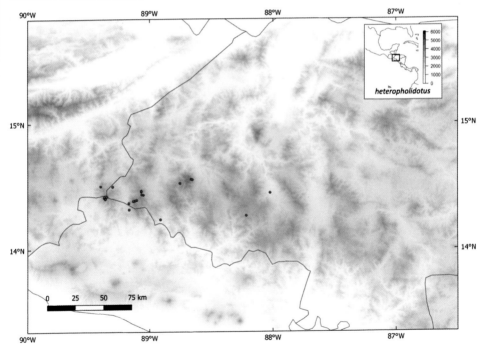

— ANOLIS HOBARTSMITHI — PLATE 1.92, 2.10

DESCRIPTION — Nieto Montes de Oca, A. 1995. Key to the species of the *Anolis schiedii* group south and east of the Isthmus of Téhuantépec. In Flores, O., F. Mendoza Quijano, G. González Porter. *Recopilación de Claves para la Determinación de Anfibios y Reptiles de México*. Publicaciones Especiales de la Museo Zoologia Facultad de Ciencias, UNAM 10, 160.
TYPE SPECIMEN — MZFC 6371.
TYPE LOCALITY — "Parador Selva Negra on Mexico Highway 195 (approximately 0.5 km W Pinabeto), Municipality of Rayon, at the border with Municipality of Pueblo Nuevo Solistahuacan, Chiapas, Mexico; 17 12 58 N, 92 57 47 W; 1992 m el."
MORPHOLOGY — Body length to 48 M, 51 F; ventral scales smooth to faintly keeled; middorsal scales gradually enlarged in several rows; interparietal smaller than tympanum; scales across the snout between the second canthals 9–12; toe lamellae 15–18; suboculars and supralabials in contact or separated by a row of scales; scales separating supraorbital semicircles 1–3; supraocular scales gradually enlarged; hindlimbs long, appressed toe to between eye and tip of snout; body color brown; male and female dewlaps purple-pink, reddish at edge.
SIMILAR SPECIES — *Anolis hobartsmithi* is a *small brown* highland anole. *Anolis laeviventris* frequently is sympatric with *A. hobartsmithi* and differs from it in possessing short hindlimbs (adpressed toe reaches approximately to ear; to between eye and tip of snout in *A. hobartsmithi*) and strongly keeled ventral scales (mostly smooth in *A. hobartsmithi*). Other geographically proximal small brown anoles differ from *A. hobartsmithi* in male dewlap color (orange in *A. rodriguezii*; yellow in *A. dollfusianus*; orange with blue central spot in *A. sericeus*; pink in *A. hobartsmithi*) and body scalation (abruptly enlarged keeled middorsal scale rows, strongly keeled ventrals, puncturelike axillary pocket in *A. compressicauda*, *A. tropidonotus*, *A. uniformis*; these traits absent in *A. hobartsmithi*). *Anolis matudai* is a similar highland species found south of the range of *A. hobartsmithi* distinguishable by its undifferentiated scales in the nasal region (elongate anterior nasal scale in *A. hobartsmithi*). *Anolis campbelli*,

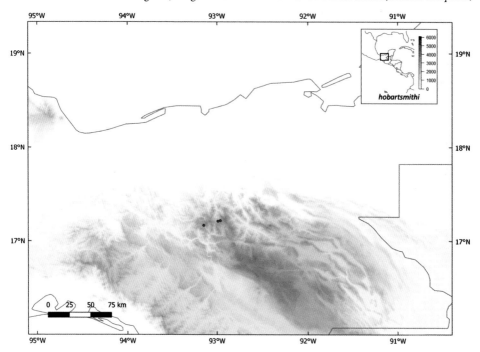

SPECIES ACCOUNTS

a similar highland anole found east of the range of *A. hobartsmithi*, differs from *A. hobartsmithi* in possessing distinctly keeled ventral scales. *Anolis parvicirculatus*, a similar highland anole found southwest of the range of *A. hobartsmithi*, differs from *A. hobartsmithi* in its red-orange male dewlap.

RANGE — From 1500 to 2200 meters in northern Chiapas State, southern Mexico.

NATURAL HISTORY — Common in forest and edge habitats; active diurnally on low vegetation, rocks, or in leaf litter; sleeps low, usually below 1.5 meters above ground on leaves, often ferns.

— *ANOLIS HUILAE* — PLATE 1.93, 2.135

DESCRIPTION — Williams, E. E. 1982. Three new species of the *Anolis punctatus* complex from Amazonian and inter-Andean Colombia, with comments on the eastern members of the *punctatus* species group. *Breviora* 467:9–16.

TYPE SPECIMEN — ICN 3725.

TYPE LOCALITY — "Herberto Herrera's coffee plantation, Palestina, Huila, Colombia." I have been unable to locate Herberto Herrera's coffee plantation (Palestina is celebrated for its coffee). The municipality of Palestina is at the southern end of Huila Department, at approximately lat 1.7 lon −76.2, ~1500–2000 m.

MORPHOLOGY — Body length to 80 M, 69 F; ventral scales smooth; middorsal scales enlarged in 0–2 rows; interparietal larger or smaller than tympanum; scales across the snout between the second canthals 7–11; toe lamellae 20–23; suboculars and supralabials in contact; supraorbital semicircles in contact; supraocular scales gradually enlarged; adpressed toe to between ear and eye or to eye; body color green, changeable to brown, with close-set yellow spots; male dewlap light greenish-yellow with black specks; female dewlap absent.

SIMILAR SPECIES — *Anolis huilae* is a *large green* highland anole that may be distinguished from geographically proximal large anoles by having the supraorbital semicircles in contact and displaying

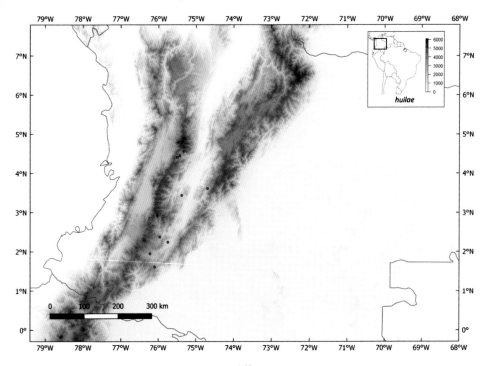

168

Anolis humilis

densely packed yellow spots on its flanks (supraorbital semicircles separated by scales and yellow spots lacking in *A. apollinaris*, *A. frenatus*, *A. ventrimaculatus*, *A. anoriensis*, *A. fitchi*, *A. limon*).
RANGE — From 1550 to 1950 meters in the Cordillera Occidental and Cordillera Oriental along the Magdalena River valley, Tolima and Huila Departments, Colombia.
NATURAL HISTORY — Common in cloud forest and edge habitat; active diurnally from 1 to 7 meters above ground on trees and bushes, with peak of activity around midday; sleeps on twigs, leaves, or vines above 3 meters above ground. See Bejarano-Bonilla and Bernal-Bautista (2019).

— *ANOLIS HUMILIS* — PLATE 1.94, 2.42

DESCRIPTION — Peters, W.C.H. 1863. Derselbe machte eine Mittheilung über einige neue Arten der Saurier-Gattung *Anolis*. *Monatsberichte der Königlich Preussischen Akademie der Wissenschaften zu Berlin*.
TYPE SPECIMEN — Syntypes ZMB 500, 55223.
TYPE LOCALITY — "Veragua." Veragua, New Granada, corresponds to modern Panama west of the Azuero peninsula.
MORPHOLOGY — Body length to 44 M, 48 F; ventral scales keeled; middorsal scales greatly enlarged in several rows; interparietal smaller than tympanum; scales across the snout between the second canthals 6–11; toe lamellae 11–14; suboculars and supralabials in contact or separated by rows of scales; scales separating supraorbital semicircles 1–4; supraocular scales gradually enlarged; hindlimbs long, addressed toe to between eye and tip of snout; body color brown; male dewlap red with yellow distal edge; puncturelike axillary pocket.
SIMILAR SPECIES — *Anolis humilis* is a widespread, predominantly lowland, largely terrestrial *small brown* anole. Its combination of strongly keeled ventral scales, abruptly enlarged middorsal scale rows, and deep (puncturelike) axillary pocket will distinguish *A. humilis* from most potentially

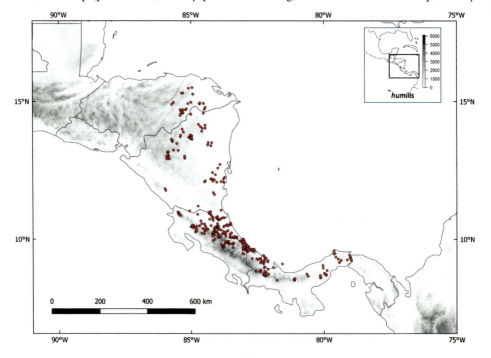

sympatric species. *Anolis humilis* may be distinguished from other Central American species that possess these traits by male dewlap color (reddish-orange with yellow edge and dark central streaking in *A. tropidonotus*; pink with dark central blotch in *A. uniformis*; pinkish purple with yellow edge in *A. compressicauda*; dull red with dark central streaks in *A. marsupialis*; red with yellow edge in *A. humilis*).

RANGE — From 0 to 1700 meters from eastern Honduras to central Panama; absent from central to southern Pacific Costa Rica, where it is replaced by similar species *Anolis marsupialis*.

NATURAL HISTORY — Common or very common in forest; active diurnally in leaf litter and low on tree trunks; sleeps on twigs or leaves below 0.5 meters above ground and in leaf litter. See Echelle et al. (1971a); Fitch (1973a, b, 1975); Andrews (1976); Talbot (1976, 1977, 1979); Corn (1981); Berkum (1986); Guyer (1986a, b; 1988a, b); Pounds (1988); Parmalee and Guyer (1995); Savage (2002); Lattanzio (2009); Paemelaere et al. (2011); Baruch et al. (2016); Curlis et al. (2017); Thompson et al. (2018); Hilje et al. (2020); Perez-Martinez et al. (2021).

COMMENT — I consider *Anolis quaggulus* (type locality: "San Juan River, Nicaragua") to be a junior synonym of *A. humilis* (see appendix 1 for justification). For those who wish to recognize *A. quaggulus*, it is usually recognized as the form of *A. humilis* found on the Caribbean versant from eastern Honduras to northern Costa Rica.

— *ANOLIS HYACINTHOGULARIS* — PLATE 1.95, 2.128

DESCRIPTION — Torres-Carvajal, O., F. P. Ayala-Varela, S. E. Lobos, S. Poe, A. Narvaez. 2017. Two new Andean species of *Anolis* lizard (Iguanidae: Dactyloinae) from southern Ecuador. *Journal of Natural History* (online): 8–12.
TYPE SPECIMEN — QCAZ 14136.

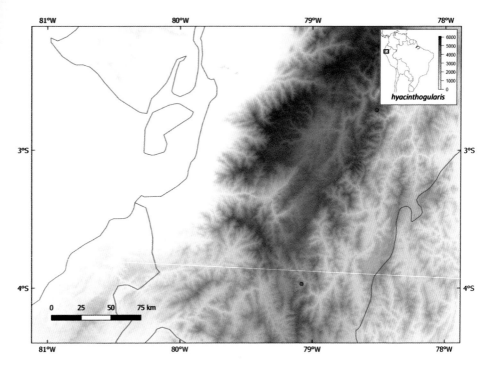

Anolis ibanezi

TYPE LOCALITY — "Ecuador, Provincia Zamora Chinchipe, San Francisco Research Station, 03.971°S, 79.082°W, WGS84, 1441 m"
MORPHOLOGY — Body length to 57 M, 52 F; ventral scales smooth; middorsal scales enlarged in 0–2 rows; interparietal larger than tympanum; scales across the snout between the second canthals 6–9; toe lamellae 16–17; suboculars and supralabials in contact; scales separating supraorbital semicircles 2–3; supraocular scales gradually enlarged; hindlimbs short, adpressed toe to posterior to ear; body color grayish-green to brown, sometimes with faint lateral banding and/or speckling; male dewlap sky blue; female dewlap cream with blue blotches.
SIMILAR SPECIES — *Anolis hyacinthogularis* is a *small green* or *brown* highland anole known to be sympatric with *A. soinii* and *A. lososi*. It is distinguishable from both species (and other nearby species) via its sky blue male dewlap (yellow in *A. soinii*, white in *A. lososi*) and cream with blue blotches female dewlap (female dewlap absent in *A. soinii*, orange with black blotches in *A. lososi*).
RANGE — From 1400 to 2200 meters on the Amazonian Andean slope of Zamora Chinchipe and Morona Santiago Provinces in southern Ecuador.
NATURAL HISTORY — Rare in forest and forest edge; sleeps on twigs or leaves above 2 meters above ground.
COMMENT — The spelling of this species name at its first use was *hyacintogularis*. I here apply Article 32.5.1 of the International Code of Zoological Nomenclature in accepting our (Torres-Carvajal et al. 2017) preferred spelling of *hyacinthogularis*.

— *ANOLIS IBANEZI* — PLATE 1.96, 2.66

DESCRIPTION — Poe, S., I. M. Latella, M. J. Ryan, E. W. Schaad. 2009. A new species of *Anolis* lizard (Squamata, Iguania) from Panama. *Phyllomedusa* 8:91–97.
TYPE SPECIMEN — MSB 72574.

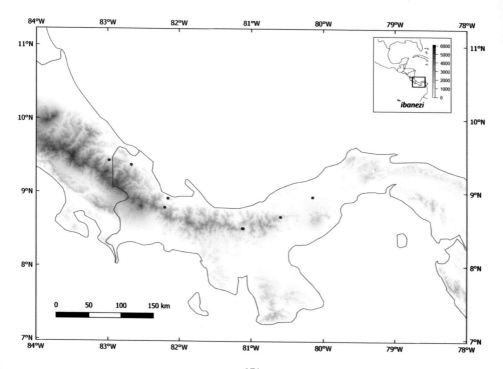

SPECIES ACCOUNTS

TYPE LOCALITY — "The trails of Parque National General de División Omar Torrijos Herrera, 5 km North of El Copé, Coclé Province, Panama, 8 40' 18.9" S, 80 35' 31.08" W (SA69 datum)," at approximately 800 m.

MORPHOLOGY — Body length to 81 M, 74 F; ventral scales smooth; middorsal scales enlarged in 0–2 rows; interparietal usually smaller than tympanum; scales across the snout between the second canthals 9–12; toe lamellae 17–19; suboculars and supralabials in contact; scales separating supraorbital semicircles 2–3; supraocular scales gradually enlarged; hindlimbs long, adpressed toe to eye (body also is long); body color green with thin black oblique lateral lines on posterior flanks; male dewlap red-orange to yellow-orange with green scales; female dewlap dark orange with green and white scales.

SIMILAR SPECIES — *Anolis ibanezi* is a *large green* anole, distinguishable from commonly sympatric large green anoles by its smooth ventral scales (keeled in *A. biporcatus*), thin oblique black lines on the flanks (rows of ocelli in *A. frenatus*), and expanded toe pads (narrow toes, 12–17 lamellae in *A. kunayalae*; broad toepads, 17–19 lamellae in *A. ibanezi*). Larger species (SVL > 90 mm; to 81 mm in *A. ibanezi*) *Anolis woodi, A. insignis, A. brooksi, A. savagei, A. ginaelisae, A. microtus, A. casildae,* and *A. kathydayae* may appear green but are strongly dorsally patterned with broad bands, spots, and/or ocelli. *Anolis ibanezi* is very similar to *A. purpurescens* and *A. maia*, which appear to geographically replace it in eastern Panama and Colombia, respectively, differing mainly in lateral body pattern (lateral ocelli or spots in these species, lines in *A. ibanezi*).

RANGE — From 400 to 900 meters. Specimens have been collected in Veraguas, Coclé, and Bocas del Toro Provinces in Panama and in neighboring Caribbean Costa Rica.

NATURAL HISTORY — Rare in forest; described as "trunk-crown" (Savage 2002, as *A. chocorum*); sleeps on twigs, vines, or (most frequently) leaves, above 2 meters above ground.

— *ANOLIS IMMACULOGULARIS* — PLATE 1.97; MAP opposite

DESCRIPTION — Köhler, G., R. Gómez Trejo Pérez, C.B.P. Peterson, F. R. Méndez de la Cruz. 2014. A revision of the Mexican *Anolis* (Reptilia, Squamata, Dactyloidae) from the Pacific versant west of the Isthmus de Téhuantépec in the states of Oaxaca, Guerrero, and Puebla, with the description of six new species. Zootaxa 3862:96–103.

TYPE SPECIMEN — SMF 96266.

TYPE LOCALITY — "Puerto Escondido, Punta Colorada (15.87038°N, 97.10152°W, WGS84), 40 m, Estado de Oaxaca, Mexico."

MORPHOLOGY — Body length to 43 M, 50 F; ventral scales keeled; middorsal scales greatly enlarged in several rows; interparietal larger or smaller than tympanum; scales across the snout between the second canthals 6–9; toe lamellae 12–15; suboculars and supralabials in contact; scales separating supraorbital semicircles 0–1; supraocular scales abruptly enlarged; adpressed toe to eye or just posterior to eye; body color grayish-brown, sometimes with light lateral stripe; male dewlap red, usually with yellowish distal edge.

SIMILAR SPECIES — *Anolis immaculogularis* is a *small brown* lowland anole. Among geographically proximal lowland brown anoles, *A. sericeus* and *A. nebulosus* differ from *A. immaculogularis* in hindlimb length (adpressed toe reaches to ear in these species; at least to eye in *A. immaculogularis*); *A. nebuloides* and *A. megapholidotus* have larger middorsal scales (middorsals larger than ventral scales; ventrals larger than middorsals in *A. immaculogularis*) and shorter hindlimbs (adpressed toe usually reaches to ear in these species). *Anolis immaculogularis* differs from parapatric forms *A. subocularis* (to the west) and *A. boulengerianus* (to the east) in male dewlap color pattern (*A. subocularis*: pinkish-red, paler around dewlap scale rows [*A. immaculogularis* lacks pale tint around scale rows]; *A. boulengerianus*: yellow-orange).

RANGE — From 0 to 250 meters on the Pacific versant of western Oaxaca State, Mexico.

NATURAL HISTORY — Very common in open and forest edge habitats; sleeps on twigs or leaves up to 2 meters above ground.

COMMENT — The status of *Anolis immaculogularis* should be evaluated with reference to *A. subocularis*.

Anolis inderenae

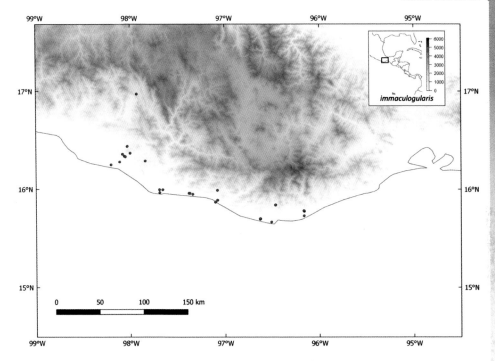

— *ANOLIS INDERENAE* — PLATE 1.98, 2.143; MAP overleaf

DESCRIPTION — Rueda-Almonacid, J. V., J. L. Hernández-Camacho. 1988. *Phenacosaurus inderenae* (Sauria: Iguanidae), Nueva especie gigante, proveniente de la cordillera Oriental de Colombia. *Trianea* 2:339–350.

TYPE SPECIMEN — IND 3213.

TYPE LOCALITY — "Departamento de Cundinamarca, Municipio de Gutiérrez, Vereda el Carmen, Finca 'El Carmen', vertiente Este de la Cordillera Oriental de Colombia. 4 15' latitud N, 74 01' W de Greenwich, ca. 2350 m.s.n.m."

MORPHOLOGY — Body length to 98 M, 118 F; ventral scales smooth; middorsal scales form a crest of triangular plates; interparietal larger than tympanum; scales across the snout between the second canthals 2; toe lamellae 23–26; suboculars and supralabials in contact; supraorbital semicircles in contact; hindlimbs short, adpressed toe to ear or posterior to ear; body gray to black with red; male dewlap cream with brown distally; female dewlap mostly brown with cream; heterogeneous lateral scutellation.

SIMILAR SPECIES — *Anolis inderenae* is a large highland *twig* anole, unlikely to be confused with sympatric forms due to its very large smooth head scales, crested middorsum, casqued head, and heterogeneous lateral scutellation. *Anolis heterodermus*, *A. quimbaya*, *A. tequendama*, and *A. richteri* shares these traits but are smaller (SVL to 76 mm; to 118 mm in *A. inderenae*) and have fewer toe lamellae (16–24; 23–26 in *A. inderenae*).

RANGE — From 1850 to 2350 m on the Amazonian Andean slope of central Colombia.

NATURAL HISTORY — Only information is from the type description: Found in remnant scrub vegetation near streams 2 to 3 meters above ground.

SPECIES ACCOUNTS

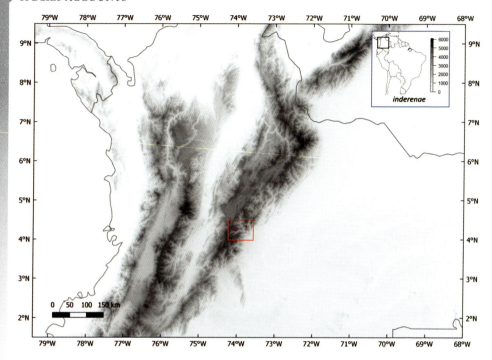

― *ANOLIS INSIGNIS* ― PLATE 1.99, 2.70; MAP opposite

DESCRIPTION — Cope, E. D. 1871. Ninth contribution to the herpetology of tropical America. *Proceedings of the Academy of Natural Sciences of Philadelphia* 23:213–214.
TYPE SPECIMEN — Lost, according to Taylor (1956:81; also Savage and Talbot [1978]), "*fide* Barbour."
TYPE LOCALITY — "San José." Savage (1974) states that Cope's *Anolis* collections labeled San José "may have been taken from … the Atlantic slopes of the Cordillera Central between Alto La Palma and Carrillo or near La Palma on the Pacific slope." Savage and Talbot (1978) state the specimen is "probably from near La Palma." Savage's (1974) La Palma is at approximately lat 10.07 lon –84.00, 1500 m.
MORPHOLOGY — Body length to 157 M, 140 F; ventral scales smooth; middorsal scales enlarged in 0–2 rows; interparietal smaller than tympanum; scales across the snout between the second canthals 9–11; toe lamellae 25–27; suboculars and supralabials in contact; scales separating supraorbital semicircles 2–3; supraocular scales subequal to gradually enlarged; hindlimbs short, appressed toe to ear, or posterior to ear; body color gray, brown, and/or green, usually with some banding; male and female dewlaps red-orange.
SIMILAR SPECIES — *Anolis insignis* is a Costa Rican *giant* anole that differs from other giant Costa Rican anoles as follows: *Anolis frenatus* has long hindlimbs (appressed toe reaches past eye; to ear in *A. insignis*) and a white dewlap in males and females (dewlap red-orange in *A. insignis*); *A. capito* and *A. biporcatus* have keeled ventral scales (smooth in *A. insignis*); *A. savagei*, which geographically replaces *A. insignis* on the Pacific southeast of Costa Rica, has a pink dewlap with dark streaks in males and females (dewlaps solid red-orange in *A. insignis*) and a prominent blotch posterior to the eye (blotch absent in *A. insignis*).
RANGE — From 400 to 1500 meters in northern Costa Rica.
NATURAL HISTORY — Rare in forest; Savage (2002) and Pounds (1988) state that this species inhabits forest canopy. I have found this species sleeping on narrow branches in edge and forest habitat above 3 meters above ground. See Fitch (1975); Pounds (1988); Savage (2002).

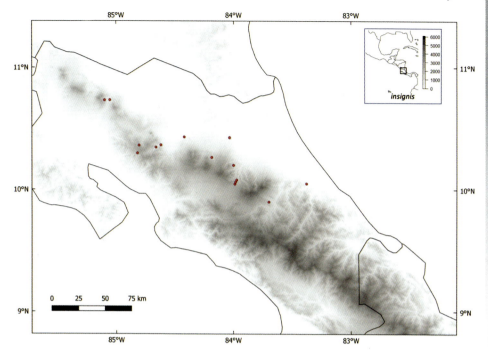

— *ANOLIS JACARE* — PLATE 1.100, 2.130; MAP overleaf

DESCRIPTION — Boulenger, G. A. 1903. On some batrachians and reptiles from Venezuela. *Annals and Magazine of Natural History* 7:482–483.
TYPE SPECIMEN — Syntypes BMNH 1946.8.12.80-84.
TYPE LOCALITY — "Merida, Venezuela, at an altitude of 1600 metres." Merida, Venezuela is at approximately lat 8.6 lon −71.1.
MORPHOLOGY — Body length to 75 M, 70 F; ventral scales smooth; middorsal scales enlarged in 0–2 rows; interparietal equal to or larger than tympanum; scales across the snout between the second canthals 5–9; toe lamellae 18–25; suboculars and supralabials in contact; scales separating supraorbital semicircles 0–2; supraocular scales gradually enlarged; hindlimbs short, adpressed toe approximately to ear; body olive green to brown with dark flecks, a yellow to orange stripe extending posteriorly over the ear from the mouth; male dewlap tan; female dewlap tan with black blotches.
SIMILAR SPECIES — *Anolis jacare* is a *large green* or *brown* highland anole known to be sympatric only with *A. nicefori*, with which it differs in its smaller head scales (large interparietal in contact with supraorbital semicircles in *A. nicefori*; interparietal separated from supraorbital semicircles by rows of scales in *A. jacare*) and lack of middorsal cresting (raised triangular plates usually present middorsally in *A. nicefori*; absent in *A. jacare*). *Anolis tetarii* and *A. euskalerriari*, highland twig anoles found near the range of *A. jacare*, both have larger head scales than *A. jacare* (4 scales across the snout in *A. tetarii*, 2–3 in *A. euskalerriari*; 5–9 in *A. jacare*).
RANGE — From 1400 to 2200 meters in the eastern Andes in extreme northeastern Colombia and northwestern Venezuela.
NATURAL HISTORY — Common in forest, forest edge, and disturbed areas; active diurnally on tree trunks and branches; sleeps on vegetation. See Williams et al. (1970); Ugueto et al. (2007); Perdoma and LaMarca (2016).

SPECIES ACCOUNTS

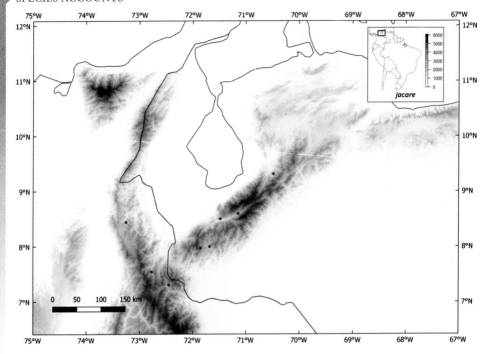

— *ANOLIS JOHNMEYERI* — PLATE 1.101, 2.63; MAP opposite top

DESCRIPTION — Wilson, L. D., J. R. McCranie. 1982. A new cloud forest *Anolis* (Sauria: Iguanidae) of the *schiedei* group from Honduras. *Transactions of the Kansas Academy of Science* 85: 133–141.
TYPE SPECIMEN — LSUMZ 37834.
TYPE LOCALITY — "El Cusuco (15° 30' N, 88° 13' W), 5.6 km WSW Buenos Aires (the latter locality about 19 km N Cofradia), 1580 meters elevation, Sierra de Omoa, Depto. Cortes, Honduras."
MORPHOLOGY — Body length to 73 M, 68 F; ventral scales smooth; middorsal scales enlarged in 0–2 rows or gradually enlarged in several rows; interparietal smaller than tympanum; scales across the snout between the second canthals 7–10; toe lamellae 15–18; suboculars and supralabials usually separated by a row of scales; scales between supraorbital semicircles 1–5; supraocular scales gradually enlarged; adpressed toe to eye or between eye and tip of snout; body color brown; male dewlap red with blue center; female dewlap yellow with blue center.
SIMILAR SPECIES — *Anolis johnmeyeri* is a *large brown* highland anole, unlikely to be confused with other anoles due to its striking male (red with blue central spot) and female (yellow with blue central spot) dewlaps.
RANGE — From 1300 to 2000 meters in northwestern Honduras.
NATURAL HISTORY — Common in forest; active diurnally on low vegetation and ground; sleeps on leaves, especially ferns, up to 2 meters above ground; McCranie and Köhler (2015) reported *A. johnmeyeri* to sleep on tree trunks. See Townsend and Wilson (2008); McCranie and Köhler (2015).

— *ANOLIS KATHYDAYAE* — PLATE 1.102, 2.73; MAP opposite bottom

DESCRIPTION — Poe, S., M. J. Ryan. 2017. Description of two new species similar to *Anolis insignis* (Squamata: Iguanidae) and resurrection of *Anolis (Diaphoranolis) brooksi*. *Amphibian & Reptile*

Anolis kathydayae

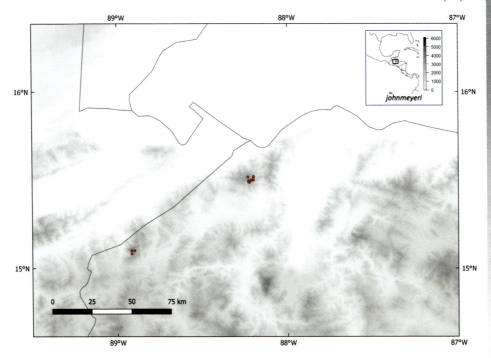
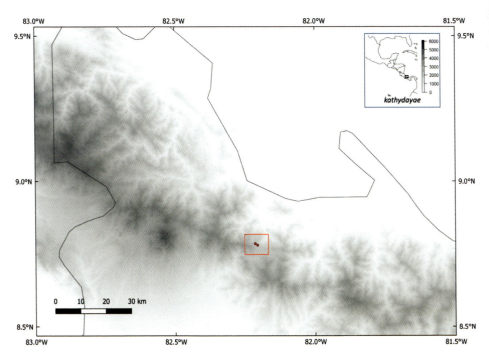

SPECIES ACCOUNTS

Conservation 11 [General Section]: 1–16.
TYPE SPECIMEN — MSB 96614.
TYPE LOCALITY — "Panama, Chiriquí, trail from paved road near Chiriquí/Bocas del Toro province boundary at Fortuna pass; 8.78533, –82.21434, 1178 m."
MORPHOLOGY — Body length to 148 M, 136 F; ventral scales smooth; middorsal scales enlarged in 0–2 rows; interparietal smaller than tympanum; scales across the snout between the second canthals 9–11; toe lamellae 23–27; suboculars and supralabials in contact; scales separating supraorbital semicircles 3–4; supraocular scales subequal to gradually enlarged; hindlimbs short, addressed toe to ear, or posterior to ear; body color gray, brown, and/or green, usually with some banding; male and female dewlaps white with greenish and/or grayish tint.
SIMILAR SPECIES — *Anolis kathydayae* is a *giant* anole differentiated from other potentially sympatric giant anoles by its short hindlimbs (addressed toe reaches to past eye in *A. frenatus*, *A. casildae*, *A. capito*; to ear in *A. kathydayae*). *Anolis biporcatus* and *A. capito* differ from *A. kathydayae* in their keeled ventral scales (smooth in *A. kathydayae*). *Anolis kunayalae* has a narrow toepad and a large claw on the fourth toe and fewer than 18 toe lamellae (typical broad anoline toepad, 23–27 lamellae in *A. kathydayae*). *Anolis microtus* and *A. ginaelisae* have larger head scales than *A. kathydayae* (5–9 scales across the snout between the second canthals; 9–11 in *A. kathydayae*) and a pink dewlap in males and females (white in *A. kathydayae*). *Anolis kathydayae* is replaced geographically to the east and southwest by similar anoles *A. brooksi* and *A. savagei*, respectively; these species differ from *A. kathydayae* mainly in dewlap color pattern (tan in males, white with dark streaks in female *A. brooksi*; pink with dark lines in males and females of *A. savagei*).
RANGE — From 1000 to 1200 meters in the Fortuna pass area of Panama.
NATURAL HISTORY — Rare in forest; sleeps above 3 meters above ground on narrow branches.

ANOLIS KEMPTONI — PLATE 1.103, 2.34; MAP opposite

DESCRIPTION — Dunn, E. R. 1940. New and noteworthy herpetological material from Panamá. *Proceedings of the Academy of Natural Sciences of Philadelphia* 92:111–112.
TYPE SPECIMEN — ANSP 21708.
TYPE LOCALITY — "Finca Lerida, 5300 feet, above Boquete, Chiriquí." Finca Lerida (lat 8.813 lon –82.483) is a coffee farm that now also functions as a hotel.
MORPHOLOGY — Body length to 53 M, 55 F; ventral scales smooth; middorsal scales enlarged in 0–2 rows; interparietal larger than tympanum; scales across the snout between the second canthals 7–11; toe lamellae 14–18; suboculars and supralabials in contact; scales separating supraorbital semicircles 0–2; supraocular scales gradually enlarged; hindlimbs short, addressed toe to ear or posterior to ear; body color mostly brown, sometimes with faint yellowish-brown crossbands; male dewlap orange-red anteriorly, pink posteriorly; female dewlap absent or dirty white.
SIMILAR SPECIES — *Anolis kemptoni* is a *small brown* highland anole with short hindlimbs (addressed toe reaches to ear) and smooth ventral scales. Among geographically proximal species that share these traits, the male dewlap color of *A. kemptoni* (pink posteriorly, orange-red anteriorly) is distinctive (male dewlap solid red in *A. salvini* and *A. fungosus*; purple-pink in *A. pentaprion*; orange in *A. carpenteri*, *A. gruuo*, *A. elcopeensis*, *A. altae*; orange-red anteriorly, yellow posteriorly in *A. fortunensis*). *Anolis pseudokemptoni* is found east of the range of *A. kemptoni* and is reported to differ from *A. kemptoni* in its bilobed hemipenes (unilobed in *A. kemptoni*; Köhler et al. 2007).
RANGE — From 1100 to 1800 meters in the Cordillera de Talamanca of Costa Rica and Panama between the Fortuna forest reserve, Panama, and Las Cruces, Costa Rica.
NATURAL HISTORY — Very common in forest, edge, and (especially) disturbed areas; sleeps on twigs or leaves at all observable heights above ground. See Poe and Armijo (2014).

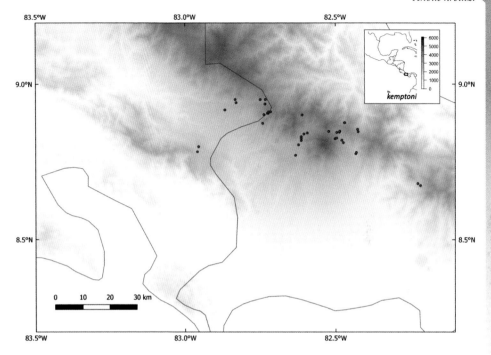

— *ANOLIS KREUTZI* — PLATE 1.104; MAP overleaf

DESCRIPTION — McCranie, J. R., G. Köhler, L. D. Wilson. 2000. Two new species of anoles from northwestern Honduras related to *Norops laeviventris* (Wiegmann 1834) (Reptilia, Squamata, Polychrotidae). *Senckenbergiana biologica* 80:218–222.
TYPE SPECIMEN — SMF 78844.
TYPE LOCALITY — "2.5 airline km NNE La Fortuna (15° 26' N, 87° 18' W), 1670–1690 m elevation, Departamento de Yoro, Honduras."
MORPHOLOGY — Body length to 48 M, 51 F; ventral scales keeled; middorsal scales gradually enlarged in several rows; interparietal larger than tympanum; scales across the snout between the second canthals 6–7; toe lamellae 16–18; suboculars and supralabials in contact; supraorbital semicircles usually in contact; supraocular scales gradually enlarged; hindlimbs short, appressed toe to ear or posterior to ear; body color grayish-brown; male dewlap pale yellow.
SIMILAR SPECIES — *Anolis kreutzi* is a *small brown* to whitish highland anole. The combination of keeled ventral scales, gray to white coloration, and short hindlimbs (appressed toe reaches to ear) will distinguish *A. kreutzi* from potentially sympatric anoles. *Anolis cusuco* and *A. laeviventris* are very similar parapatric species to *A. kreutzi* that differ in size of male dewlap (larger, well onto chest in *A. cusuco*; smaller, just past arms in *A. kreutzi*) or color of male dewlap (white in *A. laeviventris*, pale yellow in *A. kreutzi*).
RANGE — From 950 to 1700 meters in northwestern Honduras.
NATURAL HISTORY — Common in forest edge; active diurnally on tree trunks; sleeps on twigs or leaves. See McCranie and Köhler (2015).

SPECIES ACCOUNTS

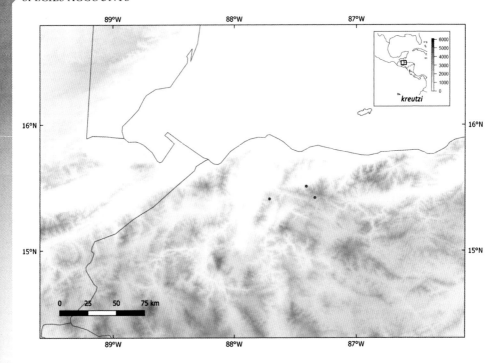

— *ANOLIS KUNAYALAE* — FIGURE 1.1; PLATE 1.105, 2.65; MAP opposite

DESCRIPTION — Hulebak, E., S. Poe, R. Ibáñez, E. E. Williams. 2007. A striking new species of *Anolis* lizard (Squamata, Iguania) from Panama. *Phyllomedusa* 6:5-10.
TYPE SPECIMEN — MSB 72605.
TYPE LOCALITY — "Trails of Parque Nacional General de División Omar Torrijos Herrera, 5 km north of El Copé, Coclé Province, Panama (08°40.315'N, 80°35.518'W)." (elevation ~800 m).
MORPHOLOGY — Body length to 109 M, 95 F; ventral scales smooth; middorsal scales enlarged in 0-2 rows; interparietal smaller than tympanum; scales across the snout between the second canthals 12-17; toe lamellae 11-15; suboculars and supralabials in contact or separated by rows of scales; at least four scales separating supraorbital semicircles; supraocular scales subequal to gradually enlarged; adpressed toe to between ear and eye; body color green sometimes with blue anteriorly, usually with some lateral transverse lines, changeable to dark brown; male dewlap white with orange-yellow distal edge and greenish scales; female dewlap green with yellow edge; iris reddish-brown.
SIMILAR SPECIES — *Anolis kunayale* is a *large green* to *giant* anole distinguishable from all potentially sympatric large anoles by its toe structure: a narrow, indistinct toepad with relatively few lamellae (11-15) and a large claw are present on the longest toe of the hindfoot. Other large potentially sympatric anoles display the typical broad, distinct anoline toepad and greater than 15 lamellae. The male body (anteriorly blue) and dewlap color (white with orange-yellow distally) of *A. kunayalae* also are locally distinctive.
RANGE — From 300 to 1100 meters from the Darién of Panama west to El Copé National Park on the Pacific slope of the Cordillera Central and Bocas del Toro on the Caribbean slope.
NATURAL HISTORY — Rare in forest and edge habitats; sleeps on twigs or leaves above 1.5 meters above ground.

Anolis laevis

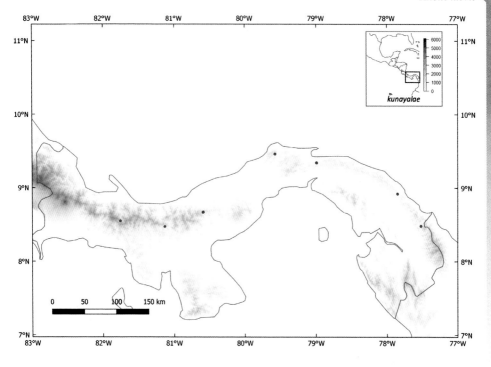

— *ANOLIS LAEVIS* — PLATE 1.106, 2.151

DESCRIPTION — Cope, E. D. 1876. Report on the reptiles brought by Professor James Orton from the middle and upper Amazon, and western Peru. *Journal of the Academy of Natural Sciences of Philadelphia* 8:165–166.
TYPE SPECIMEN — ANSP 11368.
TYPE LOCALITY — "Between Moyabamba and Balsa Puerto, on the river Huallaga in Eastern Peru." There is a historically well-known trail that, previous to the road graded from Tarapoto to Yurimaguas, was a standard passageway between the Andes and the Amazon. Professor Orton's voyage along this path is detailed in Orton (1876: 388): "It is difficult to conceive how such a path, daily used for the traffic of the great city of Moyobamba, can be tolerated." Presumably the specimen was taken during this journey. The Huallaga river passes some 50 km east of Balsa Puerto at 200 meters elevation, and this species seems likely to be a highland form based on morphology and recent collections. Thus, Orton's river may be a higher-elevation tributary of the Huallaga. The trail between Moyobamba and Balsapuerto is very roughly located at lat −6.0 lon −76.7.
MORPHOLOGY — Body length to at least 60 M (females unknown); ventral scales smooth; middorsal scales form a crest of triangular plates; interparietal larger than tympanum; scales across the snout between the second canthals 4; suboculars and supralabials in contact; supraorbital semicircles in contact; supraocular scales gradually enlarged; hindlimbs short; male dewlap white; snout protrudes anteriorly over lower jaw.
SIMILAR SPECIES — *Anolis laevis* is a *twig* anole. The strongly overlapping rostral scale of *A. laevis* should distinguish it from all nearby anoles. Two other twig anoles are found geographically proximal to *A. laevis*, *A. williamsmittermeierorum* and *A. peruensis*. These differ from *A. laevis* in male dewlap color (yellow in *A. peruensis*, white with tan distally in *A. williamsmittermeierorum*, solid white in *A. laevis*) and in lacking a serrated middorsal crest (crest present in *A. laevis*).

SPECIES ACCOUNTS

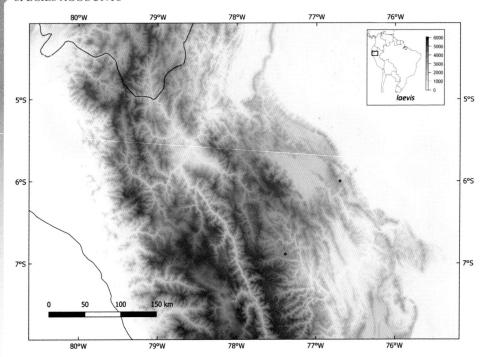

RANGE — In northern Amazonian Andean Peru, known from the type locality and a nearby high-elevation locality.
NATURAL HISTORY — Unknown. Rare. Based on its morphology, I would expect it to sleep high above ground on twigs.

— *ANOLIS LAEVIVENTRIS* — PLATE 1.107, 2.44

DESCRIPTION — Wiegmann, A.F.A. 1834. *Herpetologia Mexicana, seu Descriptio Amphibiorum Novae Hispaniae, quae Itineribus comitis de Sack, Ferdinandi Deppe et Chr. Guil. Schiede in Museum Zoologicum Berolinense Pervenerunt. Pars Prima, Saurorum Species Amplectens. Adiecto Systematis Saurorum Prodromo, Additisque multis in hunc Amphibiorum Ordinem Observationibus*. Berlin: Lüderitz. 47.
TYPE SPECIMEN — ZMB 525.
TYPE LOCALITY — "Mexico." Restricted to Jalapa, Veracruz, Mexico (lat 19.55 lon −96.92, 1500 m) by Smith and Taylor (1950).
MORPHOLOGY — Body length to 46 M, 49 F; ventral scales keeled; middorsal scales appear uniform or gradually enlarged in several rows; interparietal larger than tympanum; scales across the snout between the second canthals 6–10; toe lamellae 11–19; suboculars and supralabials usually in contact, occasionally separated by a row of scales; scales separating supraorbital semicircles 0–1; supraocular scales gradually enlarged; hindlimbs short, adpressed toe approximately to ear; body color grayish-brown to whitish; male dewlap white; female dewlap usually absent.
SIMILAR SPECIES — *Anolis laeviventris* is a *small brown* to light gray or whitish highland anole sympatric with many forms throughout its large range. The white male dewlap of *A. laeviventris* separates this species from all small highland forms except *A. cusuco*. For females, the combination of keeled ventral scales, lack of abruptly enlarged rows of middorsal scales, and short hindlimbs (adpressed toe reaches

Anolis lamari

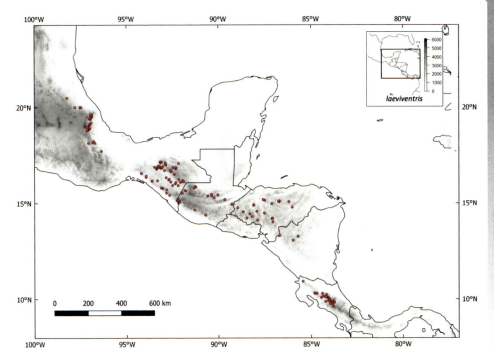

to ear) will distinguish *A. laeviventris* from potentially sympatric small highland anoles. *Anolis cusuco* is the version of *A. laeviventris* found at Cusuco Park, Honduras; it differs from *A. laeviventris* in its larger male dewlap (extending well on to chest; just beyond forelimbs in *A. laeviventris*). *Anolis kreutzi* is the version of *A. laeviventris* in Yoro Department, Honduras; it differs from *A. laeviventris* in its pale yellow male dewlap.

RANGE — From 800 to 2050 meters from Mexico to Costa Rica.

NATURAL HISTORY — Common in forest, edge, and disturbed habitats; diurnally often on tree trunks and branches; sleeps on twigs and leaves from 1 to 6 meters above ground. See Echelle et al. (1971; as *Anolis intermedius*); Fitch (1973a, b; 1975; as *A. intermedius*); Berkum (1986; as *A. intermedius*); Pounds (1988; as *A. intermedius*); Savage (2002; as *A. intermedius*).

COMMENT — Some authors (e.g., Savage 2002; Köhler and Vargas 2019) recognize the Costa Rican populations of *Anolis laeviventris* as *A. intermedius* (type locality "Veragua"). I suspect this inference is correct, but I have not seen published information delimiting these forms and I have not evaluated pertinent material.

— *ANOLIS LAMARI* — PLATE 1.108, 2.144

DESCRIPTION — Williams, E. E. 1992. New or problematic *Anolis* from Colombia. VII. *Anolis lamari*, a new anole from the Cordillera Oriental of Colombia, with a discussion of *tigrinus* and *punctatus* species group boundaries. *Breviora* 495:1–24.

TYPE SPECIMEN — ICN 6762.

TYPE LOCALITY — "Portachuelo, about 2 miles (by air) north of Manzanares, a police inspection station in the Municipio de Acacias, Meta, Colombia … Elevation ca. 1,600 m." Portachuelo is at approximately lat 4.166 lon −73.819.

SPECIES ACCOUNTS

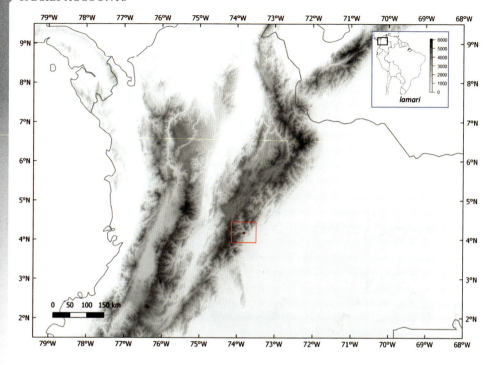

MORPHOLOGY — Body length to 52 M, 56 F; ventral scales smooth; middorsal scales enlarged in 0–2 rows; interparietal larger than tympanum; scales across the snout between the second canthals 7–8; toe lamellae 19–21; suboculars and supralabials in contact; supraorbital semicircles in contact; supraocular scales gradually enlarged; hindlimbs short, appressed toe to ear or short of ear; body color tan and green; tail banded; male dewlap yellow-orange distally, white proximally; female dewlap black.

SIMILAR SPECIES — *Anolis lamari* is a highland *twig* anole. Comparably sized, potentially sympatric forms differ in condition of ventral scales (keeled in *A. scypheus*, smooth in *A. lamari*) and size of head scales (supraorbital semicircles separated by at least one row of scales in *A. fuscoauratus, A. scypheus*; supraorbital semicircles in contact with each other and interparietal in *A. lamari*). *Anolis lamari* is replaced geographically by *A. ruizii* to the north, with which it is reported to differ in its possession of a posterior protruding knob on the skull (absent in *A. ruizii*; Williams 1992; but see below), and by *A. orcesi* to the south, which differs in its larger head scales (4–5 scales across the snout in *A. orcesi*; 7–8 in *A. lamari*).

RANGE — Known only from the type locality.

NATURAL HISTORY — Common to rare. Sleeps on twigs or leaves, especially ferns, above 1.5 meters above ground. My group was able to find *Anolis lamari* in considerable numbers across a 100-meter stretch of road at its exact type locality, but failed to find a single additional individual during intensive search of surrounding areas. See Barnett et al. (2022).

COMMENT — The skull trait purported to distinguish *Anolis lamari* from the geographically proximal species *A. ruizii* does not in fact separate these forms (Barnett et al. 2022). The status of *A. lamari* relative to *A. ruizii* should be investigated.

— ANOLIS LATIFRONS — PLATE 1.109, 2.132

DESCRIPTION — Berthold, A. A. 1845. Über verschiedene neue oder seltene Reptilien aus Neu-Granada und Crustaceen aus China. *Abhandlungen der Königlichen Gesellschaft der Wissenschaften in Göttingen* 3:6–7.
TYPE SPECIMEN — ZFMK 21342.
TYPE LOCALITY — "Provinz Popayan." At the time of description, Popayan Province in the Republic of New Granada encompassed a vast area stretching from the Andes south of Cali east to cover much of the southern Amazonian part of Colombia. The known range of *Anolis latifrons* west of the Andes restricts its presence in historical Popayan Province to the lower western slope of the Andes in current Cauca Department.
MORPHOLOGY — Body length to 133 M, 110 F; ventral scales smooth; middorsal scales enlarged in 0–2 rows; interparietal smaller than tympanum; scales across the snout between the second canthals 10–15; toe lamellae 19–26; suboculars and supralabials in contact or separated by rows of scales; scales separating supraorbital semicircles 3–5; supraocular scales subequal; hindlimbs long, toe of adpressed hindlimb to beyond eye; body color green with lateral bands of ocelli; tail banded; male and female dewlaps dull white.
SIMILAR SPECIES — *Anolis latifrons* is a *giant* green lowland anole. Its long limbs (adpressed toe reaches to beyond eye), oblique lateral rows of ocelli on the body, and all-white dewlap in males and females will differentiate *A. latifrons* from all potentially sympatric large green species except *A. frenatus* and *A. princeps* (see these species for additional comparisons). *Anolis latifrons* differs from these species in the presence of squarish scales along the dorsolateral edge of the orbit (a single elongate superciliary scale in *A. frenatus*, *A. princeps*).
RANGE — From 0 to 950 meters in forest from the Darién of Panama south to Valle del Cauca in Pacific Colombia.

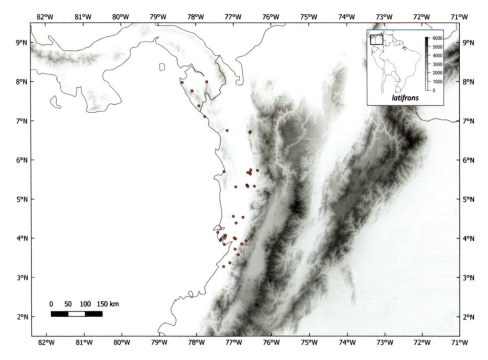

SPECIES ACCOUNTS

NATURAL HISTORY — Common in forest; sleeps on vegetation, usually above 1 meter above ground, usually on twigs or vines. See Castro-Herrera (1988); Vargas S. and Bolaños-L. (1999); Rengifo et al. (2014, 2015, 2019, 2021); Moreno-Arias et al. (2020).
COMMENT — See Comment under *Anolis frenatus*.

— *ANOLIS LEMURINUS* — PLATE 1.110, 2.58

DESCRIPTION — Cope, E. D. 1861. Notes and descriptions of anoles. *Proceedings of the Academy of Natural Sciences of Philadelphia* 13:213.
TYPE SPECIMEN — "Types, originally in ANS[P], now apparently lost," according to Stuart (1955).
TYPE LOCALITY — "Veragua, New Grenada." Veragua province in New Granada corresponds to modern Panama west of the Azuero peninsula.
MORPHOLOGY — Body length to 79 M, 78 F; ventral scales keeled; middorsal scales enlarged in 0–2 rows; interparietal larger or smaller than tympanum; scales across the snout between the second canthals 7–11; toe lamellae 15–20; suboculars and supralabials in contact or separated by rows of scales; scales separating supraorbital semicircles 0–2; supraocular scales gradually enlarged; hindlimbs long, adpressed toe to well anterior to eye; body color brown, sometimes appearing whitish (especially when sleeping), often with light lateral stripe and middorsal blotches and a dark bar dorsally between the eyes; male dewlap red; female dewlap usually white with dark flecks.
SIMILAR SPECIES — *Anolis lemurinus* is a widespread *large brown* lowland anole. Its combination of keeled ventral scales, long hindlimbs (adpressed toe reaches anterior to eye), and distinctive small male and female dewlaps (red in males, blue or white in females, reaching barely past forelimbs) serve to differentiate it from similar potentially sympatric species. Potentially confusing species *A. lionotus*, *A. oxylophus*, *A. barkeri*, *A. purpuronectes*, *A. robinsoni*, *A. riparius*, and *A. aquaticus* differ from *A. lemurinus* in their semiaquatic lifestyles, always being found within 2 meters of a stream or river,

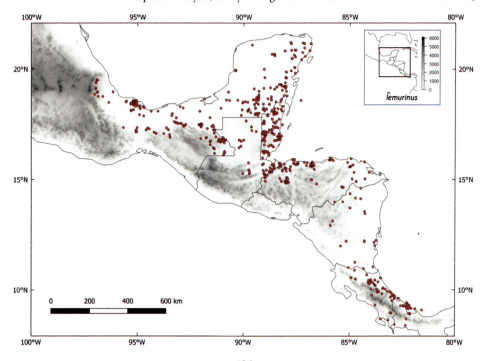

and in male dewlap color (*A. lionotus, A. oxylophus*: orange; *A. barkeri, A. aquaticus, A. riparius*: red with yellow; *A. purpuronectes*: purple; *A. robinsoni*: dark brown) and in usually lacking a dewlap in females. *Anolis capito* has small dewlaps in male and female, but these are differently colored than the dewlaps of *A. lemurinus* (yellow-brown in males and females). Also, *A. capito* tends to have a greater number of supralabial scales counted from the rostral to the center of the eye (8–11 in *A. capito*; 5–8 in *A. lemurinus*). *Anolis woodi* has blue eyes (brown or red in *A. lemurinus*) and tends to live at higher elevations than *A. lemurinus*. Similar anole *A. vittigerus* replaces *A. lemurinus* to the south/east and differs in male and female dewlap (dark central spot in both sexes; spot absent in *A. lemurinus*) and in possessing a greater number of more strongly enlarged rows of middorsal scales (more than 5 noticeably enlarged middorsal rows in *A. vittigerus*; middorsals only slightly enlarged in *A. lemurinus*).
RANGE — From 0 to 1150 meters from Mexico to western Panama.
NATURAL HISTORY — Common, occasionally rare, in forest, edge, and disturbed habitat; diurnally often on tree trunks, also low vegetation; sleeps on twigs, leaves, or vines from 0.5 to 4 meters above ground. See Fitch (1975); Henderson and Fitch (1975); Corn (1981); Berkum (1986); Villareal-Benítez (1997); Savage (2002); D'Cruze (2005); D'Cruze and Stafford (2006); Urbina-Cardona et al. (2006); Luja et al. (2008); Logan et al. (2012); Logan et al. (2015); McCranie and Köhler (2015); Russildi et al. (2016); Cardona and Reynoso (2017); Hilje et al. (2020); Perez-Martinez et al. (2021).
COMMENT — Work by Prado-Irwin (2022) suggests multiple species exist within *Anolis lemurinus* as currently recognized.

— *ANOLIS LIMIFRONS* — PLATE 1.111, 2.26

DESCRIPTION — Cope, E. D. 1862. Contributions to Neotropical saurology. *Proceedings of the Academy of Natural Sciences of Philadelphia* 14:178–179.
TYPE SPECIMEN — Syntypes ANSP 7900–01.

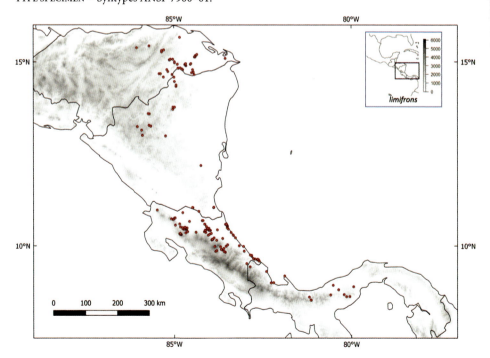

SPECIES ACCOUNTS

TYPE LOCALITY — "Veragua." Veragua, Republic of New Granada, included all of Panama west of the Azuero peninsula at time of description.
MORPHOLOGY — Body length to 48 M, 46 F; ventral scales smooth; middorsal scales enlarged in 0–2 rows; interparietal larger or smaller than tympanum; scales across the snout between the second canthals 9–16; toe lamellae 13–17; suboculars and supralabials in contact; scales separating supraorbital semicircles 1–3; supraocular scales gradually enlarged; hindlimbs long, adpressed toe to between eye and tip of snout; body color brown, often with a light lateral stripe and middorsal rectangles; tail banded; male dewlap white with or without basal yellow spot; female dewlap absent or small and white.
SIMILAR SPECIES — *Anolis limifrons* is a widespread *small brown* lowland anole. Its combination of smooth ventral scales and long hindlimbs (adpressed toe reaches to between eye and tip of snout) will differentiate *A. limifrons* from most potentially sympatric small brown anoles. Among lowland species that share these traits, *A. polylepis* and *A. cristatellus* differ from *A. limifrons* in male dewlap (*A. polylepis*: solid orange; *A. cristatellus*: red distally, green proximally; *A. limifrons*: white) and larger body size (SVL to 57 mm in *A. polylepis*, to 73 mm in *A. cristatellus*; to 48 mm in *A. limifrons*). *Anolis biscutiger* differs from *A. limifrons* in its slightly shorter hindlimbs (adpressed toe usually reaches to eye) and darker male dewlap (ashy gray) (see comments under *A. biscutiger*). *Anolis zeus* essentially is the northernmost version of *A. limifrons* (see comments under that species). *Anolis rodriguezii* is similar to *A. limifrons* and found north of its range; it differs from *A. limifrons* in its orange male dewlap. *Anolis apletophallus* replaces *A. limifrons* to the south/east at approximately the Panama Canal and differs from *A. limifrons* in its solid yellow male dewlap.
RANGE — From 0 to 1350 meters from eastern Honduras to western Panama.
NATURAL HISTORY — Very common in forest, edge, and disturbed habitats; active diurnally on the ground and low on vegetation; sleeps on twigs, leaves, or vines at all visible vertical levels; often the most abundant anole at any locality it inhabits; called a "weed" or gap species by Savage (2002), as it commonly occupies open or disturbed areas formerly covered by forest. See Ballinger et al, (1970); Echelle et al. (1971a); Sexton et al. (1972); Fitch (1973a, b, 1975); Andrews and Rand (1974, 1982); Fitch et al. (1976); Andrews (1976); Talbot (1976, 1977, 1979); Jenssen and Hover (1976); Scott et al. (1976); Andrews (1979b, 1982, 1989); Corn (1981); Andrews and Sexton (1981); Andrews et al. (1983); Berkum (1986); Andrews and Nichols (1990); Savage (2002); Curlis et al. (2017); Barquero and Bolaños (2018); Thompson et al. (2018); Hilje et al. (2020); Perez-Martinez et al. (2021).
COMMENT — I consider *Anolis cryptolimifrons* (type locality: "Cerro Brujo [9° 11' 16.4" N, 82° 11' 25.4" W], 10 m, Bocas del Toro Province, Panama") to be a junior synonym of *A. limifrons* (see appendix 1 for justification). For those who wish to recognize *A. cryptolimifrons*, it is the version of *A. limifrons* found near Almirante, Bocas del Toro, Panama.

— *ANOLIS LIMON* — PLATE 1.112

DESCRIPTION — Velasco, J. A., J. P. Hurtadao-Gomez. 2014. A new green anole lizard of the "Dactyloa" clade (Squamata: Dactyloidae) from the Magdalena river valley of Colombia. Zootaxa 3785:201–216.
TYPE SPECIMEN — MHUA-R 11760.
TYPE LOCALITY — "Department of Antioquia, Gomez Plata municipality, Vereda La Clara, Hacienda Vegas de la Clara; 6 34' 53" N, 75 11' 43" W (1093 m)."
MORPHOLOGY — Body length to 79 M, 82 F; ventral scales smooth; middorsal scales enlarged in 0–2 rows; interparietal larger or smaller than tympanum; scales across the snout between the second canthals 8–11; toe lamellae 20–22; suboculars and supralabials in contact; scales separating supraorbital semicircles 2–3; supraocular scales gradually enlarged; adpressed toe probably to eye (specimens not checked for this trait); body color solid green or green with thick bands or dark markings; male dewlap white.
SIMILAR SPECIES — *Anolis limon* is a *large green* anole with smooth ventral scales. Among geographically proximal similar green anoles, *A. antioquiae* has smaller head scales (16–19 scales across the snout; 8–11 in *A. limon*); *A. frenatus*, *A. latifrons*, and *A. apollinaris* are larger than *A. limon* (>100 mm

Anolis lineatus

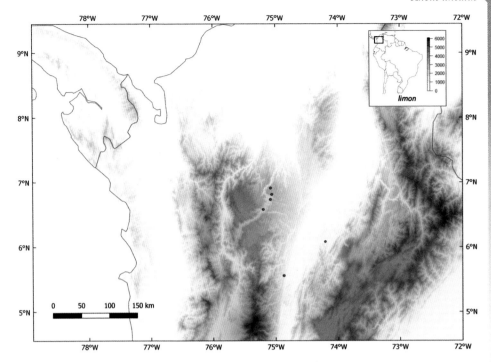

SVL) and have greater numbers of toe lamellae (usually 23–28; 20–22 in *A. limon*). *Anolis frenatus* and *A. latifrons* further differ from *A. limon* in lateral body pattern (oblique rows of ocelli in *A. frenatus*, *A. latifrons*; solid green or green with bands, lacking ocelli in *A. limon*). *Anolis purpurescens* is very similar to *A. limon* and replaces it geographically to the northwest; it differs from *A. limon* in its yellow-orange male dewlap (white in *A. limon*).
RANGE — From 500 to 1550 meters in the northern Magdalena River valley of Colombia.
NATURAL HISTORY — Only information is from the type description: "In secondary forest and forest edges, near streams and ponds, mostly sleeping on leaves at night."

— *ANOLIS LINEATUS* — PLATE 1.113, 2.157

DESCRIPTION — Daudin, F. M. 1802. *Histoire Naturelle, Générale et Particulière des Reptiles*, vol. 4. F. Dufart, Paris, 66–69.
TYPE SPECIMEN — MNHN 795.
TYPE LOCALITY — "Il existe dans diverses parties de l'Amérique méridionale peut-être même dans les îles Antilles" (i.e., various parts of South America, possibly the West Indies). Duméril and Bibron (1837) suggested Martinique as an original locality based on supposedly known provenance of one of their two specimens, an almost certainly erroneous proposal tentatively followed by Boulenger (1885; "Leeward Islands; Martinique?"). Curaçao or Aruba are more likely type localities.
MORPHOLOGY — Body length to 73 M, 62 F; ventral scales keeled; middorsal scales enlarged in 0–2 rows; interparietal larger or smaller than tympanum; scales across the snout between the second canthals 4–9; toe lamellae 16–24; suboculars and supralabials in contact; supraorbital semicircles in contact; body color brown, with two parallel longitudinal lines on each flank, sometimes with faint lateral bars or crossbanding; male dewlap yellow-orange with dark central blotch; female dewlap like males but smaller.

SPECIES ACCOUNTS

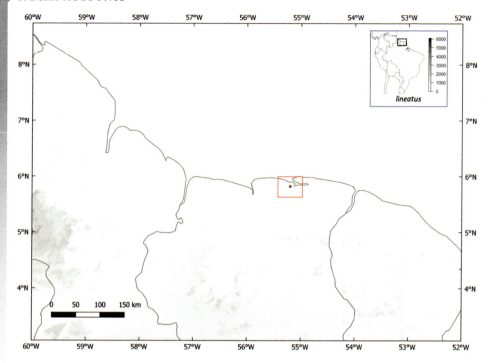

SIMILAR SPECIES — *Anolis lineatus* is a *large brown* anole. The unique male dewlap color pattern of orange-yellow with a dark diffuse central blotch, distinctive lateral body pattern of two parallel longitudinal dark lines, and a restricted mainland distribution of a small highly disturbed area in Suriname should render identification straightforward.
RANGE — On the mainland, known from a population established in a garden in Rainville, Paramaribo, Suriname (Hoogmoed 1981). Native to the islands Curaçao and Aruba.
NATURAL HISTORY — On "wooden and concrete poles of a fence and trunks of trees" in Suriname. Very common and associated with trees on Curaçao (Rand and Rand 1967). See Hoogmoed (1981).

— *ANOLIS LIOGASTER* — PLATE 1.114, 2.1

DESCRIPTION — Boulenger, G. A. 1905. Descriptions of new reptiles discovered in Mexico by Dr. H. Gadow, F.R.S. *Proceedings of the Zoological Society of London* 1905:245–246.
TYPE SPECIMEN — Syntypes BMNH 1946.8.8.53-54.
TYPE LOCALITY — "Omilteme, Guerrero, 7600 ft." Omiltemi, Guerrero, Mexico, is at lat 17.556 lon −99.687, 2160 m.
MORPHOLOGY — Body length to 66 M, 59 F; ventral scales smooth; middorsal scales gradually enlarged in several rows; interparietal larger or smaller than tympanum; scales across the snout between the second canthals 5–7; toe lamellae 14–18; suboculars and supralabials in contact; scales separating supraorbital semicircles 0–1; supraocular scales abruptly enlarged; adpressed toe to eye or just beyond; body color brown, usually with light lateral stripe; male dewlap pink.
SIMILAR SPECIES — *Anolis liogaster* is a *small brown* highland anole with smooth ventral scales. *Anolis liogaster* is distinguished from other northern Pacific Mexico anoles with smooth ventral scales by its solid pink dewlap and band of at least 10 rows of enlarged middorsal scales (*Anolis omiltemanus*

Anolis lionotus

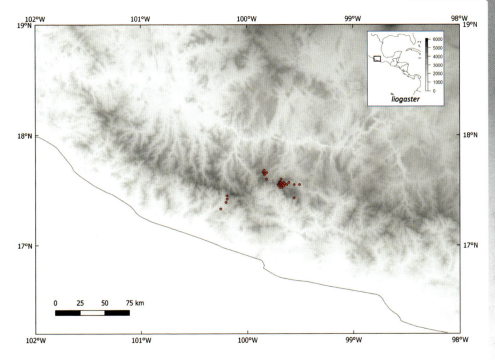

and *A. peucephilus* have a solid orange male dewlap; *Anolis taylori* and *A. gadovii* have patterned pink-purple male dewlaps; *A. dunni* has an orange-red male dewlap; all five species have fewer than four enlarged middorsal rows). *Anolis brianjuliani* essentially is an eastern version of *A. liogaster* that differs in its slightly larger middorsal scales.
RANGE — From 1800 to 2500 meters on the Pacific versant of central Guerrero, Mexico.
NATURAL HISTORY — Common in open habitats with trees; active diurnally on ground or low on tree trunks; sleeps on twigs or leaves. See Davis and Dixon (1961); Muñoz-Alonso (1988); Muñoz-Alonso and Flores-Villela (1990).

— *ANOLIS LIONOTUS* — PLATE 1.115, 2.83

DESCRIPTION — Cope, E. D. 1861. Notes and descriptions of anoles. *Proceedings of the Academy of Natural Sciences of Philadelphia* 13:210–211.
TYPE SPECIMEN — ANSP 7909.
TYPE LOCALITY — "Cocuyas de Veraguas, New Grenada." Myers (1974) states that Cocuyos de Veraguas refers to a gold mine in Veraguas, Panama, at "8° 45' N 81° 00' W."
MORPHOLOGY — Body length to 77 M, 65 F; ventral scales keeled; middorsal scales greatly enlarged in several rows; interparietal larger or smaller than tympanum; scales across the snout between the second canthals 9–13; toe lamellae 13–18; suboculars and supralabials separated by one or two rows of scales; scales separating supraorbital semicircles 1–2; supraocular scales abruptly or gradually enlarged; addressed toe to eye or anterior to eye; body color brown with light lateral stripe; male dewlap yellow-orange; female dewlap absent.
SIMILAR SPECIES — *Anolis lionotus* is a *semiaquatic* anole. Its streamside lifestyle, keeled ventral scales, and abruptly enlarged rows of 10–15 smooth middorsal scales will distinguish this form from all

SPECIES ACCOUNTS

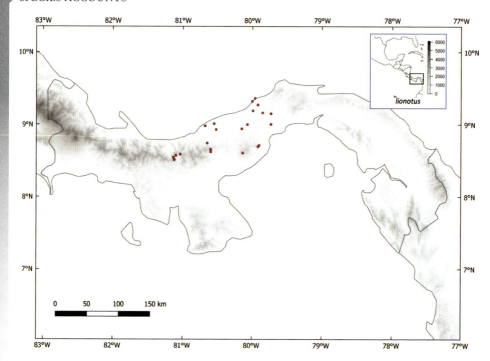

geographically proximal congeners. In particular, the enlarged middorsal scale rows distinguish *A. lionotus* from all nearby large brown anoles and semiaquatic species except *A. oxylophus*, which geographically replaces *A. lionotus* to the northwest in Costa Rica. In *A. oxylophus*, the enlarged middorsal scale rows are smaller, approximately equal to the ventrals, and usually keeled (middorsal rows much larger than ventrals, smooth in *A. lionotus*). *Anolis poecilopus*, which geographically and ecologically replaces *A. lionotus* as a large brown semiaquatic anole to the southeast approximately at the Panama Canal, has only weakly enlarged (and slightly keeled) middorsal scale rows and small head scales (14–24 scales across the snout; 9–13 scales across the snout in *A. lionotus*).
RANGE — From 0 to 900 meters from the Panama Canal area west to near the Costa Rican border.
NATURAL HISTORY — Common in forest; active diurnally in and along streams on boulders, logs, and low vegetation; sleeps on twigs, leaves, or boulders or rock walls up to 1 meter above ground. See Campbell (1973); Andrews (1976); Montgomery et al. (2011); Muñoz et al. (2015).

— *ANOLIS LOSOSI* — PLATE 1.116, 2.148

DESCRIPTION — Torres-Carvajal, O., F. P. Ayala-Varela, S. E. Lobos, S. Poe, A. E. Narváez. 2017. Two new Andean species of *Anolis* lizard (Iguanidae: Dactyloinae) from southern Ecuador. *Journal of Natural History*, https://doi.org/10.1080/00222933.2017.1391343.
TYPE SPECIMEN — QCAZ 10173.
TYPE LOCALITY — "Ecuador, Provincia Zamora Chinchipe, Romerillos Alto, 4.227°S, 78.939°W, WGS84, 1550 m."
MORPHOLOGY — Body length to 61 M, 60 F; ventral scales smooth; middorsal scales enlarged in 0–2 rows; interparietal larger than tympanum; scales across the snout between the second canthals 4–6; toe lamellae 18–21; suboculars and supralabials in contact; supraorbital semicircles in contact; supraocular

Anolis loveridgei

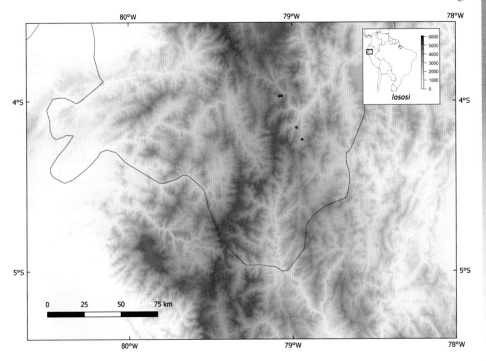

scales abruptly enlarged; hindlimbs short, adpressed toe to ear or posterior to ear; body color brown or lichenous white, green, and gray; male dewlap white; female dewlap orange with black blotches.
SIMILAR SPECIES — *Anolis lososi* is a highland *twig* anole that may be sympatric with *A. soinii*, *A. podocarpus*, and *A. hyacinthogularis*. The small body size and short hindlimbs of *A. lososi* will distinguish it from large species *A. soinii* and *A. podocarpus* (maximum SVL > 80, adpressed hindlimb reaches at least to eye; SVL to 61, adpressed hindlimb to ear in *A. lososi*). *Anolis lososi* may be distinguished from *A. hyacinthogularis* by male and female dewlap color (male: sky blue, female: cream with blue blotches; male: white, female: orange with black blotches in *A. lososi*). Similar twig anole *A. williamsmittermeierorum* is found south and north of the range of *A. lososi* (the northern population may represent a distinct species) and differs in dewlap colors (male: white with tan edge; female: white with black blotches).
RANGE — From 1550 to 2000 meters on the Amazonian Andean slope of southern Ecuador.
NATURAL HISTORY — Common in forest and in open areas; usually sleeps on twigs, from 1 to 8 meters above ground.

— *ANOLIS LOVERIDGEI* — PLATE 1.117

DESCRIPTION — Schmidt, K. P. 1936. New amphibians and reptiles from Honduras in the Museum of Comparative Zoology. *Proceedings of the Biological Society of Washington* 49:47–48.
TYPE SPECIMEN — MCZ 38700.
TYPE LOCALITY — "Portillo Grande, 4100 feet altitude, Yoro, Honduras." Monroe (1965) places Portillo Grande, Yoro, at "5 miles northwest of Yorito." That is, at approximately lat 15.10 lon −87.35.
MORPHOLOGY — Body length to 118 M, 116 F; ventral scales weakly keeled; middorsal scales enlarged in 0–2 rows; interparietal smaller than tympanum; scales across the snout between the second canthals

SPECIES ACCOUNTS

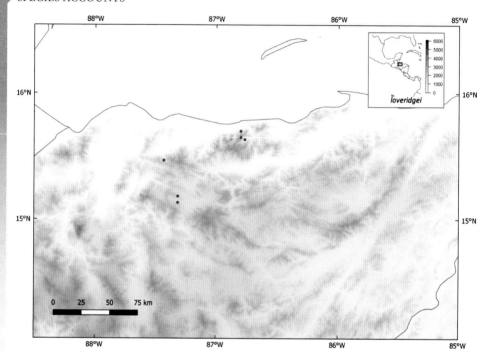

8–15; toe lamellae 24–28; scales between suboculars and supralabials 0–1; scales between supraorbital semicircles 4–8; supraocular scales subequal; hindlimbs long, adpressed toe to between eye and tip of snout; body color brown, possibly with light lateral bands or striations; male dewlap orange with broad red streaks; female dewlap red.

SIMILAR SPECIES — *Anolis loveridgei* is a giant anole in a region of the anole distribution with few other large anoles. Large anoles *A. petersii* and *A. biporcatus* differ from *A. loveridgei* in their shorter hindlimbs (adpressed toe reaches to ear or just past ear; to between eye and tip of snout in *A. loveridgei*). *Anolis capito* has fewer scales separating the supraorbital semicircles than *A. loveridgei* (0–3; 4–8 in *A. loveridgei*).

RANGE — From 550 to 1600 meters in northern Honduras.

NATURAL HISTORY — In forest; active diurnally low on vegetation; sleeps on vegetation. McCranie and Köhler (2015) note individuals sleeping on tree trunks and on a log. See McCranie and Köhler (2015).

— *ANOLIS LYNCHI* — PLATE 1.118, 2.154

DESCRIPTION — Miyata, K. 1985. A new *Anolis* of the *lionotus* group from northwestern Ecuador and southwestern Colombia (Sauria: Iguanidae). *Breviora* 481:1–13.

TYPE SPECIMEN — MCZ 124406.

TYPE LOCALITY — "Santo Domingo de los Colorados, 600 m elevation, Provincia del Pichincha, Ecuador." Santo Domingo, Pichincha, Ecuador is at lat −0.25 lon −79.18.

MORPHOLOGY — Body length to 62 M, 59 F; ventral scales keeled; middorsal scales enlarged in 0–2 rows; interparietal usually equal to or smaller than tympanum; scales across the snout between the second canthals 16–29; toe lamellae 14–18; suboculars and supralabials separated by a row of scales; scales separating supraorbital semicircles 3–5; supraocular scales subequal; adpressed toe to between

Anolis macrinii

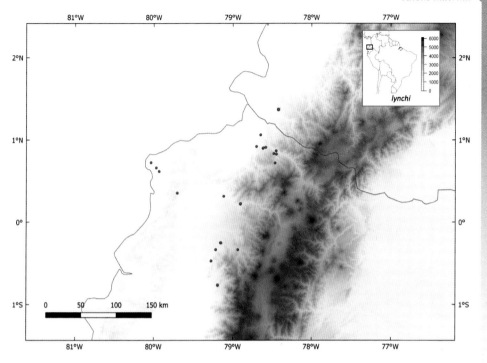

eye and tip of snout; body color brown with white lateral stripe along body; male dewlap yellow-orange; female dewlap absent.

SIMILAR SPECIES — *Anolis lynchi* is a *semiaquatic* lowland anole. Its tiny head scales (16–29 scales across the snout) and semiaquatic habits will separate it from geographically proximal brown anoles. Potentially confusing species may be separated from *A. lynchi* as follows: *Anolis vittigerus* has a red dewlap in males, blue or white dewlap in females (male dewlap yellow-orange, female dewlap absent in *A. lynchi*); *Anolis maculiventris*, *A. granuliceps*, and *A. festae* have smooth or weakly keeled ventral scales (strongly keeled in *A. lynchi*); *A. binotatus* and *A. gracilipes* have larger head scales (8–17 scales across the snout; 16–29 in *A. lynchi*). *Anolis macrolepis* is a semiaquatic anole found north of the range of *A. lynchi* that differs from *A. lynchi* in its 10–15 rows of enlarged middorsal scales and larger head scales (7–12 scales across the snout between the second canthals).

RANGE — From 100 to 1150 meters in northern Pacific Ecuador and Southern Colombia.

NATURAL HISTORY — Rare in forest and disturbed areas in and along streams; active on ground, in streams, and on low vegetation; sleeps on boulders or low (<1.0 meters above ground) vegetation within or near streams; stays underwater by "rebreathing" via an air bubble (Bocci et al. 2021). See Narváez (2017); Torres-Carvajal et al. (2019); Arteaga et al. (2023).

— *ANOLIS MACRINII* — PLATE 1.119, 2.25

DESCRIPTION — Smith, H. M. 1968b. Two new lizards, one new, of the genus *Anolis* from Mexico. *Journal of Herpetology* 2:143–6.
TYPE SPECIMEN — MCZ 46202.
TYPE LOCALITY — "Cafetal Santa Hedvigis, Pochutla, Oaxaca." I am unable to locate this cafetal. The central city of Pochutla District, Oaxaca, is San Pedro Pochutla (lat 15.75 lon −96.47, 160 m).

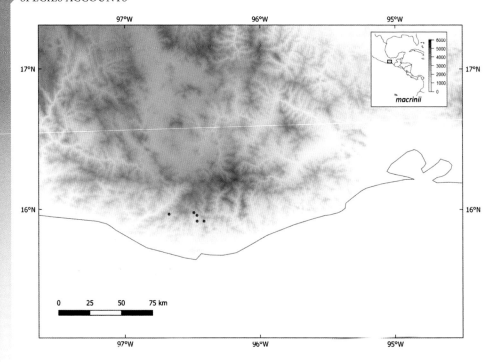

MORPHOLOGY — Body length to 95 M, 96 F; ventral scales smooth to weakly keeled; middorsal scales enlarged in 0–2 rows; interparietal larger than tympanum; scales across the snout between the second canthals 6–8; toe lamellae approximately 25; suboculars and supralabials in contact; scales separating supraorbital semicircles 0–1; supraocular scales gradually enlarged; hindlimbs short, adpressed toe to ear or posterior to ear; body color solid green, changeable to dark brown under stress; male dewlap dark orange distally, light proximally; female dewlap dark.
SIMILAR SPECIES — As the only *large green* anole in Pacific Oaxaca, *Anolis macrinii* is unlikely to be confused with other anole species.
RANGE — From 350 to 1400 meters on the Pacific slope of central Oaxaca, Mexico.
NATURAL HISTORY — Common to rare in forest and edge habitat; active diurnally on bushes and trees, including coffee plants; sleeps on twigs or leaves. See Köhler et al. (2013, 2014a).

— *ANOLIS MACROLEPIS* — PLATE 1.120, 2.153

DESCRIPTION — Boulenger, G. A. 1911. Descriptions of new reptiles from the Andes of South America, preserved in the British Museum. *Annals and Magazine of Natural History* 8:21–22.
TYPE SPECIMEN — Syntypes BMNH 1946.8.13.2, 1946.8.5.95–96.
TYPE LOCALITY — "A female [i.e., BMNH 1946.8.13.2] from Novita, Rio Tamaná, Choco, S. W. Colombia [lat 4.995 lon −76.606], 150–200 feet, and two young males [i.e., BMNH 1946.8.5.95–96] from Condoto [lat 5.091 lon −76.650], in the same district, 150 feet."
MORPHOLOGY — Body length to 62 M, 55 F; ventral scales keeled; middorsal scales greatly enlarged in several rows; interparietal larger than tympanum; scales across the snout between the second canthals 7–12; toe lamellae 13–19; suboculars and supralabials in contact or separated by a row of scales; scales separating supraorbital semicircles 0–1; supraocular scales gradually or abruptly enlarged; hindlimbs

Anolis macrophallus

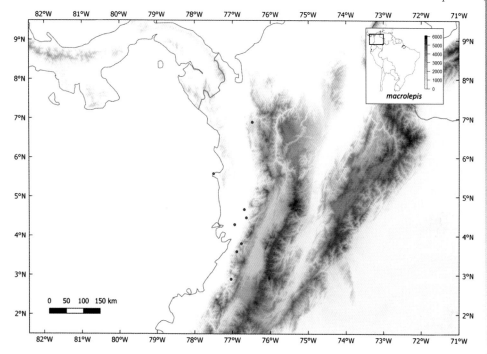

long, adpressed toe approximately to tip of snout; body color brown, reddish ventrolaterally; male dewlap orange-yellow; female dewlap absent.

SIMILAR SPECIES — *Anolis macrolepis* is a *semiaquatic* lowland anole with keeled ventral scales. Potentially confusing geographically proximal brown species with keeled ventral scales may be separated from *A. macrolepis* as follows: *Anolis vittigerus* lacks greatly enlarged middorsal scale rows (several rows of abruptly enlarged middorsal scales in *A. macrolepis*); *Anolis auratus*, *A. notopholis*, *A. gaigei*, and *A. tropidogaster* are smaller (usually under 50 mm SVL) and have different male dewlaps (blue-black in *A. auratus*; red and yellow in *A. tropidogaster*, *A. gaigei*; orange-red in *A. notopholis*; orange-yellow in *A. macrolepis*); *A. binotatus* and *A. gracilipes* have solid orange male dewlaps and greater numbers of scales separating supraorbital semicircles (usually 2–4; 0–1 in *A. macrolepis*); *Anolis rivalis*, *A. poecilopus*, and *A. maculigula* are semiaquatic anoles found within the range of *A. macrolepis*. *Anolis maculigula* and *A. poecilopus* lack greatly enlarged middorsal scales. *Anolis rivalis* has smaller head scales than *A. macrolepis* (13–20 scales across the snout; 7–12 in *A. macrolepis*). *Anolis lynchi* is a semiaquatic anole found south of the range of *A. macrolepis* and differing from it in possessing minute head scales (usually greater than 20 scales across the snout) and lacking enlarged middorsal scale rows.
RANGE — From 0 to 800 meters in Pacific coastal Colombia.
NATURAL HISTORY — Rare in and along streams. Sleeps on low (<1.0 meters) twigs or leaves along streams in forest. See Castro-Herrera (1988); Velasco and Herrel (2007).

— *ANOLIS MACROPHALLUS* — PLATE 1.121

DESCRIPTION — Werner, F. 1917. Über einige neue Reptilien und einen neuen Frosch des Zoologischen Museums in Hamburg. *Mitteilungen aus dem Zoologischen Museum*, 2. Beiheft zum Jahrbuch der Hamburgischen Wissenschaftlichen Anstalten 34:31–32.

SPECIES ACCOUNTS

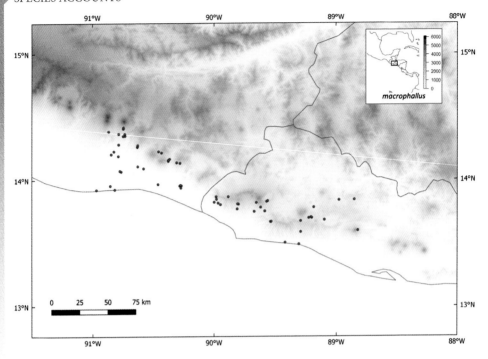

TYPE SPECIMEN — Destroyed from ZMH (Fitch et al. 1972). Neotype: SMF 79035 designated by Köhler and Kreutz (1999).
TYPE LOCALITY — "S. José de Guatemala." Presumably Port of San Jose, Escuintla Department, Guatemala (Köhler and Kreutz 1999; lat 13.92 lon −90.82, 0 m). Neotype is from "18.2 km from Puerto San José on the road to Escuintla, Guatemala, 14°04.35' N, 90°46.59' W, 20 m a.s.l."
MORPHOLOGY — Body length to 48 M, 47 F; ventral scales keeled; middorsal scales gradually enlarged in several rows; interparietal larger than tympanum; scales across the snout between the second canthals 7–11; toe lamellae 14–15; suboculars and supralabials in contact or separated by a row of scales; scales between supraorbital semicircles 1–3; supraocular scales gradually enlarged; hindlimbs long, appressed toe to eye or just beyond eye; body color brown; male dewlap orange-brown proximally, pale pink distally.
SIMILAR SPECIES — *Anolis macrophallus* is a *small brown* lowland anole. Geographically proximal brown anoles differ from *A. macrophallus* as follows: *A. dollfusianus* is smaller (usually less than 40 mm SVL; to 48 mm in *A. macrophallus*) and has a yellow male dewlap (male dewlap orange-brown proximally, pale pink distally in *A. macrophallus*); *A. laeviventris*, *A. sericeus*, and *A. cristifer* have short hindlimbs (adpressed toe reaches approximately to ear; to eye or just beyond in *A. macrophallus*); *A. serranoi* is larger than *A. macrophallus* (to 85 mm SVL) with greater numbers of toe lamellae (16–18; 14–15 in *A. macrophallus*). *Anolis crassulus*, *A. caceresae*, and *A. heteropholidotus* inhabit higher elevations than *A. macrophallus* and differ from it in their more abruptly enlarged middorsal scales (middorsals gradually enlarged in *A. macrophallus*) and heterogeneous mix of large and small lateral body scales (lateral body scales homogeneous, granular in *A. macrophallus*). *Anolis cupreus* is very similar to *A. macrophallus* and geographically replaces it to the south. *Anolis cupreus* differs from *A. macrophallus* in possessing stouter lobes of the hemipenes and a larger basal orange blotch on the male dewlap.
RANGE — From 0 to 1350 meters on the Pacific slope of Guatemala and El Salvador.
NATURAL HISTORY — Common in forest and disturbed areas, including coffee plantations; active diurnally low on vegetation, including bushes and small tree trunks; sleeps on twigs or leaves up to 2 meters above ground. See Köhler and Acevedo (2004).

— ANOLIS MACULIGULA — PLATE 1.122, 2.155

DESCRIPTION — Williams, E. E. 1984b. New or problematic *Anolis* from Colombia. III. Two new semi-aquatic anoles form Antioquia and Choco, Colombia. *Breviora* 478:2–7.

TYPE SPECIMEN — LACM 42150.

TYPE LOCALITY — "Quebrada San Lorenzo, tributary of the Rio Arquía near the small town of Belén (6 15 N 76 39 W), about 10 to 15 km upstream from the junction of the Rio Arquía with the Rio Atrato, western Antioquia, Colombia." The GPS point maps to an area without stream or river visible, and I am unable to locate a "Belén" that fits geographically with the described area. Fortunately, the LACM catalog gives a more informative description of the type locality than does the species description: "Rio Arquía, Belén (town), ca. 2hr 45min upstream from Pta. Palacios, 5 min upstream from Vegaes by canoe." This point (i.e., where a stream enters Rio Arquía just east of Vegaez, Antioquia) is at approximately lat 6.195 lon −76.545, 60 m.

MORPHOLOGY — Body length to 107 M, 75 F; ventral scales smooth to weakly keeled; middorsal scales enlarged in 0–2 rows; interparietal larger or smaller than tympanum; scales across the snout between the second canthals 12–19; toe lamellae 16–22; suboculars and supralabials separated by a row of scales; scales separating supraorbital semicircles 1–4; supraocular scales subequal to gradually enlarged; hindlimbs long, appressed toe to between eye and tip of snout; body color greenish-brown, often with bluish tint, with dorsal blotches; strong mottling in gular area; male dewlap anteriorly pale pink-peach to red, posterior pale blue to nearly white, with blue or brown scales; female dewlap dark gray anteriorly, lighter posteriorly.

SIMILAR SPECIES — *Anolis maculigula* is a *semiaquatic* anole with large body size and nearly or completely smooth ventral scales. *Anolis frenatus*, *A. latifrons*, *A. biporcatus*, and *A. purpurescens* are similarly sized to *A. maculigula* but are predominantly green and differ in dorsal pattern and male dewlap color (*A. purpurescens*: lateral ocelli on body, orange male dewlap; *A. biporcatus*: solid green

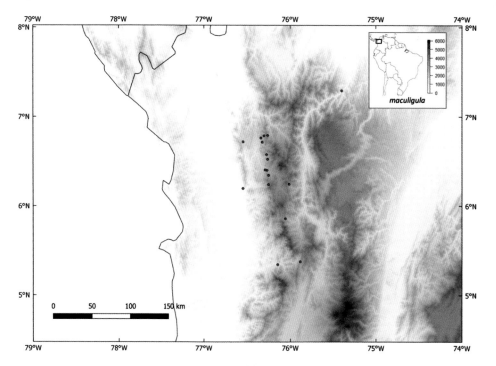

SPECIES ACCOUNTS

body, red, blue, and yellow male dewlap; *A. frenatus, A. latifrons*: lateral ocelli on body, white male dewlap; *A. maculigula*: predominantly bluish-brown and lacking lateral ocelli, pale pink and blue male dewlap). *Anolis ventrimaculatus* is smaller than *A. maculigula* (SVL to 80 mm; to 107 mm in *A. maculigula*) and differs in male dewlap color (dirty orange). *Anolis eulaemus* has a pale brown male dewlap and grayish-brown body. *Anolis mirus* has fewer toe lamellae than *A. maculigula* (12–15; 16–22 in *A. maculigula*). Several large nonaquatic Pacific Colombian species share small head scales and narrow toepads with *A. maculigula* but tend to have smaller posterior head scales (usually 4–5 scales between supraorbital semicircles in *A. antioquiae, A. megalopithecus, A. mirus*; usually 1–3 in *A. maculigula*). The other western Colombian semiaquatic anoles (*A. lynchi, A. macrolepis, A. poecilopus, A. rivalis*) differ from *A. maculigula* in possessing strongly keeled ventral scales.

RANGE — From 50 to 2200 meters in northern Pacific Colombia. The type and paratype series of this species are from near sea level but recent collections assigned to this species are from high elevations. I have observed confirmed *Anolis maculigula* at 1400 (near Pueblo Rico, Risaralda, Colombia) and 1600 meters (near Nutibara, Antioquia, Colombia), and iNaturalist photographs of individuals from still higher elevations appear to jibe with Williams's (1965) description of the species. I do not know whether the apparent broad elevational range of this species indicates extraordinary environmental tolerance in this form, cryptic taxonomic diversity, or some other factor.

NATURAL HISTORY — Common along streams in forest and disturbed areas. The type description states that individuals were observed on large boulders in the splash zones of a 10-meter-wide stream. I have observed individuals sleeping on boulders and rock walls, occasionally vegetation, in and around streams and rivers, often in splash zones, sometimes in great densities; found syntopically with fellow semiaquatic anole *A. rivalis*. Stays underwater by "rebreathing" via an air bubble (Bocci et al. 2021).

— *ANOLIS MACULIVENTRIS* — PLATE 1.123, 2.89

DESCRIPTION — Boulenger, G. A. 1898. An account of the reptiles and batrachians collected by Mr. W.F.H. Rosenberg in western Ecuador. *Proceedings of the Zoological Society of London* 1898:111.
TYPE LOCALITY — "Paramba." Rosenberg collected at Hacienda Paramba, in northern Imbabura on the south bank of Rio Mira (Paynter 1993), lat 0.817 lon −78.350, 700 m. However, Ayala and Williams (1988), citing unsuitable habitat at Hacienda Paramba, suggested the actual collection must have occurred downstream from Hacienda Paramba along the Rio Mira.
TYPE SPECIMEN — Syntypes: BMNH 1946.8.13.33–34.
MORPHOLOGY — Body length to 46 M, 49 F; ventral scales smooth; middorsal scales enlarged in 0–2 rows; interparietal equal to or smaller than tympanum; scales across the snout between the second canthals 10–16; toe lamellae 12–19; suboculars and supralabials in contact; scales separating supraorbital semicircles 2–5; supraocular scales gradually enlarged; hindlimbs long, adpressed toe to between eye and tip of snout; body color grayish-brown, sometimes with light and/or dark lateral ocelli; male dewlap pink to red posteriorly, orange distally; female dewlap usually absent.
SIMILAR SPECIES — *Anolis maculiventris* is a *small brown* lowland anole with smooth ventral scales and long hindlimbs (adpressed toe reaches to between eye and tip of snout). Among potentially sympatric species that fit this description, *A. granuliceps* has a small, orangeish-yellow male dewlap (large pink and orange dewlap in *A. maculiventris*) and usually has a light lateral longitudinal stripe (*A. maculiventris* frequently is transversely banded with lines, bars, or rows of ocelli); *A. vicarius* has smaller head scales than *A. maculiventris* (17 scales across the snout; 10–16 in *A. maculiventris*) and dark blotching on the nape (blotching absent in *A. maculiventris*). *Anolis apletophallus* differs from *A. maculiventris* in its solid yellow male dewlap. *Anolis fuscoauratus* and several high-elevation Andean species are very similar to *A. maculiventris* but unlikely to overlap *A. maculiventris* in range (see accounts and maps of *A. antonii, A. urraoi, A. mariarum, A. tolimensis, A. fuscoauratus*). Aside from geography, I know of no traits to distinguish *A. medemi* (Gorgona Island) and *A. maculiventris* (Pacific mainland Colombia and Ecuador). *Anolis urraoi* lives at higher elevations than *A. maculiventris* (1700–2250 meters; to 1800 meters in *A. maculiventris*) and has paler pink in the posterior aspect of the male dewlap.

Anolis magnaphallus

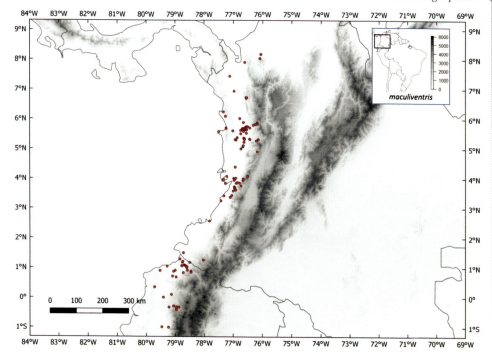

RANGE — From 0 to 1900 meters in Pacific Colombia and northern Ecuador.
NATURAL HISTORY — Very common in forest and disturbed areas; often the most abundant *Anolis* species in its area; uses a variety of microhabits and a range of vertical arboreal perches diurnally and nocturnally. See Fitch et al. (1976); Ayala et al. (1988); Castro-Herrera (1988); Moreno et al. (2007); Ríos et al. (2011); Carvajal-Cogollo and Urbina-Cardona (2015); Pinilla-Renteria et al. (2015); Narváez (2017); Viteri (2015); Rengifo et al. (2014, 2015); Rengifo et al. (2019, 2021); Pinto-Erazo et al. (2020); Moreno et al. (2020); Torres-Carvajal et al. (2019); Arteaga et al. (2023).
COMMENT — Poe et al.'s (2015b) suggestion that *Anolis maculiventris* may occur in Darién Panama was based on collection of fuscoauratid anoles near Yaviza with *maculiventris*-like dewlaps. I am uncertain of the proper taxonomic status of this population.

— *ANOLIS MAGNAPHALLUS* — PLATE 1.124, 2.49

DESCRIPTION — Poe, S., R. Ibañez. 2007. A new species of *Anolis* lizard from the Cordillera de Talamanca of western Panama. *Journal of Herpetology* 41:263–70.
TYPE SPECIMEN — MSB 72579.
TYPE LOCALITY — "Eastern entrance to Sendero Quetzales, 8 km North of Boquete, approximately 08°49.0 N, 82° 28.6 W, Chiriquí Province, Panama." The holotype and much of the paratype series were collected between the above GPS point (elevation 1670 m) and lat 8.82979 lon −82.47943, 1575 m, along the Bajo Mono loop road north of Boquete.
MORPHOLOGY — Body length to 55 M, 57 F; ventral scales keeled; middorsal scales greatly enlarged in several rows; interparietal larger than tympanum; scales across the snout between the second canthals 10–13; toe lamellae 13–16; suboculars and supralabials in contact or separated by rows of scales; scales separating supraorbital semicircles 2–4; supraocular scales gradually enlarged; hindlimbs long,

SPECIES ACCOUNTS

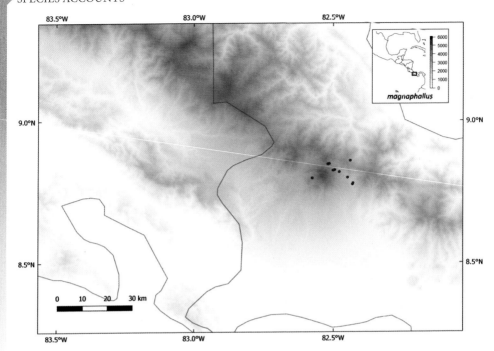

adpressed toe to between eye and tip of snout; body color mostly brown with some reddish accents, a dark line back from eye and a cream line under eye; dorsum sometimes with chevrons pointing posteriorly; a dark interorbital bar is usually present; tail banded; male dewlap red; female dewlap absent or white; males have a bulging hemipenial area.

SIMILAR SPECIES — *Anolis magnaphallus* is a *small brown* highland anole, distinguishable from potentially sympatric similar anoles by its long hindlimbs (adpressed toe to anterior to eye; posterior to eye in *A. salvini*, *A. kemptoni*, *A. pseudokemptoni*, *A. fungosus*, *A. gruuo*, *A. fortunensis*). *Anolis datzorum* differs from *A. magnaphallus* in its larger head scales (7–9 scales across the snout, 0–2 scales between the supraorbital semicircles; 10–13 scales across the snout, 2–4 scales between the supraorbital semicircles in *A. magnaphallus*). Other geographically proximal small brown anoles (e.g., *A. polylepis*, *A. limifrons*, *A. biscutiger*, *A. auratus*) are found at lower elevations than *A. magnaphallus*. *Anolis pseudopachypus*, *A. benedikti*, and *A. pachypus* are highland anoles that are very similar to *A. magnaphallus* and found east of, syntopic to, and west of the range of *A. magnaphallus*, respectively. These species differ from *A. magnaphallus* in male dewlap color (orange-red with yellow in *A. pachypus*, *A. benedikti*; dull yellow in *A. pseudopachypus*; dark red in *A. magnaphallus*).

RANGE — From 1450 to 2600 meters in the Cordillera de Talamanca of Panama near Volcán Baru.

NATURAL HISTORY — Common in forest, edge, and disturbed habitat; sleeps on low vegetation, often shrubs, usually less than 1 meter above ground. See Poe and Armijo (2014).

— *ANOLIS MAIA* — PLATE 1.125, 2.68

DESCRIPTION — Batista, A., M. Vesely, K. Mebert, S. Lotzkat, G. Köhler. 2015. A new species of *Dactyloa* from eastern Panama, with comments on other *Dactyloa* species present in the region. *Zootaxa* 4039:65–73.

TYPE SPECIMEN — SMF 97268.

Anolis maia

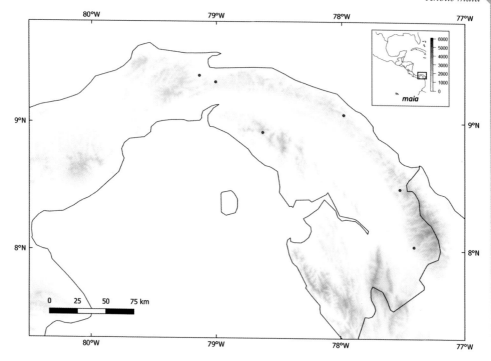

TYPE LOCALITY — "The ridge of the Serranía de Darién (fig. 1) along the trail that connects the Comarca Wargandí and the Comarca Guna Yala, about 10 km northeast of the village Nurra, 9.06142° N, 77.97961° W, 344 m asl., Corregimiento de Nurra, Comarca Wargandí, Panama."

MORPHOLOGY — Body length to 78 (sexual size dimorphism unknown); ventral scales smooth; mid-dorsal scales enlarged in 0–2 rows; interparietal larger or smaller than tympanum; scales across the snout between the second canthals 10–16; toe lamellae 17–22; suboculars and supralabials in contact; scales separating supraorbital semicircles 2–3; supraocular scales gradually enlarged; adpressed toe approximately to eye; body color green with lateral broad dark bands or bands broken into blotches; male dewlap orange with white distal edge; female dewlap dark orange.

SIMILAR SPECIES — *Anolis maia* is a *large green* anole. Among geographically proximal, comparably sized green anoles, *A. frenatus*, *A. latifrons*, and *A. brooksi* are larger than *A. maia* (>100 mm SVL; to 78 mm in *A. maia*) and have greater numbers of toe lamellae (usually 23–28; 17–22 in *A. maia*); *A. biporcatus* and *A. chloris* have keeled ventral scales (smooth in *A. maia*); *A. kunayalae* differs in its narrow fourth (largest) toe with large claw and fewer than 17 lamellae (broad, usual anoline toepad in *A. maia*). *Anolis purpurescens* and *A. ibanezi* are very similar to *A. maia* and replace it geographically to the east and west, respectively. These forms may differ from *A. maia* in their lateral body patterns (dense arrangement of ocelli and reticulations on green background in *A. purpurescens*; widely separated rows of blotches or thick bands in *A. maia*; widely separated narrow lines in *A. ibanezi*).

RANGE — From 300 to 900 meters in eastern Panama.

NATURAL HISTORY — Rare in forest. Sleeps on leaves or (less frequently) twigs at least 1 meter above ground.

SPECIES ACCOUNTS

— ANOLIS MARIARUM — PLATE 1.126, 2.91

DESCRIPTION — Barbour, T. 1932. New anoles. *Proceedings of the New England Zoological Club* 11:100–101.
TYPE SPECIMEN — MCZ 32303.
TYPE LOCALITY — "Sampedro, a village some 45 kilometers north of Medellín, Department of Antioquia, Colombia." I am unable to locate an appropriate "Sampedro" in Antioquia. The locality may refer to San Pedro de los Milagros (lat 6.46 lon −75.56, 2470 m), which is approximately 25 km straight north of Medellín, perhaps 35 km by road.
MORPHOLOGY — Body length to 52 M, 51 F; ventral scales smooth to weakly keeled; middorsal scales enlarged in 0–2 rows; interparietal larger or smaller than tympanum; scales across the snout between the second canthals 7–12; toe lamellae 14–18; suboculars and supralabials in contact; scales separating supraorbital semicircles 1–3; supraocular scales gradually enlarged; adpressed toe to between eye and tip of snout; body color brown; male dewlap red-orange anteriorly, yellow posteriorly, sometimes appearing overall red-orange; female dewlap absent.
SIMILAR SPECIES — *Anolis mariarum* is a *small brown* highland anole. Similar Colombian Andean anoles are best distinguished by male dewlap color and locality: The male dewlap of *A. urraoi* is pale pink posteriorly, orange anteriorly; *A. urraoi* is found in the northern Cordillera Occidental. The male dewlap of *A. tolimensis* is solid pinkish red; *A. tolimensis* is found in the Cordillera Central in upper regions of the southern Magdalena River valley. The male dewlap of *A. antonii* is red-orange to orange-red anteriorly, pale pink posteriorly; *A. antonii* is found in the southern Cordillera Occidental. The male dewlap of *A. fuscoauratus* usually is pale pink in Colombia; *A. fuscoauratus* is found on the Amazonian slope of the Cordillera Oriental and in the northern Magdalena River valley. The male dewlap of *A. mariarum* is red-orange anteriorly, yellow posteriorly; *A. mariarum* is found in the northern Cordillera Central.

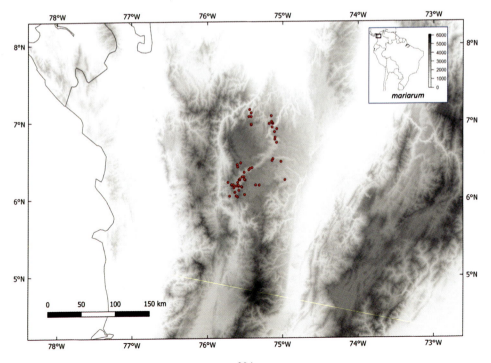

RANGE — From 1300 to 2700 meters in the northern Cordillera Central of the Colombian Andes.
NATURAL HISTORY — Common in forest and disturbed habitats; sleeps on a variety of substrates across a range of vertical perches. See Molina Zuluaga and Gutiérrez-Cárdenas (2007); Bock et al. (2009); Bock et al. (2010); Rubio-Rocha et al. (2011); Carvajal-Cogollo and Urbina-Cardona (2015).
COMMENT — Work remains to be done to establish the distributional limits of *Anolis mariarum* relative to *A. urraoi* to the west, *A. tolimensis* and *A. antonii* to the south, and *A. fuscoauratus* and *A. maculiventris* to the north. See Espitia Sanabria (2023) for issues of species boundary and geographic range in *A. mariarum*, and Comment under *A. antonii*.

— *ANOLIS MARMORATUS* — PLATE 1.127, 2.160

DESCRIPTION — Duméril, A.M.C., G. Bibron. 1837. *Erpétologie Générale ou Histoire Naturelle Complete des Reptiles*, vol. 4. Librairie Encyclopedique de Roret.
TYPE SPECIMEN — Lectotype MNHN 0794.
TYPE LOCALITY — "Martinique." Lazell (1972) designated "Capesterre, La Guadeloupe," which is located at lat 16.05 lon −61.60, as a new type locality.
MORPHOLOGY — Body length to 82 M, 57 F; ventral scales smooth; middorsal scales enlarged in 0–2 rows; interparietal equal to or larger than tympanum; scales across the snout between the second canthals 4–8; toe lamellae 20–30; suboculars and supralabials in contact; scales separating supraorbital semicircles 0–1; supraocular scales gradually enlarged; appressed toe to eye or just past eye; body color pattern is highly variable on its native Guadeloupe, possibly including greens, browns, orange, blue, spotting, and reticulations (Skip Lazell [1964: 373]) called this species "the most beautiful anole I have ever seen"). In my experience with *A. marmoratus* on Guadeloupe, green is the commonest resting (i.e., sleeping) background color. Male dewlap is yellow to orange, female dewlap absent.

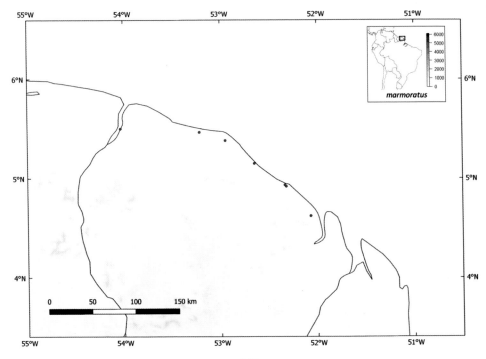

SPECIES ACCOUNTS

SIMILAR SPECIES — *Anolis marmoratus* is a *large green* anole. The only other green anoles in French Guiana or Suriname are *A. punctatus*, which is a forest anole unlikely to be found in the same disturbed areas inhabited by *A. marmoratus*, and *A. callainus*. *Anolis callainus* differs from *A. marmoratus* in its shorter hindlimbs (adpressed toe reaches approximately to ear; to eye in *A. marmoratus*) and bluish-gray and black male dewlap (yellow to orange in *A. marmoratus*).
RANGE — On the mainland, along the coast of French Guiana. Native to Guadeloupe, Lesser Antilles.
NATURAL HISTORY — Unstudied in French Guiana. See Schwartz and Henderson (1991) and references therein for ecology on Guadeloupe.

— *ANOLIS MARSUPIALIS* — PLATE 1.128

DESCRIPTION — Taylor, E. H. 1956. A review of the lizards of Costa Rica. *University of Kansas Science Bulletin* 38:97–100.
TYPE SPECIMEN — KU 40893.
TYPE LOCALITY — "About 15 km WSW of San Isidro del General along the Dominical Road." This marshy spot (lat 9.3135 lon −83.7724, 900 m) is a classic Costa Rican herping site, a type locality for four other species (Savage 1974).
MORPHOLOGY — Body length to 48 M, 45 F; ventral scales keeled; middorsal scales greatly enlarged in several rows; interparietal smaller than tympanum; scales across the snout between the second canthals 9–11; toe lamellae 11–14; suboculars and supralabials in contact or separated by a row of scales; scales separating supraorbital semicircles 2–4; supraocular scales gradually enlarged; adpressed toe to between eye and tip of snout; male dewlap dull red with dark red central streaks; puncturelike axillary pocket.
SIMILAR SPECIES — *Anolis marsupialis* is a small brown mainly terrestrial anole. Its combination of strongly keeled ventral scales, abruptly enlarged keeled middorsal scales, and a puncturelike axillary

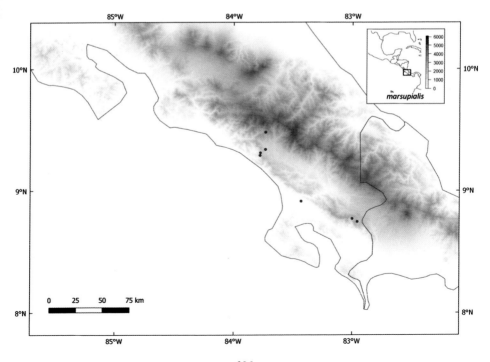

pocket will distinguish *A. marsupialis* from all potentially sympatric anoles. *Anolis humilis* is geographically proximal, parapatric, and shares these traits, but differs in male dewlap color (red with yellow distal edge in *A. humilis*; dull red with dark red central streaks in *A. marsupialis*).
RANGE — From 600 to 1000 meters in Pacific southeastern Costa Rica.
NATURAL HISTORY — Common in forest; active diurnally in leaf litter and low vegetation; sleeps on vegetation less than 1 meter above ground. See Köhler et al. (2015).

— *ANOLIS MATUDAI* — PLATE 1.129

DESCRIPTION — Smith, H. M. 1956. A new anole (Reptilia: Squamata) from Chiapas, Mexico. *Herpetologica* 12:1–2.
TYPE SPECIMEN — UIMNH 34199.
TYPE LOCALITY — "Region de Soconusco, Chiapas." The town of Soconusco (elevation 30 m) is too low in elevation to support *Anolis matudai*. Nieto Montes de Oca (1994a) suggested the holotype was collected "at some point intermediate among ... Finca Prusia [lat 15.690 lon −92.802], Monte Ovando [lat 15.412 lon −92.614], and Volcán Tacana [lat 15.13 lon −92.1]." (Parenthetical coordinates mine).
MORPHOLOGY — Body length to 51 M, 51 F; ventral scales smooth to faintly keeled; middorsal scales gradually enlarged in several rows; interparietal smaller than tympanum; scales across the snout between the second canthals 8–10; toe lamellae 15–17; suboculars and supralabials usually separated by a row of scales; scales separating supraorbital semicircles 2–3; supraocular scales gradually enlarged; adpressed toe to between eye and tip of snout; body color brown; male dewlap pink with light edge.
SIMILAR SPECIES — *Anolis matudai* is a *small brown* highland anole. Similar and potentially sympatric forms *A. crassulus*, *A. dollfusianus*, *A. laeviventris*, *A. serranoi*, and *A. sericeus* differ from *A. matudai*

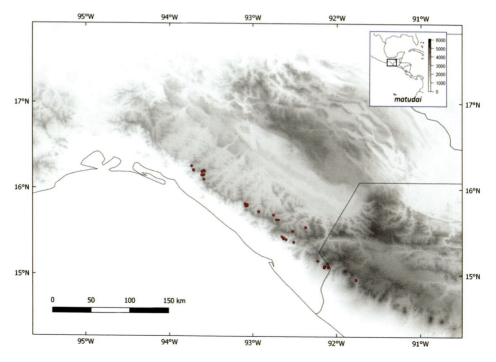

SPECIES ACCOUNTS

in possessing strongly keeled ventral scales (ventrals usually smooth in *A. matudai*). *Anolis cuprinus*, *A. parvicirculatus*, *A. hobartsmithi*, *A. campbelli*, and *A. cobanensis* are very similar highland species found to the west, northwest, north, northeast, and east of the range of *A. matudai*, respectively. *Anolis hobartsmithi*, *A. campbelli*, and *A. cobanensis* have an elongate anterior nasal scale (scales of nasal region undifferentiated in *A. matudai*), and *A. campbelli* and *A. cuprinus* have distinctly keeled ventral scales. *Anolis parvicirculatus* differs from *A. matudai* in its orange-red male dewlap (pink in *A. matudai*).

RANGE — From 1200 to 2150 meters on the Atlantic and Pacific slopes of Sierra Madre de Chiapas, Chiapas, Mexico.

NATURAL HISTORY — In forest; active diurnally on low vegetation or in leaf litter. See Nieto Montes de Oca (1994a).

— *ANOLIS MEDEMI* — PLATE 1.130

DESCRIPTION — Ayala, S. C., E. E. Williams. 1988. New or problematic *Anolis* from Colombia, VI: two fuscoauratid anoles from the Pacific lowlands, *A. maculiventris* Boulenger, 1898 and *A. medemi*, a new species from Gorgona Island. *Breviora* 490:1–16.

TYPE SPECIMEN — ICN 4371.

TYPE LOCALITY — "Isla Gorgona (2 59 N 76 12 W), La Esperanza, Cauca, Colombia."

MORPHOLOGY — Body length to 52 M, 50 F; ventral scales smooth; middorsal scales enlarged in 0–2 rows; interparietal equal to or smaller than tympanum; scales across the snout between the second canthals 9–18; toe lamellae 13–17; suboculars and supralabials in contact; scales separating supraorbital semicircles 2–4; supraocular scales gradually enlarged; hindlimbs long, adpressed toe to between eye and tip of snout; body color tan; male dewlap orange, slightly pinkish posteriorly; female dewlap absent.

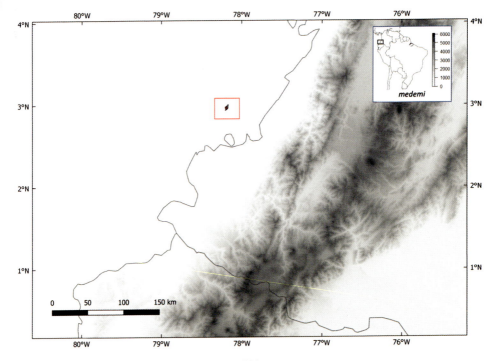

SIMILAR SPECIES — *Anolis medemi* is a *small brown anole*, the only small brown anole found on Gorgona Island. Other anole species on Gorgona are much larger and green (*A. princeps, A. purpurescens, A. parvauritus*) or blue (*A. gorgonae*).
RANGE — Throughout Gorgona Island (and see Comment below).
NATURAL HISTORY — Common on lower regions of tree trunks. See Castro-Herrera et al. (2012); Phillips et al. (2019).
COMMENT — Some authors (e.g., Grisales et al. 2017) have considered some Pacific Colombian lowland small brown anoles with smooth ventral scales to be *Anolis medemi*, whereas others (e.g., Phillips et al. 2019) consider *A. medemi* to be restricted to Gorgona Island. *Anolis medemi* originally was diagnosed from *A. maculiventris* based on subtleties of color pattern in a sample of 15 *A. maculiventris* and 23 *A. medemi*, including an "orange-brown" dorsum and a "well-defined pattern of darker brown bars and spots" on the body in *A. medemi* (Ayala and Williams 1988). I have collected individuals of *A. maculiventris* from near its type locality that fit this color pattern description of *A. medemi*. Because it seems likely that the population of small brown anoles on Gorgona, 55 km over water from the Colombian mainland, is evolving with some independence relative to the mainland population, I continue to recognize *A. medemi*, but consider its range restricted to Gorgona Island.

— *ANOLIS MEGALOPITHECUS* — PLATE 1.131, 2.121

DESCRIPTION — Rueda Almonacid, J.V.R. 1989. Un nuevo y extraordinario saurio de color rojo (Iguanidae: *Anolis*) para la cordillera Occidental de Colombia. *Trianea* 3:85–92.
TYPE SPECIMEN — ICN 6703.
TYPE LOCALITY — "Departamento de Antioquia, municipio de Frontino, Km 10 carretera entre la cabecera del corregimiento de Nutibara y Murri (=La Blanquita), ca 6 18 N, 76 17 W, alto de las

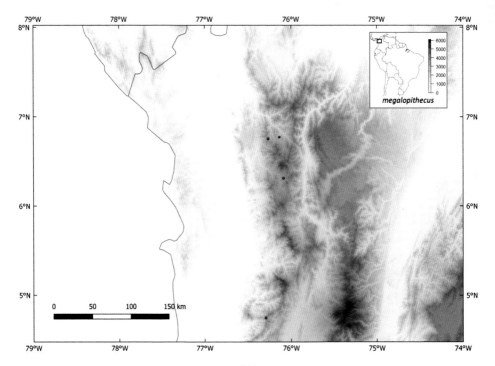

> SPECIES ACCOUNTS

Cuevas, vertiente occidental de la Cordillera Occidental de Colombia. 2000 m.s.n.m." Las Cuevas along the Nutibara-Murri road is at lat 6.752 lon −76.265.

MORPHOLOGY — Body length to 88 M, 78 F; ventral scales smooth; middorsal scales enlarged in 0–2 rows; interparietal usually absent; scales across the snout between the second canthals 13–21; toe lamellae 21–24; suboculars and supralabials separated by a row of scales; scales separating supraorbital semicircles 4–5; supraocular scales subequal to gradually enlarged; hindlimbs long, adpressed toe approximately to tip of snout; body color dark brown patterned with red, tan and green; male dewlap red to red-orange with black markings.

SIMILAR SPECIES — *Anolis megalopithecus* is a *large brown* highland anole. Its distinctive dorsal colors (red with dark brown) and male dewlap (red to red-orange with black) render males of this form unlikely to be confused with potentially sympatric species; females may be green, patterned like males, or otherwise highly variable (see below). Comparably sized, geographically proximal potentially highland anoles that share smooth ventral scales, narrow toepads, and long hindlimbs with *A. megalopithecus* include *A. antioquiae*, which is predominantly green and has an orange or white female dewlap (see Comment below); *A. eulaemus* and *A. anoriensis*, which lack red on the body and have brown male dewlaps; *A. ventrimaculatus*, which is smaller than *A. megalopithecus* [male to 80 mm, female to 62 mm; to 88 mm in *A. megalopithecus*], lacks red on the body, and has a dirty orange male dewlap; and semiaquatic *A. maculigula*, which has larger posterior head scales (usually 1–3 scales between supraorbital semicircles; usually 4–5 scales between supraorbital semicircles in *A. megalopithecus*) and a pale pink and blue male dewlap. *Anolis danieli* and lowland forms *A. frenatus* and *A. latifrons* are predominantly green forms with broad toepads and white male dewlaps. *Anolis purpurescens* is green with an orange male dewlap.

RANGE — From 1900 to 2350 meters on the Pacific Andean slope of central Colombia.

NATURAL HISTORY — Rare in cloud forest and cloud forest edge; sleeps on leaves, especially ferns, above 1 meter above ground.

COMMENT — *Anolis megalopithecus* may represent the male version of *A. antioquiae* (insight due to D. L. Mahler, personal communication 2021), which would render the name *megalopithecus* a junior synonym of *antioquiae*. Evidence for this inference: *Anolis antioquiae* was described from only females; *A. megalopithecus* was described without comparison to *A. antioquiae*; the two species are identical in scalation; close relatives of *A. antioquiae* and *A. megalopithecus* are highly variable in color pattern, especially in females; and my group has found green females identified as *A. antioquiae* sympatric with dark brown and red males identified as putative *A. megalopithecus*.

— *ANOLIS MEGAPHOLIDOTUS* — PLATE 1.132, 2.13

DESCRIPTION — Smith, H. M. 1933. Notes on some Mexican lizards of the genus *Anolis*, with the description of a new species, *A. megapholidotus*. *Transactions of the Kansas Academy of Science* 36:315–320.

TYPE SPECIMEN — FMNH 100105.

TYPE LOCALITY — "Between Rincon and Cajones (about 40–45 kilometers south of Chilpancingo), Guerrero, Mexico." This site is likely to be at approximately lat 17.32 lon −99.47, 950 m.

MORPHOLOGY — Body length to 39 M, 38 F; ventral scales keeled; middorsal scales greatly enlarged in several rows; interparietal equal to or larger than tympanum; scales across the snout between the second canthals 6–9; toe lamellae 12–17; suboculars and supralabials usually in contact, occasionally separated by one row of scales; scales separating supraorbital semicircles 0–2; supraocular scales abruptly enlarged; hindlimbs short, adpressed toe to ear, or posterior to ear; body color brown, sometimes with light lateral stripe; male dewlap pink.

SIMILAR SPECIES — *Anolis megapholidotus* is a (very) *small brown* anole. In the region where *A. megapholidotus* lives, its combination of keeled ventral scales, abruptly enlarged middorsal scale rows, and short hindlimbs (adpressed toe reaches to ear) is distinctive. Among geographically proximal species that might be confused with it, *A. microlepidotus*, *A. sericeus*, and *A. nebulosus* lack abruptly enlarged middorsal scale rows; *A. nebuloides*, found east of *A. megapholidotus*, has a yellow edge to its pink male dewlap (male dewlap solid pink in *A. megapholidotus*).

Anolis menta

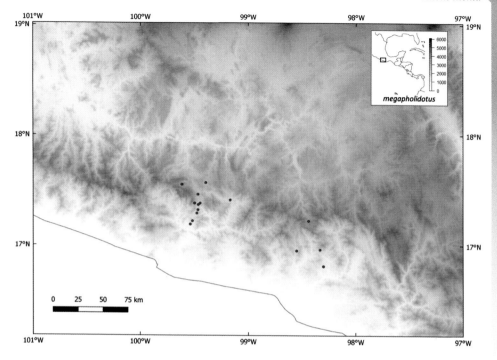

RANGE — From 550 to 1200 meters on the Pacific slope of central and southern Guerrero.
NATURAL HISTORY — In forest and edge habitat; active diurnally on or near ground; sleeps on twigs or leaves, usually below 2 meters above ground. See Fitch et al. (1976); Köhler et al. (2014a).
COMMENT — I consider *Anolis nietoi* (type locality: "Cascada Iliatenco (17.06753°N, 98.77796°W, WGS84), 1185 m, Estado de Guerrero, Mexico") to be a junior synonym of *A. megapholidotus* (see appendix 1 for justification). For those who wish to recognize *A. nietoi*, it was described as the version of *A. megapholidotus* found in the region of Iliatenco, Guerrero, Mexico. Köhler et al. (2014a) list a male individual of *A. nietoi* at 50 mm SVL, which is considerably larger than my known observations of other *A. megapholidotus*.

— *ANOLIS MENTA* — PLATE 1.133, 2.145

DESCRIPTION — Ayala, S. C., D. M. Harris, E. E. Williams. 1984. *Anolis menta*, sp. n. (Sauria, Iguanidae), a new *tigrinus* group anole from the west side of the Santa Marta mountains, Colombia. *Papéis Avulsos de Zoologia* 35:135–145.
TYPE SPECIMEN — MCZ 159013.
TYPE LOCALITY — "Cuchilla, Hierbabuena, 4 km southeast of San Pedro de la Sierra in the Sierra Nevada de Santa Marta Mountains, Magdalena Department, Colombia (10 53 N; 74 1 W)." Elevation at this locality is approximately 1700 m.
MORPHOLOGY — Body length to 56 M, 54 F; ventral scales smooth; middorsal scales enlarged in 0-2 rows; interparietal larger than tympanum; scales across the snout between the second canthals 5–6; toe lamellae 16–21; suboculars and supralabials in contact; supraorbital semicircles in contact; supraocular scales gradually or abruptly enlarged; hindlimbs short, adpressed toe to ear, or posterior to ear; body color brown to olive with broad dorsal bands in male, brown to olive with dorsal blotches

SPECIES ACCOUNTS

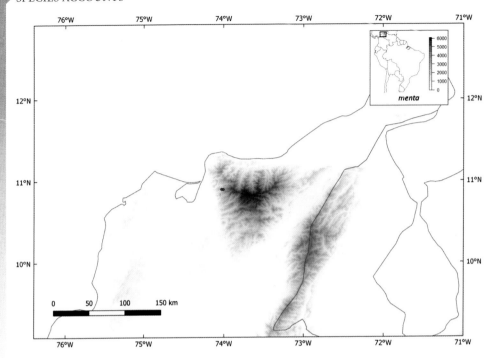

and lateral spots in females; tail banded; male dewlap white with yellow distally; female dewlap light brown with dark blotches.

SIMILAR SPECIES — *Anolis menta* is a *twig* or *small brown* or *green* highland anole. We have found this species to lack sympatric congeners (i.e., a "solitary" anole sensu Williams 1970) at the high-elevation localities it inhabits. It is geographically replaced by *A. solitarius* to the north, from which it differs by female dewlap (solid dark brown, lacking dark blotches in *A. solitarius*; light brown with dark blotches in *A. menta*), and *A. santamartae* to the south, from which it differs by male dewlap (all-white in *A. santamartae*; white with yellow distally in *A. menta*). Geographically proximal lowland species have keeled ventral scales (*A. gaigei, A. biporcatus, A. auratus*; ventrals smooth in *A. menta*).

RANGE — From 1650 to 2250 meters on the southwestern slope of the Santa Marta mountains in northern Colombia.

NATURAL HISTORY — Common in forest and open forest edge and disturbed areas; sleeps above 0.5 meters above ground on twigs, leaves, or (often) ferns. We have found this form to be patchily distributed, microlocally extremely abundant in some areas and absent in other proximal areas that do not appear to differ substantially in habitat.

— *ANOLIS MERIDIONALIS* — PLATE 1.134

DESCRIPTION — Boettger, O. 1885. Liste von Reptilien und Batrachiern aus Paraguay. *Zeitschrift für Naturwissenschaften* 58:215–216, 437. (Former pages refer to description of "*Anolis* (Draconura) *chrysolepis*," which Boettger later [p. 437] identifies as members of a new species according to Boulenger).

TYPE SPECIMEN — Lost (Barbour 1934). Neotype (Motte and Cacciali 2009) MNHNP 6608.

TYPE LOCALITY — Original type locality assumed to be Paraguay based on the title of Boettger's paper. Neotype locality is "Colonia Ybycui, Estancia Ybycui (aproximadamente 5 Km NE de la casa principal), Departamento Canindeyú ... 23° 43' S, 55° 30' O)" [Paraguay].

MORPHOLOGY — Body length to 58 M, 58 F; ventral scales keeled; middorsal scales greatly enlarged in several rows; interparietal usually equal to or larger than tympanum; scales across the snout between the second canthals 6–12; toe lamellae 11–16; suboculars and supralabials in contact or separated by a row of scales; scales separating supraorbital semicircles 0–3; supraocular scales gradually enlarged; adpressed toe to between ear and eye; body color brown, sometimes with lateral stripe and/or mid-dorsal broken bands; male dewlap greenish-blue.

SIMILAR SPECIES — *Anolis meridionalis* is a *small brown* lowland anole. Among geographically proximal species, *A. meridionalis* is unusual in its narrow indistinct toepad (toepad expanded and distinct in *A. fuscoauratus*, *A. ortonii*, *A. punctatus*, *A. brasiliensis*, *A. tandai*, *A. transversalis*). In addition, *A. fuscoauratus* and *A. ortonii* have smooth ventral scales (keeled in *A. meridionalis*). *Anolis punctatus* is large (SVL to 89 mm; to 58 mm in *A. meridionalis*) and green. *Anolis transversalis* is large (SVL to 79 mm) and has blue eyes (brown in *A. meridionalis*). *Anolis brasiliensis* and *A. tandai* differ from *A. meridionalis* in hindlimb length (adpressed toe reaches beyond snout; to between ear and eye in *A. meridionalis*). All of the above compared species tend to be forest forms, whereas *A. meridionalis* inhabits open areas. *Anolis auratus* is similar to *A. meridionalis* in morphology and preference for open grassy areas, but is found north of the range of *A. meridionalis* and differs in possessing fewer supralabial scales counted from the rostral to the center of the eye (*A. auratus*: 4–6; *A. meridionalis*: 6–9).

RANGE — From 100 to 1050 meters in southern Brazil, Bolivia, and Paraguay.

NATURAL HISTORY — In open areas; diurnally terrestrial but associated with vegetation and termite nests. See Smith et al. (1972); Vitt (1991); Vitt and Caldwell (1993); Colli et al. (2002); Nogueira et al. (2005, 2009); Langsroth (2006); Giraldelli (2007) ; Vaz-Silva et al. (2007); Veras and Santos (2007); Motte and Caciella (2009); Souza et al. (2010); Veludo (2011); Cassel et al. (2012); Costa et al. (2013); Gainsbury and Colli (2014); Araujo et al. (2014); Uetanabaro et al. (2017).

COMMENT — Work by Guarnizo et al. (2016) suggests multiple species exist within *Anolis meridionalis*.

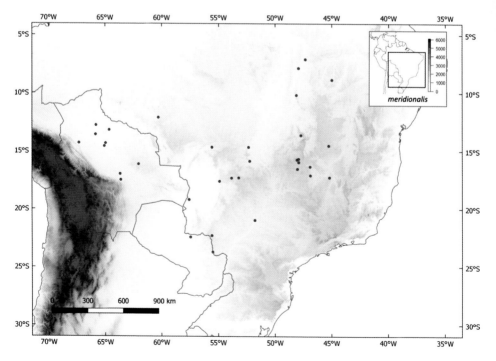

SPECIES ACCOUNTS

— *ANOLIS MICROLEPIDOTUS* — PLATE 1.135, 2.20

DESCRIPTION – Davis, W. B. 1954. Three new anoles from Mexico. *Herpetologica* 10:4–5.
TYPE SPECIMEN — TCWC 10276.
TYPE LOCALITY — "Four miles west of Chilpancingo, 5800 feet, Guerrero." I.e., approximately lat 17.543 lon −99.554.
MORPHOLOGY — Body length to 44 M, 48 F; ventral scales keeled; middorsal scales gradually enlarged in several rows; interparietal larger than tympanum; scales across the snout between the second canthals 4–7; toe lamellae 11–16; suboculars and supralabials in contact; supraorbital semicircles usually in contact; supraocular scales abruptly enlarged; hindlimbs short, adpressed toe to ear or posterior to ear; body color brown; male dewlap red and yellow.
SIMILAR SPECIES — *Anolis microlepidotus* is a *small brown* highland anole. In the Pacific Mexico area where *A. microlepidotus* is found, many small brown anole species have smooth ventral scales (keeled in *A. microlepidotus*). Geographically proximal species with keeled ventral scales differ from *A. microlepidotus* in male dewlap color: *A. nebuloides*: pink with yellow; *A. laeviventris*: white; *A. quercorum*, *A. nebuloides*: pink; *A. microlepidotus*: red with yellow). *Anolis megapholidotus* and *A. nebuloides* have abruptly enlarged middorsal scales (middorsals gradually, not abruptly enlarged in *A. microlepidotus*). *Anolis laeviventris* lacks abruptly enlarged supraocular scales (usually 3 supraocular scales abruptly enlarged in *A. microlepidotus*). *Anolis nebulosus* usually is found at lower elevations than *A. microlepidotus* and subtly differs from it in male dewlap color (usually solid yellow to orange, sometimes with darker striping/markings between scale rows, often with white edge; red with yellow blotching/striping that does not follow dewlap scale rows, edge not white in *A. microlepidotus*).
RANGE — From 1100 to 1900 meters in central to eastern Guerrero and southwestern Puebla States, Mexico.

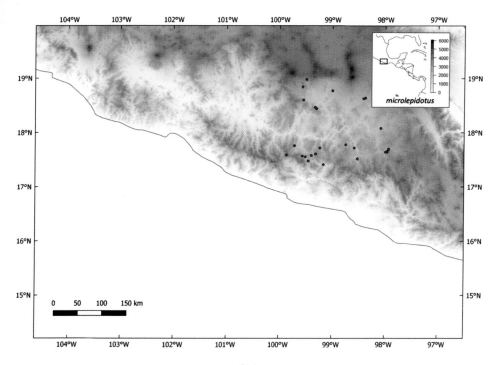

214

Anolis microtus

NATURAL HISTORY — Common in forest, edge, and disturbed habitats in both xeric and (less commonly) mesic areas; sleeps on twigs or leaves from 0.5 to 3 meters above ground. See Fitch et al. (1976); Köhler et al. (2014a).

— *ANOLIS MICROTUS* — PLATE 1.136, 2.75

DESCRIPTION — Cope, E. D. 1871. Ninth contribution to the herpetology of tropical America. *Proceedings of the Academy of Natural Sciences of Philadelphia* 23:214–215.
TYPE SPECIMEN — USNM 31282.
TYPE LOCALITY — "San José." Savage (1974) states that Cope's *Anolis* collections labeled San José "may have been taken from … the Atlantic slopes of the Cordillera Central between Alto La Palma and Carrillo or near La Palma on the Pacific slope." Savage and Talbot (1978) state the specimen is "probably from near La Palma." Savage's (1974) La Palma is at approximately lat 10.07 lon −84.00, 1500 m.
MORPHOLOGY — Body length to 120 M, 106 F; ventral scales smooth to weakly keeled; middorsal scales enlarged in 0–2 rows; interparietal usually smaller than tympanum; scales across the snout between the second canthals 5–9; toe lamellae 19–24; suboculars and supralabials in contact; scales separating supraorbital semicircles 2–3; supraocular scales gradually enlarged; adpressed toe to between ear and eye; body color variable, green with brown, banded; tail banded; male and female dewlaps pink.
SIMILAR SPECIES — *Anolis microtus* is a *giant* highland anole that differs from most other large geographically proximal species by its short hindlimbs (adpressed toe past eye in *A. frenatus*, *A. casildae*, *A. capito*; to between ear and eye in *A. microtus*). *Anolis biporcatus* has keeled ventral scales (usually smooth in *A. microtus*) and tends to be solid green in color (*A. microtus* may be green but usually is patterned). *Anolis kunayalae* has narrow toes with a large claw and few lamellae (<17; 19–24 in *A. microtus*). *Anolis brooksi*, *A. savagei*, and *A. kathydayae* have smaller head scales (usually at least

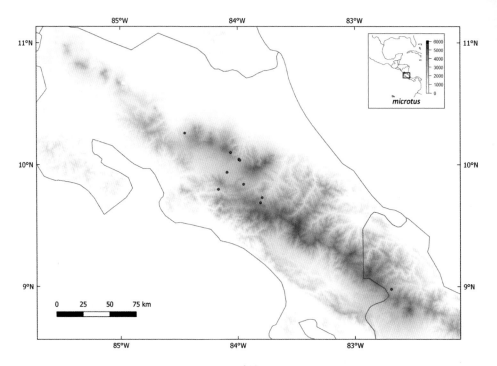

SPECIES ACCOUNTS

10 scales across the snout; 5–9 in *A. microtus*). *Anolis microtus* is replaced geographically to the east by similar species *A. ginaelisae*, which is reported to differ from *A. microtus* in its slightly longer hindlimbs (adpressed toe to cye).
RANGE — From 900 to 2100 meters in extreme western Panama and northern Costa Rica.
NATURAL HISTORY — In forest; sleeps on branches or leaves above 1.5 meters above ground. See Savage (2002); Lotzkat et al. (2013).
COMMENT — The apparent distributional gap between the northern populations and the Panama population of *A. microtus* (including *A. ginaelisae*, which is nearly identical to *A. microtus* as diagnosed by Lotzkat et al. 2013) suggests the potential for species status for these populations.

— *ANOLIS MILLERI* — PLATE 1.137, 2.7

DESCRIPTION — Smith, H. M., E. H. Taylor. 1950. An annotated checklist and key to the reptiles of Mexico exclusive of the snakes. *Bulletin of the US National Museum* 199:64. (The name is listed as "*Anolis milleri* Smith, new species"; i.e., the authors associate only Smith with the species name).
TYPE SPECIMEN — USNM 120957.
TYPE LOCALITY — "Quetzaltepec, Oaxaca." San Miguel Quetzaltepec, Oaxaca, Mexico, is at lat 16.975 lon −95.763 (1200 meters). According to Wilson and McCranie (1982:138) "The holotype came from the outskirts of Quetzaltepec in the region of Cerro Zempoaltepec."
MORPHOLOGY — Body length to 54 M, 50 F; ventral scales smooth; middorsal scales enlarged in 0–2 rows or gradually enlarged in several rows; interparietal usually smaller than tympanum; scales across the snout between the second canthals 8–10; toe lamellae 15–17; suboculars and supralabials in contact; scales separating supraorbital semicircles 1–3; supraocular scales gradually enlarged; adpressed toe to between eye and tip of snout; body color brown; male dewlap pink.

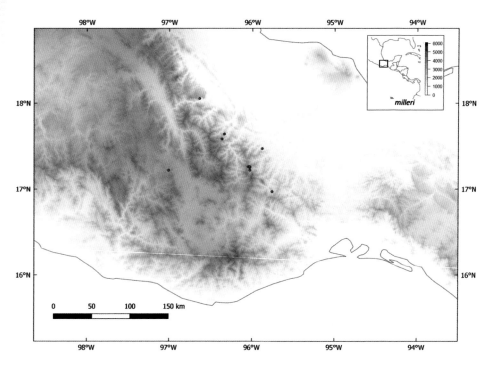

Anolis mirus

SIMILAR SPECIES — *Anolis milleri* is a *small brown* highland anole that is potentially sympatric with the following similar forms (Nieto Montes de Oca 1994a): *Anolis laeviventris, A. sericeus, A. tropidonotus,* and *A. uniformis* have keeled ventral scales (smooth in *A. milleri*); *A. rodriguezii* lacks enlarged mid-dorsal scale rows (*A. milleri* has approximately 4 rows of gradually enlarged middorsal scales). *Anolis rubiginosus* is very similar to *A. milleri* and found parapatrically southwest of its range. The dewlap of *A. rubiginosus* is pink but paler than the pink dewlap of *A. milleri*, appearing orange distally. *Anolis schiedii* and *A. cymbops* are similar species found northwest of the range of *A. milleri*. *Anolis schiedii* can be distinguished from *A. milleri* by male dewlap color (orange-red) and having a greater number of scale rows on the male dewlap (~8, vs. ~5 in *A. milleri*). *Anolis cymbops* can be distinguished from *A. milleri* by its greater number of scale rows on the male dewlap (~7, vs. ~5 in *A. milleri*) and greater number of scales per row at the mid-dewlap level (~18, vs. ~24 in *A. milleri*). Geographically proximal forms *Anolis quercorum, A. nebuloides,* and *A. megapholidotus* have keeled ventral scales.
RANGE — From 400 to 1900 meters in northern Oaxaca, Mexico.
NATURAL HISTORY — Common in forest and edge habitat; active diurnally in leaf litter; sleeps on twigs or leaves up to 2 meters above ground. See Nieto Montes de Oca (1994a). According to Nieto Montes de Oca (1994a), Campbell's (1989) studied specimens of "*Anolis milleri*" are *A. rubiginosus*.

— *ANOLIS MIRUS* — PLATE 1.138, 2.115

DESCRIPTION — Williams, E. E. 1963. Studies of South American anoles. Description of *Anolis mirus*, new species, from Rio San Juan, Colombia, with comment on digital dilation and dewlap as generic and specific characters in anoles. *Bulletin of the Museum of Comparative Zoology* 129:463–480.
TYPE SPECIMEN — BMNH 1910.7.11.5.
TYPE LOCALITY — "Rio San Juan, S. W. Colombia." The San Juan is a major river in Chocó that flows

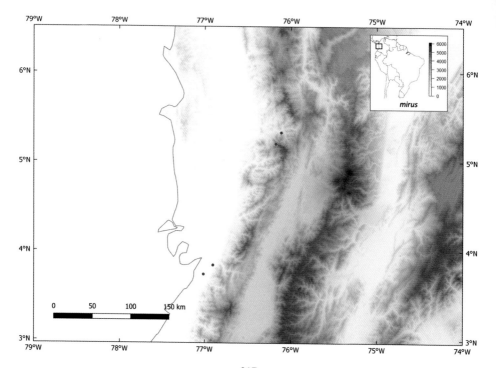

SPECIES ACCOUNTS

380 km from the western Andes to the Pacific about 50 km northwest of Buenaventura. Williams and Duellman (1984) implied that the "Rio San Juan" locality of M. G. Palmer, the collector of the *Anolis mirus* type specimen, is Pueblo Rico in Risaralda Department (lat 5.22 lon −76.03, 1550 m), a town on the edge of Tatama national park, an area where I have found *A. mirus*.

MORPHOLOGY — Body length to 116 M, 94 F; ventral scales weakly keeled; middorsal scales enlarged in 0–2 rows; interparietal equal to or smaller than tympanum; scales across the snout between the second canthals 10–13; toe lamellae 12–15; suboculars and supralabials in contact or separated by a row of scales; scales separating supraorbital semicircles 4–5; supraocular scales subequal to gradually enlarged; weakly expanded toepads, clawed (ultimate) phalange not distinct from expanded scales of toepad, i.e., the toepad is not "raised" (Williams 1963); greatly enlarged claw on fourth (longest) toe; adpressed toe to ear or just beyond; body green, patterned with light and dark blotches, sometimes with lateral white bars; female dewlap dark yellow, greenish proximally; male dewlap white with faint blotching and yellow scales; iris reddish-brown.

SIMILAR SPECIES — *Anolis mirus* is a *giant* anole, distinguishable from all potentially sympatric similar anoles by its unusual morphology of the fourth (longest) toe, including very few expanded toe lamellae (at least 20 in all comparably sized species; 12–15 in *A. mirus*), narrow toepad not discrete from distal toe scales, and large claw. *Anolis parilis*, found south of the range of *A. mirus*, shares the unusual toe morphology of *A. mirus* but has smaller head scales (14–18 scales across the snout; 10–13 in *A. parilis*) and smooth ventral scales (weakly keeled in *A. mirus*).

RANGE — From 0 to 1550 meters in Pacific Colombia.

NATURAL HISTORY — The two individuals of this form I have found were sleeping on twigs, one at 4 meters above ground in edge habitat and the other at 3 meters above ground within forest.

— *ANOLIS MONTEVERDE* — PLATE 1.139, 2.37

DESCRIPTION — Köhler, G. 2009. New species of *Anolis* formerly referred to as *Anolis altae* from Monteverde, Costa Rica (Squamata: Polychrotidae). *Journal of Herpetology* 43:11–20.

TYPE SPECIMEN — SMF 86920.

TYPE LOCALITY — "2 km east of Santa Elena (10° 20' 30.5" N, 84° 48' 16.0" W), 1,550 m, Puntarenas Province, Costa Rica."

MORPHOLOGY — Body length to 46 M, 50 F; ventral scales weakly keeled; middorsal scales enlarged in 0–2 rows; interparietal equal to or larger than tympanum; scales across the snout between the second canthals 6–9; toe lamellae 14–15; suboculars and supralabials in contact; scales separating supraorbital semicircles 1–3; supraocular scales gradually enlarged; adpressed toe approximately to ear; body color brown; male dewlap orange.

SIMILAR SPECIES — *Anolis monteverde* is a *small brown* highland anole that may be distinguished from potentially sympatric similar forms as follows: *A. humilis*, *A. laeviventris*, *and A. sericeus* have strongly keeled ventral scales (weakly keeled in *A. monteverde*); *A. pachypus*, *A. tropidolepis*, *A. cupreus*, *A. limifrons*, *A. biscutiger*, and *A. polylepis* have longer hindlimbs (adpressed toe reaches to beyond eye; to ear in *A. monteverde*); *A. kemptoni* has smooth ventral scales (weakly keeled in *A. monteverde*) and a bicolor orange and pink male dewlap (orange in *A. monteverde*). *Anolis tenorioensis* and *A. altae* are very similar to *A. monteverde* but have been found only around Volcán Tenorio and south of Monteverde, respectively, allopatric to *A. monteverde*. *Anolis tenorioensis* differs from *A. monteverde* in its darkly spotted male dewlap; *A. altae* is reported to differ in its bilobed hemipenes (unilobed in *A. monteverde*) and slightly longer tail (Köhler 2009). *Anolis arenal* is similar to *A. monteverde* but is a lowland form (known only from below 600 m) with a differently colored male dewlap (dark centrally, orange-red distally).

RANGE — 1500 to 1600 meters, known only from at and near the type locality east of Santa Elena in the Monteverde area, Costa Rica.

NATURAL HISTORY — Formerly common, apparently now rare (Pounds et al. 1999) in forest, edge, and disturbed areas; active diurnally and sleeps on low vegetation. See Pounds (1988; as *Anolis altae*); Pounds and Fogden (2000; as *Anolis altae*).

Anolis morazani

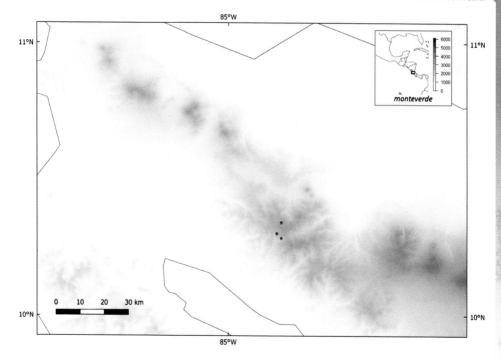

COMMENT — *Anolis monteverde* essentially constitutes the northernmost population of *A. altae*. The stated differences between these forms are modal, their male dewlaps are identical (compare fig. 2E and fig. 2F in Köhler [2011a]), and their genetic divergence is compatible with conspecificity (e.g., 1.6% in the mitochondrial gene 16S; see table 1 in Köhler [2011]). The status of *A. monteverde* relative to *A. altae* should be investigated.

— *ANOLIS MORAZANI* — PLATE 1.140

DESCRIPTION — Townsend, J. H., L. D. Wilson. 2009. New species of cloud forest *Anolis* (Squamata: Polychrotidae) in the *crassulus* group from Parque Nacional Montaña de Yoro, Honduras. *Copeia* 2009:62–70.
TYPE SPECIMEN — SMF 87153.
TYPE LOCALITY — "Honduras, Departamento de Francisco Morazán, Municipio de Marale, Parque Nacional Montaña de Yoro, Cataguana, 15° 01' N, 87° 06' W, 1910 m elevation."
MORPHOLOGY — Body length to 52 M, 59 F; ventral scales keeled; middorsal scales greatly enlarged in several rows; interparietal larger than tympanum; scales across the snout between the second canthals 4–8; toe lamellae about 15; suboculars and supralabials in contact; scales separating supraorbital semicircles 0–2; supraocular scales gradually enlarged; adpressed toe approximately to eye; body color brown; male dewlap orange-red with white scales.
SIMILAR SPECIES — *Anolis morazani* is a *small brown* anole found at high elevations in Yoro National Park, Honduras. Among geographically proximal small brown highland anoles, *A. yoroensis*, *A. laeviventris*, *A. cusuco*, and *A. kreutzi* differ from *A. morazani* in their lack of abruptly enlarged middorsal scale rows (present in *A. morazani*). *Anolis caceresae* is similar to *A. morazani* but found well outside of its range, at high elevations in southern Honduras. *Anolis muralla*, found just east of the range of

SPECIES ACCOUNTS

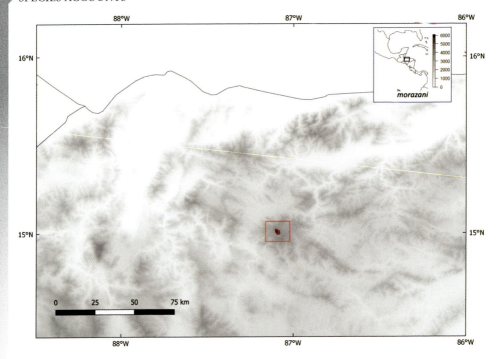

A. *morazani* in Muralla Reserve, is similar to A. *morazani* but differs in possessing smooth ventral scales (keeled in A. *morazani*). *Anolis rubribarbaris*, found just west of the range of A. *morazani* in Montaña de Yoro National Park, apparently differs from A. *morazani* in hemipenial structure (asulcate process of hemipenes divided in A. *morazani*, undivided in A. *rubribarbaris*; McCranie and Köhler 2015). Small brown anoles found in the general region of A. *morazani* but at lower elevations generally lack the abruptly enlarged middorsal scale rows of A. *morazani* (A. *sericeus*, A. *ustus*, A. *zeus*, A. *limifrons*) or have a puncturelike axillary pocket (A. *tropidonotus*, A. *uniformis*; absent in A. *morazani*).

RANGE — From 1250 to 2150 meters in the area of Montaña de Yoro National Park in central Honduras.

NATURAL HISTORY — Common in forest and edge habitat; sleeps on low vegetation. See McCranie and Köhler (2015).

— *ANOLIS MURALLA* — PLATE 1.141

DESCRIPTION — Köhler, G., J. R. McCranie, L. D. Wilson. 1999. Two new species of anoles of the *Norops crassulus* group from Honduras (Reptilia: Sauria: Polychrotidae). *Amphibia-Reptilia* 20:285–296.

TYPE SPECIMEN — SMF 78093.

TYPE LOCALITY — "Honduras, Departamento de Olancho, Parque Nacional La Muralla, along trail to Cerro de Enmedio, 1500 m." The USNM catalog georeferences this point to lat 15.08 lon −86.73.

MORPHOLOGY — Body length to 48 M, 57 F; ventral scales smooth; middorsal scales greatly enlarged in several rows; interparietal equal to or smaller than tympanum; scales across the snout between the second canthals 5–7; toe lamellae 13–17; suboculars and supralabials in contact; scales between

Anolis nasofrontalis

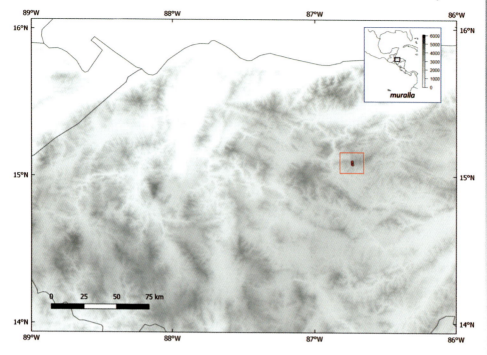

supraorbital semicircles 0–2; supraocular scales gradually enlarged; adpressed toe to between ear and eye; body color brown; male dewlap red; iris "metallic green."
SIMILAR SPECIES — *Anolis muralla* is a *small brown* highland anole found in Muralla Refuge, Honduras. Among potentially sympatric geographically proximal small brown highland anoles, *Anolis yoroensis*, *A. laeviventris*, and *A. kreutzi* differ from *A. muralla* in their keeled ventral scales (smooth in *A. muralla*). *Anolis morazani*, found just west of the range of *A. muralla* in Yoro National Park, is similar to *A. morazani* but differs in possessing keeled ventral scales. Small brown anoles found in the general region of *A. muralla* but at lower elevations generally lack the abruptly enlarged middorsal scale rows of *A. muralla* (*A. sericeus*, *A. zeus*, *A. limifrons*, *A. rodriguezii*, *A. lemurinus*, *A. sagrei*) or have a puncturelike axillary pocket (absent in *A. muralla*) and keeled ventral scales (*A. tropidonotus*, *A. uniformis*).
RANGE — From 1400 to 1750 meters in Muralla Refuge, north-central Honduras.
NATURAL HISTORY — In forest; active diurnally on low vegetation and ground; sleeps on low vegetation. See McCranie and Köhler (2015).

— *ANOLIS NASOFRONTALIS* — PLATE 1.142

DESCRIPTION — Amaral, A. D. 1933. Estudios sobre lacertilios neotropicos. I. Novos gêneros e espécies de lagartos do Brasil. *Memorias do Instituto Butantan* 7:58–59.
TYPE SPECIMEN — MZUSP 440.
TYPE LOCALITY — "Estado do Espírito Santo, Brasil." Ernesto Garbe, collector of the type specimen of *Anolis nasofrontalis*, is known to have collected along the Rio Doce in 1906, the listed year of collection for the type specimen. However, at least some of Garbe's listings of his Rio Doce collections explicitly referenced that river (see, e.g., *A. pseudotigrinus*). Thus, Garbe's listed type locality for

SPECIES ACCOUNTS

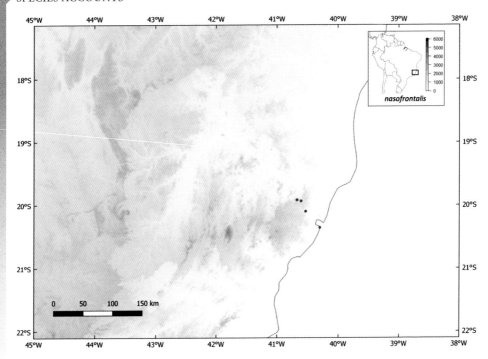

A. nasofrontalis may refer to Vila Velha (lat −20.36 lon −40.30), the coastal town located in Espírito Santo State that was itself formerly known as Vila do Espírito Santo.

MORPHOLOGY — Body length to 38 M (n = 1), 55 F; ventral scales smooth; middorsal scales enlarged in 0–2 rows; interparietal larger than tympanum; scales across the snout between the second canthals 4–7; toe lamellae 22–25; subocular and supralabial scales in contact; supraorbital semicircles in contact; supraocular scales abruptly or gradually enlarged; body color lichenous, white to tan with green, yellow, black, and brown lines and spots; male dewlap red or pink; female dewlap tan, pale posteriorly.

SIMILAR SPECIES — *Anolis nasofrontalis* is a *twig anole*. Prates et al. (2017) state this species is sympatric with *A. punctatus*, a large green species with which it is unlikely to be confused, and *A. fuscoauratus* and *A. pseudotigrinus*, from which it differs amply. *Anolis fuscoauratus* differs from *A. nasofrontalis* in its smaller head scales (1–3 scales between the supraorbital semicircles; supraorbital semicircles in broad contact in *A. nasofrontalis*). *Anolis pseudotigrinus* differs from *A. nasofrontalis* in female dewlap color (white; tan, pale posteriorly in *A. nasofrontalis*) and male dewlap scalation (12–14 rows of scales; 5–7 in *A. nasofrontalis*). *Anolis neglectus*, found south of the range of *A. nasofrontalis*, is very similar to *A. nasofrontalis* but differs in female dewlap color (yellow-orange). *Anolis ortonii* may be sympatric with *A. nasofrontalis*, and differs in its smaller head scales (7–11 scales across the snout between the second canthals; 4–7 in *A. nasofrontalis*). *Anolis brasiliensis*, *A. sagrei*, and *A. meridionalis* differ from *A. nasofrontalis* in their keeled ventral scales (smooth in *A. nasofrontalis*).

RANGE — From 50 to 750 meters in coastal Brazil, Espíritu Santo.

NATURAL HISTORY — One specimen was found sleeping 3 meters above ground on a narrow branch. See Prates et al. (2017).

— ANOLIS NAUFRAGUS — PLATE 1.143, 2.11

DESCRIPTION — Campbell, J. A., D. M. Hillis, W. W. Lamar. 1989. A new lizard of the genus *Norops* (Sauria: Iguanidae) from the cloud forest of Hidalgo, Mexico. *Herpetologica* 45:232–242.
TYPE SPECIMEN — UTA 11514.
TYPE LOCALITY — "10.1 km NE Tlanchinol, 1237 m, Hidalgo, Municipio de Tlanchinol, Mexico." This locality is inferred to be at approximately lat 21.02 lon −98.61.
MORPHOLOGY — Body length to 50 M, 53 F; ventral scales smooth to weakly keeled; middorsal scales enlarged in 0–2 rows; interparietal smaller than tympanum; scales across the snout between the second canthals 7–10; toe lamellae approximately 15; suboculars and supralabials usually separated by a row of scales; scales separating supraorbital semicircles 1–2; supraocular scales gradually enlarged; hindlimbs long, adpressed toe approximately to tip of snout; body color brown; male dewlap orange-red.
SIMILAR SPECIES — *Anolis naufragus* is a *small brown* highland anole. Nieto Montes de Oca (1994a) lists *A. laeviventris* and *A. sericeus* as sympatric with *A. naufragus*. These species differ from *A. naufragus* in their short hindlimbs (adpressed toe usually reaches to ear; to tip of snout in *A. naufragus*). *Anolis schiedii* is very similar to *A. naufragus* and found just southeast of its range. *Anolis schiedii* is reported to differ from *A. naufragus* in its multicarinate dorsal scales on the limbs (dorsal limb scales unicarinate in *A. naufragus*; Nieto Montes de Oca 1994a; also see below).
RANGE — From 400 to 2200 meters on the Atlantic slope of the Sierra Madre Oriental from Hidalgo to northern Puebla States, Mexico.
NATURAL HISTORY — Rare in cloud forest and edge habitat; active diurnally on ground or low vegetation, or in rock crevices; sleeps on low vegetation. See Nieto Montes de Oca (1994a); Canseco-Márquez et al. (2006); Ramirez-Bautista and Cruz-Elizalde (2013).
COMMENT — The status of this name should be investigated with reference to *Anolis schiedii*. *Anolis naufragus* was described before *A. schiedii* was "rediscovered," and these species share similar ranges, morphology, and dewlap colors (see Nieto Montes de Oca [1994b] for additional discussion).

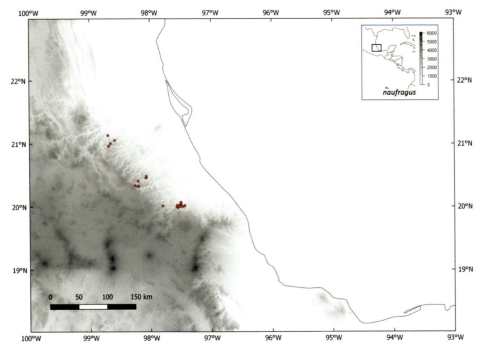

SPECIES ACCOUNTS

— *ANOLIS NEBLININUS* — PLATE 1.144, 2.141

DESCRIPTION — Myers, C. M., E. E. Williams, R. W. McDiarmid. 1993. A new anoline lizard (*Phenacosaurus*) from the highland of Cerro de la Neblina, southern Venezuela. *American Museum Novitates* 3070:1–15.

TYPE SPECIMEN — AMNH 129241.

TYPE LOCALITY — "Camp 7, 1850 m, Cerro de la Neblina, Amazonas, Venezuela (0 50' 40" N, 65 58' 10" W)."

MORPHOLOGY — Body length to 64 M, 57 F; ventral scales smooth; middorsal scales form a serrated crest, larger nuchally than on dorsum and larger in males than females; interparietal larger than tympanum; scales across the snout between the second canthals 4–6; toe lamellae 17–23; suboculars and supralabials in contact; supraorbital semicircles in contact; supraocular scales abruptly or gradually enlarged; hindlimbs short, adpressed toe to posterior to ear; body color greenish brown, lichenous; male dewlap bluish-gray proximally, pale yellow distally, or solid whitish; female dewlap pale gray, orange, or brown, with black spots.

SIMILAR SPECIES — *Anolis neblininus* is a highland *twig* anole. Geographically proximal anoles are unlikely to be found at the high elevations where *A. neblininus* lives. Among proximal lowland species, *A. auratus*, *A. planiceps*, *A. bombiceps*, and *A. scypheus* differ from *A. neblininus* in their keeled ventral scales (smooth in *A. neblininus*); *Anolis fuscoauratus*, *A. ortonii*, and *A. punctatus* have smaller head scales than *A. neblininus* (at least 7, usually 9 or more scales across the snout at the second canthals; 4–6 in *A. neblininus*).

RANGE — From 1650 to 2100 meters in Cerro de Neblina, southern Venezuela, and adjacent northern Brazil.

NATURAL HISTORY — In open areas with shrubs; active diurnally on leaves; sleeps from 1 to 4 meters up on vegetation, often ferns. See Prates (2017).

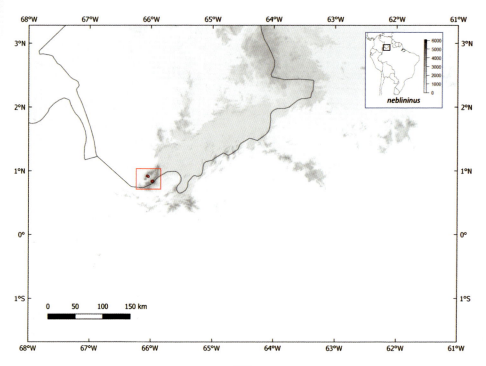

— ANOLIS NEBULOIDES — PLATE 1.145, 2.16

DESCRIPTION — Bocourt, M. F. 1873. Etudes sur les reptiles. In: Duméril, M. A., M. F. Bocourt, M. Mocquard, eds., *Mission Scientifique au Mexique et dans l'Amérique Centrale*, livr. 3, 74–75. Paris: Imprimerie impériale.
TYPE SPECIMEN — Syntypes MNHN 2494, 1994.0984-86.
TYPE LOCALITY — "Putla, province d'Oaxaca (Mexique)." Putla, Oaxaca, is at approximately lat 17.02 lon −97.93, 730 m.
MORPHOLOGY — Body length to 51 M, 42 F; ventral scales keeled; middorsal scales greatly enlarged in several rows; interparietal larger or smaller than tympanum; scales across the snout between the second canthals 6–9; toe lamellae 13–16; suboculars and supralabials in contact; scales separating supraorbital semicircles 0–1; supraocular scales abruptly enlarged; adpressed toe to ear or between ear and eye; body color brown with light lateral stripe; male dewlap purple-pink or red-pink with yellow edge.
SIMILAR SPECIES — *Anolis nebuloides* is a *small brown* anole. Its combination of keeled ventral scales and 10–12 rows of abruptly enlarged keeled middorsal scales will distinguish *A. nebuloides* from most geographically proximal similar species. Among nearby small brown anoles with keeled ventral scales, *A. boulengerianus*, *A. microlepidotus*, *A. nebulosus*, *A. immaculogularis*, *A. quercorum*, and *A. subocularis* differ from *A. nebuloides* in their smaller dorsal scales (middorsal scales smaller than ventral scales in these species; middorsals larger than ventrals in *A. nebuloides*). *A. megapholidotus*, found just west of the range of *A. nebuloides*, is similar to *A. nebuloides* but lacks a yellow edge to its pink male dewlap.
RANGE — From 350 to 2050 meters on the Pacific slope of western and central Oaxaca and extreme eastern Guerrero, Mexico.
NATURAL HISTORY — Common in forest and edge habitat; active diurnally low on tree trunks and ground; sleeps on twigs or leaves up to 2.5 meters above ground. See Köhler et al. (2014a).

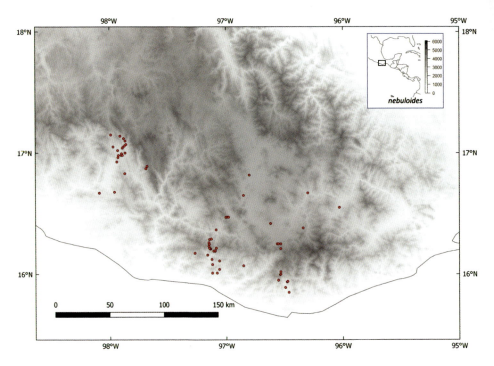

SPECIES ACCOUNTS

COMMENT — I consider *Anolis stevepoei* (type locality: "about 15.8 airline km NNW San Gabriel Mixtepec on road to El Vidrio (16.2205°N, 97.14883°W), 1900 m, Estado de Oaxaca, Mexico") and *A. zapotecorum* (type locality: "Pluma Hidalgo (15.942410°N, 96.430440°W), 1350 m, Estado de Oaxaca, Mexico") to be junior synonyms of *A. nebuloides* (see appendix 1 for justification). For those who wish to recognize these forms, *A. zapotecorum* composes the eastern end of the distribution of *A. nebuloides*, near Pluma Hidalgo in Oaxaca State. *Anolis stevepoei* is reported to range from just west of there, north of Puerto Escondido in Oaxaca.

— *ANOLIS NEBULOSUS* — PLATE 1.146, 2.19

DESCRIPTION — Wiegmann, A.F.A., et al. 1834. *Herpetologia Mexicana, seu Descriptio Amphibiorum Novae Hispaniae, quae Itineribus comitis de Sack, Ferdinandi Deppe et Chr. Guil. Schiede in Museum Zoologicum Berolinense Pervenerunt. Pars Prima, Saurorum Species Amplectens. Adiecto Systematis Saurorum Prodromo, Additisque multis in hunc Amphibiorum Ordinem Observationibus*, 47. Berlin: Lüderitz.
TYPE SPECIMEN — ZMB 527.
TYPE LOCALITY — "Mexico." Restricted to Mazatlan, Sinaloa, Mexico (lat 23.25 lon −106.42, 10 m) by Smith and Taylor (1950).
MORPHOLOGY — Body length to 47 M, 44 F; ventral scales keeled; middorsal scales gradually enlarged in several rows; interparietal larger than tympanum; scales across the snout between the second canthals 5–7; toe lamellae 11–15; suboculars and supralabials in contact; supraorbital semicircles in contact; supraocular scales abruptly enlarged; hindlimbs short, appressed toe to ear; body color grayish-brown; male dewlap yellow-orange to red-orange, sometimes with white edge distally (variable by population), sometimes with darker striping/blotching between scale rows; female dewlap absent.

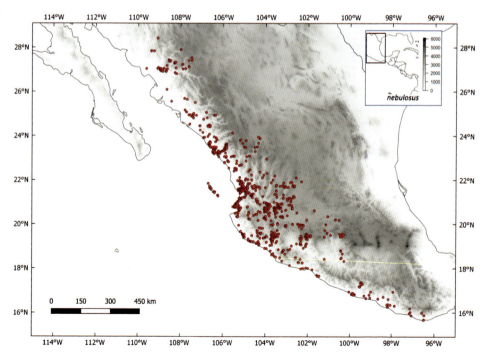

SIMILAR SPECIES — *Anolis nebulosus* is a *small brown* anole. Its combination of keeled ventral scales and short hindlimbs (adpressed toe reaches to ear) will separate it from all similar geographically proximal species except *A. microlepidotus*, *A. sericeus*, *A. megapholidotus*, and *A. nebuloides*. *Anolis sericeus* differs from *A. nebulosus* in its lack of abruptly enlarged supraocular scales (usually three abruptly enlarged supraocular scales in *A. nebulosus*); *A. megapholidotus* and *A. nebuloides* both display large, abruptly enlarged rows of middorsal scales (small, gradually enlarged middorsal scale rows in *A. nebulosus*). *Anolis microlepidotus* is similar to *A. nebulosus* but is not found at low elevations and differs in male dewlap color pattern (red with jumbled yellow blotching/striping; yellow-orange to red-orange, sometimes with darker striping/blotching between scale rows, often with white edge in *A. nebulosus*)

RANGE — From 0 to 2100 meters on the Pacific slope of Mexico from Sonora to Oaxaca.

NATURAL HISTORY — Very common in forest, edge, and disturbed areas; active diurnally on low vegetation or ground; sleeps on twigs or leaves up to 2 meters above ground, or in leaf litter (Jenssen 1970). See Jenssen (1969, 1970a, b; 1971); Lister and Aguayo (1992); Ramírez-Bautista and Vitt (1997); Ramírez-Bautista and Benabib (2001); Ramírez-Bautista (2002); García (2008); Boyd et al. (2007); Köhler et al. (2014a); Hernández-Salinas et al. (2014); Siliceo-Cantero and García (2014, 2015); Woolrich-Piña et al. (2015a, b); Hernández-Salinas et al. (2016); Siliceo-Cantero et al. (2016, 2017); Ramírez-Bautista et al. (2017); Hernández-Salinas et al. (2019).

COMMENT — The extraordinary recorded elevational (and latitudinal) range of *Anolis nebulosus* warrants taxonomic and physiological studies.

— *ANOLIS NEGLECTUS* — PLATE 1.147, 2.140

DESCRIPTION — Prates, I., P. R. Melo-Sampaio, K. de Queiroz, A. C. Carnaval, M. T. Rodrigues, L. O. Drummond. 2019. Discovery of a new species of *Anolis* lizard from Brazil and its implications

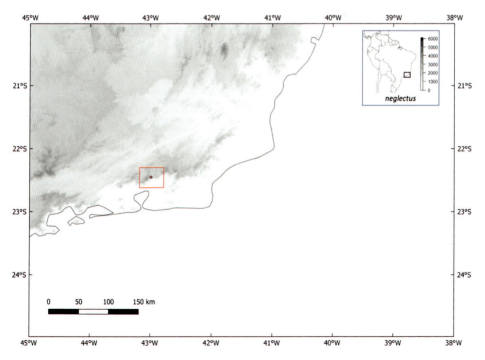

SPECIES ACCOUNTS

for the historical biogeography of montane Atlantic Forest endemics. *Amphibia-Reptilia.* https://doi.org/10.1163/15685381-20191179.
TYPE SPECIMEN — MNRJ 26927.
TYPE LOCALITY — "Parque Nacional da Serra dos Órgãos (22°26 55.9 S, 42°59 09.9 W; 981 m above sea level), municipality of Teresópolis, state of Rio de Janeiro, Brazil."
MORPHOLOGY — Body length to 55 M, 63 F; ventral scales smooth; middorsal scales enlarged in 0–2 rows; interparietal larger than tympanum; scales across the snout between the second canthals 5–9; toe lamellae unknown; suboculars and supralabials in contact; supraorbital semicircles in contact or separated by 1 row of scales; supraocular scales abruptly enlarged; hindlimbs short, appressed toe to insertion of forelimb, well posterior to ear; body color lichenous, may appear white or brown, with black and green lines as banding and reticulations; male dewlap red; female dewlap orange-yellow.
SIMILAR SPECIES — *Anolis neglectus* is a *twig* anole, found near fellow twig anoles *A. nasofrontalis* and *A. pseudotigrinus*. *Anolis nasofrontalis* is nearly indistinguishable from *A. neglectus*, differing in female dewlap color (yellow-orange in *A. neglectus*; tan, fading posteriorly to pale in *A. nasofrontalis*); these species are best differentiated by geography (they appear genetically distinct [Prates et al. 2019]). *Anolis pseudotigrinus* differs from *A. neglectus* in its larger male dewlap (well on to chest; to posterior insertion of forelimbs in *A. neglectus*), and greater number of dewlap scale rows (12–14; 4–7 in *A. neglectus*). Among other geographically proximal species, *A. punctatus* is larger than *A. neglectus* (SVL to 89 mm; to 63 mm in *A. neglectus*) and tends to be solid green in body color (lichenous white or brown with green in *A. neglectus*); *A. fuscoauratus* is brown (never green) and has smaller head scales (8–14 scales across the snout; 5–9 in *A. neglectus*). *A. ortonii* differs in male and female dewlap color (red with yellow-orange in males and females); *A. brasiliensis*, *A. sagrei*, and *A. meridionalis* differ in their keeled ventral scales (smooth in *A. neglectus*).
RANGE — Known only from the type locality.
NATURAL HISTORY — Only information is from the type description: All specimens were found in forest edge habitat sleeping at night on "the tips of twigs or other narrow plant structures at a height of 1.5 to 8.0 m" (Prates et al. 2019:11).

— *ANOLIS NEMONTEAE* — PLATE 1.148, 2.117

DESCRIPTION — Ayala-Varela, F., S. Valverde, S. Poe, A. Narváez, M Yánez-Muñoz, O. Torres-Carvajal. 2021. A new giant anole (Squamata: Iguanidae: Dactyloinae) from southwestern Ecuador. *Zootaxa* 4991:295–317.
TYPE SPECIMEN — QCAZ 14595.
TYPE LOCALITY — "Ecuador, Provincia El Oro, Reserva Buenaventura, 3.654 S, 79.777 W, WGS84, 417 m."
MORPHOLOGY — Body length to 115 M, 103 F; ventral scales weakly keeled; middorsal scales enlarged in 0–2 rows; interparietal smaller than tympanum; scales across the snout between the second canthals 7–11; scales separating supraorbital semicircles 0–2; toe lamellae 21–23; suboculars and supralabials in contact; scales separating supraorbital semicircles 3–4; supraocular scales gradually enlarged; hindlimbs short, appressed toe to posterior to ear; body color variable but often green with broad dark bands, fading to brown; male dewlap bluish-white with yellowish scales, female dewlap colored as male's but with black blotches and yellow stripes.
SIMILAR SPECIES — *Anolis nemonteae* is a *giant* anole in a region, southern Pacific Ecuador, where no other giant anole species are found. *Anolis fraseri* is very similar to *A. nemonteae* and geographically replaces it to the north. *Anolis fraseri* differs most obviously from *A. nemonteae* in male and female dewlap color (male dewlap cream to yellowish with greenish-yellow scales, female dewlap pale brown and lacking black blotches; male dewlap bluish-white, female dewlap bluish-white with black blotches in *A. nemonteae*).
RANGE — From 370 to 1000 meters on the Pacific Andean slope of southern Ecuador.

Anolis nicefori

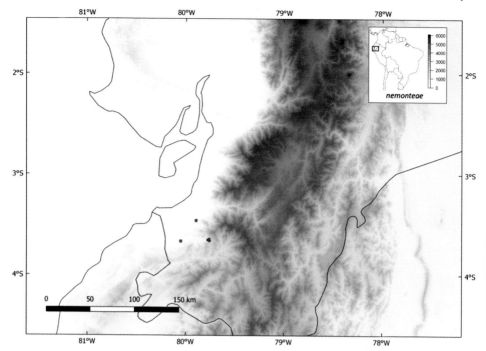

NATURAL HISTORY — Common in edge and forest habitat; sleeps on twigs, branches, and leaves from 2.5 to 7 meters above ground. See Torres-Carvajal et al. (2019); Arteaga et al. (2023).

— *ANOLIS NICEFORI* — PLATE 1.149

DESCRIPTION — Dunn, E. R. 1944. The lizard genus *Phenacosaurus*. *Caldasia* 3:100.
TYPE SPECIMEN — ILS 64 according to Lazell (1969). Description states "Largest male in the Institute de La Salle series, collected by Hmno. Niceforo Maria." (Dunn 1944:59). Lazell (1969) states that ILS 64 is an adult female.
TYPE LOCALITY — "Vicinity of Pamplona, Norte de Santander (2340 m)." This locality is at lat 7.38 lon −72.65.
MORPHOLOGY — Body length to 63 M, 58 F; ventral scales smooth; middorsal scales form a crest of triangular scales; interparietal larger than tympanum; scales across the snout between the second canthals 3–8; toe lamellae 15–22; suboculars and supralabials in contact; supraorbital semicircles in contact; supraocular scales gradually enlarged; hindlimbs short, adpressed toe to posterior to ear; dewlap "creamy white with pale orange stripes" (Barros et al. 1996).
SIMILAR SPECIES — *Anolis nicefori* is a highland *twig* anole. The combination of short hindlimbs (adpressed toe reaches to posterior to ear), large smooth head scales, heterogeneous lateral scales, and crested middorsum will distinguish *A. nicefori* from all nearby anoles except *A. richteri* and *A. tetarii*, found south and north of the range of *A. nicefori*, respectively. *Anolis tetarii* differs from *A. nicefori* in its larger body size (*A. tetarii*: SVL to 86 mm; *A. nicefori*: SVL to 63 mm) and different male dewlap colors (yellow; white with orange stripes in *A. nicefori*). *Anolis richteri* differs from *A. nicefori* in its larger body size (SVL to 76 mm) and possessing more extreme heterogeneity in flank scales (large round scales separated by small interstitial scales; heterogeneity slight, with large scales at most 2x the size of smaller scales in *A. nicefori*).

229

SPECIES ACCOUNTS

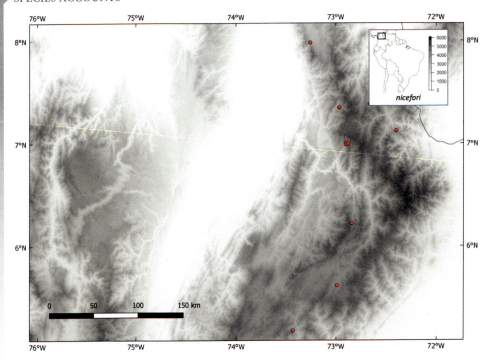

RANGE — From 1800 to 2400 meters along the northern Colombia-Venezuela border.
NATURAL HISTORY — In forest and pasture; has been observed on bushes and moss. See Pérez-Rojas et al. (2020).

— *ANOLIS NOTOPHOLIS* — PLATE 1.150, 2.106

DESCRIPTION — Boulenger, G. A. 1896. Descriptions of new reptiles and batrachians from Colombia. *Annals and Magazine of Natural History* 17:17.
TYPE SPECIMEN — Syntypes 1946.8.12.85-91.
TYPE LOCALITY — "Near Buenaventura." Buenaventura, Valle del Cauca, Colombia is at lat 3.88 lon −77.03, 0 m.
MORPHOLOGY — Body length to 49 M, 52 F; ventral scales keeled; middorsal scales greatly enlarged in several rows; interparietal equal to or smaller than tympanum; scales across the snout between the second canthals 7–10; toe lamellae 11–15; suboculars and supralabials in contact or separated by a row of scales; scales separating supraorbital semicircles 0–2; supraocular scales gradually enlarged; hindlimbs long, appressed toe to anterior to tip of snout; body color brown dorsally, yellow-green ventrolaterally, with middorsal gray diamonds; male dewlap orange-red; puncturelike axillary pocket.
SIMILAR SPECIES — *Anolis notopholis* is a *small brown* lowland anole, the only anole in its region that combines strongly keeled ventral scales, abruptly enlarged rows of keeled middorsal scales, and a puncturelike axillary pocket. The most similar potentially sympatric form is *A. auratus*, which differs from *A. notopholis* in its lack of a puncturelike axillary pocket.
RANGE — From 0 to 1400 meters in Pacific Colombia, rarely above 500 meters.
NATURAL HISTORY — Common in open areas and edge habitat; active diurnally on ground and low vegetation, sleeps on grass blades or other low vegetation, including tree stumps, usually below 1.5 meters

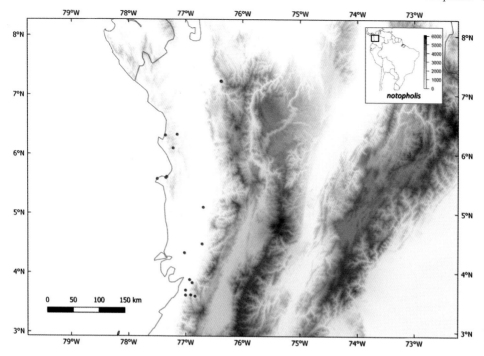

above ground. See Castro-Herrera (1988); Van den Elzen and Schuchmann (1980); Ríos et al. (2011); Rengifo et al. (2014, 2015); Rengifo et al. (2019, 2021).

— ANOLIS OCELLOSCAPULARIS — PLATE 1.151, 2.52

DESCRIPTION — Köhler, G., J. R. McCranie, L. D. Wilson. 2001. A new species of anole from western Honduras (Squamata: Polychrotidae). *Herpetologica* 57:247–255.
TYPE SPECIMEN — SMF 78841.
TYPE LOCALITY — "Honduras, Departamento de Copán, near Quebrada Grande off a trail to Laguna del Cerro, 15° 04.82' N, 88° 55.45' W, 1200 m elevation."
MORPHOLOGY — Body length to 43 M, 49 F; ventral scales weakly keeled; middorsal scales enlarged in 0–2 rows; interparietal equal to or larger than tympanum; scales across the snout between the second canthals 9–12; toe lamellae 16–17; suboculars and supralabials in contact; scales between supraorbital semicircles 1–3; supraocular scales gradually enlarged; addressed toe to between eye and tip of snout; body color brown, with dark shoulder spot, usually with light lateral stripe; male dewlap orange, lighter orange distally.
SIMILAR SPECIES — *Anolis ocelloscapularis* is a *small brown* highland anole. Its combination of keeled ventral scales, an ocellated shoulder spot and undifferentiated (i.e., not abruptly enlarged) middorsal scales will distinguish *A. ocelloscapularis* from potentially confusing nearby species. In addition, *A. cusuco*, *A. kreutzi*, *A. laeviventris*, *A. beckeri*, *A. sericeus*, and *A. ustus* differ from *A. ocelloscapularis* in having shorter hindlimbs (adpressed toe reaches approximately to ear; to between eye and tip of snout in *A. ocelloscapularis*). *Anolis cupreus* differs from *A. ocelloscapularis* in male dewlap color (orange-brown proximally, pale pink-peach distally) and in usually having a row of scales separating the suboculars and supralabials (suboculars and supralabials in broad contact in *A. ocelloscapularis*).

SPECIES ACCOUNTS

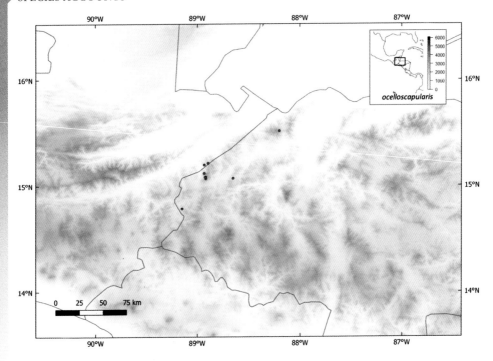

Anolis ocelloscapularis, A. rodriguezii, and *A. yoroensis* share similar body proportions and an orange male dewlap, and overlap in most scale counts. *Anolis ocelloscapularis* is the only one of these three that usually has an ocellated shoulder spot. Also, *A. rodriguezii* has smooth ventral scales and *A. yoroensis* tends to have a row of scales separating suboculars and supralabials.
RANGE — From 1000 to 1550 meters in northwestern Honduras and eastern Guatemala.
NATURAL HISTORY — Common in forest and forest edge; active on low vegetation and ground; sleeps on twigs or leaves up to 3 meters above ground. See Townsend et al. (2006); Townsend and Wilson (2008); McCranie and Köhler (2015).

— *ANOLIS OMILTEMANUS* — PLATE 1.152, 2.3

DESCRIPTION — Davis, W. B. 1954. Three new anoles from Mexico. *Herpetologica* 10:2–3.
TYPE SPECIMEN — TCWC 10278.
TYPE LOCALITY — "Two miles west of Omiltemi, 7800 feet, Guerrero." Assuming the distance is by road, this location is at lat 17.55 lon −99.70.
MORPHOLOGY — Body length to 44 M, 46 F; ventral scales smooth; middorsal scales enlarged in 0–2 rows; interparietal larger than tympanum; scales across the snout between the second canthals 4–6; toe lamellae 14–17; suboculars and supralabials in contact; supraorbital semicircles in contact; supraocular scales abruptly enlarged; hindlimbs short, adpressed toe to ear; body color grayish-brown; male dewlap orange; female dewlap white.
SIMILAR SPECIES — *Anolis omiltemanus* is a *small* gray to *brown* highland anole. Its combination of smooth ventral scales, short hindlimbs (adpressed toe reaches ear), and orange male dewlap separates it from all other species in its region except *A. peucephilus,* which is allopatric to *A. omiltemanus* to

Anolis onca

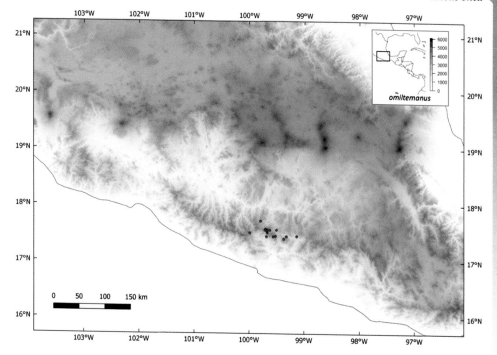

the east. *A. peucephilus* differs from *A. omiltemanus* in its especially short hindlimbs (adpressed toe to between axilla and ear).
RANGE — From 1850 to 2550 meters on the Pacific slope of central Guerrero.
NATURAL HISTORY — Common in open areas with pine trees; the species has been stated to be "strictly arboreal" (Köhler et al. 2013b) and "on the ground in leaf litter" (Flores-Villela and Muñoz 1990); sleeps on twigs of pine trees from 1.5 to at least 6 meters above ground. The dozens of *Anolis omiltemanus* I have collected were all sleeping in trees. See Flores-Villela and Muñoz (1990, 1993); Köhler et al. (2013b, 2014a).

— *ANOLIS ONCA* — PLATE 1.153

DESCRIPTION — O'Shaughnessy, A.W.E. 1875. List and revision of the species of Anolidae in the British-Museum Collection, with descriptions of new species. *Annals and Magazine of Natural History* 4:280–281.
TYPE SPECIMEN — Syntypes BMNH 1961.1067-8.
TYPE LOCALITY — "Venezuela and Dominica."
MORPHOLOGY — Body length to 92 M, 74 F; ventral scales keeled; middorsal scales enlarged in 0–2 rows; interparietal larger or smaller than tympanum; scales across the snout between the second canthals 8–12; toes narrow, no distinct expanded toe lamellae; suboculars and supralabials in contact or separated by a row of scales; scales separating supraorbital semicircles 1–3; supraocular scales abruptly or gradually enlarged; adpressed toe to between ear and eye; body color tan with light and dark brown markings; male dewlap intermixed yellow and red.
SIMILAR SPECIES — *Anolis onca* is a *large brown* to gray anole that is distinctive in its xeric, usually coastal terrestrial habitat use and unique anoline toe structure. The lack of expanded subdigital

SPECIES ACCOUNTS

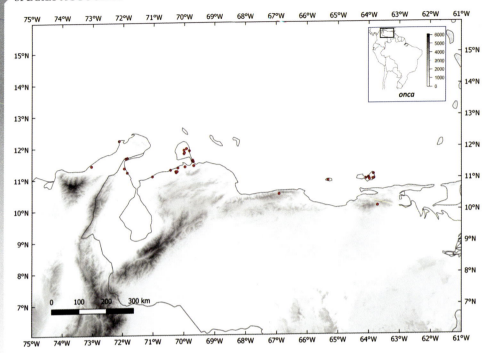

lamellae and presence of ventral keeling on all toe scales distinguishes *A. onca* from all other anoles. *Anolis annectens* also lacks expanded subdigital scales but differs in having few keeled ventral toe scales (all ventral toe scales keeled in *A. onca*) and possessing obviously enlarged rows of middorsal scales (middorsals not or slightly enlarged in *A. onca*).
RANGE — From 0 to 100 meters in northern coastal areas of Venezuela and extreme northeastern Colombia.
NATURAL HISTORY — Common in open arid areas; usually terrestrial, or on low vegetation; has been observed to escape underground. See Williams (1974a); Barros et al. (2007); González et al. (2007); Velásquez et al. (2011).

— *ANOLIS ORCESI* — PLATE 1.154, 2.147

DESCRIPTION — Lazell, J. D. 1969. The genus *Phenacosaurus* (Sauria: Iguanidae). *Breviora* 325:14–18.
TYPE SPECIMEN — MCZ 38937.
TYPE LOCALITY — "Mt. Sumaco, Napo, Pastaza Province, Ecuador." Volcán Sumaco is at lat −0.54 lon −77.63.
MORPHOLOGY — Body length to 59 M, 59 F; ventral scales smooth; middorsal scales form a crest of triangular scales; interparietal larger than tympanum; scales across the snout between the second canthals 4–5; toe lamellae 17–19; suboculars and supralabials in contact; supraorbital semicircles in contact; supraocular scales abruptly or gradually enlarged; hindlimbs short, appressed toe to ear or just posterior to ear; body color gray with green, black, and brown, appearing lichenous; male and female dewlaps yellow.
SIMILAR SPECIES — *Anolis orcesi* is a small highland *twig* anole that generally is sympatric with *A. fuscoauratus* and one large to giant brown anole (usually *A. fitchi*). *Anolis orcesi* is distinguished from

Anolis ortonii

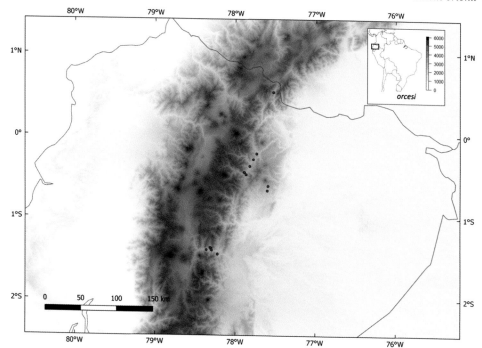

A. fuscoauratus by its larger head scales (supraorbital semicircles in contact in *A. orcesi*; 1–3 scales between supraorbital semicircles in *A. fuscoauratus*), and from *A. fitchi* by its diminutive body size (SVL to 59 mm in *A. orcesi*, 91 mm in *A. fitchi*). Similar twig anoles *A. williamsmittermeierorum* and *A. lososi* are found south of the range of *A. orcesi*. These differ from *A. orcesi* most obviously in dewlap colors (male dewlap white with tan edge, female dewlap white with black blotches in *A. williamsmittermeierorm*; male dewlap white, female dewlap orange with black blotches in *A. lososi*; male and female dewlaps yellow in *A. orcesi*) and lacking a serrated middorsal crest.

RANGE — From 1500 to 2200 meters on the Amazonian Andean slope of Ecuador.

NATURAL HISTORY — Common in forest and edge habitat; sleeps on twigs above 1 meter above ground. See Narváez (2017); Torres-Carvajal et al. (2019); Arteaga et al. (2023).

COMMENT — The paratype of *A. orcesi* (USNM 16533) and other recent collections from near the Ecuador-Colombia border probably represent an undescribed species.

— *ANOLIS ORTONII* — PLATE 1.155, 2.97

DESCRIPTION — Cope, E. D. 1868. An examination of the Reptilia and Batrachia obtained by the Orton expedition to Equador and the upper Amazon, with notes on other species. *Proceedings of the Academy of Natural Sciences of Philadelphia* 20:97–98.

TYPE SPECIMEN — ANSP 11404.

TYPE LOCALITY — "Napo or Upper Marañon." Cope (1868:96) described a part of the Orton expedition, during which the specimen was collected, as proceeding "to Napo on the River Napo; thence by canoe down the Napo to the Marañon and Amazons." Presumably the specimen was collected some time during this leg of the journey. The Napo enters the Marañon (=Amazon) just east of Iquitos, at approximately lat −3.46 lon −72.74, 100 m.

SPECIES ACCOUNTS

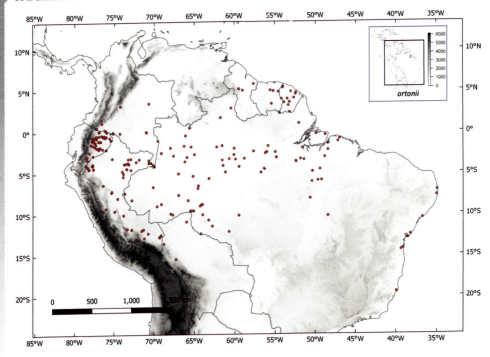

MORPHOLOGY — Body length to 59 M, 50 F; ventral scales smooth; middorsal scales enlarged in 0–2 rows; interparietal equal to or larger than tympanum; scales across the snout between the second canthals 7–11; toe lamellae 15–20; suboculars and supralabials in contact; supraorbital semicircles usually in contact; supraocular scales gradually enlarged; hindlimbs short, adpressed toe to ear or posterior to ear; body color generally gray with brown, black, and white; tail banded; male and female dewlaps red with yellow-orange along scale rows.

SIMILAR SPECIES — *Anolis ortonii* is a *small brown* to gray anole that is a member of the complement of anoles usually found at lowland Amazonian sites. Among these species, *A. ortonii* differs from *A. transversalis* and *A. punctatus* in its smaller body size (SVL to 89 mm in *A. punctatus*, 79 mm in *A. transversalis*; to 59 mm in *A. ortonii*). Also, *A. punctatus* is green (white to brown in *A. ortonii*) and *A. transversalis* has blue eyes (brown in *A. ortonii*). *Anolis chrysolepis*, *A. bombiceps*, *A. tandai*, *A. planiceps*, and *A. scypheus* differ from *A. ortonii* in their keeled ventral scales (smooth in *A. ortonii*). *Anolis fuscoauratus* is smaller than *A. ortonii* (SVL to 47 mm) and differs in male and female dewlap color (pink or white in male *A. fuscoauratus*, dewlap usually absent in females; red with yellow in males and females of *A. ortonii*). *Anolis trachyderma* has longer hindlimbs than *A. ortonii* (adpressed toe reaches at least to eye; to ear in *A. ortonii*). Among lowland species with restricted or non-Amazonian distributions, *A. sagrei*, *A. auratus*, *A. meridionalis*, *A. dissimilis*, and *A. caquetae* differ from *A. ortonii* in their keeled ventral scales; *A. phyllorhinus* differs in green body coloration; *A. neglectus*, *A. nasofrontalis*, and *A. pseudotigrinus* differ in female dewlap color (*A. neglectus*: orange-yellow, *A. nasofrontalis*: tan fading posteriorly to white, *A. pseudotigrinus*: white).

RANGE — From 50 to 1100 meters in Amazonia.

NATURAL HISTORY — Rare in forest and open areas with tall trees, including disturbed areas; diurnally often on tree trunks and large branches; sleeps over 2 meters above ground on vegetation, frequently on large leaves. See Crump (1971); Hoogmoed (1973); Dixon and Soini (1975, 1986); Duellman (1978); Zimmerman and Rodrigues (1990); Vitt and Zani (1996b); Ávila-Pires (1995); Vitt et al.

(1999); Vitt (2000); Macedo et al. (2008); Silva et al. (2011); Pinto Aguirre (2014); de Oliveira et al. (2014); Araújo-Nieto (2017); Moreno-Arias et al. (2020); Torres-Carvajal et al. (2019); Arteaga et al. (2023); Pinto and Torres-Carvajal (2023).

— *ANOLIS OTONGAE* — PLATE 1.156, 2.111

DESCRIPTION — Ayala-Varela, F. P., J. A. Velasco. 2010. A new species of dactyloid anole (Squamata: Iguanidae) from the western Andes of Ecuador. *Zootaxa* 2577:46–56.
TYPE SPECIMEN — QCAZ 2051.
TYPE LOCALITY — "Ecuador, Provincia Cotopaxi, Cantón Sigchos, Reserva de Bosque Integral Otonga, near San Francisco de Las Pampas, 0 25'8.04"S, 79 0'14.04"W, 2000–2200 m."
MORPHOLOGY — Body length to 67 M, 63 F; ventral scales smooth; middorsal scales enlarged in 0–2 rows; interparietal smaller than tympanum; scales across the snout between the second canthals 12–17; toe lamellae 16–23; suboculars and supralabials in contact; scales separating supraorbital semicircles 3–5; supraocular scales gradually enlarged; hindlimbs long, adpressed toe to eye or just posterior to eye; body color mix of browns and greens, with lateral bands of cream ocelli, female with transverse dark middorsal saddles; tail banded; male dewlap pale yellow, greenish-yellow anteriorly; female dewlap absent; iris blue.
SIMILAR SPECIES — *Anolis otongae* is a *small green* highland anole, often sympatric with *A. aequatorialis* and parapatric with *A. gemmosus*. *Anolis aequatorialis* differs from *A. otongae* in its larger body size (SVL to 92 mm; to 67 mm in *A. otongae*) and male and female dewlap color (dark with black, green, and blue in male and female; male dewlap pale yellow, greenish-yellow anteriorly, female dewlap absent in *A. otongae*). *Anolis gemmosus* is very similar to *A. otongae*, and most easily distinguished by female dorsal pattern: females of *A. otongae* have a series of dark brown transverse saddles

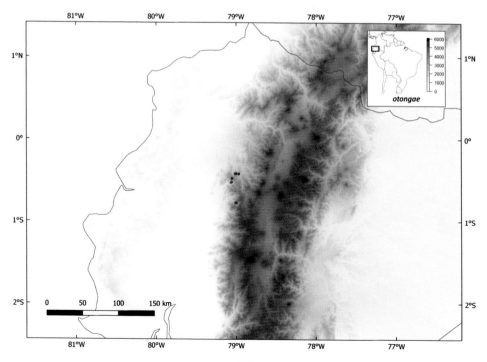

dorsally (pattern absent in *A. gemmosus*). Other green anoles of Pacific Ecuador live only at lower elevations than *A. otongae*, excepting *A. poei*. *Anolis poei* is allopatric to the south of the range of *A. otongae* and differs only subtly in morphology (it appears to be molecularly distinct [Ayala-Varela et al. 2010]).
RANGE — From 1750 to 2500 meters on the Pacific Andean slope of Ecuador.
NATURAL HISTORY — Common in cloud forest and disturbed areas; sleeps on leaves, especially ferns, occasionally twigs, from 0.5 to 3 meters. In areas where both *A. otongae* and *A. gemmosus* are present, *A. otongae* occurs at higher elevations. See Torres-Carvajal et al. (2019); Arteaga et al. (2023).

— *ANOLIS OXYLOPHUS* — PLATE 1.157, 2.84

DESCRIPTION — Cope, E. D. 1876. On the Batrachia and Reptilia of Costa Rica with notes on the herpetology and ichthyology of Nicaragua and Peru. *Journal of the Academy of Natural Sciences of Philadelphia* 8:123–124.
TYPE SPECIMEN — Syntypes USNM 30556–7.
TYPE LOCALITY — "Costa Rica." Taylor (1956:141) proposed to "fix or restrict the type locality to Sipurio, Limon Province, C. R." (lat 9.53 lon −82.948, 70 m).
MORPHOLOGY — Body length to 76 M, 67 F; ventral scales keeled; middorsal scales greatly enlarged in several rows; interparietal larger or smaller than tympanum; scales across the snout between the second canthals 7–14; toe lamellae 12–19; suboculars and supralabials separated by a row of scales; scales separating supraorbital semicircles 1–2; supraocular scales gradually enlarged; appressed toe to eye; body color brown with light lateral stripe; male dewlap yellow; female dewlap absent.
SIMILAR SPECIES — *Anolis oxylophus* is a widespread *semiaquatic* form. The combination of semi-aquatic lifestyle (almost always found within one meter of a stream or river), keeled ventral scales,

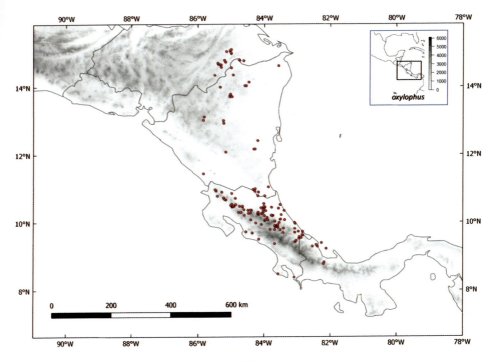

and enlarged rows of 10–15 keeled middorsal scales will distinguish *A. oxylophus* from geographically proximal congeners. The enlarged middorsal scale rows separate this form from all nearby semiaquatic species except *A. lionotus*, which geographically replaces *A. oxylophus* to the south in Panama. In *A. lionotus*, the enlarged middorsal scale rows are smooth and much larger than the ventrals (usually keeled, approximately equal in size to ventrals in *A. oxylophus*). The other semiaquatic anoles in the range of *A. oxylophus* (*A. aquaticus*, *A. riparius*, *A. robinsoni*) lack enlarged middorsal scale rows.
RANGE — From 0 to 1300 meters from central Honduras to western Panama.
NATURAL HISTORY — Common in forest; active diurnally on boulders, logs, and tree buttresses in and around streams; sleeps on leaves, twigs, or boulders or rock walls, up to 1 meter above ground; stays underwater by "rebreathing" via an air bubble (Bocci et al. 2021). See Fitch (1970, 1973a, b; 1975); Corn (1981; as *Anolis lionotus*); Fitch and Seigel (1984); Berkum (1986; as *A. lionotus*); Vitt et al. (1995); Savage (2002); Dappen (2003); Guyer and Donnelly (2004); Muñoz et al. (2009); Ream and Reider (2013); Muñoz et al. (2015); McCranie and Köhler (2015); Herrmann (2017); Lara-Resendiz et al. (2017); Hilje et al. (2020); Perez-Martinez et al. (2021).

— *ANOLIS PACHYPUS* — PLATE 1.158, 2.50

DESCRIPTION — Cope, E. D. 1876. On the Batrachia and Reptilia of Costa Rica with notes on the herpetology and ichthyology of Nicaragua and Peru. *Journal of the Academy of Natural Sciences of Philadelphia* 8:122–123.
TYPE SPECIMEN — USNM 30683.
TYPE LOCALITY — "Slope of Pico Blanco." Savage (1974, 2002) states the type specimen was collected on the slope of Cerro Utyum, Canton de Talamanca, Limon Province. Cerro Utyum is at lat 9.327 lon −83.178.

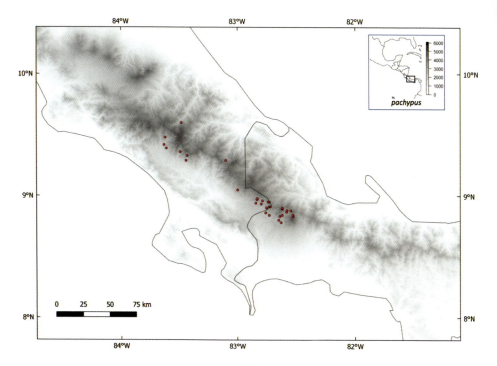

SPECIES ACCOUNTS

MORPHOLOGY — Body length to 54 M, 50 F; ventral scales weakly keeled to smooth; middorsal scales gradually enlarged in several rows; interparietal smaller than tympanum; scales across the snout between the second canthals 10–14; toe lamellae 12–16; suboculars and supralabials in contact or separated by rows of scales; scales separating supraorbital semicircles 2–4; supraocular scales gradually enlarged; hindlimbs long, adpressed toe to eye or just beyond; body color brown, sometimes with middorsal chevrons; male dewlap yellow with reddish-orange; female dewlap white or absent; males have a bulging hemipenial area.

SIMILAR SPECIES — *Anolis pachypus* is a *small brown* highland anole with long hindlimbs (adpressed toe reaches to eye or beyond). Among geographically proximal species with long hindlimbs, *Anolis datzorum* has larger head scales than *A. pachypus* (7–9 scales across the snout; 10–14 in *A. pachypus*). *Anolis magnaphallus* and *A. benedikti* are very similar highland anoles found east of the range of *A. pachypus*, and *A. tropidolepis* appears geographically to replace *A. pachypus* to the west. Both *A. magnaphallus* and *A. tropidolepis* may be distinguished from *A. pachypus* by their solid red male dewlaps (yellow with red-orange in *A. pachypus*). *Anolis benedikti* has smaller head scales than *A. pachypus* (11–18 scales across the snout, 4–5 scales separating supraorbital semicircles; 2–4 scales separating supraorbital semicircles in *A. pachypus*) and an orange posterior, yellow anterior male dewlap (in *A. pachypus*, orange may appear to surround yellow, or there may be multiple patches of orange on the male dewlap).

RANGE — From 1700 to 2600 meters in western Panama and eastern Costa Rica. See Comment.

NATURAL HISTORY — Common in forest, edge, and disturbed areas on or near the ground; associated with moss, fog, and rainfall; sleeps on low vegetation, often shrubs, usually below 1 meter above ground.

COMMENT — Savage (2002) considered the range of *Anolis pachypus* to extend north through Costa Rica to the Nicaraguan border. I have tentatively adopted the view of Köhler et al. (2014c) that this species is replaced by *A. tropidolepis* in northern Costa Rica. See additional comments under *A. tropidolepis*.

— ANOLIS PARILIS — PLATE 1.159, 2.116; MAP opposite

DESCRIPTION — Williams, E. E. 1975. South American *Anolis*: *Anolis parilis*, new species, near *A. mirus* Williams. Breviora 434:1–8.

TYPE SPECIMEN — UIMNH 82901.

TYPE LOCALITY — "Rio Baba, 2.4 km S Sto Domingo de los Colorados, Pichincha, Ecuador." The actual wording on the specimen tag appears to be "24 km S" of Santo Domingo (Chun 2010: fig. 8), which places the locality approximately at the Río Palenque Reserve (lat −0.588 lon −79.363, elevation 170 m).

MORPHOLOGY — Body length to 82 M, 80 F; ventral scales smooth; middorsal scales enlarged in 0–2 rows; interparietal smaller than tympanum; scales across the snout between the second canthals 14–18; toe lamellae 13–16; suboculars and supralabials in contact or separated by a row of scales; scales separating supraorbital semicircles 4–5; supraocular scales subequal; hindlimbs short, adpressed toe reaches approximately to ear; body color patterned green and brown; male dewlap orange-yellow distally, white proximally; female dewlap yellow-orange; iris reddish-brown.

SIMILAR SPECIES — *Anolis parilis* is a *large green* lowland anole that is distinguishable from potentially sympatric large anoles by its distinctive fourth (longest) toe morphology (narrow, indistinct toepad with large claw and <17 lamellae). *Anolis mirus*, found to the north of the range of *A. parilis*, shares the unusual toe morphology of *A. parilis* but has larger head scales (10–13 scales across the snout; 14–18 in *A. parilis*) and weakly keeled ventral scales (smooth in *A. parilis*).

RANGE — From 0 to 700 meters in Pacific Ecuador.

NATURAL HISTORY — Rare in forest and forest edge; active diurnally on arboreal perches; sleeps on twigs, leaves, or vines above 1.5 meters. See Torres-Carvajal et al. (2019); Arteaga et al. (2023).

Anolis parvauritus

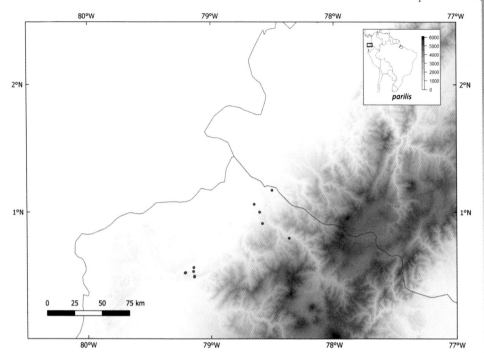

— *ANOLIS PARVAURITUS* — PLATE 1.160, 2.129; MAP overleaf

DESCRIPTION — Williams, E. E. 1966. South American anoles: *Anolis biporcatus* and *Anolis fraseri* (Sauria, Iguanidae) compared. *Breviora* 239:7–11.
TYPE SPECIMEN — MCZ 78935
TYPE LOCALITY — "Banana plantation, woods and penal colony camp, northern Gorgona Island, Cauca, Colombia, 5–45 meters altitude." Gorgona Island is located at lat 2.97 lon −78.18, 0–330 m.
MORPHOLOGY — Body length to 88 M, 95 F; ventral scales keeled; middorsal scales enlarged in 0–2 rows; interparietal approximately equal to tympanum; scales across the snout between the second canthals 9–14; toe lamellae 20–25; suboculars and supralabials separated by a row of scales; scales separating supraorbital semicircles 1–3; supraocular scales gradually enlarged; appressed toe to eye; body color usually solid green, turning dark brown under stress; male dewlap red distally, yellow and blue proximally, with black scales; female dewlap sky blue or white with black flecks.
SIMILAR SPECIES — *Anolis parvauritus* is a *large green* lowland anole that essentially is the southern Colombian and Ecuadorean version of *A. biporcatus*, from which it differs in having black scales on the red, yellow, and blue male dewlap (scales white to green in *A. biporcatus*). The combination of large body size (SVL to 95 mm), keeled ventral scales, patternless green dorsum (unless stressed), and distinctive male and female dewlaps (red, blue, and yellow in male; sky blue or white with dark flecks in female) will differentiate *A. parvauritus* from all other potentially sympatric large green anoles.
RANGE — From 0 to 850 meters in Pacific Ecuador and southern Colombia.
NATURAL HISTORY — Common in disturbed areas and edge habitat; active arboreally above 2 meters above ground; sleeps above 2 meters above ground, usually on or among green leafy vegetation. See Castro-Herrera (1988; as *A. biporcatus*); Castro-Herrera et al. (2012; as *A. biporcatus*); Armstead et al. (2017); Narváez (2017); Narváez et al. (2019); Torres-Carvajal et al. (2019); Pinto-Erazo et al. (2020); Arteaga et al. (2023).

SPECIES ACCOUNTS

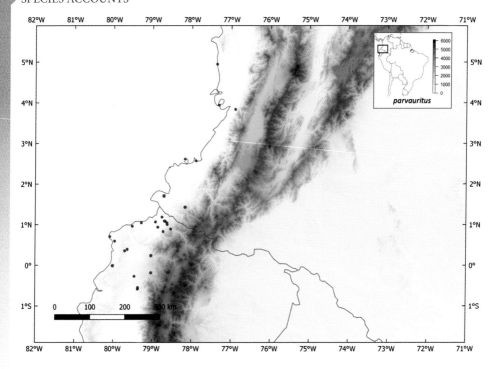

— *ANOLIS PARVICIRCULATUS* — PLATE 1.161, 2.8

DESCRIPTION — Álvarez del Toro, M. and H. M. Smith. 1956. Notulae Herpetologicae. *Herpetologica* 12:5–7.

TYPE SPECIMEN — UIMNH 38045.

TYPE LOCALITY — "El Suspiro, Chiapas, 1200 m." Earlier in the type description paper, Álvarez del Toro and Smith locate El Suspiro 9 km northwest of Berriozabal. Wake and Johnson (1989) state that El Suspiro is an abandoned coffee finca. Fallingrain.com, an online gazetteer, lists nine El Suspiros in Chiapas. The El Suspiro that best fits the additional information is at lat 16.844 lon −93.302.

MORPHOLOGY — Body length to 50 M, 50 F; ventral scales smooth; middorsal scales gradually enlarged in several rows; interparietal smaller than tympanum; scales across the snout between the second canthals 6–9; toe lamellae 15–18; suboculars and supralabials usually separated by a row of scales; scales separating supraorbital semicircles 1–2; supraocular scales gradually enlarged; hindlimbs long, appressed toe to between eye and tip of snout; body color brown with some grayish markings; male dewlap orangeish-red centrally, light orange distally; small female dewlap with some color like males.

SIMILAR SPECIES — *Anolis parvicirculatus* is a *small brown* anole. Its long hindlimbs and smooth ventral scales will distinguish *A. parvicirculatus* from most sympatric congeners. Potentially confusing species *A. rodriguezii* has smaller dorsal head scales than *A. parvicirculatus* (8–13 scales across the snout between the second canthals; 6–9 in *A. parvicirculatus*) and suboculars and supralabials in contact (usually separated in *A. parvicirculatus*). *Anolis parvicirculatus* is similar to geographically surrounding highland species *A. hobartsmithi* (northeast), *A. cuprinus* (southwest), and *A. matudai* (southeast). It differs from these in either ventral keeling (*A. cuprinus*: keeled) or male dewlap color (*A. matudai*, *A. hobartsmithi*: pink; *A. parvicirculatus*: orangeish-red centrally, light orange distally).

RANGE — From 600 to 1150 meters in west-central Chiapas, Mexico.

Anolis pentaprion

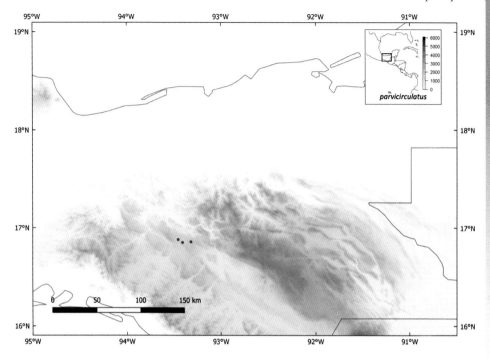

NATURAL HISTORY — Common in forest; active diurnally on ground or low vegetation; escapes into limestone cracks; sleeps on twigs or leaves. See Nieto Montes de Oca (1994a); Fitch et al. (1976); Álvarez del Toro (1982); Luna-Reyes et al. (2017).

— *ANOLIS PENTAPRION* — PLATE 1.162, 2.80

DESCRIPTION — Cope, E. D. 1863. Contributions to Neotropical saurology. *Proceedings of the Academy of Natural Sciences of Philadelphia* 14:178.
TYPE SPECIMEN — Neotype SMF 83608.
TYPE LOCALITY — The species originally was described from "New Granada, near the river Truando" (i.e., northwest Colombia). The holotype is said to be lost (Köhler 2010:12). Köhler (2010) designated a neotype from "San Rafael, ca. 15 km S Los Chiles, 10.73719°N, 84.49378°W, 60 m elevation, Alajuela Province, Costa Rica."
MORPHOLOGY — Body length to 77 M, 64 F; ventral scales smooth; middorsal scales enlarged in 0–2 rows; interparietal larger than tympanum; scales across the snout between the second canthals 5–9; toe lamellae 19–23; suboculars and supralabials in contact; scales separating supraorbital semicircles 0–2; supraocular scales gradually enlarged; hindlimbs short, appressed toe to ear or posterior to ear; body color usually whitish with dark flecks, lichenous; male and female dewlaps reddish-purple.
SIMILAR SPECIES — *Anolis pentaprion* is a widespread lowland *twig*, or *small* to *large brown* anole. Its combination of short hindlimbs, lichenous white color, and smooth body scales will distinguish it from many potentially sympatric similar species. Among potentially sympatric short-legged forms with smooth ventrals, the white lichenous dorsal color of *A. pentaprion* will most easily distinguish it from *A. fortunensis*, *A. elcopeensis*, *A. gruuo*, *A. kemptoni*, *A. arenal*, and *A. pseudokemptoni*, which are

SPECIES ACCOUNTS

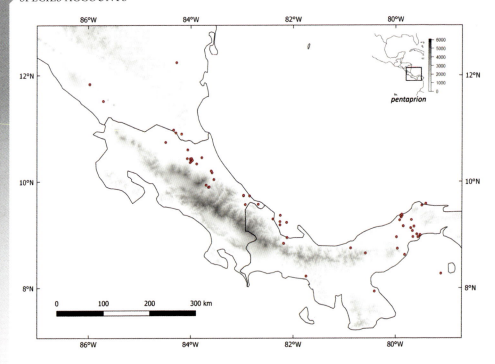

brown. Highland species *A. fungosus* is similar to *A. pentaprion* but smaller (SVL to 48 mm; to 77 mm in *A. pentaprion*) and with fewer toe lamellae (13–16; 19–23 in *A. pentaprion*). *Anolis charlesmyersi*, parapatric to *A. pentaprion* in Caribbean Costa Rica, differs from *A. pentaprion* in its red male and female dewlaps (dewlaps reddish-purple in *A. pentaprion*) and widely spaced dewlap scales (4–9 scales per dewlap scale row, vs. 17–25 close-set scales in *A. pentaprion*). *Anolis triumphalis*, parapatric to *A. pentaprion* to the east, differs from *A. pentaprion* in its red and yellow dewlap.
RANGE — From 0 to 900 meters from central Panama to central Nicaragua, except Pacific versant of Costa Rica.
NATURAL HISTORY — Common to rare in forest and edge habitats, and on large trees in disturbed habitat; active diurnally on tree trunks and branches; sleeps on twigs or (more frequently) large leaves usually above 2 meters above ground. See Echelle et al. (1971a); Corn (1981); Pérez-Higareda et al. (1997); Savage (2002); Hilje et al. (2020); Perez-Martinez et al. (2021).

— *ANOLIS PERACCAE* — PLATE 1.163, 2.100

DESCRIPTION — Boulenger, G. A. 1898. An account of the reptiles and batrachians collected by Mr. W.F.H. Rosenberg in western Ecuador. *Proceedings of the Zoological Society of London* 1898:108–109.
TYPE SPECIMEN — I have been unable to confirm the proper type material for *Anolis peraccae*. Unpublished notes from Ernest Williams recognize BMNH 1946.13.13-17 as syntypes.
TYPE LOCALITY — "Chimbo." Boulenger (1898:108) described Chimbo as "Puente del Chimbo, the railway terminus about 70 miles from Guayaquil, at an elevation of about 1000 feet." This locality appears to be at the town of General Elizalde, Bucay Province, Ecuador, just west of the town of Cumandá, Chimborazo Province, in the western Andean foothills, at lat −2.202 lon −79.141, 290 m (see Paynter 1993).

Anolis peruensis

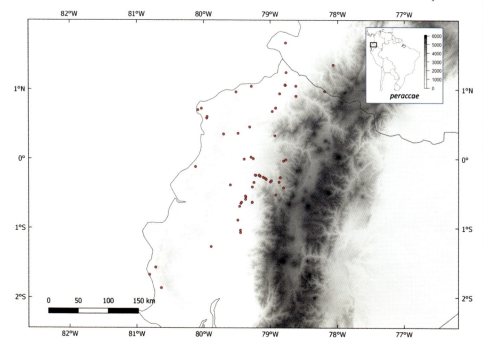

MORPHOLOGY — Body length to 52 M, 48 F; ventral scales smooth; middorsal scales enlarged in 0–2 rows; interparietal larger or smaller than tympanum; scales across the snout between the second canthals 6–14; toe lamellae 13–19; suboculars and supralabials in contact; scales separating supraorbital semicircles 0–2; supraocular scales gradually enlarged; adpressed toe to eye; body color brown with dorsal chevrons; tail banded; male dewlap dirty white; female dewlap absent; iris may appear blue, green, or brown.

SIMILAR SPECIES — *Anolis peraccae* is a *small brown* to olive-greenish lowland anole. Among potentially sympatric species that may fit this description, *A. festae* has shorter hindlimbs than *A. peraccae* (adpressed toe reaches approximately to ear; to eye in *A. peraccae*) and an unusually shallow male dewlap (typical anole dewlap depth in *A. peraccae*). *Anolis granuliceps* has a small, orangeish-yellow male dewlap (dirty white in *A. peraccae*) and usually has a light lateral longitudinal stripe (lateral stripe absent in *A. peraccae*). *Anolis gracilipes*, *A. binotatus*, *A. vittigerus*, *A. chloris*, and *A. lynchi* have strongly keeled ventral scales (smooth in *A. peraccae*). *Anolis anchicayae* is similar to *A. peraccae* but larger (SVL to 63 mm; to 52 mm in *A. peraccae*) and with a yellow-green male dewlap.

RANGE — From 0 to 1400 meters in Pacific central Ecuador and southern Colombia.

NATURAL HISTORY — Common in forest and forest edge; active diurnally on tree trunks; often observed low on vegetation but may inhabit higher reaches of trees; sleeps on twigs, leaves, or vines, usually below 2 meters above ground. See Fitch et al. (1976); Miyata (2013); Narváez (2017); Viteri (2015); Torres-Carvajal et al. (2019); Arteaga et al. (2023).

— ***ANOLIS PERUENSIS*** — PLATE 1.164, 2.150

DESCRIPTION — Poe, S., I. M. Latella, F. Ayala-Varela, C. Yañez-Miranda, O. Torres-Carvajal. 2015a. A new species of phenacosaur *Anolis* (Squamata: Iguanidae) from Peru and a comprehensive phylogeny of *Dactyloa*-clade *Anolis* based on new DNA sequences and morphology. *Copeia* 103:639–650.

SPECIES ACCOUNTS

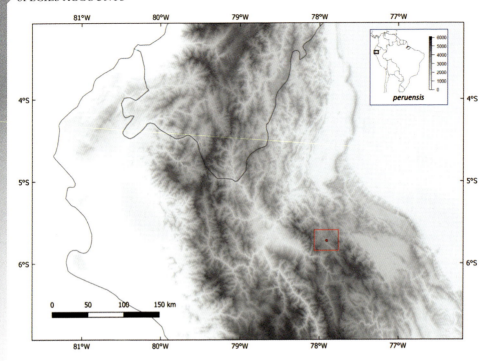

TYPE SPECIMEN — MSB 72532. The type description erroneously lists MSB 72521 as Holotype and MSB 72522 as paratype (both are *A. williamsmittermeierorum*; MSB 72533 was scored as the MSB paratype of *A. peruensis*).
TYPE LOCALITY — "2.4 km west of Esperanza, Amazonas Province, Peru, 05' 43.553 S 77' 54.328 W, 1857 m."
MORPHOLOGY — Body length to 56 M, 51 F; ventral scales smooth; middorsal scales enlarged in 0–2 rows; interparietal larger than tympanum; scales across the snout between the second canthals 4–5; toe lamellae 18–19; subocular and supralabial scales in contact; supraorbital semicircles in contact; supraocular scales abruptly or gradually enlarged; hindlimbs short, adpressed toe to ear or posterior to ear; body color gray, sometimes with brown and green, may appear lichenous; male dewlap orange-yellow to yellow; female dewlap black.
SIMILAR SPECIES — *Anolis peruensis* is a highland *twig* anole that may be found sympatrically with *A. fuscoauratus*, *A. soinii*, *A. williamsmittermeierorum*, an anole similar to *A. fitchi*, and possibly *A. laevis*. Among these species, *A. peruensis* might only be confused with *A. williamsmittermeierorum* and *A. laevis* due to their similar body size, scalation and body proportions (SVL < 60 mm, short limbs and tail, large smooth head scales). *Anolis williamsmittermeierorum* differs from *A. peruensis* most obviously in male and female dewlap color (male: white with band of dark peach at distal edge; female: white with black blotches; male dewlap yellow, female dewlap black in *A. peruensis*). *Anolis laevis* differs from *A. peruensis* in its greatly protruding rostral scale (scale slightly or not protruding in *A. peruensis*), white male dewlap, and middorsal crest of triangular scales (crest absent in *A. peruensis*).
RANGE — Known only from the vicinity of the type locality.
NATURAL HISTORY — Rare in edge and disturbed habitat; sleeps on twigs above 2 meters above ground.

— *ANOLIS PETERSII* — PLATE 1.165, 2.69

DESCRIPTION — Bocourt, M. F. 1873. Études sur les reptiles et les Batraciens. In Duméril, M. A., M. Bocourt, M. Mocquard, eds., *Recherches Zoologiques* vol. 3, section 1, 79–80. Paris: Mission Scientifique au Mexique et dans L'Amerique Centrale.
TYPE SPECIMEN — Syntypes MNHN 2479, 2479a.
TYPE LOCALITY — "Haute Vera Paz (Guatemala)." Alta Verapaz Department is central along the northern border of Guatemala. According to collector Bocourt's map in Duméril et al. (1873), Bocourt only went as far north as San Pedro, just north of Coban, so the southern aspect of Alta Verapaz seems a slightly more precise type region.
MORPHOLOGY — Body length to 118 M, 108 F; ventral scales keeled; middorsal scales enlarged in 0–2 rows; interparietal larger or smaller than tympanum; scales across the snout between the second canthals 8–12; toe lamellae 24–30; suboculars and supralabials in contact or separated by a row of scales; scales separating supraorbital semicircles 1–2; supraocular scales subequal; hindlimbs short, adpressed toe to ear or between ear and eye; body color light brown with black and green markings; tail banded; male dewlap pink with black blotches and pale yellow edge; female dewlap pale with black blotches.
SIMILAR SPECIES — *Anolis petersii* is a highland *giant* anole. In most areas of its range, it is the only possible giant anole, especially at high elevations. *Anolis biporcatus* tends to be solid green in dorsal color (patterned brown and green in *A. petersii*). Lowland form *A. capito* has very long hindlimbs and a short snout (adpressed toe reaches beyond snout; to ear or just beyond in *A. petersii*). *Anolis loveridgei* has longer hindlimbs than *A. petersii* (adpressed toe to between eye and tip of snout).
RANGE — From 1100 to 1550 meters from Mexico to northwestern Honduras.
NATURAL HISTORY — Rare in forest, edge, and disturbed areas; diurnally arboreal; sleeps on twigs or leaves, from 1.5 meters up to the limits of observability. See Villareal-Benítez (1997); Townsend and Wilson (2008); McCranie and Köhler (2015).

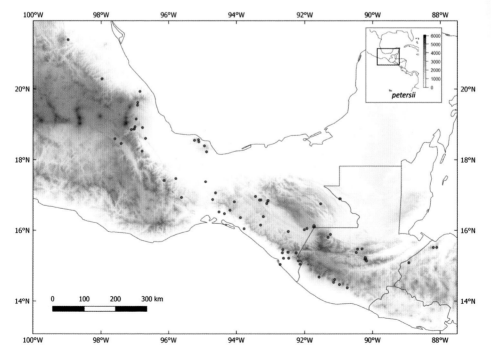

SPECIES ACCOUNTS

— *ANOLIS PEUCEPHILUS* — PLATE 1.166

DESCRIPTION — Köhler, G., R. Gómez Trejo Pérez, C.B.P. Petersen, F. Méndez de la Cruz. 2014b. A new species of pine anole from the Sierra Madre del Sur in Oaxaca, Mexico (Reptilia, Dactyloidae: *Anolis*). *Zootaxa* 3753:453–468.
TYPE SPECIMEN — SMF 96368.
TYPE LOCALITY — "Ca. 27 km on road N San Gabriel Mixtepec (16.19135°N, 97.09820°W, WGS84), 1325 m, Estado de Oaxaca, Mexico."
MORPHOLOGY — Body length to 41 M, 43 F; ventral scales smooth; middorsal scales enlarged in 0–2 rows; interparietal larger than tympanum; scales across the snout between the second canthals 6–7; toe lamellae 17–18; suboculars and supralabials in contact; supraorbital semicircles in contact; supraocular scales abruptly enlarged; hindlimbs short, adpressed toe to posterior to ear; body color brown; male dewlap orange; female dewlap off-white.
SIMILAR SPECIES — *Anolis peucephilus* is a *small brown* to gray highland anole. Its combination of smooth ventral scales, short hindlimbs, and an orange male dewlap distinguishes it from all other species in its region except *A. omiltemanus*, which is allopatric to *A. peucephilus* to the west. *Anolis omiltemanus* differs from *A. peucephilus* in its slightly longer hindlimbs (adpressed toe to ear; short of ear in *A. peucephilus*).
RANGE — From 1300 to 1950 meters along the road (131) north of San Gabriel Mixtepec, Oaxaca State, Mexico.
NATURAL HISTORY — In open areas with pine trees; sleeps above 1.5 meters, mainly in pine trees.

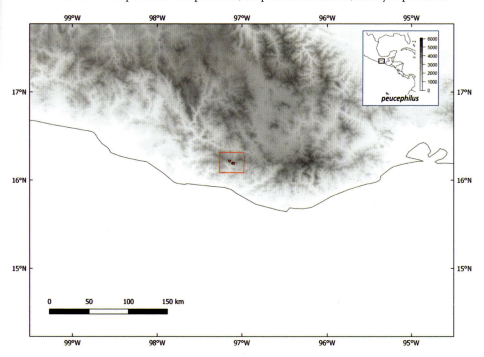

— *ANOLIS PHYLLORHINUS* — PLATE 1.167

DESCRIPTION — Myers, G. S., A. Leitão de Carvalho. 1945. A strange new leaf-nosed lizard of the genus *Anolis* from Amazonia. *Boletim do Museu Nacional* 43:1–14.
TYPE SPECIMEN — MNRJ 1804.
TYPE LOCALITY — "Borba, lower Rio Madeira, State of Amazonas, Brazil." Borba, Amazonas, Brazil is at lat −4.39 lon −59.59, 20 m.
MORPHOLOGY — Body length to 87 M, 72 F; ventral scales smooth; middorsal scales enlarged in 0–2 rows; interparietal smaller than tympanum; scales across the snout between the second canthals 8–10; toe lamellae 24–25; subocular and supralabial scales in contact; scales separating supraorbital semicircles 1–2; body color green; male dewlap mostly red, light blue posteriorly; female dewlap small or absent; male with scaly extension anterior to snout.
SIMILAR SPECIES — *Anolis phyllorhinus* is a *large green* lowland Amazonian anole. Males of *A. phyllorhinus* are unmistakable due to their possession of a scaly snout extension. Females are unlikely to be confused with most potentially sympatric species due to green dorsal color and/or smooth ventral scales. *Anolis dissimilis* is smaller than *A. phyllorhinus* (SVL to 58 mm; to 72 mm in females of *A. phyllorhinus*) with a well-developed female dewlap (absent or small in *A. phyllorhinus*). *Anolis transversalis* is predominantly brown with blue eyes (brown eyes in *A. phyllorhinus*). Females of *A. punctatus* are not straightforwardly distinguishable from females of *A. phyllorhinus*.
RANGE — From 0 to 250 meters in Amazonian Brazil.
NATURAL HISTORY — In edge habitat near primary forest; diurnally from 1 to 5 meters above ground on vegetation, especially tree trunks. See Ávila-Pires (1995); Rodriguez et al. (2002).

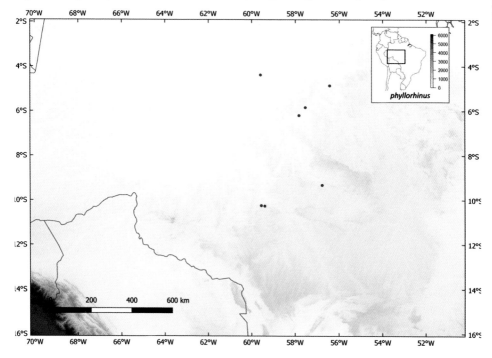

SPECIES ACCOUNTS

— *ANOLIS PIJOLENSE* — PLATE 1.168

DESCRIPTION — McCranie, J. R., L. D. Wilson, K. L. Williams. 1993b. Another new species of lizard of the *Norops schiedei* group (Sauria: Polychrotidae) from northern Honduras. *Journal of Herpetology* 27:393–399.
TYPE SPECIMEN — USNM 322871.
TYPE LOCALITY — "East slope of Pico Pijol (15° 10' N, 87° 33' W), Montaña de Pijol, northwest of Tegucigalpita, 2050 m elevation, Departamento de Yoro, Honduras."
MORPHOLOGY — Body length to 59 M, 60 F; ventral scales smooth; middorsal scales enlarged in 0–2 rows; interparietal smaller than tympanum; scales across the snout between the second canthals 6–12; toe lamellae 16–17; suboculars and supralabials in contact or separated by a row of scales; scales beween supraorbital semicircles 2–4; supraocular scales subequal to gradually enlarged; adpressed toe to between eye and tip of snout; body color brown; male dewlap pink, purple centrally.
SIMILAR SPECIES — *Anolis pijolense* is a small brown highland anole with smooth ventral scales. Among geographically proximal species that share these traits, *A. muralla*, *A. heteropholidotus*, *A. sminthus*, and *A. amplisquamosus* differ from *A. pijolense* in having abruptly and greatly enlarged rows of keeled middorsal scales; *A. limifrons*, *A. zeus*, *A. beckeri*, and *A. rodriguezii* are lowland forms that are smaller than *A. pijolense* (SVL to 49 mm; to 60 mm in *A. pijolense*) and differ in male dewlap color (*A. limifrons*, *A. zeus*: white; *A. rodriguezii*: orange; *A. beckeri*: reddish-pink; *A. pijolense*: pink with purple centrally). *Anolis purpurgularis* is similar to *A. pijolense* and found northeast of its range; it differs from *A. pijolense* in its uniformly purple male dewlap and multicarinate snout scales (unicarinate in *A. pijolense*).
RANGE — From 1150 to 2050 meters in northwestern Honduras.
NATURAL HISTORY — Common in forest; active diurnally on ground or low on tree trunks or other vegetation; sleeps on twigs or leaves. See McCranie and Köhler (2015).

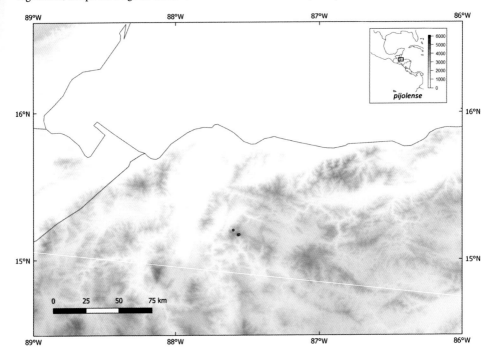

— ANOLIS PINCHOTI — PLATE 1.169

DESCRIPTION — Cochran, D. 1931. A new lizard (*Anolis pinchoti*) form Old Providence Island. *Journal of the Washington Academy of Sciences* 21 (15):354–355.
TYPE SPECIMEN — USNM 76945.
TYPE LOCALITY — "Old Providence Island, Colombia." Providencia Island is at lat 13.35 lon −81.37, 0–360 m.
MORPHOLOGY — Body length to 52 M, 46 F; ventral scales keeled; middorsal scales gradually enlarged in several rows; interparietal larger or smaller than tympanum; scales across the snout between the second canthals 8–13; toe lamellae 16–21; suboculars and supralabials in contact; scales separating supraorbital semicircles 0–2; supraocular scales gradually enlarged; adpressed toe to eye; body color brown; male dewlap red.
SIMILAR SPECIES — *Anolis pinchoti*, a *small brown* anole, is the only anole species recorded from Providence Island.
RANGE — Widespread on Providence Island and surrounding cayes, east of Nicaragua in the Caribbean Sea.
NATURAL HISTORY — Common on bushes and trees. See Tamsitt and Valdivieso (1963); Corn and Dalby (1973); Grote (2003); Calderón-Espinosa and Barragán-Forero (2011).

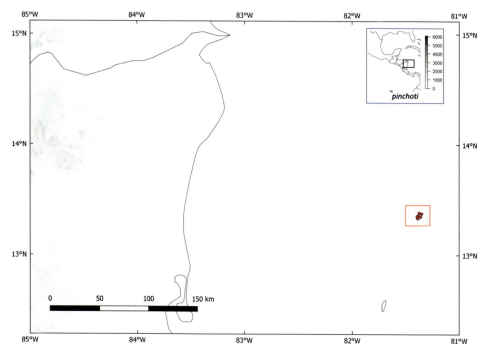

— *ANOLIS PLANICEPS* — PLATE 1.170

DESCRIPTION — Troschel, F. H. 1848. Amphibien. In: R. Schomburgk, *Reisen in British Guiana, in den Jahren 1840–1844. Im Auftrag Sr. Majestat des Konigs von Preussen ausgefuhrt von Richard Schomburgk*, vol. 3: *Versuch einer fauna und flora von British Guiana*, 649–650.
TYPE SPECIMEN — ZMB 529.
TYPE LOCALITY — "Caracas." Caracas, Venezuela is at lat 10.48 lon −66.90, 900 m.
MORPHOLOGY — Body length to 76 M 69 F; ventral scales keeled; middorsal scales gradually enlarged in several rows; interparietal usually larger than tympanum; scales across the snout between the second canthals 8–14; toe lamellae 15–18; subocular and supralabial scales separated by a row of scales; scales separating supraorbital semicircles 1–3; supraocular scales gradually enlarged; hindlimbs long, adpressed toe to beyond tip of snout; body color brown, sometimes banded or with middorsal diamonds; male and female dewlaps orangeish-red.
SIMILAR SPECIES — *Anolis planiceps* is a *large brown* lowland anole from northern Amazonia. Its keeled ventral scales distinguish it from smooth-scaled *A. fuscoauratus, A. ortonii, A. trachyderma, A. transversalis*, and *A. tigrinus*. *Anolis auratus* is smaller than *A. planiceps* (SVL to 54 mm; to 76 mm in *A. planiceps*) and has abruptly enlarged rows of keeled middorsal scales (middorsals gradually enlarged in *A. planiceps*). *Anolis onca* and *A. annectens* lack the typical anoline toepad (expanded toepad present in *A. planiceps*). Similar species *A. chrysolepis, A. brasiliensis, A. tandai, A. bombiceps*, and *A. scypheus* differ from *A. planiceps* in female dewlap color (red distally, blue proximally in *A. scypheus*; blue with cream distally in *A. tandai*; blue in *A. brasiliensis, A. bombiceps*; cream in *A. chrysolepis*; red in *A. planiceps*).
RANGE — From 0 to 1200 meters in Amazonian Venezuela, Trinidad, Guyana, and northern Brazil.
NATURAL HISTORY — Common in forest; diurnally usually on ground; sleeps on low vegetation. See Beebe (1944; as *Anolis nitens*); Hoogmoed (1973); Vitt and Zani (1996a; as *A. chrysolepis*); Myers and Donnely (2008); de Oliveira et al. (2014).

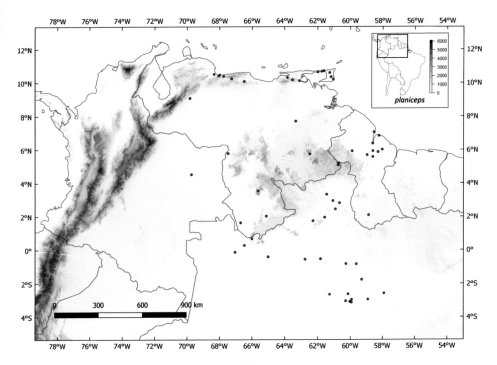

— *ANOLIS PODOCARPUS* — PLATE 1.171, 2.122

DESCRIPTION — Ayala-Varela, F. P., O. Torres-Carvajal. 2010. A new species of dactyloid anole (Iguanidae, Polychrotinae, *Anolis*) from the southeastern slopes of the Andes of Ecuador. *Zookeys* 53:59–73.
TYPE SPECIMEN — QCAZ 10126.
TYPE LOCALITY — "Ecuador, Provincia Zamora-Chinchipe, Romerillos Alto, 04 13' 35.6" S, 78 56' 23.0" W, 1550 m."
MORPHOLOGY — Body length to 96 M, 89 F; ventral scales smooth; middorsal scales enlarged in 0–2 rows; interparietal smaller than tympanum; scales across the snout between the second canthals 14–20; toe lamellae 20–25; suboculars and supralabials in contact or separated by a row of scales; scales separating supraorbital semicircles 1–3; supraocular scales subequal; hindlimbs long, adpressed toe to tip of snout; body color usually greenish-brown with dark bands and light ocelli; tail banded; male dewlap brown with light brown along scale rows; female dewlap darker brown with light brown along scale rows; iris blue.
SIMILAR SPECIES — *Anolis podocarpus* is a *large brown* to *giant* highland anole frequently in sympatry with *A. soinii* and *A. fuscoauratus*, and sometimes found with *A. lososi* and *A. hyacinthogularis*. *Anolis soinii* differs from *A. podocarpus* in its green dorsum and predominantly white male dewlap (brown body and dewlap in *A. podocarpus*). Small anoles *A. fuscoauratus*, *A. lososi*, and *A. hyacinthogularis* have larger head scales than *A. podocarpus* (scales across the snout 9–13 in *A. fuscoauratus*, 4–6 in *A. lososi*, 6–9 in *Anolis hyacinthogularis*; 14–20 in *A. podocarpus*). *Anolis fitchi* is very similar to *A. podocarpus* and is found at comparable elevations north of the range of *A. podocarpus*. *Anolis podocarpus* and *A. fitchi* are separable by male dewlap (*A. fitchi*: yellow-tan, with scattered scales; *A. podocarpus*: dark brown, with separated rows of light scales).
RANGE — From 1400 to 1950 meters on the Amazonian Andean slope of southern Ecuador.
NATURAL HISTORY — Rare in forest; often sleeps along streams, usually on leaves, especially ferns, 0.5 to 2.5 meters above ground; see Narváez (2017); Torres-Carvajal et al. (2019); Arteaga et al. (2023).

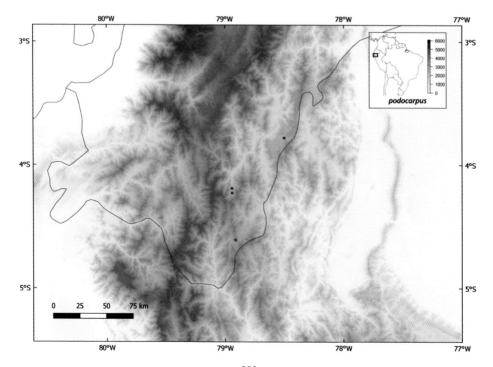

SPECIES ACCOUNTS

— *ANOLIS POECILOPUS* — PLATE 1.172, 2.152

DESCRIPTION — Cope, E. D. 1863. Contributions to Neotropical saurology. *Proceedings of the Academy of Natural Sciences of Philadelphia* 14:170.
TYPE SPECIMEN — Syntypes USNM 4320, 4331. Lost, according to Barbour (1934).
TYPE LOCALITY — "Near Carthagena, and on the Truando, New Granada." The Truando River is a tributary of the Atrato River, which flows north across Chocó to enter the Gulf of Uraba at the current Panama–Colombia border. The Truando enters the Atrato nearly 400 km southwest of Cartagena near the town of Riosucio (lat 7.43 lon −77.11, 0 m). The Truando was explored and collected extensively in the mid-1800s as a potential site for an Atlantic to Pacific waterway (e.g., Cassin 1860/1861).
MORPHOLOGY — Body length to 72 M, 68 F; ventral scales keeled; middorsal scales may be interpreted as slightly enlarged in 13–24 rows or as more or less uniform (see Williams 1984b); interparietal approximately equal to tympanum; scales across the snout between the second canthals 14–24; toe lamellae 13–22; suboculars and supralabials in contact or separated by rows of scales; scales separating supraorbital semicircles 2–5; supraocular scales gradually enlarged; adpressed toe to eye or just beyond; body color brown with light lateral stripe; male dewlap yellow-orange; female dewlap absent.
SIMILAR SPECIES — *Anolis poecilopus* is a *semiaquatic* lowland anole. Among comparably sized, potentially sympatric brown anoles: *Anolis vittigerus* differs from *A. poecilopus* in its larger head scales (scales across the snout 6–13; 14–24 in *A. poecilopus*); *A. capito*, *A. gaigei*, *A. gracilipes*, and *A. tropidogaster* have longer hindlimbs (adpressed toe reaches past eye; to eye in *A. poecilopus*); *A. notopholis* has a deep, puncturelike axillary pocket (absent in *A. poecilopus*). *Anolis poecilopus* seems most likely to be confused with the other western Colombian semiaquatic anoles *A. maculigula*, *A. macrolepis*, and *A. rivalis*, and *A. lionotus*, which geographically replaces *A. poecilopus* to the west at the Panama Canal. *Anolis rivalis*, *A. macrolepis*, and *A. lionotus* have several (10+) rows of abruptly enlarged middorsal scales (middorsals barely enlarged in *A. poecilopus*), and *A. macrolepis* and *A. lionotus* have

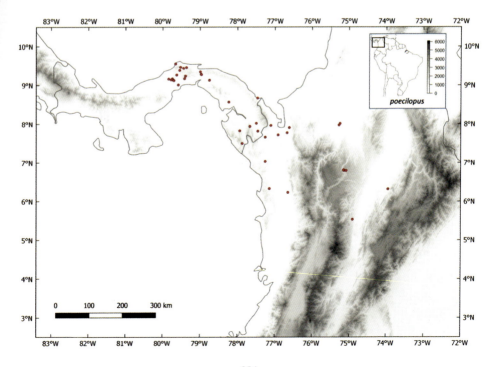

larger head scales than *A. poecilopus* (7–14 scales across the snout). *Anolis maculigula* differs from *A. poecilopus* in its smooth ventral scales, blue-greenish body color (brown in *A. poecilopus*), and larger body size (SVL to over 100 mm).
RANGE — From 0 to 750 meters, from central Panama to northern Pacific and southwestern Caribbean Colombia.
NATURAL HISTORY — Common along streams; perches on roots and trunks; sleeps on vegetation or rock surfaces less than 1 meter above ground. See Campbell (1973); Andrews (1976); Rand and Myers (1990); Muñoz et al. (2015).
COMMENT — The specimens of purported *Anolis poecilopus* from the Magdalena River valley may represent an undescribed species.

— *ANOLIS POEI* — PLATE 1.173, 2.112

DESCRIPTION — Ayala-Varela, F. P., D. Troya-Rodríguez, X. Talero-Rodríguez, O. Torres-Carvajal. 2014. A new Andean anole species of the Dactyloa clade (Squamata: Iguanidae) from western Ecuador. *Amphibian and Reptile Conservation* [Special Section] 8:8–24.
TYPE SPECIMEN — QCAZ 3449.
TYPE LOCALITY — "Ecuador, Provincia Bolívar, Telimbela, 01.65789° S, 79.15334° W, WGS84 1,354 m."
MORPHOLOGY — Body length to 60 M, 59 F; ventral scales smooth; middorsal scales enlarged in 0–2 rows; interparietal smaller than tympanum; scales across the snout between the second canthals 11–14; toe lamellae 18–19; suboculars and supralabials in contact; scales separating supraorbital semicircles 1–3; supraocular scales gradually enlarged; adpressed toe approximately to eye; body color nearly solid green or with lateral green and black spots; male dewlap bluish-white with green anteriorly; female dewlap absent; iris brown.

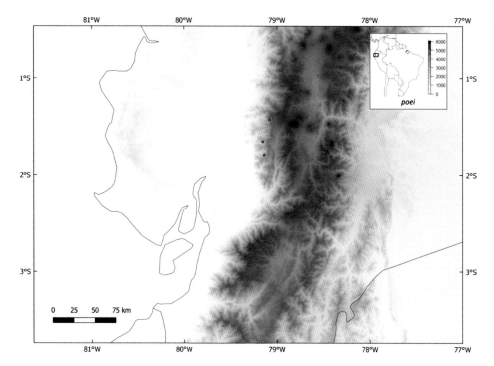

SPECIES ACCOUNTS

SIMILAR SPECIES — *Anolis poei* is a *small green* highland anole. *Anolis poei* is sympatric with *A. aequatorialis*, from which it differs in smaller body size (maximum SVL to 92 mm; to 60 mm in *A. poei*) and male and female dewlap (large and dark, black and green with blue, brown, and/or orange in *A. aequatorialis*; bluish-white with green in *A. poei*). Most green anoles potentially to be confused with *A. poei* are found at lower elevations (e.g., *A. festae*, *A. peraccae*, *A. parilis*, *A. fasciatus*, *A. biporcatus*, *A. parvauritus*, *A. purpurescens*, *A. chloris*). *Anolis poei* essentially is the southern version of *A. gemmosus*, nearly indistinguishable in external morphology and found at comparable elevations (these species appear genetically distinct [Ayala-Varela et al. 2014]). *Anolis gemmosus* tends to have fewer toe lamellae (14–18; 18–19 in *A. poei*) and greater numbers of scales across the snout than *A. poei* (mean = 15; mean = 12 in *A. poei*)
RANGE — From 1300 to 2650 meters on the Pacific Andean slope of central Ecuador.
NATURAL HISTORY — In forest and disturbed areas; sleeps on twigs or leaves from 0.5 to 4.5 meters above ground. See Torres-Carvajal et al. (2019); Arteaga et al. (2023).

— *ANOLIS POLYLEPIS* — PLATE 1.174, 2.54

DESCRIPTION — Peters, W.C.H. 1874. Über neue Saurier (*Spæriodactylus, Anolis, Phrynosoma, Tropidolepisma, Lygosoma, Ophioscincus*) aus Centralamerica, Mexico und Australien. *Monatsberichte der Koniglichen Preussischen Akademie der Wissenschaften zu Berlin*.
TYPE SPECIMEN — Syntypes ZMB 7825–26, 7830, 58002–09, MCZ 21962–3, 171299. MCZ 171299 previously was a duplicate of MCZ 21962, which explains the discrepancy in specimen numbers noted in Köhler et al. (2010).
TYPE LOCALITY — "Chiriqui." Presumably Chiriquí Province, Panama.
MORPHOLOGY — Body length to 57 M, 53 F; ventral scales smooth; middorsal scales enlarged in 0–2 rows; interparietal larger or smaller than tympanum; scales across the snout between the second

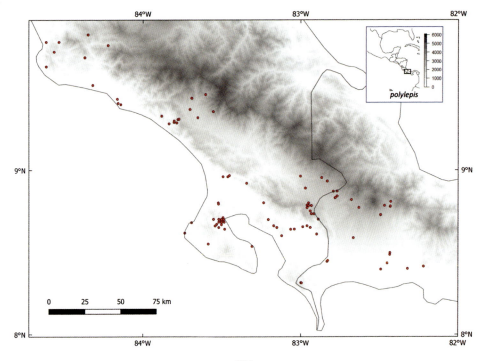

canthals 9–15; toe lamellae 13–17; suboculars and supralabials usually in contact, occasionally separated by a row of scales; scales separating supraorbital semicircles 2–4; supraocular scales gradually enlarged; hindlimbs long, adpressed toe to eye or anterior to eye; body color mostly brown with a narrow white line extending posteriorly from shoulder along flank, fading posteriorly; a dark interorbital bar usually present; male dewlap orange; female dewlap absent.

SIMILAR SPECIES — *Anolis polylepis* is a *small brown* lowland anole with smooth ventral scales and long hindlimbs (adpressed toe reaches to eye or anterior to eye). *Anolis limifrons* and *A. biscutiger* share these traits and differ from *A. polylepis* in male dewlap color (white; orange in *A. polylepis*) and smaller body size (SVL to 48 mm; to 57 mm in *A. polylepis*).

RANGE — From 0 to 1500 meters in Pacific western Panama and southern Costa Rica.

NATURAL HISTORY — Very common in forest, edge, and disturbed areas; active diurnally on low vegetation, often leaves; sleeps on twigs or leaves, usually below 2 meters above ground. See Andrews (1971, 1976, 1983); Clark (1973); Hertz (1974); Perry (1996); Savage (2002); Socci et al. (2005); Schlaepfer (2006); Barquero and Arguedas (2009); Steffen (2010); Suárez and Gutiérrez (2013); Ryan and Poe (2014); Jiménez and Rodríguez-Rodríguez (2015).

COMMENT — I consider *Anolis osa* (type locality: "about 6.3 km WSW Rincón de Osa, 8°40'36.7"N, 83°32'7.9"W, about 150 m elevation, Puntarenas Province, Costa Rica.") to be a junior synonym of *A. polylepis* (see appendix 1 for justification). For those who wish to recognize *A. osa*, it is the form of *A. polylepis* found on the Osa Peninsula, Costa Rica.

— *ANOLIS PRINCEPS* — PLATE 1.175, 2.133

DESCRIPTION — Boulenger, G. A. 1902. Description of new batrachians and reptiles from northwestern Ecuador. *Annals and Magazine of Natural History* 7(9):54–55.

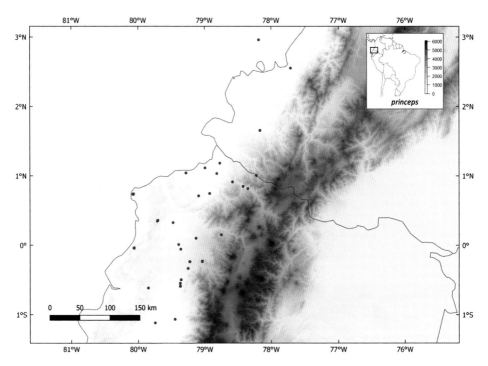

SPECIES ACCOUNTS

TYPE SPECIMEN — Syntypes BMNH 1901.3.29.97, 1946.8.8.45, 1946.8.12.99, 1946.8.13.4, 1946.9.7.62-63; MCZ 70226 (formerly BMNH 1946.8.8.46).
TYPE LOCALITY — "Several specimens from St. Javier (60 feet), Salidero (350 feet), R. Lita (3000 feet), and Paramba (3500 feet)." The elevation for "R. Lita" matches the "Lita" locality of Paynter (1993), which is located at the junction of Esmeraldas, Carchi, and Imbabura Provinces (approximately lat 0.87 lon −78.45). Salidero is at approximately lat 1.08 lon −78.67. "St. Javier" may refer to San Javier, Esmeraldas of Paynter (1993), which is located at lat 1.07 lon −78.78. Paramba is likely Hacienda Paramba (lat 0.817 lon −78.350).
MORPHOLOGY — Body length to 117 M, 110 F; ventral scales smooth; middorsal scales enlarged in 0–2 rows; interparietal smaller than tympanum; scales across the snout between the second canthals 12–17; toe lamellae 20–25; suboculars and supralabials separated by a row of scales; scales separating supraorbital semicircles 2–5; supraocular scales subequal; hindlimbs long, adpressed toe to between eye and tip of snout; body color green with lateral oblique rows of ocelli; male and female dewlaps dirty white.
SIMILAR SPECIES — *Anolis princeps* is a *giant* green lowland anole. Its combination of long hindlimbs (adpressed toe reaches to between eye and tip of snout), lateral rows of ocelli, and white dewlap in males and females will differentiate it from all potentially sympatric giant green species except *A. frenatus*. I know of no consistent trait to separate *A. princeps* from *A. frenatus*, which is found in northern Colombia and Central America (see discussion under *A. frenatus*). Commonly sympatric large green lowland anoles differ from *A. princeps* in having keeled ventral scales (*A. parvauritus*; smooth in *A. princeps*), narrow indistinct toepads (*A. parilis*; broad overlapping toepad in *A. princeps*), and suboculars and supralabials in broad contact (*A. purpurescens*, *A. fasciatus*; suboculars and supralabials separated by a row of scales in *A. princeps*).
RANGE — From 0 to 1000 meters from southern Pacific Colombia into Ecuador.
NATURAL HISTORY — Common in forest and edge habitats; diurnally often on tree trunks; sleeps on vegetation, usually above 1 meter above ground, usually on twigs or vines. See Fitch et al. (1976); Castro-Herrera et al. (2012); Miyata (2013); Narváez (2017); Pinto-Erazo et al. (2020); Torres-Carvajal et al. (2019); Arteaga et al. (2023).
COMMENT — See comments under *Anolis frenatus*.

— *ANOLIS PROBOSCIS* — FIGURE 1.1; PLATE 1.176

DESCRIPTION — Peters, J. A., G. Orcés-V. 1956. A third leaf-nosed species of the lizard genus *Anolis* from South America. *Breviora* 62:1–8.
TYPE SPECIMEN — MCZ 54300.
TYPE LOCALITY — "Neighborhood of Cunuco, a small town at 1200 meters elevation, five kilometers northwest of Mindo, on the south bank of the Rio Mindo, a northern tributary of the upper Rio Blanco, in Pichincha Province, Ecuador." Cunuco is located at approximately lat −0.26 lon −79.81.
MORPHOLOGY — Body length to 75 M, 73 F; ventral scales smooth; middorsal scales form a low crest of triangular plates; interparietal larger than tympanum; scales across the snout between the second canthals 8–10; toe lamellae 18–21; suboculars and supralabials in contact; scales separating supraorbital semicircles 2–3; supraocular scales subequal to gradually enlarged; hindlimbs short, adpressed toe to ear or just posterior to ear; body color green, tan, and brown, often with dorsal crossbands, female dorsum often solid olive; male and female dewlaps greenish-white; male (but not female) with scaly anterior snout extension.
SIMILAR SPECIES — Males of *twig* anole *Anolis proboscis* are unlikely to be confused with any potentially sympatric form due to their possession of a scaly anterior extension of the snout. In the areas where *A. proboscis* occurs, females are distinctive in their short limbs (adpressed toe reaches to ear) and tail and large (for a female) greenish-white dewlap. The serrated dorsal crest of both males and females also is differentiating.
RANGE — From 1100 to 1800 meters in the vicinity of Mindo, on the Pacific Andean slope of central Ecuador.

Anolis propinquus

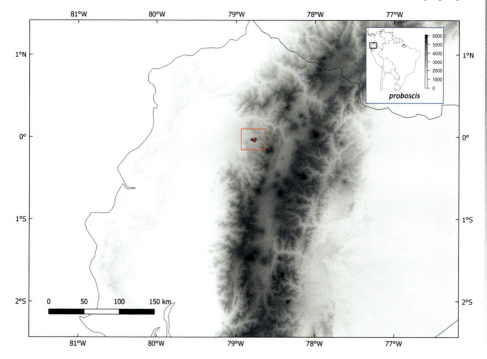

NATURAL HISTORY — Common in edge habitat and tree islands in pasture. Sleeps on twigs or vines above 1.5 meters above ground. See Yánez-Muñoz et al (2010); Poe et al. (2012); Losos et al. (2012); Arteaga et al. (2013); Narváez (2017); Quirola et al. (2017); Torres-Carvajal et al. (2019); Arteaga et al. (2023).

— *ANOLIS PROPINQUUS* — PLATE 1.177

DESCRIPTION — Williams, E. E. 1984a. New or problematic *Anolis* from Colombia. II. *Anolis propinquus*, another new species from the cloud forest of western Colombia. *Breviora* 477:1–7.
TYPE SPECIMEN — KU 169833.
TYPE LOCALITY — "Rio Calima, 1.5 km W Lago Calima, Valle, Colombia." This locality is at approximately lat 3.8818 lon −76.5740, 1300 m.
MORPHOLOGY — Maximum body length unknown, as the only known specimen is a juvenile of SVL 41; ventral scales smooth; middorsal scales enlarged in 0–2 rows; interparietal smaller than tympanum (only known specimen lacks a differentiated interparietal scale); scales across the snout between the second canthals 10; toe lamellae 25; suboculars and supralabials separated by a row of scales; scales separating supraorbital semicircles 3. Body and dewlap color in life of adults unknown.
SIMILAR SPECIES — *Anolis propinquus* is known from a single juvenile specimen, but is likely to be a *giant* anole as an adult, as it has been estimated to be a phylogenetically close relative of giant species (Poe et al. 2017a) and its number of toe lamellae suggests large body size. Among large, high-elevation, geographically proximal species, *A. eulaemus*, *A. dracula*, and *A. ventrimaculatus* possess smaller head scales (approximately 15 scales across the snout; 10 in *A. propinquus*) and fewer toe lamellae than *A. propinquus* (16–23; 25 in *A. propinquus*). *Anolis propinquus* is likely to resemble its close phylogenetic relatives *A. danieli* and *A. apollinaris* as an adult (see accounts for those species).

SPECIES ACCOUNTS

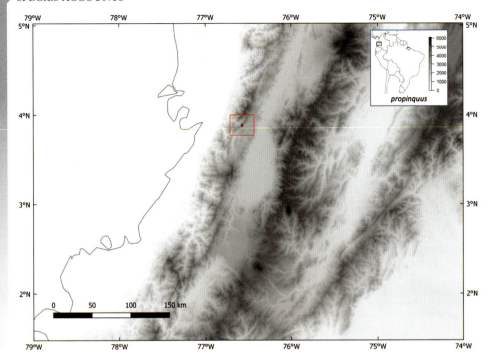

RANGE — Known only from the type locality.
NATURAL HISTORY — Only information is from the type description: type specimen was collected sleeping 0.5 meters above ground on vegetation.

— *ANOLIS PSEUDOKEMPTONI* — PLATE 1.178, 2.35

DESCRIPTION — Köhler, G., M. Ponce, J. Sunyer, A. Batista. 2007. Four new species of anoles (genus *Anolis*) from the Serranía de Tabasará, West-Central Panama (Squamata: Polychrotidae). *Herpetologica* 63:380–383.
TYPE SPECIMEN — SMF 85420.
TYPE LOCALITY — "La Nevera, 8 29 45 N, 81 46 35 W, 1600 m elevation, Serranı́a de Tabasara, Comarca Ngöbe Bugle, Distrito de Nole Düima, Corregimiento de Jadeberi, Panama." Corrected by Lotzkat et al. (2010) to 8° 30' N, 81° 46' 20" W.
MORPHOLOGY — Body length to 55 M, 55 F; ventral scales smooth; middorsal scales enlarged in 0–2 rows; interparietal larger than tympanum; scales across the snout between the second canthals 9–10; toe lamellae 14–17; suboculars and supralabials in contact; scales separating supraorbital semicircles 1; supraocular scales gradually enlarged; hindlimbs short, adpressed toe to ear or posterior to ear; body color grayish-brown; tail banded; male dewlap pink posteriorly blurring to orange-red anteriorly; female dewlap pale orange.
SIMILAR SPECIES — *Anolis pseudokemptoni* is a *small brown* highland anole with smooth ventral scales and short hindlimbs (adpressed toe reaches to ear). Geographically proximal species that share these traits are most easily distinguished by male dewlap color: orange in *A. gruuo*, *A. elcopeensis*, *A. carpenteri*; orange anteriorly, yellow posteriorly in *A. fortunensis*; red in *A. fungosus*, *A. salvini*. Anolis

Anolis pseudopachypus

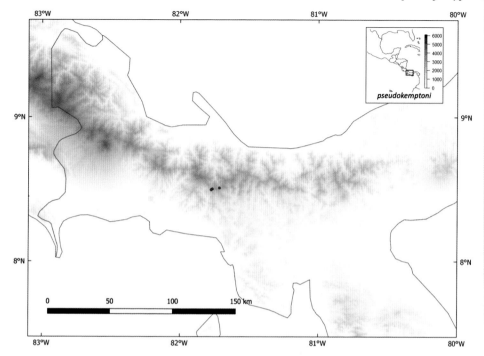

kemptoni is nearly indistinguishable from *A. pseudokemptoni*, differing subtly in male dewlap color shades (compare dewlap plates, and note geography).
RANGE — From 1100 to 2050 meters in western Panama from Cerro Santiago to Cerro Sagui.
NATURAL HISTORY — Common in forest and forest edge; sleeps on twigs or leaves below 3 meters above ground.

— *ANOLIS PSEUDOPACHYPUS* — PLATE 1.179

DESCRIPTION — Köhler, G., M. Ponce, J. Sunyer, A. Batista. 2007. Four new species of anoles (genus *Anolis*) from the Serranía de Tabasará, West-Central Panama (Squamata: Polychrotidae). *Herpetologica* 63:383–385.
TYPE SPECIMEN — SMF 85153.
TYPE LOCALITY — "La Nevera, 8 29 45 N, 81 46 35 W, 1600 m elevation, Serranı́a de Tabasara, Comarca Ngöbe Bugle, Distrito de Nole Düima, Corregimiento de Jadeberi, Panama." Corrected by Lotzkat et al. (2010) to 8° 30' N, 81° 46' 20" W.
MORPHOLOGY — Body length to 46 M, 47 F; ventral scales keeled; middorsal scales enlarged in 0–2 rows; interparietal smaller than tympanum; scales across the snout between the second canthals 15–21; toe lamellae 12–14; suboculars and supralabials usually separated by rows of scales; scales separating supraorbital semicircles 6–9; supraocular scales subequal to gradually enlarged; appressed toe to between eye and tip of snout; body color mostly brown, patterned; tail banded; male dewlap orange-yellow.
SIMILAR SPECIES — *Anolis pseudopachypus* is a *small brown* highland anole that is distinctive in its tiny head scales (15–21 scales across the snout between the second canthals; 6–9 scales between supraorbital semicircles). Similar highland anoles *A. magnaphallus*, *A. benedikti*, and *A. pachypus* are found

SPECIES ACCOUNTS

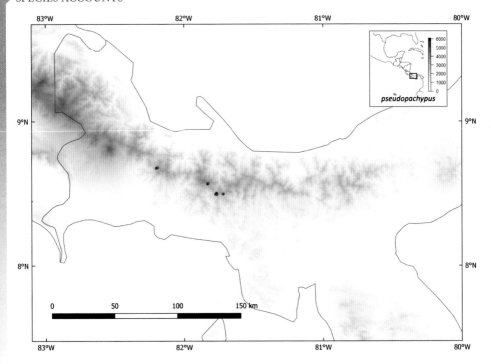

west of the range of *A. pseudopachypus*. These species differ from *A. pseudopachypus* in male dewlap color (red, or red-orange with yellow; dull orange-yellow in *A. pseudopachypus*) and in their larger head scales (usually fewer than 16 scales across the snout, fewer than 6 scales between the supraorbital semicircles).
RANGE — From 1550 to 2050 meters in western Panama around Cerro Santiago.
NATURAL HISTORY — In forest; active diurnally on ground; sleeps on low vegetation.

— *ANOLIS PSEUDOTIGRINUS* — PLATE 1.180

DESCRIPTION — Amaral, A. 1933. Estudos sobre lacertilios neotropicos I. Novos generos e especies de lagartos do Brasil. *Memorias do Instituto Butantan* 7:60.
TYPE SPECIMEN — MZUSP 721.B
TYPE LOCALITY — "Regiao do rio Doce, Espirito Santo, Brasil." Ernesto Garbe, collector of the type specimen, traveled the Rio Doce in Espirito Santo State from the border of Minas Gerais to Linhares in 1906 (Hamilton 2021). Linhares, Espirito Santo, Brazil is at lat −19.40 lon −40.06, 15 m.
MORPHOLOGY — Body length to 53 M, 54 F; ventral scales smooth; middorsal scales enlarged in 0–2 rows; interparietal larger than tympanum; scales across the snout between the second canthals 5–6; toe lamellae 17–20; subocular and supralabial scales in contact; supraorbital semicircles in contact; female dewlap white.
SIMILAR SPECIES — *Anolis pseudotigrinus* is a *twig anole*. Prates et al. (2017) state this species is sympatric with *A. punctatus*, a large green species with which it is unlikely to be confused, and *A. fuscoauratus* and *A. nasofrontalis*, from which it differs amply. *Anolis fuscoauratus* differs from *A. pseudotigrinus* in its smaller head scales (1–3 scales between the supraorbital semicircles; supraorbital semicircles in contact in *A. pseudotigrinus*). *Anolis nasofrontalis* differs from *A. pseudotigrinus* in female dewlap

262

Anolis punctatus

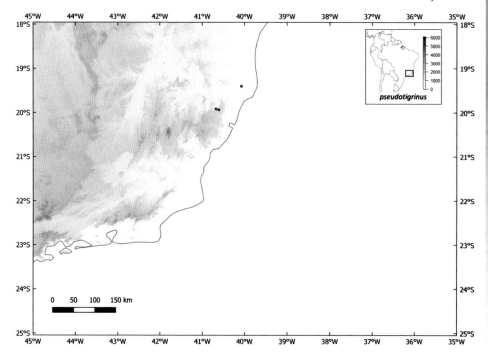

color (tan fading to white posteriorly; white in *A. pseudotigrinus*) and dewlap scalation (12–14 rows of dewlap scales in *A. pseudotigrinus*; 5–7 in *A. nasofrontalis*). *Anolis ortonii* may be sympatric with *A. pseudotigrinus* and differs in its smaller head scales (7–11 scales across the snout; 5–6 in *A. pseudotigrinus*) and female dewlap color (red with yellow-orange along scale rows). *Anolis sagrei*, *A. brasiliensis*, and *A. meridionalis* differ from *A. pseudotigrinus* in their keeled ventral scales (smooth in *A. pseudotigrinus*).

RANGE — From 700 to 750 meters in coastal Brazil, Espirito Santo.

NATURAL HISTORY — In forest or forest edge; sleeps on twigs 2–3 meters above ground. See Prates et al. (2017).

— *ANOLIS PUNCTATUS* — FIGURE 3.22; PLATE 1.181

DESCRIPTION — Daudin, F. M. 1802. *Histoire Naturelle, Générale et Particulière des Reptiles*, vol. 4. F. Dufart, Paris. 84.

TYPE SPECIMEN — MNHN 2340.

TYPE LOCALITY — "L'Amérique Méridionale." That is, South America.

MORPHOLOGY — Body length to 89 M, 77 F; ventral scales smooth to weakly keeled; middorsal scales enlarged in 0–2 rows; interparietal larger or smaller than tympanum; scales across the snout between the second canthals 8–14; toe lamellae 22–32; suboculars and supralabials in contact; scales separating supraorbital semicircles 0–2; supraocular scales gradually enlarged; appressed toe to ear or to between ear and eye; body color green, changing to brown under stress; male dewlap yellow-orange; female dewlap absent or small, colored similar to male's; male snout enlarged, often bulbous.

SIMILAR SPECIES — *Anolis punctatus* is a *large green* lowland Amazonian anole. *Anolis punctatus* is unlikely to be confused with most potentially sympatric species due to large size (SVL to 89 mm),

SPECIES ACCOUNTS

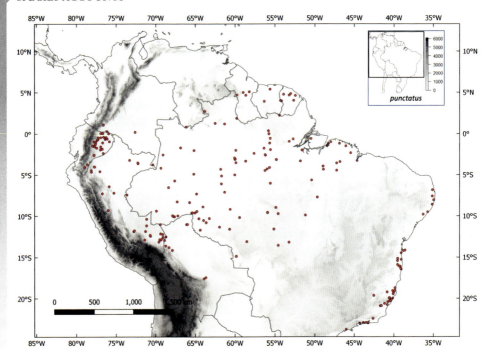

dorsal color, or ventral scales (smooth to weakly keeled). *Anolis bombiceps, A. scypheus, A. planiceps, A. tandai, A. brasiliensis,* and *A. chrysolepis* are predominantly brown with strongly keeled ventral scales. *Anolis fuscoauratus, A. trachyderma,* and *A. ortonii* are smaller than *A. punctatus* (SVL < 59 mm) and brown or gray. *Anolis transversalis* is predominantly brown with blue eyes (brown in *A. punctatus*). The rare green species *Anolis dissimilis* is smaller than *A. punctatus* (SVL to 58 mm) and has a differently colored male dewlap (white; yellow-orange in *A. punctatus*). The rare green species *Anolis phyllorhinus* has a scaly leaf-like snout extension in males (lacking in *A. punctatus*) and a red and blue male dewlap. *Anolis caquetae,* known from a single specimen from one Eastern Colombian locality, apparently is smaller than *A. punctatus* (SVL in mature male holotype 58 mm) and distinctive in its large interparietal scale contacting the supraorbital semicircles (scales separate interparietal and supraorbital semicircles in *A. punctatus*). Its male dewlap coloration may be distinctive (see accounts of *A. caquetae, A. vaupesianus*).
RANGE — From 0 to 1450 meters in Amazonia.
NATURAL HISTORY — Rare to common in forest and forest edge; active diurnally on tree trunks and other vegetation, potentially at very high abundance (Arteaga et al. 2023); sleeps above 1.5 meters above ground on twigs or leaves. See Beebe (1944); Rand and Humphrey (1968); Crump (1971); Hoogmoed (1973); Gasnier et al. (1994); Ávila-Pires (1995); Vitt and Zani (1996b); Duellman (1978); Martins (1991; as *A. philopunctatus*); Vitt et al. (1999; Vitt (2000); Vitt et al. (2003a); Pinto-Aguirre (2014); Landauro and Morales (2007); Macedo et al. (2008); Silva et al. (2011); de Oliveira et al. (2014); Araújo-Nieto (2017); Faria et al. (2019); Moreno-Arias et al. (2020); Oliveira et al. (2021); Torres-Carvajal et al. (2019); Arteaga et al. (2023); Pinto and Torres-Carvajal (2023).
COMMENT — I consider *Anolis philopunctatus* (type locality: "Brasil: Amazonas: INPA/WWF Reserves (2206 Dimona), 90 km NW Manaus") to be a junior synonym of *A. punctatus* based on Prates et al. (2015).

— *ANOLIS PURPURESCENS* — PLATE 1.182, 2.67

DESCRIPTION — Cope, E. D. 1899. Contributions to the herpetology of New Granada and Argentina with descriptions of new forms. *The Philadelphia Museums Scientific Bulletin* 1:7–9.
TYPE SPECIMEN — USNM 4321.
TYPE LOCALITY — "Truando River, new Granada." Possibly lat 7.43 lon −77.11, elevation 0. See comments regarding type locality for *A. poecilopus*.
MORPHOLOGY — Body length to 79 M, 76 F; ventral scales smooth; middorsal scales enlarged in 0–2 rows; interparietal usually smaller than tympanum; scales across the snout between the second canthals 10–15; toe lamellae 17–22; suboculars and supralabials in contact; scales separating supraorbital semicircles 1–5; supraocular scales gradually enlarged; adpressed toe to eye; body color green with lateral rows of ocelli; male dewlap yellow-orange; female dewlap dark olive.
SIMILAR SPECIES — *Anolis purpurescens* is a *large green* lowland anole. Among geographically proximal, comparably sized green anoles, *A. frenatus*, *A. latifrons*, and *A. princeps* have greater numbers of toe lamellae (usually 23–28; 17–22 in *A. purpurescens*) and frequently have a row of scales separating suboculars and suprabials (suboculars and supralabials in broad contact in *A. purpurescens*); *A. biporcatus*, *A. chloris*, and *A. parvauritus* differ from *A. purpurescens* in their keeled ventral scales (smooth in *A. purpurescens*). *Anolis brooksi* and *A. fraseri* are larger than *A. purpurescens* (>100 mm) and have shorter hindlimbs (adpressed toe reaches approximately to ear; to eye in *A. purpurescens*). *Anolis festae*, *A. anchicayae*, *A. fasciatus*, and *A. peraccae* may have blue eyes (brown in *A. purpurescens*) and differ in male dewlap color (white or yellow green; yellow-orange in *A. purpurescens*). *Anolis kunayalae*, *A. mirus*, and *A. parilis* differ from *A. purpurescens* in their narrow fourth (longest) toe with a large claw and few toe lamellae (<17; broad, distinct toepad and usual anoline claw, 17–22 lamellae in *A. purpurescens*). *Anolis limon* is similar to *A. purpurescens* but found northeast of its range; it is differentiable from *A. purpescens* by its white male dewlap (orange

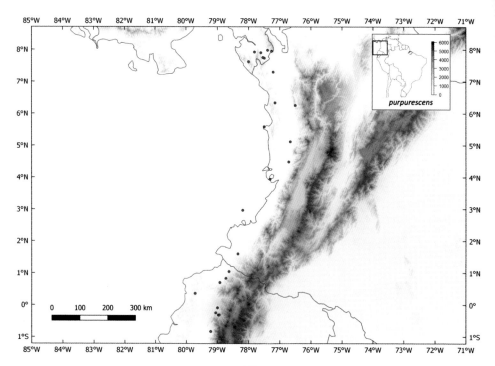

SPECIES ACCOUNTS

in *A. purpurescens*). *Anolis maia* is similar to *A. purpurescens* and replaces it geographically to the west. *Anolis maia* differs from *A. purpurescens* in its lateral body pattern (widely separated rows of ocelli or thick bands in *A. maia*; denser arrangement of ocelli and reticulations on green background in *A. purpurescens*).

RANGE — From 0 to 1100 meters, from the Darién of Panama east and south to Pacific Colombia and northern Ecuador.

NATURAL HISTORY — Rare in forest interior; active diurnally on trees; sleeps on twigs and (more frequently) leaves, usually over 1.5 meters above ground. See Williams and Duellman (1967; description of *Anolis chocorum*); Castro-Herrera (1988; as *A. chocorum*); Moreno et al. (2007); Rengifo et al. (2014, 2015; as *A. chocorum*); Narváez (2017); Viteri (2015; as *A. chocorum*); Rengifo et al. (2019); Torres-Carvajal et al. (2019); Arteaga et al. (2023).

COMMENT — Literature before 2016 that references this species calls it *A. chocorum* (see Chun 2010; Batista et al. 2015)

— *ANOLIS PURPURGULARIS* — PLATE 1.183

DESCRIPTION — McCranie, J. R., G. A. Cruz, P. A. Holm. 1993a. A new species of cloud forest lizard of the *Norops schiedei* group (Sauria: Polychrotidae) from northern Honduras. *Journal of Herpetology* 27:386–392.

TYPE SPECIMEN — USNM 322885.

TYPE LOCALITY — "2.5 airline km NNE La Fortuna (15° 26' N, 87° 18' W), 1690 m elevation, Cordillera Nombre de Dios, Departamento de Yoro, Honduras."

MORPHOLOGY — Body length to 59 M, 58 F; ventral scales smooth to weakly keeled; middorsal scales gradually enlarged in 2 to several rows; interparietal smaller than tympanum; scales across the

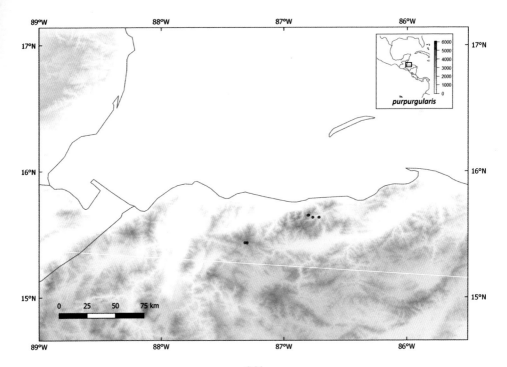

snout between the second canthals 6–9; toe lamellae 15–18; suboculars and supralabials separated by a row of scales; scales between supraorbital semicircles 1–4; supraocular scales gradually enlarged; adpressed toe to between eye and tip of snout; body color brown; male dewlap pinkish purple.

SIMILAR SPECIES — *Anolis purpurgularis* is a *small brown* highland anole. Potentially sympatric species *A. laeviventris, A. kreutzi, A. cusuco, A. beckeri,* and *A. sericeus* have shorter hindlimbs than *A. purpurgularis* (adpressed toe reaches to ear; to between eye and tip of snout in *A. purpurgularis*). *Anolis tropidonotus, A. uniformis, A. crassulus, A. caceresae, A. morazani, A. muralla, A. heteropholidotus, A. sminthus, A. amplisquamosus,* and *A. rubribarbaris* differ from *A. purpurgularis* in having more abruptly and greatly enlarged rows of keeled middorsal scales (middorsals slightly enlarged in *A. purpurgularis*). *Anolis cupreus* has strongly keeled ventral scales (smooth to weakly keeled in *A. purpurgularis*). *Anolis pijolense* is similar to *A. purpurgularis* and found southwest of its range; it differs from *A. purpurgularis* in male dewlap color (pink with purple center; solid pinkish-purple in *A. purpurgularis*) and its unicarinate snout scales (multicarinate in *A. purpurgularis*).

RANGE — From 1550 to 2050 meters in north central Honduras.

NATURAL HISTORY — In forest; active diurnally on ground or low vegetation; sleeps on twigs or leaves. See McCranie and Köhler (2015).

— *ANOLIS PURPURONECTES* — PLATE 1.184, 2.87

DESCRIPTION — Gray, L., R. Meza-Lazaro, S. Poe, A. Nieto-Montes de Oca. 2016. A new species of semiaquatic *Anolis* (Dactyloidae) from Oaxaca and Veracruz, Mexico. *The Herpetological Journal* 26:253–262.

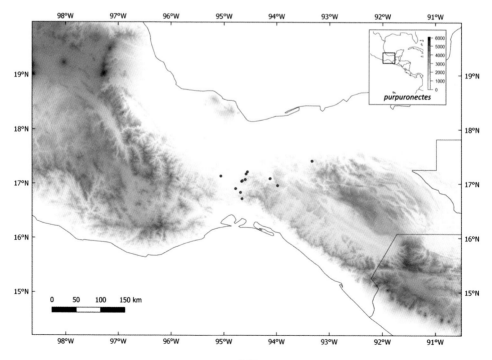

SPECIES ACCOUNTS

TYPE SPECIMEN — MZFC 28961.
TYPE LOCALITY — "Approximately 1.6 km N of Chalchijapa, Municipality of Santa María Chimalapa, Oaxaca, Mexico, 17.04377 N, 94.66586 W, 268 m."
MORPHOLOGY — Body length to 92 M, 72 F; ventral scales keeled; middorsal scales enlarged in 0–2 rows; interparietal usually smaller than tympanum; scales across the snout between the second canthals 9–12; toe lamellae 15–18; suboculars and supralabials usually separated by a row of scales; scales separating supraorbital semicircles 1–4; supraocular scales gradually enlarged; adpressed toe to between ear and eye (body is long); body color brown with some reddish tint, especially ventrally, and weak banding and a light lateral stripe; male dewlap purple; female dewlap usually absent.
SIMILAR SPECIES — *Anolis purpuronectes* is a *semiaquatic* anole. Among comparably sized geographically proximal brown anoles, *Anolis capito*, *A. alvarezdeltoroi*, and *A. lemurinus* have long hindlimbs (adpressed toe reaches anterior to snout in *A. capito*, *A. alvarezdeltoroi*, anterior to eye in *A. lemurinus*; to between ear and eye in *A. purpuronectes*); *A. petersii* has a distinctively colored small dewlap in males and females (pink with black blotches in males; pale with black blotches in females; large and purple in males, absent in female of *A. purpuronectes*); *Anolis cristatellus* and *A. sagrei* have larger head scales (4–9 scales across the snout; 9–12 in *A. purpuronectes*) and tend to be found only in disturbed areas. Semiaquatic *A. barkeri* is very similar to *A. purpuronectes* but found north and east of its range and differing from *A. purpuronectes* in its red with yellow male dewlap.
RANGE — From 150 to 1200 meters in eastern Oaxaca and Veracruz, Mexico.
NATURAL HISTORY — Common in forest along clear streams; sleeps on streamside boulders or vegetation up to 1 meter above ground. See Meyer (1968; part of sample, as *Anolis barkeri*); Powell and Birt (2001; part of sample, as *A. barkeri*).
COMMENT — I am not confident of the geographical boundaries of *Anolis purpuronectes* and similar species *A. barkeri* displayed on the range maps (compare maps for these species). More work is needed to delimit the geographic boundaries and species status of these forms.

— *ANOLIS PYGMAEUS* — PLATE 1.185

DESCRIPTION — Álvarez del Toro, M., H. M. Smith. 1956. Notulae Herpetologicae Chiapasiae I. *Herpetologica* 12:7.
TYPE SPECIMEN — UIMNH 37975.
TYPE LOCALITY — "El Ocote, 600 m." Earlier in the description this locality is placed at "35 km. NW Ocozocoautla." The "old road" (Johnson et al. 1976) that heads northwest from Ocozocoautla de Espinosa reaches 600 meters approximately 35 km northwest of the city, near the eastern edge of El Ocote reserve at approximately lat 16.96 lon −93.47.
MORPHOLOGY — Body length to 35 M (unknown for females); ventral scales keeled; middorsal scales greatly enlarged in several rows; interparietal larger or smaller than tympanum; scales across the snout between the second canthals 7–8; toe lamellae 11–13; suboculars and supralabials in contact; scales separating supraorbital semicircles 1–2; supraocular scales gradually enlarged; adpressed toe to between ear and eye; body color brown; male dewlap red.
SIMILAR SPECIES — *Anolis pygmaeus* is a *small brown* lowland anole. Its abruptly enlarged rows of middorsal scales will separate *A. pygmaeus* from other geographically proximal small brown anoles except *A. tropidonotus*, *A. uniformis*, and *A. compressicauda*. These species differ from *A. pygmaeus* in their possession of a puncturelike axillary pocket (absent in *A. pygmaeus*). *Anolis duellmani* is very similar to *A. pygmaeus* and found west of its range. These species may differ in male dewlap color (pink in *A. duellmani*, red in *A. pygmaeus*).
RANGE — From 50 to 700 meters in western Chiapas and eastern Veracruz and Oaxaca, central Mexico.
NATURAL HISTORY — I am not aware of published natural history information on this species, and I have not observed it in life. Based on anecdotal reports and the ecology of its congener and potential

Anolis quercorum

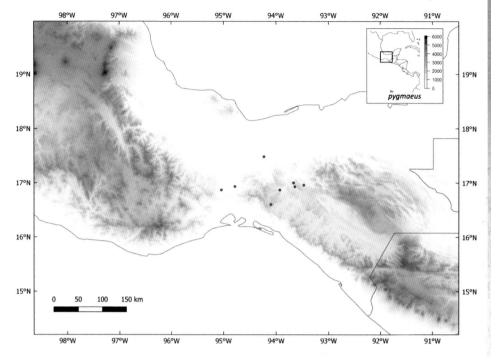

conspecific *Anolis duellmani*, I expect *A. pygmaeus* to be active on and near the ground and sleep on low vegetation such as bushes.
COMMENT — See discussion under *Anolis duellmani*.

— *ANOLIS QUERCORUM* — PLATE 1.186, 2.18

DESCRIPTION — Fitch, H. S. 1978. Two new anoles (Reptilia: Iguanidae) from Oaxaca with comments on other Mexican species. *Contributions in Biology and Geology*, Milwaukee Public Museum 20:6–8.
TYPE SPECIMEN — KU 176050.
TYPE LOCALITY — "26 km SE Nochixtlán (2.5 km NW Cuesta Blanca, Highway 190), Oaxaca, Mexico." This location is at lat 17.331 lon −97.161, 2250 m.
MORPHOLOGY — Body length to 45 M, 41 F; ventral scales keeled; middorsal scales gradually or abruptly enlarged in several rows; interparietal smaller than tympanum; scales across the snout between the second canthals 5–8; toe lamellae 10–16; suboculars and supralabials in contact; scales separating supraorbital semicircles 0–1; supraocular scales abruptly enlarged; adpressed toe to between ear and eye; body color brown; male dewlap pink.
SIMILAR SPECIES — *Anolis quercorum* is a *small brown* highland anole. Among geographically proximal small brown species, *A. milleri* and *A. rubiginosus* have smooth ventral scales (keeled in *A. quercorum*); *A. laeviventris*, *A. microlepidotus*, and *A. nebulosus* have shorter hindlimbs (adpressed toe reaches to ear; to between ear and eye in *A. quercorum*). In addition, *A. milleri*, *A. rubiginosus*, and *A. laeviventris* lack the abruptly enlarged supraocular scales present in *A. quercorum* (supraocular scales grade in size in these species). *Anolis nebuloides* and *A. megapholidotus* differ from *A. quercorum* in their relatively larger middorsal scales (larger than ventral scales; midddorsals smaller than ventrals in *A. quercorum*).

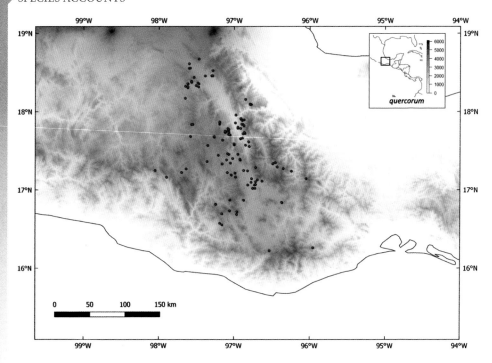

RANGE — From 650 to 2550 meters in central and western Oaxaca, Mexico.
NATURAL HISTORY — In open oak woodlands; active diurnally low (<1 meter) on trees or in leaf litter; sleeps on twigs or leaves or (as observed by Julian Davis) ensconced in hanging moss. See Ramírez-Bautista et al. (2002a, b); Ramírez-Bautista (2003); Peterson et al. (2004); Canseco-Márquez & Gutiérrrez-Mayén (2010).
COMMENT — I consider *Anolis carlliebi* (type locality: "Ixtlán de Juárez, university campus (17.316350°N, 96.483470°W, WGS84), 1945 m, Estado de Oaxaca, Mexico) and *A. sacamecatensis* ("Cerro Sacamecates (16.549440°N, 95.819820°W, WGS84), 2035 m, Estado de Oaxaca, Mexico") to be junior synonyms of *A. quercorum* (see appendix 1 for justification). For those who wish to recognize these forms, *A. carlliebi* is composed of the northern individuals of *A. quercorum*; *A. sacamecatensis* corresponds to the eastern members of *A. quercorum*.

— *ANOLIS QUIMBAYA* — PLATE 1.187

DESCRIPTION — Moreno-Arias, R. A., M. A. Méndez-Galeano, I. Beltrán, M. Vargas-Ramírez. 2023. Revealing the anole diversity in the highlands of Northern Andes: New and resurrected species of the *Anolis heterodermus* species group. *Vertebrate Zoology* 73:161–188.
TYPE SPECIMEN — MHUA-R 12691.
TYPE LOCALITY — "Medellín municipality, Antioquia department, Colombia (6.2688°N −75.4992°W, 2400 m asl)."
MORPHOLOGY — Body length to 73 M, 69 F; ventral scales smooth; middorsal scales form a crest of triangular plates (see Lazell 1969: fig. 1); interparietal larger than tympanum; scales across the snout between the second canthals 3–7; toe lamellae 20–24; suboculars and supralabials in contact; supraorbital semicircles in contact; supraocular scales abruptly enlarged; hindlimbs short, adpressed

Anolis quimbaya

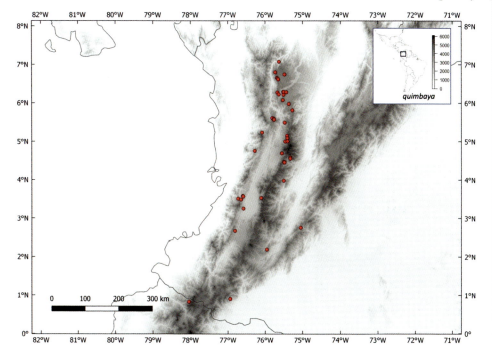

toe to ear or posterior to ear; body color brown or green with white and/or gray, with light stripe from supralabial area back to shoulder; dewlap orange in male, green in female; heterogeneous lateral scutellation.

SIMILAR SPECIES — *Anolis quimbaya* is a distinctive highland *twig* anole. The heterogeneous lateral scalation, wherein large smooth oval scales are separated by granular interstitial scales, is differentiating in *A. quimbaya*, especially in combination with its short hindlimbs and tail and large smooth head scales, throughout its geographic range in the southern, central and western Andes of Colombia. Among eastern Andean anoles that share these traits, *A. vanzolinii* and *A. inderenae* are larger than *A. quimbaya* (to >100 mm SVL; to 73 mm in *A. quimbaya*) and possess greater numbers of toe lamellae (23–28; 20–24 in *A. quimbaya*); *A. heterodermus*, *A. richteri*, and *A. tequendama* differ from *A. quimbaya* in male and female dewlap color (*A. heterodermus*: purple, red, or black in males and females; *A. richteri*: cream to yellow in males and females, sometimes with red distally; *A. tequendama*: brown with or without spotting in male, yellow with dark stripes in female; *A. quimbaya*: orange in male, green in female).

RANGE — From 1800 to 3100 meters on the Pacific Andean slope of extreme northern Ecuador to the western and central Andes of Colombia (i.e., high Colombian localities excluding the northeast Andean ranges of similar species *Anolis richteri*, *A. heterodermus*, *A. tequendama*, and *A. inderenae*).

NATURAL HISTORY — Common in cloud forest, edge habitat, and páramo; diurnally active on vegetation above 0.5 meters above ground; sleeps on twigs, leaves, or vines above 1.5 meters above ground (presumably lower in páramo). See Dunn (1944; as *Phenacosaurus heterodermus*); Torres-Carvajal et al. (2010; as *Anolis heterodermus*); Arteaga et al. (2023).

SPECIES ACCOUNTS

— *ANOLIS RICHTERI* — PLATE 1.188

DESCRIPTION — Dunn, E. R. 1944. The lizard genus *Phenacosaurus. Caldasia* 3 (11):60.
TYPE SPECIMEN — ICN-R 5974.
TYPE LOCALITY — "Tabio (2645 m), Cundimarca." Georeferenced to lat 4.9167 lon −75.1000, 2645 m by Moreno-Arias et al. (2023).
MORPHOLOGY — Body length to 73 M, 69 F; ventral scales smooth; middorsal scales form a crest of triangular plates (see Lazell 1969: fig. 1); interparietal larger than tympanum; scales across the snout between the second canthals 4–5; toe lamellae 21–22; suboculars and supralabials in contact; scales separating supraorbital semicircles 0–1; supraocular scales abruptly enlarged; hindlimbs short, adpressed toe to ear or posterior to ear; body color green or brown with yellow and black, light line from snout back to shoulder; male and female dewlaps cream to yellow, sometimes with red distally; heterogeneous lateral scutellation.
SIMILAR SPECIES — *Anolis richteri* is a distinctive highland *twig* anole. The heterogeneous lateral scalation, wherein large smooth oval scales are separated by granular interstitial scales, is differentiating in *A. richteri*, especially in combination with its short hindlimbs and tail and large smooth head scales. Among potential eastern Andean forms that share these traits, *A. vanzolinii* and *A. inderenae* are larger than *A. richteri* (to >100 mm SVL; to 73 mm in *A. richteri*) and possess greater numbers of toe lamellae (23–28; 21–22 in *A. richteri*); *A. heterodermus*, *A. quimbaya*, and *A. tequendama* differ from *A. richteri* in male and female dewlap color (*A. heterodermus*: purple, red, or black in males and females; *A. quimbaya*: orange in male, green in females; *A. tequendama*: brown with or without spotting in male, yellow with dark stripes in female; *A. richteri*: cream to yellow in males and females, sometimes with red distally).
RANGE — From 2500 to 3500 meters in the Eastern Andes of Colombia from Cundimarca north to Santander Department. "Northern populations of *A. heterodermus* have been found in sympatry

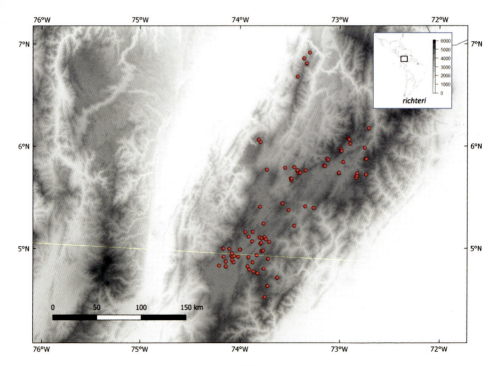

with southern populations of *A. richteri*" (Moreno-Arias et al. 2023). See range discussion under *A. heterodermus*.

NATURAL HISTORY — In forest and scrublands; uses narrow perches such as twigs. See Dunn (1944; as *Phenacosaurus richteri*); Osorno and Osorno (1946; as *P. richteri*); Jenssen (1975; as *P. heterodermus*); Miyata (1983; as *P. heterodermus*); Lombo (1989; as *P. heterodermus*); Moreno-Arias and Urbina-Cardona (2013; as *Anolis heterodermus*); Méndez-Galeano and Calderón -Espinosa (2017, 2020; as *A. heterodermus*); Beltrán and Barragán-Contreras (2019; as *A. heterodermus*); Moreno-Arias et al. (2023).

— *ANOLIS RIPARIUS* — PLATE 1.189

DESCRIPTION — Chaves, G., M. J. Ryan, F. Bolaños, C. Márquez, G. Köhler, S. Poe. 2023. Two new species of semiaquatic *Anolis* (Squamata: Dactyloidae) from Costa Rica. *Zootaxa* 5319:249–262.
TYPE SPECIMEN — UCR 5579.
TYPE LOCALITY — "Pacuarito river, Pacuarito 10 Km S-SW of Pérez Zeledón, 9.31390 N, 83.77220 W, 880 masl, San José Province, Costa Rica."
MORPHOLOGY — Body length to 73 M, 65 F; ventral scales keeled; middorsal scales enlarged in 0–2 rows; interparietal smaller than tympanum; scales across the snout between the second canthals 15–19; toe lamellae 16–18; suboculars and supralabials usually separated by a row of scales; scales separating supraorbital semicircles 4–5; supraocular scales subequal; addpressed toe to eye or between eye and tip of snout; body color mainly brown with some green, with darker dorsal bands and a light lateral stripe; male dewlap orange-red with yellow streaks.
SIMILAR SPECIES — *Anolis riparius* is a *semiaquatic* anole. The combination of semiaquatic ecology and large brown body should distinguish *A. riparius* from all potentially sympatric congeners except *A. oxylophus* and *A. woodi*. *Anolis woodi* possesses a different male dewlap (orange in *A. woodi*;

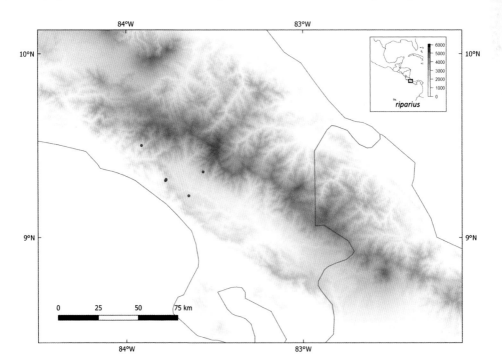

SPECIES ACCOUNTS

orange-red with yellow in *A. riparius*) and has larger head scales than *A. riparius* (9–14 scales across the snout; 15–19 in *A. riparius*). *Anolis oxylophus*, another semiaquatic anole, has 10–15 rows of enlarged middorsal scales (middorsals undifferentiated in *A. riparius*). *Anolis riparius* is similar in ecology and morphology to *A. robinsoni* and *A. aquaticus*, but these species are not found sympatrically with *A. riparius* (*A. robinsoni*: north of *A. riparius*; *A. aquaticus*: south of *A. riparius*) and differ in either male dewlap color (dark brown in *A. robinsoni*) or size of head scales (larger in *A. aquaticus*: 8–16 scales across the snout between the second canthals). Other large brown anoles in the region of *A. riparius* have smooth ventral scales (*A. pentaprion*, *A. charlesmyersi*, *A. cristatellus*; keeled in *A. riparius*) or larger head scales than *A. riparius* (5–13 scales across the snout in *A. lemurinus*, *A. capito*, *A. sagrei*; 15–19 in *A. riparius*).

RANGE — From 100 to 1450 meters in south central Pacific Costa Rica including the Pacuar, Guabo, and Savegre River basins.

NATURAL HISTORY — Common in forests and occasionally open areas; active diurnally on boulders along streams; sleeps on moss, leafy vegetation, boulders, or rock walls by streams, often in and near splash zones.

ANOLIS RIVALIS — PLATE 1.190

DESCRIPTION — Williams, E. E. 1984b. New or problematic *Anolis* from Colombia. III. Two new semiaquatic anoles from Antioquia and Choco, Colombia. *Breviora* 478:7–20.
TYPE SPECIMEN — LACM 42124.
TYPE LOCALITY — "Belen, Rio Arquia, Antioquia, Colombia (6 15 N, 76 39 W)." Belén along the Río Arquía in Antioquia is located at approximately lat 6.195 lon −76.545, 60 m. See discussion in species account for *Anolis maculigula*.

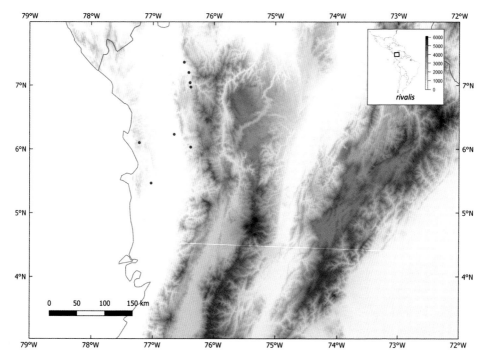

MORPHOLOGY — Body length to 70 M, F known to at least 40 mm, almost certainly larger; ventral scales keeled; middorsal scales gradually or greatly enlarged in several rows; scales across the snout between the second canthals 13–20; toe lamellae 12–18; suboculars and supralabials in contact or separated by a row of scales; scales separating supraorbital semicircles 1–3; supraocular scales abruptly or gradually enlarged; hindlimbs long, adpressed toe to eye or anterior to eye; body color brown; male dewlap yellow-orange; female dewlap absent.
SIMILAR SPECIES — *Anolis rivalis* is a *semiaquatic* anole with a zone of about 13 noticeably enlarged rows of middorsal scales. Geographically proximal similar species may be separated from *A. rivalis* as follows: *Anolis auratus*, *A. tropidogaster*, *A. gaigei*, and *A. notopholis* are smaller (usually under 55 mm SVL; to 70 mm in *A. rivalis*) and have different male dewlaps (blue-black in *A. auratus*; red and yellow in *A. tropidogaster*, *A. gaigei*; orange-red in *A. notopholis*; yellow-orange in *A. rivalis*); *A. gracilipes* has a small interparietal (smaller than tympanum, separated from supraorbital semicircles by several scales; larger than tympanum, often in contact with supraorbital semicircles in *A. rivalis*); *A. binotatus* has larger head scales (8–13 scales across the snout; 13–20 in *A. rivalis*). *Anolis rivalis* seems most likely to be confused with the other western Colombian semiaquatic anoles *A. maculigula*, *A. macrolepis*, and *A. poecilopus*. *Anolis poecilopus* has approximately uniform dorsal scales. *Anolis maculigula* has smooth ventral scales and large body size (SVL to over 100 mm; ventrals keeled in *A. rivalis*). *Anolis macrolepis* has larger head scales than *A. rivalis* (7–12 scales across the snout between the second canthals). *Anolis lynchi* is a semiaquatic anole found south of the range of *A. rivalis* and differing from it in possessing minute head scales (usually greater than 20 scales across the snout).
RANGE — From 200 to 1100 meters in northern Pacific Colombia.
NATURAL HISTORY — Strongly associated with streams; found syntopically with fellow semiaquatic anole *Anolis maculigula*; active diurnally and sleeps on rocks, rock walls, and vegetation within 1 meter of streams.

— *ANOLIS ROATANENSIS* — PLATE 1.191; MAP overleaf

DESCRIPTION — Köhler, G., J. R. McCranie. 2001. Two new species of anoles from northern Honduras (Reptilia, Squamata, Polychrotidae). *Senckenbergiana biologica* 81:240–243.
TYPE SPECIMEN — SMF 79953.
TYPE LOCALITY — "Between West End Point and Flowers Bay (16° 17.98' N, 86° 34.82' W), 30 m elevation, Isla de Roatán, Departamento de Islas de la Bahía, Honduras."
MORPHOLOGY — Body length to 65 M, 62 F; ventral scales keeled; middorsal scales enlarged in 0–2 rows; interparietal about equal to tympanum; scales across the snout between the second canthals 8–11; toe lamellae 16–18; suboculars and supralabials separated by a row of scales; scales between supraorbital semicircles 1–4; supraocular scales gradually enlarged; adpressed toe to eye; body color brown, usually with light lateral stripe; male dewlap red with white scales.
SIMILAR SPECIES — *Anolis roatanensis* is a *large brown* anole. It differs from the other two anole species known from Roatán as follows: *A. allisoni* is green, with a pink male dewlap (red in *A. roatanensis*); *A. sagrei* has larger head scales (usually 6–8 scales across the snout, suboculars and supralabials in contact; 8–11 scales across the snout, suboculars and supralabials usually separated by a row of scales in *A. roatanensis*) and a red male dewlap with yellow edge.
RANGE — Throughout Roatán Island and Islas de la Bahía, Honduras.
NATURAL HISTORY — Common in forest, forest edge, and disturbed areas; active diurnally low on tree trunks and ground; sleeps on twigs or leaves; McCranie and Köhler (2015) report individuals sleeping on cave walls and human trash. See Logan et al. (2013); McCranie and Köhler (2015).

SPECIES ACCOUNTS

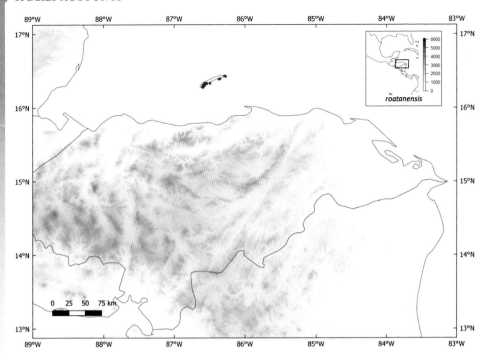

— *ANOLIS ROBINSONI* — PLATE 1.192

DESCRIPTION — Chaves, G., M. J. Ryan, F. Bolaños, C. Márquez, G. Köhler, S. Poe. 2023. Two new species of semiaquatic *Anolis* (Squamata: Dactyloidae) from Costa Rica. *Zootaxa* 5319:249–262.

TYPE SPECIMEN — UCR 2463.

TYPE LOCALITY — "Palma stream bridge 5.1 km south of Santa Marta de Santiago de Puriscal, 9.79230 N,—84.39530 W, ca. 800 masl, San José Province, Costa Rica."

MORPHOLOGY — Body length to 74 M, 56 F; ventral scales keeled; middorsal scales enlarged in 0–2 rows; interparietal smaller than tympanum; scales across the snout between the second canthals 15–20; toe lamellae 15–17; suboculars and supralabials usually separated by a row of scales; scales between supraorbital semicircles 5–7; supraocular scales subequal; hindlimbs long, adpressed toe to between eye and tip of snout; body color mainly brown with some green, with darker dorsal bands and a light lateral stripe; male dewlap chocolate brown; females usually with small dewlap similar to male's in color.

SIMILAR SPECIES — *Anolis robinsoni* is a *semiaquatic* anole. The combination of semiaquatic ecology and large brown body should distinguish *A. robinsoni* from all potentially sympatric congeners except *A. oxylophus* and *A. woodi*. *Anolis woodi* possesses a different male dewlap (orange in *A. woodi*; dark brown in *A. robinsoni*) and larger head scales than *A. robinsoni* (9–14 scales across the snout; 15–20 in *A. robinsoni*) and sleeps on vegetation (*A. robinsoni* usually sleeps on boulders or rock walls in splash zones of streams). *Anolis oxylophus* has 10–15 rows of enlarged middorsal scales (middorsals undifferentiated in *A. robinsoni*). *Anolis robinsoni* is similar in ecology and morphology to *A. riparius* and *A. aquaticus*, but these species are not found sympatrically with *A. aquaticus* and differ in male dewlap color (red with yellow in both species). Other large brown anoles with keeled ventral scales in the region have larger head scales than *A. robinsoni* (5–13 scales across the snout in *A. lemurinus*, *A. capito*, *A. sagrei*).

276

Anolis rodriguezii

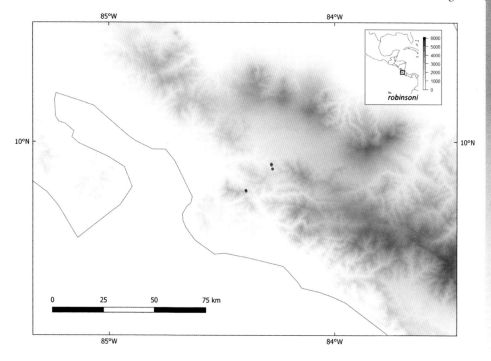

RANGE — From 500 to 1100 meters in central Pacific Costa Rica, between Tarcoles and Candelaria Rivers.

NATURAL HISTORY — Common in forest and occasionally open areas; active diurnally on boulders along and in streams; sleeps on boulders or moss in and near streams, often in splash zones. See Márquez–Baltán (1994); Márquez et al. (2005); Márquez and Márquez (2009).

ANOLIS RODRIGUEZII — PLATE 1.193, 2.30

DESCRIPTION — Bocourt, M. F. 1873. Etudes sur les reptiles et les Batraciens. In Duméril, M. A., M. Bocourt, M. Mocquard, eds., *Recherches Zoologiques* vol. 3, section 1, 62–63. Paris: Mission Scientifique au Mexique et dans L'Amerique Centrale.

TYPE SPECIMEN — MNHN 2411.

TYPE LOCALITY — "Pansos sur le Polochic (Amerique centrale)." Panzos, Alta Verapaz, Guatemala is located at lat 15.399 lon −89.640, 20 m. The Polochic River passes about 1 km east of Panzos.

MORPHOLOGY — Body length to 46 M, 49 F; ventral scales smooth; middorsal scales enlarged in 0–2 rows; interparietal larger or smaller than tympanum; scales across the snout between the second canthals 8–13; toe lamellae 15–17; suboculars and supralabials in contact; scales separating supraorbital semicircles 1–2; supraocular scales gradually enlarged; appressed toe to between eye and tip of snout; body color brown; tail banded; male dewlap yellow-orange proximally, orange distally; female dewlap absent.

SIMILAR SPECIES — *Anolis rodriguezii* is a widespread *small brown* lowland anole with smooth ventral scales. Among geographically proximal small brown species that share these traits, *A. cobanensis* has very long hindlimbs (appressed toe reaches to beyond tip of snout; to between eye and tip of snout in *A. rodriguezii*) and a pinkish-purple male dewlap (orange in *A. rodriguezii*); *A. beckeri* has short

SPECIES ACCOUNTS

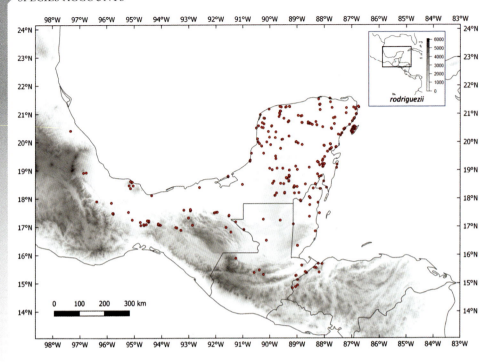

hindlimbs (adpressed toe reaches to ear) and a dark red male dewlap; *A. heteropholidotus* has shorter hindlimbs (adpressed toe reaches to eye) and greatly enlarged middorsal scales (middorsals uniform in *A. rodriguezii*); *A. parvicirculatus* has larger head scales (usually 6–9 scales across the snout; 8–13 in *A. rodriguezii*) and suboculars and supralabials usually separated by a row of scales (in contact in *A. rodriguezii*). *Anolis zeus*, *A. limifrons*, and *A. cupreus* are similar to *A. rodriguezii* and found southeast of its range; these species differ from *A. rodriguezii* in male dewlap color (*A. limifrons*, *A. zeus*: predominantly white; *A. cupreus*: orange-brown proximally, pale pink-peach distally).

RANGE — From 0 to 1400 meters from Mexico to northwestern Honduras.

NATURAL HISTORY — Very common in forest, edge, and disturbed habitats; active diurnally on a variety of arboreal perches across a broad vertical range; sleeps on twigs or leaves. See Lee (1996); López-González and González-Romero (1997); Villarreal Benítez (1997); D'Cruze (2005); D'Cruze and Stafford (2006); McCranie and Köhler (2015); Cardona and Reynoso (2017).

COMMENT — I consider *Anolis microlepis* (type locality: "El Ocote, Chiapas, 600 m") to be a junior synonym of *A. rodriguezii* (see appendix 1 for justification).

— *ANOLIS RUBIGINOSUS* — PLATE 1.194, 2.14

DESCRIPTION — Bocourt, M. F. 1873. Notes erpetologiques. *Annales des Sciences Naturelles*, series 5. Zoologie et Paleontologie. Tome 17, article 2:1.

TYPE SPECIMEN — MNHN 2636.

TYPE LOCALITY — "Prov. d'Oxaca, Mexique;" that is, the state of Oaxaca, Mexico. Nieto Montes de Oca (1994a:234) compellingly implied Totontepec (lat 17.26 lon −96.03, 1850 m) as the likely type locality.

MORPHOLOGY — Body length to 49 M, 51 F; ventral scales smooth; middorsal scales enlarged in 0–2 rows; interparietal larger or smaller than tympanum; scales across the snout between the second

Anolis rubiginosus

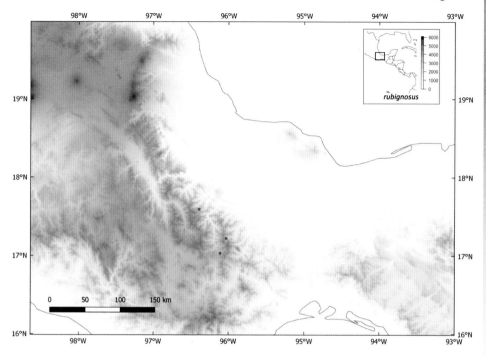

canthals 6-10; toe lamellae 16-21; suboculars and supralabials in contact or separated by a row of scales; scales separating supraorbital semicircles 0-2; supraocular scales gradually enlarged; adpressed toe to between eye and tip of snout; body color brown; male dewlap pale pink, faintly orangeish distally; female dewlap absent.

SIMILAR SPECIES — *Anolis rubiginosus* is a *small brown* anole, distinguishable from similar geographically proximal forms as follows: *A. laeviventris*, *A. sericeus*, and *A. ustus* differ from *A. rubiginosus* in their shorter hindlimbs (adpressed toe reaches to ear; to between eye and tip of snout in *A. rubiginosus*); *A. tropidonotus* and *A. uniformis* have a puncturelike axillary pocket (absent in *A. rubiginosus*); *A. nebulosus*, *A. microlepidotus*, and *A. quercorum* have approximately three abruptly enlarged supraocular scales (supraoculars gradually enlarged in *A. rubiginosus*). *Anolis milleri* is very similar to *A. rubiginosus* and found southwest of the range of *A. rubiginosus*. The male dewlap of *A. milleri* is solid purple-pink (pale pink, appearing orange distally in *A. rubiginosus*). *Anolis schiedii* and *A. cymbops* are similar species found northwest of the range of *A. rubiginosus* that can be distinguished from *A. rubiginosus* by male dewlap color (*A. schiedii*: orange-red; *A. cymbops*: solid pink). In addition, *A. rubiginosus* may be distinguished from *A. milleri*, *A. schiedii*, and *A. cymbops* by its multicarinate dorsal head scales (most head scales unicarinate to smooth in these species).

RANGE — From 1400 to 2200 meters on the northeastern slope of Sierra de Juarez in north-central Oaxaca, Mexico.

NATURAL HISTORY — Common in forest, edge, and disturbed habitat; sleeps on bushes or other low vegetation. See Nieto Montes de Oca (1994a).

— *ANOLIS RUBRIBARBARIS* — PLATE 1.195

DESCRIPTION — Köhler, G., J. R. McCranie, L. D. Wilson. 1999. Two new species of anoles of the *Norops crassulus* group from Honduras (Reptilia: Sauria: Polychrotidae). *Amphibia-Reptilia* 20:280–285.

TYPE SPECIMEN — UF 90206.

TYPE LOCALITY — "Honduras, Departamento de Santa Bárbara, N slope of Montaña de Santa Bárbara, 4 km S of San Luís de los Planes, elevation 1700 m." Assuming the "4 km S" refers to road rather than straight line distance (which would result in an elevation of 2200+ meters), this locality is located at approximately lat 14.95 lon −88.11.

MORPHOLOGY — Body length to 52 M, 52 F; ventral scales keeled; middorsal scales greatly enlarged in several rows; interparietal larger or smaller than tympanum; middorsal scales enlarged in 0–2 rows; interparietal about equal to tympanum; scales across the snout between the second canthals 4–7; toe lamellae 15–18; suboculars and supralabials in contact; scales separating supraorbital semicircles 0–3; supraocular scales gradually enlarged; appressed toe to between ear and eye; body color brown; male dewlap red with white scales.

SIMILAR SPECIES — *Anolis rubribarbaris* is a *small brown* highland anole. Among geographically proximal small brown highland anoles, *Anolis yoroensis*, *A. ocelloscapularis*, *A. laeviventris*, *A. cusuco*, and *A. kreutzi* differ from *A. rubribarbaris* in their lack of abruptly enlarged middorsal scale rows (middorsals greatly enlarged in several rows in *A. rubribarbaris*). *Anolis caceresae* is similar to *A. rubribarbaris* but found well outside of its range, at high elevations in southern Honduras. *Anolis morazani*, found just east of the range of *A. rubribarbaris*, is similar to *A. rubribarbaris* but reported to differ in hemipenial structure (asulcate process of hemipenes divided in *A. morazani*, undivided in *A. rubribarbaris*; McCranie and Köhler 2015). Small brown anoles found in the general region of *A. rubribarbaris* but at lower elevations generally have homogeneous lateral

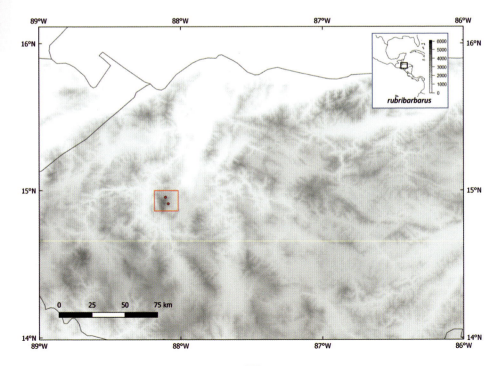

scales (heterogeneous lateral scales in *A. rubribarbaris*) and lack the abruptly enlarged middorsal scale rows of *A. rubribarbaris* (*A. sericeus, A. ustus, A. zeus, A. limifrons, A. rodriguezii*) or have a puncturelike axillary pocket (*A. tropidonotus, A. uniformis*; puncturelike axillary pocket absent in *A. rubribarbaris*).
RANGE — From 1600 to 2200 meters in western Honduras.
NATURAL HISTORY — In forest; active diurnally on low vegetation or ground; sleeps on vegetation. See McCranie and Köhler (2015).

— *ANOLIS RUIZII* — PLATE 1.196

DESCRIPTION — Rueda, J. V., E. E. Williams. 1986. Una nueva especie de saurio para la cordillera oriental de Colombia (Sauria; Iguanidae). *Caldasia* 15:511–24.
TYPE SPECIMEN — ICN 6189.
TYPE LOCALITY — "Quebrada 'La Limonita,' tributaria del Rio Cusiana, Inspeccion de Policia de Corinto, Municipio de Pajarito, Departamento de Boyaca, flanco oriental de la Cordillera Oriental de Colombia, 1620 m s.n.m, 5 30 N, 72 17 W." A quebrada in Pajarito that feeds into the Cusiana river crosses the main road just north of Corinto at lat 5.420 lon −72.723, 1630 m (the GPS point listed in the description of the type locality, if taken precisely, identifies a site east of Pajarito municipality at approximately 500 meters elevation).
MORPHOLOGY — Body length to 57 M, 58 F; ventral scales smooth; middorsal scales enlarged in 0–2 rows; interparietal larger than tympanum; scales across the snout between the second canthals 5–8; toe lamellae 20; suboculars and supralabials in contact; supraorbital semicircles in contact; supraocular scales gradually enlarged; hindlimbs short, appressed toe to ear or posterior to ear; body color light brown, greenish yellow on head and neck; male dewlap pale yellow.

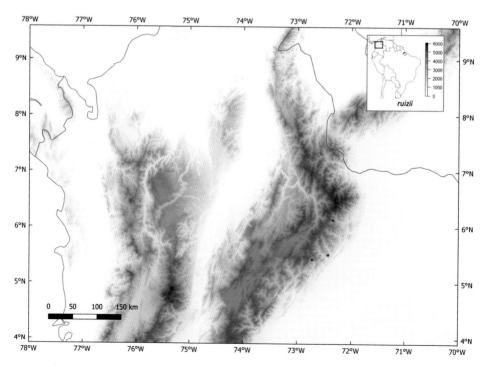

SPECIES ACCOUNTS

SIMILAR SPECIES — *Anolis ruizii* appears to be a highland *twig* anole (Rueda and Williams 1986). *Anolis ruizii* may be sympatric with large green species *A. huilae* and *A. punctatus*, with which it is unlikely to be confused based on size (SVL to at least 80 mm; to 58 mm in *A. ruizii*) and dorsal color pattern (mostly solid green in *A. huilae*, *A. punctatus*; lichenous in *A. ruizii*) and *A. scypheus*, which differs from *A. ruizii* in its keeled ventral scales (smooth in *A. ruizii*). *Anolis fuscoauratus* differs from *A. ruizii* in its smaller head scales (8–14 scales across the snout; 5–8 in *A. ruizii*). *Anolis lamari* is very similar to *A. ruizii* and replaces it geographically to the south; it is reported to differ from *A. ruizii* in its possession of a bony knob on the posterior aspect of the skull (Williams 1992; see discussion of *A. lamari*).

RANGE — From 1500 to 1650 meters, on the Amazonian Andean slope in central Colombia.

NATURAL HISTORY — Only information is from the type description: One individual found active diurnally 1 meter above ground on a tree trunk; others found sleeping on ferns or bushes 0.7 to 1.8 meters above ground; associated with streams.

— *ANOLIS SAGREI* — PLATE 1.197, 2.167

DESCRIPTION — Duméril, A.M.C., Bibron, G. 1837. *Erpétologie Générale ou Histoire Naturelle Complete des Reptiles*, vol. 4, 149–150. Paris: Librairie Encyclopedique de Roret.
TYPE SPECIMEN — MNHN 2430, 6797 (5 syntypes).
TYPE LOCALITY — "Cuba."
MORPHOLOGY — Body length to 70 M, 46 F; ventral scales keeled; middorsal scales enlarged in 0–2 rows; interparietal usually smaller than tympanum; scales across the snout between the second canthals 5–9; toe lamellae 15–22; suboculars and supralabials in contact; scales separating supraorbital semicircles 1–2; supraocular scales gradually enlarged; adpressed toe usually to eye, occasionally

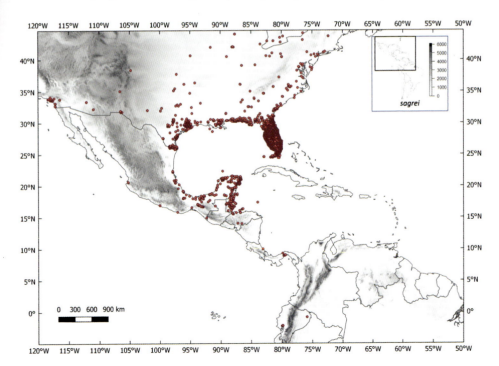

posterior to eye; body color brown, frequently with light lateral stripes and dark and light markings; male dewlap red to orange-red with yellow edge; female dewlap absent.

SIMILAR SPECIES — *Anolis sagrei* is a *small brown* lowland anole that has been introduced to multiple mainland localities. This species should be entertained as a potential anole identification at most coastal localities, especially near ports. The combination of keeled ventral scales, red male dewlap with yellow edge, and occupation of and high abundance in disturbed areas such as city parks will indicate some likelihood that the species to be identified is *A. sagrei*.

RANGE — Within the geographic purview of this book, *Anolis sagrei* is established in most of the southeastern mainland USA, Mexico, Guatemala, Belize, Honduras, Costa Rica, Panama, Ecuador, and Brazil. This species also lives in Cuba, Jamaica, several other Caribbean islands, Taiwan, Singapore, Hawaii, California, and Ascension Island. According to GBIF, *Anolis sagrei* has been observed in 31 US states and two Canadian provinces as of this writing. Although likely the northern and many of the southern of these observations do not represent breeding populations, I have included most of these observations in the range map, out of likely reader interest and because I am unable to determine whether many of the observations represent isolated observations or established populations.

NATURAL HISTORY — Very common in disturbed habitats, also in edge and (occasionally) forest; usually the most abundant, most visible lizard at its localities; active diurnally on low vegetation, human structures, and ground; a "trunk-ground" anole (Williams 1983) that frequently surveys habitat from a head-down position up to 1 meter above ground on tree trunks; sleeps on any available vegetation, usually up to 1.5 meters above ground. See Meshaka (199a, b); Amador et al. (2017); Campbell (2000); Delaney and Warner (2017); Edwards and Lailvaux (2012); Evans (1938); Fetters and McGlothlin (2017); Lee et al. (1989); Narváez et al. (2020); Norval et al. (2010); Stamps (1999); Tokarz (1985); Kamath and Losos (2017); Oliveira et al, (2018); Culbertson and Herrmann (2019); Tiatragul et al. (2019); Pruett et al. (2020); Vásquez-Cruz et al. (2020); Fisher et al. (2020); Logan et al. (2021). For natural history in its native range, see, e.g., Schwartz and Henderson (1991) and references in Rodriguez Schettino (1999).

— *ANOLIS SALVINI* — PLATE 1.198, 2.82

DESCRIPTION — Boulenger, G. A. 1885. *Catalogue of the lizards in the British Museum (Natural History)*, vol. 2. London. 75–76.
TYPE SPECIMEN — BMNH 1946.9.8.19.
TYPE LOCALITY — "Guatemala." (In error according to Köhler 2007).
MORPHOLOGY — Body length to 61 M, 66 F; ventral scales keeled; middorsal scales enlarged in 0–2 rows; interparietal larger or smaller than tympanum; scales across the snout between the second canthals 4–7; toe lamellae 15–21; suboculars and supralabials in contact; scales separating supraorbital semicircles 0–1; supraocular scales gradually enlarged; hindlimbs short, adpressed toe approximately to ear; body color usually pale gray with dark flecks, lichenous; male dewlap red to orange-red; female dewlap variable, sky blue or blackish-blue or red (Bientrue et al. 2013).
SIMILAR SPECIES — *Anolis salvini* is a *twig* or *small brown* to whitish-gray highland anole with short hindlimbs (adpressed hindlimb reaches to ear). Several short-limbed highland forms differ from *A. salvini* in being more or less uniformly brown, lacking the lichenous greens and blacks of *A. salvini*, and in male dewlap color: *A. kemptoni* (male dewlap pink posteriorly, orange anteriorly), *A. fortunensis* (male dewlap orange anteriorly, yellow posteriorly), *A. altae* (male dewlap orange), *A. monteverde* (male dewlap orange), *A. tenorioensis* (male dewlap reddish-orange with dark spots). *Anolis laeviventris* has smaller head scales than *A. salvini* (6–10 scales across the snout; 4–7 in *A. salvini*) and a white male dewlap. *Anolis carpenteri* tends to be green and has an orange male dewlap. *Anolis fungosus* is smaller than *A. salvini* (SVL to 48 mm; to 66 mm in *A. salvini*), rarely found, and has fewer toe lamellae (13–16; 15–21 in *A. salvini*). Similar species (i.e., with lichenous dorsum, short hindlimbs) *A. pentaprion* and *A. charlesmyersi* are found only at lower elevations than *A. salvini* (below 1000 m).

SPECIES ACCOUNTS

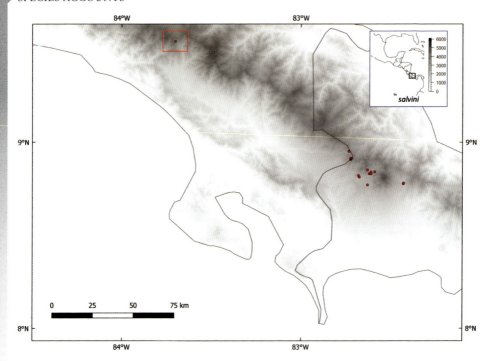

RANGE — From 1300 to 2100 meters in western Panama and eastern Costa Rica.
NATURAL HISTORY — Common in forest, open, and disturbed areas; may appear highly abundant or absent across proximal, seemingly comparable localities; sleeps on twigs or leaves, usually between 0.2 and 2 meters above ground, occasionally higher.

— *ANOLIS SANTAMARTAE* — PLATE 1.199, 2.146

DESCRIPTION — Williams, E. E. 1982. Three new species of the *Anolis punctatus* complex from Amazonian and inter-Andean Colombia, with comments on the eastern members of the *punctatus* species group. Breviora 467:16–21.
TYPE SPECIMEN — CAS 113922.
TYPE LOCALITY — "San Sebastian de Rabago, Sierra Nevada de Santa Marta, Cesar, Colombia (10 4 N, 73 16 W)." San Sebastian de Rabago is located at approximately lat 10.565 lon −73.606, 1900 m (the locality appears approximately accurately mapped in Williams [1982], but the GPS point listed for the holotype in the description, if taken precisely, is over 50 km south of San Sebastian de Rabago at 100 meters elevation).
MORPHOLOGY — Body length to 55 M, 53 F; ventral scales smooth to weakly keeled; middorsal scales gradually enlarged in several rows; interparietal larger than tympanum; scales across the snout between the second canthals 4–6; toe lamellae 19–21; suboculars and supralabials in contact; supraorbital semicircles in contact; male dewlap white.
SIMILAR SPECIES — *Anolis santamartae* may be a *twig* or *small brown* highland anole. Like its apparent close relatives to the north in the Santa Marta mountains *A. solitarius* and *A. menta*, *A. santamartae* may be a "solitary" anole that lacks sympatric congeners. *Anolis santamartae* differs from *A. menta* and *A. solitarius* in its all-white male dewlap (male dewlap predominantly yellow in *A. menta*,

Anolis savagei

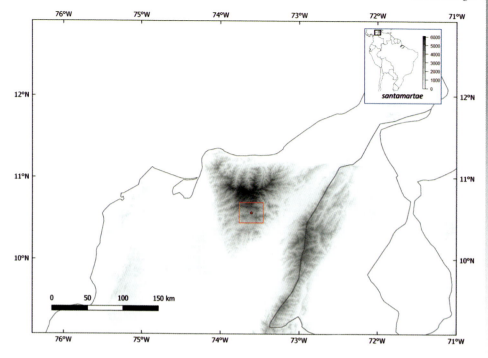

A. solitarius). Geographically proximal small brown lowland anoles *A. gaigei* and *A. auratus* have keeled ventral scales (smooth in *A. santamartae*).
RANGE — Known only from the type locality.
NATURAL HISTORY — Unknown.

— *ANOLIS SAVAGEI* — PLATE 1.200, 2.71

DESCRIPTION — Poe, S., M. J. Ryan. 2017. Description of two new species similar to *Anolis insignis* (Squamata: Iguanidae) and resurrection of *Anolis (Diaphoranolis) brooksi*. Amphibian and Reptile Conservation 11[General Section]:9-11.
TYPE SPECIMEN — MSB 96616.
TYPE LOCALITY — "Las Cruces, Puntarenas, Costa Rica; 8.78242, -82.95886, 1127 m."
MORPHOLOGY — Body length to 141 M; ventral scales smooth; middorsal scales enlarged in 0-2 rows; interparietal smaller than tympanum; scales across the snout between the second canthals 8-9; toe lamellae 25-29; suboculars and supralabials in contact; scales separating supraorbital semicircles 2; supraocular scales subequal to gradually enlarged; hindlimbs short, adpressed toe to ear or posterior to ear; body color gray, brown, and/or green, usually with some banding; prominent blotch posterior to eye; tail banded; male and female dewlaps pink with dark streaks.
SIMILAR SPECIES — *Anolis savagei* is a *giant* anole. No geographically proximal species approaches *A. savagei* in size (SVL to 141 mm), and potentially sympatric large forms are easily distinguished by their smaller size and keeled ventral scales (*Anolis biporcatus* [SVL to 108 mm], *A. capito* [SVL to 102 mm], and *A. woodi* [SVL to 92 mm]; ventrals smooth in *A. savagei*). Geographically proximal giant species differ from *A. savagei* in dewlap color (*A. microtus*, *A. ginaelisae*: dewlap pink in males and females; *A. kathydayae*: dewlap white with blue or greenish tint in males and females; *A. frenatus*,

SPECIES ACCOUNTS

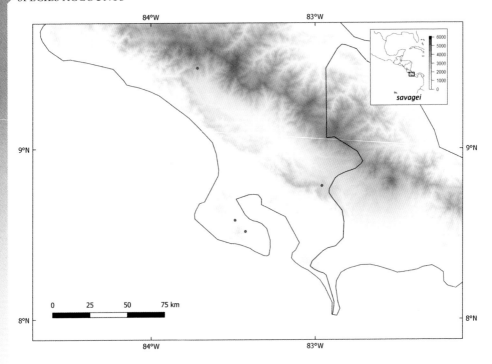

A. casildae: dewlap white in males and females; *A. savagei*: dewlap pink with dark streaks in adult males and females) and lack the dark postocular blotch of *A. savagei*.
RANGE — 0 to 1700 meters in the southeastern Pacific region of Costa Rica.
NATURAL HISTORY — Rare in forest; observed diurnally on tree branches; sleeps on narrow branches above 4 meters above ground.

— *ANOLIS SCHIEDII* — PLATE 1.201, 2.12

DESCRIPTION — Wiegmann, A.F.A. 1834. *Herpetologia Mexicana, seu Descriptio Amphibiorum Novae Hispaniae, quae Itineribus comitis de Sack, Ferdinandi Deppe et Chr. Guil. Schiede in Museum Zoologicum Berolinense Pervenerunt. Pars Prima, Saurorum Species Amplectens. Adiecto Systematis Saurorum Prodromo, Additisque multis in hunc Amphibiorum Ordinem Observationibus.* 48. Berlin: Lüderitz.
TYPE SPECIMEN — ZMB 526.
TYPE LOCALITY — None given, but presumably Mexico. Nieto Montes de Oca (1994b) restricted the type locality to Xalapa, Veracruz, Mexico (lat 19.54 lon –96.91, 1420 m).
MORPHOLOGY — Body length to 58 M, 58 F; ventral scales smooth to weakly keeled; middorsal scales enlarged in 0–2 rows; interparietal smaller than tympanum; scales across the snout between the second canthals 7–9; toe lamellae 17–19; suboculars and supralabials usually separated by a row of scales; scales separating supraorbital semicircles 1–3; supraocular scales gradually enlarged; hindlimbs long, adpressed toe to between eye and tip of snout; body color brown; male dewlap orange-red; female dewlap dull yellow.
SIMILAR SPECIES — *Anolis schiedii* is a *small brown* highland anole. Nieto Montes de Oca (1994a) listed one sympatric small brown congener: *A. laeviventris* differs from *A. schiedii* in its shorter limbs (adpressed toe reaches approximately to ear; to between eye and tip of snout in *A. schiedii*). Other

286

Anolis scypheus

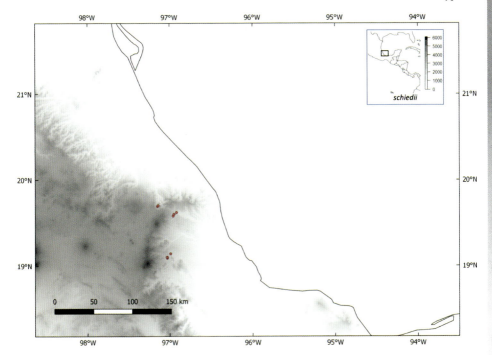

potentially sympatric forms usually found at lower elevations include *A. tropidonotus*, which differs from *A. schiedii* in its abruptly enlarged keeled middorsal scale rows (a few slightly enlarged middorsal rows in *A. schiedii*); *A. sericeus*, which differs from *A. schiedii* in its short hindlimbs (adpressed toe reaches to ear); and *A. rodriguezii*, which differs in its smaller head scales (8–13 scales across the snout; 7–9 in *A. schiedii*). *Anolis cymbops* is very similar to *A. schiedii* but differs in possessing enlarged postcloacal scales in males (absent in males of *A. schiedii*) and in male dewlap color (pink in *A. cymbops*; orange-red in *A. schiedii*). *Anolis naufragus*, found northwest of the range of *A. schiedii*, is similar to *A. schiedii* but differs in its unicarinate dorsal limb scales (multicarinate in *A. schiedii*). *Anolis milleri*, found to the southeast of the range of *A. schiedii*, can be distinguished from *A. schiedii* by male dewlap color (pink) and having fewer scale rows on the male dewlap (~5, vs. ~8 in *A. schiedii*).

RANGE — From 1300 to 2050 meters on the Atlantic slope of Sierra Madre Oriental in central Veracruz, Mexico.

NATURAL HISTORY — Common in patches of habitat in former cloud forest areas; sleeps on twigs, ferns, or leaves from 0.5 to 2 meters above ground. See Nieto Montes de Oca (1994a); Kelly-Hernandez et al. (2018).

— *ANOLIS SCYPHEUS* — PLATE 1.202, 2.95

DESCRIPTION — Cope, E. D. 1864. Contributions to the herpetology of tropical America. *Proceedings of the Academy of Natural Sciences of Philadelphia* 16:166–181.

TYPE SPECIMEN — BMNH 1946.8.8.55.

TYPE LOCALITY — None listed. "Caracas" according to Boulenger (1885); considered to be "a type locality lacking or in error" by Vanzolini & Williams (1970:85), who nevertheless considered the type

SPECIES ACCOUNTS

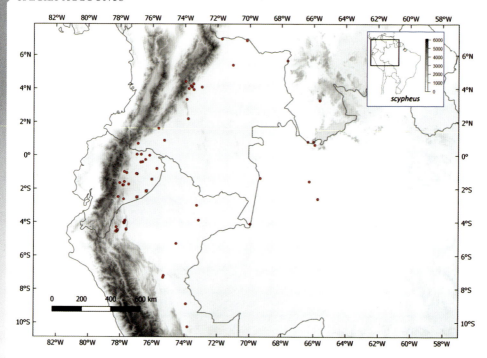

specimen to be undeniably assignable to *chrysolepis*-like ground anoles of Colombia and Peru, i.e., *Anolis scypheus*.

MORPHOLOGY — Body length to 78 M, 80 F; ventral scales keeled; middorsal scales gradually enlarged in several rows; interparietal larger or smaller than tympanum; scales across the snout between the second canthals 9–15; toe lamellae 14–21; suboculars and supralabials in contact; scales separating supraorbital semicircles 0–3; supraocular scales gradually enlarged; hindlimbs long, adpressed toe to between eye and tip of snout, sometimes past tip of snout; body color brown with middorsal markings; male and female dewlap red distally, blue proximally; short snout.

SIMILAR SPECIES — *Anolis scypheus* is a *large brown* anole, one of the common Amazonian lowland ground anoles with long hindlimbs and keeled ventral scales. *Anolis scypheus* differs from other similar Amazonian anoles with these traits in dewlap color (*A. bombiceps*: blue in male and female; *A. brasiliensis*: blue or blue-green in male and female; *A. chrysolepis*: blue in male, cream in female; *A. planiceps*: orangeish red in male and female; *A. tandai*: blue in male, blue with cream distally in female; *A. scypheus*: red distally, blue centrally in male and female). Small brown anoles that frequently are sympatric with *A. scypheus* (*A. fuscoauratus*, *A. ortonii*, *A. trachyderma*) have smooth (or very weakly keeled) ventral scales. *Anolis auratus* is smaller than *A. scypheus* (SVL to 54 mm; to 80 mm in *A. scypheus*) and has several abruptly enlarged rows of keeled middorsal scales (middorsals gradually enlarged in *A. scypheus*).

RANGE — From 50 to 1600 meters in Amazonia.

NATURAL HISTORY — Common in forest; active diurnally on ground or low vegetation; sleeps on twigs or leaves, usually below 1 meter above ground. See Fitch (1968); Hoogmoed (1973); Vanzolini and Williams (1970); Duellman (1978); Ávila-Pires (1995); Vitt and Zani (1996b; as *A. nitens*); Pinto-Aguirre (2014); Narváez (2017); Torres-Carvajal et al. (2019); Arteaga et al. (2023); Pinto and Torres-Carvajal (2023).

— *ANOLIS SERICEUS* — PLATE 1.203, 2.23

DESCRIPTION — Hallowell, E. 1856. Notes on the reptiles in the collection of the Academy of Natural Sciences of Philadelphia. *Proceedings of the Academy of Natural Sciences of Philadelphia* 8:227–228.
TYPE SPECIMEN — Lost according to Barbour (1934).
TYPE LOCALITY — "El Euceros le Jalapa, Mexico." Corrected to "El Encero de Jalapa, Vera Cruz, Mexico" by Barbour (1934). Hallowell's (1856) "El Euceros" and Barbour's (1934) "El Encero" apparently refer to El Lencero (lat 19.49 lon −96.81, 1000 m]), a town just east of Xalapa that other gringos (e.g., Oswandal 1885/2010:74) also have misidentified as "El Encero."
MORPHOLOGY — Body length to 49 M, 51 F; ventral scales keeled; middorsal scales enlarged in 0–2 rows or gradually enlarged in several rows; interparietal larger than tympanum; scales across the snout between the second canthals 5–11; toe lamellae 13–17; suboculars and supralabials in contact; scales separating supraorbital semicircles 0–2; supraocular scales gradually enlarged; hindlimbs short, adpressed toe approximately to ear; body color brown, sometimes with a light lateral stripe; male dewlap orange with blue central spot; female dewlap absent or small with male colors.
SIMILAR SPECIES — *Anolis sericeus* is a *small brown* lowland anole with keeled ventral scales and short hindlimbs (adpressed toe reaches to ear). Among potentially sympatric species with keeled ventrals and short hindlimbs, *A. microlepidotus*, *A. nebuloides*, and *A. nebulosus* have approximately three abruptly enlarged supraocular scales (supraoculars gradually enlarged in *A. sericeus*) and *A. laeviventris* tends to have heterogeneous lateral scales (homogeneous scales in *A. sericeus*) and be abundant at higher elevations than *A. sericeus*. Also, all of the above listed species lack the distinctive male dewlap colors of blue with an orange central spot in *A. sericeus*. *Anolis ustus* possesses the same distinctive dewlap colors of *A. sericeus* and is parapatric to *A. sericeus* in the Yucatan peninsula; it may be distinguished by its small dewlaps in males and females (male and female dewlaps extending approximately to arms; male dewlap large, extending on to chest, female dewlap very small or absent in *A. sericeus*).
RANGE — From 0 to 1400 meters from Mexico to Costa Rica.

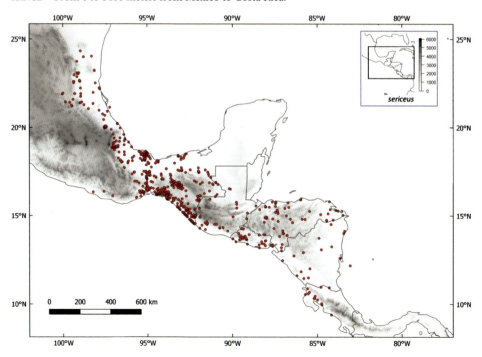

SPECIES ACCOUNTS

NATURAL HISTORY — Very common in open areas and edge habitats; active diurnally on bushes and small trees or on ground; sleeps on twigs, leaves, or vines at all visible heights. See Echelle et al. (1971a); Fitch (1973a, 1975); Lee (1996); Heras-Lara and Villareal-Benítez (1997); Savage (2002); Cardona and Reynoso (2017); Flores-Villela (2019).

COMMENT — I am treating *Anolis sericeus*, *A. wellbornae*, and *A. unilobatus* collectively as *A. sericeus*. There is conflict between the species limits of these forms suggested by Köhler and Vesely (2010) based on hemipenial variation and the molecular lineages recovered by Gray et al. (2019). However, the lineages in Gray et al. (2019) are not clearly morphologically diagnosable, their (our) southern sampling of ostensible *A. wellbornae* and *A. unilobatus* was weak, and there is significant hemipenial variation within at least one of the lineages (*A. sericeus* sensu strictu; Lara-Tufiño et al. 2016). Given these difficulties, I adopt recognition of a single species *A. sericeus*, with the acknowledgment that there certainly are distinct evolutionary lineages, i.e., species, within this form. Another option assuming the results of Gray et al. (2019) would be to recognize all *sericeus*-like anoles from Guatemala south as *A. wellbornae*, *sericeus* forms from Tabasco north along the Caribbean as *A. sericeus*, and a new species for *sericeus* anoles along Pacific Mexico. I am currently recognizing *A. ustus*, a *sericeus*-like anole from the Yucatan peninsula, in addition to *A. sericeus*. *Anolis ustus* is the only member of this "*sericeus*" complex that is both molecularly distinct (Gray et al. 2019) and morphologically diagnosable (Lara-Tufiño et al. 2016).

— *ANOLIS SERRANOI* — PLATE 1.204, 2.60

DESCRIPTION — Köhler, G. K. 1999. Eine neue Saumfingerart der Gattung *Norops* von der Pazifikseite des nördlichen Mittelamerika. *Salamandra* 35:37–52.
TYPE SPECIMEN — SMF 78834.
TYPE LOCALITY — "Wald in der Umgebung der Schmetterlingsfarm von Dr. Francisco Serrano, 13° 49,46 N, 89° 59,98 W, 225 m NN, Departamento Ahuachapan, El Salvador."

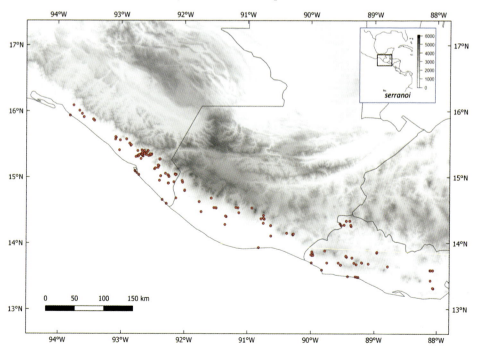

Anolis sminthus

MORPHOLOGY — Body length to 85 M, 78 F; ventral scales keeled; middorsal scales enlarged in 0–2 rows; interparietal about equal to tympanum; scales across the snout between the second canthals 6–9; toe lamellae 16–18; suboculars and supralabials usually separated by a row of scales; scales separating supraorbital semicircles 0–3; supraocular scales gradually enlarged; hindlimbs long, adpressed toe to between eye and tip of snout; body color brown; male dewlap dark red; female dewlap gray.

SIMILAR SPECIES — *Anolis serranoi* is a *large brown* lowland anole that differs from potentially sympatric brown anoles as follows: *Anolis cristifer* has smooth ventral scales (keeled in *A. serranoi*) and short hindlimbs (adpressed toe reaches to ear; to between eye and tip of snout in *A. serranoi*); *Anolis dollfusianus*, *A. laeviventris*, *A. macrophallus*, and *A. sericeus* are smaller than *A. serranoi* (SVL < 55 mm; to 85 mm in *A. serranoi*) and have different male dewlaps (*A. dollfusianus*: yellow; *A. laeviventris*: white; *A. macrophallus*: brown proximally, pink distally; *A. sericeus*: yellow with blue central spot; *A. serranoi*: dark red); *A. petersii* is larger than *A. serranoi* (SVL to 118 mm) and has a greater number of toe lamellae (24–30; 16–18 in *A. serranoi*).

RANGE — From 0 to 1100 meters on the Pacific slope of El Salvador, Guatemala, and Chiapas, Mexico.

NATURAL HISTORY — Common in forest and edge habitat; active diurnally on trees up to 2 meters above ground; sleeps on twigs, leaves, or vines from 1 to 2.5 meters above ground. See Köhler and Acevedo (2004); Luna-Reyes et al. (2012); Dor et al. (2014).

— *ANOLIS SMINTHUS* — PLATE 1.205, 2.47

DESCRIPTION — Dunn, E. R., J. T. Emlen. 1932. Reptiles and amphibians from Honduras. *Proceedings of the Academy of Natural Sciences of Philadelphia* 84:26–27.

TYPE SPECIMEN — ANSP 22946 (see notes in McCranie and Köhler 2014).

TYPE LOCALITY — "San Juancito, Honduras, 6900 feet." San Juancito is located at lat 14.22 lon –87.07. The stated elevation and Dunn and Emlen's (1932: 21) comments suggest that the holotype was

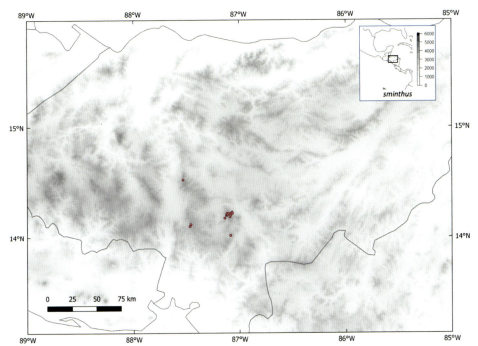

SPECIES ACCOUNTS

collected above the town, in what is now La Tigra National Park, a site where *Anolis sminthus* is common (personal observation): "San Juancito is a town situated in the mountains at 4200 ft., about fifteen miles northeast of Tegucigalpa. The camp of New York and Honduras Rosario Mining Co., about 1000 ft. above the town, offers a base from which the cloud forest area up to 7200 ft., as well as the upper pine region from 4000 ft. to 5000 ft., can be reached in a few hours' time. Six weeks were spent in this locality."

MORPHOLOGY — Body length to 52 M, 58 F; ventral scales weakly keeled; middorsal scales greatly enlarged in several rows; interparietal smaller than tympanum; scales across the snout between the second canthals 4–7; toe lamellae 17–18; suboculars and supralabials in contact; scales separating supraorbital semicircles 0–2; supraocular scales gradually enlarged; adpressed toe to eye; body color brown; male dewlap red with white scales; female dewlap yellow.

SIMILAR SPECIES — *Anolis sminthus* is a *small brown* highland anole. *Anolis laeviventris*, *A. heteropholidotus*, and *A. caceresae* are geographically proximal highland anoles that differ from *A. sminthus* in their degree of ventral keeling (*A. laeviventris*, *A. caceresae*: strongly keeled; *A. heteropholidotus*: smooth; *A. sminthus*: weakly keeled). See discussion under *A. caceresae* for comparison with other Honduran high-elevation small brown anoles.

RANGE — from 1450 to 1900 meters in southern-central Honduras.

NATURAL HISTORY — Common in forest, edge, and open areas; active diurnally on low vegetation or ground; sleeps on leaves up to 1.5 meters above ground. See McCranie and Köhler (2015).

— *ANOLIS SOINII* — FIGURE 3.22; PLATE 1.206

DESCRIPTION — Poe, S., C. Yañez-Miranda. 2008. Another new species of green *Anolis* (Squamata: Iguania) from the Eastern Andes of Peru. *Journal of Herpetology* 42:564–571.

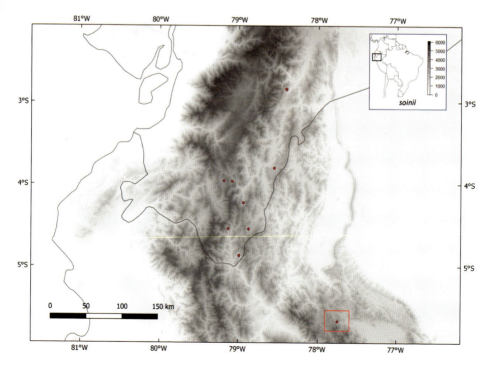

Anolis solitarius

TYPE SPECIMEN — MZUNAP 02.000179.
TYPE LOCALITY — "Peru, Department of San Martin, Venceremos, approximately 94 km west of Rioja (between old kilometer markers 390–391, near new kilometer marker 380), 05' 40.405 S 77' 45.310 W, 1,739 m."
MORPHOLOGY — Body length to 82 M, 78 F; ventral scales smooth; middorsal scales enlarged in 0–2 rows; interparietal smaller than tympanum; scales across the snout between the second canthals 9–14; toe lamellae 17–21; suboculars and supralabials in contact; supraorbital semicircles usually in contact; supraocular scales gradually enlarged; hindlimbs long, adpressed toe to eye or to between eye and snout; body color green with faint yellow lateral spots, changing to brown with distinct spots under stress; male dewlap white with dark gray spotting and streaking, with green and turquoise scales; female dewlap absent.
SIMILAR SPECIES — *Anolis soinii* is a *large green* highland anole. Its distinctive body color pattern of green with ventrolateral yellow spots will distinguish *A. soinii* from potential sympatric congeners (sympatric forms are either lichenous or predominantly brown). *Anolis fuscoauratus, A. lososi, A. williamsmittermeierorum, A. peruensis, A. laevis,* and *A. hyacinthogularis* further differ from *A. soinii* in their smaller body size (SVL to 66 mm in these species; to 82 mm in *A. soinii*) and shorter hindlimbs (adpressed toe to ear, or short of ear; to eye or past eye in *A. soinii*). Lowland green anole *A. punctatus* differs from *A. soinii* in its dorsal pattern (usually solid green; green with light lateral spots in *A. soinii*).
RANGE — From 1400 to 2000 meters on the Amazonian slope of southern Ecuador and northern Peru.
NATURAL HISTORY — Common in forest and edge habitats; diurnally arboreal on tree trunks and shrubs; sleeps from 0.5 to 5 meters above ground, usually on leaves, especially ferns. See Ayala-Varela et al. (2011); Narváez (2017); Torres-Carvajal et al. (2019); Arteaga et al. (2023).

— *ANOLIS SOLITARIUS* — FIGURE 3.21; PLATE 1.207; MAP overleaf

DESCRIPTION — Ruthven, A. G. 1916. Three new species of *Anolis* from the Santa Marta mountains of Colombia. *Occasional Papers of the Museum of Zoology*, University of Michigan 32:2–4.
TYPE SPECIMEN — UMMZ 48303.
TYPE LOCALITY — "San Lorenzo, elevation of 5,000 feet, Santa Marta Mountains, Colombia." Paynter (1997) describes collector A. G. Ruthven's San Lorenzo as Cuchilla San Lorenzo, a "semi-isolated ridge running SW to NE for ca. 15 km, ca. 25 km SE of Santa Marta." The San Lorenzo Estación Experimental (lat 11.111 lon −74.054, 2250 meters), where *Anolis solitarius* is abundant (personal observation), is a common base of operations and seems a likely type locality.
MORPHOLOGY — Body length to 52 M, 51 F; ventral scales smooth; middorsal scales enlarged in 0–2 rows; interparietal larger than tympanum; scales across the snout between the second canthals 7–9; toe lamellae 16–20; suboculars and supralabials in contact; scales separating supraorbital semicircles 0–2; body color basically green, with dark broad saddles extending laterally in males, mostly patternless green or with light and dark spots or with broad brown middorsal stripe in females; male dewlap white with yellow distal edge; female dewlap smaller, brown.
SIMILAR SPECIES — *Anolis solitarius* is a *twig* or *small green* or *brown* anole. In my experience, this is a "solitary" anole (Williams et al. 1970), i.e., not sympatric with any congeners. It is geographically replaced to the south by *A. menta*, which is very similar but differs in its spotted female dewlap (unspotted in *A. solitarius*). Nearby small brown lowland anoles *A. gaigei, A. onca,* and *A. auratus* have keeled ventral scales (smooth in *A. solitarius*).
RANGE — From 1500 to 2500 meters in cloud forest on the northern slope of the Santa Marta mountains in northern Colombia.
NATURAL HISTORY — Common in cloud forest and edge habitat; sleeps on twigs, leaves, or (often) ferns from 0.5 meters to at least 7 meters.

SPECIES ACCOUNTS

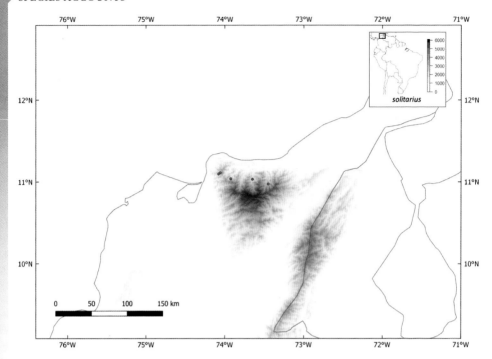

— *ANOLIS SQUAMULATUS* — PLATE 1.208; MAP opposite

DESCRIPTION — Peters, W.C.H. 1863. Über einige neue Arten der Saurier-Gattung *Anolis*. *Monatsberichte der Königlichen Preussischen Akademie der Wissenschaften zu Berlin*. 1863 (März).
TYPE SPECIMEN — Syntypes ZMB 507, 58010.
TYPE LOCALITY — Peters (1863) stated that specimens were sent from "Puerto Cabello." Presumably this is Puerto Cabello, Carabobo State, Venezuela (lat 10.47 lon −88.01, 10 m). Given that the species has verifiably been collected only at relatively high elevations, the coastal town of Puerto Cabello is an unlikely place of collection for this form. Ugueto et al. (2009) report specimens from higher elevations just inland from Puerto Cabello.
MORPHOLOGY — Body length to 95 M, 101 F; ventral scales smooth; middorsal scales enlarged in 0–2 rows; interparietal smaller than tympanum; scales across the snout between the second canthals 10–15; toe lamellae 21–27; suboculars and supralabials usually separated by a row of scales; scales separating supraorbital semicircles 3–6; supraocular scales subequal; hindlimbs long, adpressed toe to anterior to eye; body color green, often with bands; male dewlap yellow with red anterior.
SIMILAR SPECIES — *Anolis squamulatus* is a *large green* highland anole that may be sympatric with *A. planiceps*, *A. fuscoauratus*, and/or *A. tigrinus* (Ugueto et al. 2009). *Anolis squamulatus* is unlikely to be confused with these or other nearby species due to its large size (SVL < 76 mm in *A. planiceps*, *A. fuscoauratus*, *A. tigrinus*; to 101 mm in *A. squamulatus*) and green body coloration (brown or gray in *A. planiceps*, *A. fuscoauratus*, *A. tigrinus*).
RANGE — From 900 to 1650 meters in northern Venezuela, states of Cojedes, Carabobo, Aragua, and Vargas.
NATURAL HISTORY — In forest and edge habitats; active diurnally usually 1 to 2 meters above ground on trees or bushes, occasionally higher; sleeps on leaves. See Test et al. (1966); Barrio-Amorós (2006); Ugueto et al. (2009).

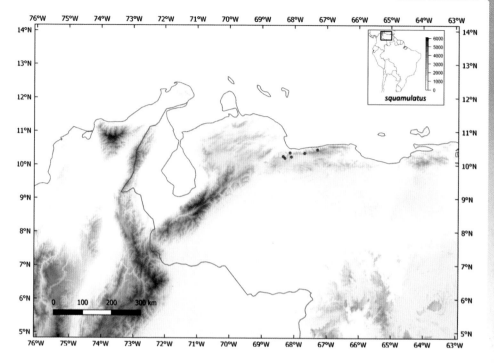

— ANOLIS SUBOCULARIS — PLATE 1.209, 2.15

DESCRIPTION — Davis, W. B. 1954. Three new anoles from Mexico. *Herpetologica* 10:3-4.
TYPE SPECIMEN — TCWC 8675.
TYPE LOCALITY — "One mile southwest of Tierra Colorada, 900 ft., Guerrero." This location is at approximately lat 17.16 lon -99.53.
MORPHOLOGY — Body length to 63 M, 48 F; ventral scales keeled; middorsal scales gradually enlarged in several rows; interparietal equal to or smaller than tympanum; scales across the snout between the second canthals 6-8; toe lamellae 12-17; suboculars and supralabials usually separated by a row of scales; scales separating supraorbital semicircles 0-2; supraocular scales abruptly enlarged; addressed toe to between ear and eye or to eye; body color grayish-brown; male dewlap pinkish-red; female dewlap usually absent.
SIMILAR SPECIES — *Anolis subocularis* is a *small brown* lowland anole with keeled ventral scales. Among geographically proximal species that share these traits, *A. sericeus* and *A. nebulosus* differ from *A. subocularis* in hindlimb length (addressed toe reaches to ear; approximately to eye in *A. subocularis*); *A. nebuloides* and *A. megapholidotus* have larger middorsal scales (middorsals larger than ventrals; ventrals larger than middorsals in *A. subocularis*), shorter hindlimbs (addressed toe usually reaches to ear), and tend to be found at higher elevations than *A. subocularis*. Parapatric eastern lowland form *A. immaculogularis* and Isthmusian form *A. boulengerianus* differ from *A. subocularis* in male dewlap color (*A. immaculogularis*: pinkish-red, without paler tint around dewlap scale rows; *A. boulengerianus*: yellow-orange; *A. subocularis*: pinkish-red, paler around dewlap scale rows).
RANGE — Sea level to 1000 meters on the Pacific slope of Mexico from the Isthmus of Téhuantépec to southeastern Guerrero.

SPECIES ACCOUNTS

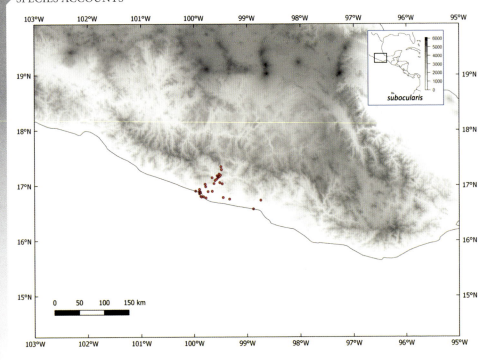

NATURAL HISTORY — Very common in open forest and edge habitat; active diurnally on rocks and low on tree trunks; sleeps on twigs or (less frequently) leaves up to 3 meters above ground. See Fitch et al. (1976); Köhler et al. (2014a).

— *ANOLIS SULCIFRONS* — PLATE 1.210, 2.108

DESCRIPTION — Cope, E. D. 1899. On a collection of batrachia and reptilia from New Granada. *The Philadelphia Museums Scientific Bulletin* 1:6–7.
TYPE SPECIMEN — AMNH 38750.
TYPE LOCALITY — "In Colombia, near Bogota ... I have not been able to ascertain the exact localities at which the specimens were obtained, but most of them, it is believed, were found in the neighborhood of Bogota." (Cope 1899:3, in the Introduction, referring to the localities of species described in the paper). Smith and Taylor (1950) restricted the type locality to Barranquilla, probably based on Barbour's (1934:145) comment that collection of the type specimen "was apparently made near Barranquilla." Dunn and Stuart (1951) rightly questioned this assignment given the known distribution of *A. sulcifrons*. They implied that the Magdalena River valley west of Bogota seems a likely collecting area for the holotype of *A. sulcifrons*, an inference with which I concur.
MORPHOLOGY — Body length to 62 M, 62 F; ventral scales smooth; middorsal scales enlarged in 0–2 rows; interparietal larger than tympanum; scales across the snout between the second canthals 7–11; toe lamellae 18–22; suboculars and supralabials in contact; supraorbital semicircles in contact; adpressed toe approximately to eye; body color gray, with lichenous white, brown, and black pattern; tail banded; male dewlap red with black markings; female dewlap absent.
SIMILAR SPECIES — *Anolis sulcifrons* is a *small brown* to gray anole, possibly interpreted as a *twig* anole. Most comparably sized geographically proximal brown anoles differ from *A. sulcifrons* by possessing

Anolis tandai

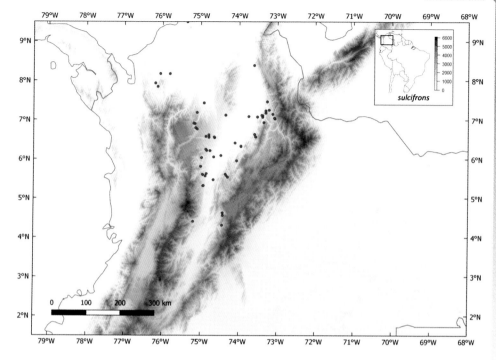

keeled ventral scales (smooth in *A. sulcifrons*): *A. auratus, A. tropidogaster, A. gaigei, A. vittigerus, A. poecilopus, A. rivalis. Anolis tolimensis, A. urraoi, A. mariarum,* and *A. maculiventris* differ from *A. sulcifrons* in possessing scales separating the supraorbital semicircles (supraorbital semicircles in contact in *A. sulcifrons*), and in their constant brown coloration (*A. sulcifrons* may appear white and/ or lichenous).

RANGE — From 100 to 1500 meters in the Magdalena River valley and farther north, Colombia.

NATURAL HISTORY — Common in forest edge and on trees in disturbed areas; active diurnally on tree trunks and branches of trees; sleeps on large leaves, usually above 3 meters above ground. See Carvajal-Cogollo and Urbina-Cordona (2015); Moreno-Arias et al. (2020).

COMMENT — I consider *Anolis ibague* (type locality "Ibague, Dto Tolima, Colombia") to be a junior synonym of *A. sulcifrons* (see appendix 1 for justification). For those who wish to recognize *A. ibague,* it is the version of *A. sulcifrons* in the region of Ibague, Colombia.

— *ANOLIS TANDAI* — PLATE 1.211, 2.96

DESCRIPTION — Ávila-Pires, T.C.S. 1995. Lizards of Brazilian Amazonia (Reptilia: Squamata). *Zoologische Verhandelingen,* Leiden 299:80–84.
TYPE SPECIMEN — MPEG 15850.
TYPE LOCALITY — "E of Porto Urucu (Petrobras station RUC-2), Rio Urucu, Amazonas State, Brazil." The Porto Urucu Petrobras station is located at lat −4.86 lon −65.30, 70 m.
MORPHOLOGY — Body length to 70 M, 61 F; ventral scales keeled; middorsal scales greatly enlarged in several rows; interparietal larger or smaller than tympanum; scales across the snout between the second canthals 10–15; toe lamellae approximately 14; suboculars and supralabials separated by a row of scales; scales separating supraorbital semicircles 1–4; supraocular scales gradually enlarged;

SPECIES ACCOUNTS

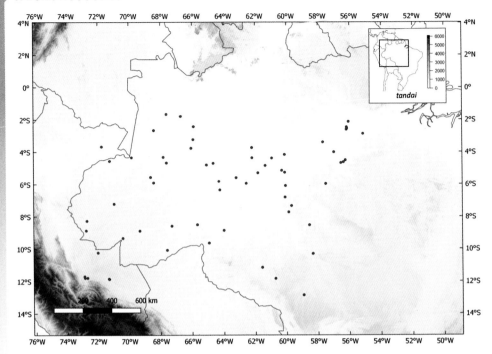

hindlimbs long, adpressed toe to beyond tip of snout; body color brown; male dewlap blue; female dewlap blue with cream distally.

SIMILAR SPECIES — *Anolis tandai* is a *small* to *large brown* lowland anole that is found in the Amazon basin with *A. bombiceps, A. scypheus, A. brasiliensis, A. planiceps,* and *A. chrysolepis*. Among these very similar species, only *A. bombiceps* may be truly sympatric with *A. tandai,* so consultation of range maps is useful for proper identification. All these species may be distinguished by female dewlap (red distally, blue proximally in *A. scypheus*; blue in *A. brasiliensis, A. bombiceps*; red in *A. planiceps*; blue in *A. chrysolepis*; blue with cream distally in *A. tandai*). The other small anoles frequently sympatric with *A. tandai* (*A. fuscoauratus, A. ortonii, A. trachyderma*) have smooth or very weakly keeled ventral scales. *Anolis auratus* is smaller than *A. tandai* (SVL to 54 mm; to 70 mm in *A. tandai*) and has shorter hindlimbs (adpressed toe reaches to between eye and snout; to beyond snout in *A. tandai*). *Anolis transversalis* has blue eyes (brown in *A. tandai*).

RANGE — From 0 to 150 meters in Amazonian Brazil, states of Amazonas and Para.

NATURAL HISTORY — Common in forest; active diurnally on ground or low vegetation; sleeps on vegetation below 1 meter above ground. See Ávila-Pires (1995); Vitt et al. (2001); Macedo et al. (2008).

— *ANOLIS TAYLORI* — PLATE 1.212, 2.5

DESCRIPTION — Smith, H. M., R. A. Spieler. 1945. A new anole from Mexico. *Copeia* 1945:165–168.
TYPE SPECIMEN — USNM 132358.
TYPE LOCALITY — "In the hills about one mile north of Acapulco, Guerrero." This area is located at approximately lat 16.9 lon −99.9, 300 m.
MORPHOLOGY — Body length to 78 M, 64 F; ventral scales smooth; middorsal scales gradually enlarged in several rows; interparietal usually larger than tympanum; scales across the snout between

Anolis tenorioensis

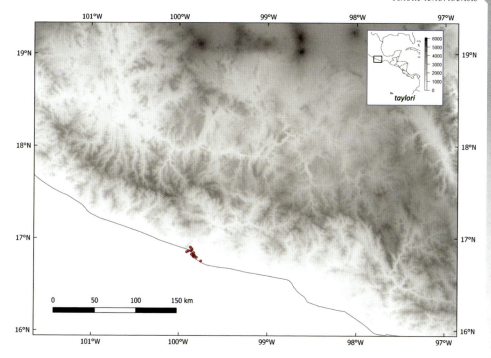

the second canthals 5–8; toe lamellae 16–21; suboculars and supralabials in contact; supraorbital semicircles in contact; supraocular scales abruptly enlarged; adpressed toe to between eye and tip of snout; body color brown, with cream lateral stripe from tip of snout to groin; male and female dewlap red with dark red and light pink markings.
SIMILAR SPECIES — *Anolis taylori* is a *large brown* anole with smooth ventral scales. Among geographically proximal species with smooth ventrals, *A. omiltemanus* and *A. peucephilus* are smaller than *A. taylori* (SVL to 48 mm; to 76 mm in *A. taylori*) and have short hindlimbs (adpressed toe reaches to ear or short of ear; between eye and tip of snout in *A. taylori*); *A. liogaster* and *A. brianjuliani* have a band of enlarged middorsal scales (middorsals uniform in *A. taylori*) and a solid pink dewlap (male and female dewlaps red with dark red and light pink markings in *A. taylori*). *Anolis dunni* is smaller than *A. taylori* (SVL to 58 mm), more catholic in its habitat use (*A. taylori* specializes on boulders), and has a different male dewlap (orange-red). *Anolis gadovii* differs from *A. taylori* mainly in color (contrasting reticulating dorsal pattern and male dewlap with dark crescents on a pinkish background in *A. gadovii*).
RANGE — From 0 to 1050 meters in the hills around Acapulco, Guerrero, Mexico.
NATURAL HISTORY — Common in forest and open areas; active diurnally on tree trunks, boulders and rock walls; sleeps on boulders or rock walls including cave interiors. See Mautz (1981); Fitch and Henderson (1976a); Fitch et al. (1976); Köhler et al. (2014a).

— *ANOLIS TENORIOENSIS* — PLATE 1.213, 2.39

DESCRIPTION — Köhler, G. 2011a. A new species of anole related to *Anolis altae* from Volcán Tenorio, Costa Rica (Reptilia, Squamata, Polychrotidae). *Zootaxa* 3120:29–42.
TYPE SPECIMEN — SMF 91985.

SPECIES ACCOUNTS

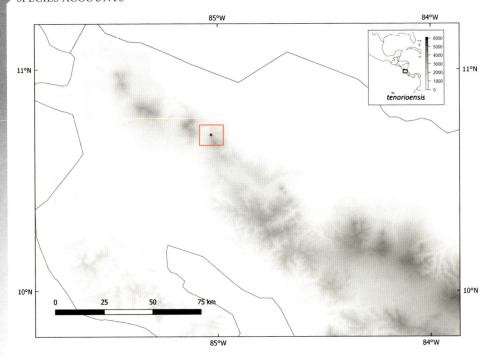

TYPE LOCALITY — "Near Laguna Danta, Volcán Tenorio (10.70521 N, 85.03068 W, WGS84), Alajuela Province, Costa Rica." Elevation at this point is 1160 m.

MORPHOLOGY — Body length to 42 M, 45 F; ventral scales smooth; middorsal scales enlarged in 0–2 rows; interparietal smaller than tympanum; scales across the snout between the second canthals 7; toe lamellae 15–16; suboculars and supralabials in contact; scales separating supraorbital semicircles 1–2; supraocular scales gradually enlarged; hindlimbs short, adpressed toe approximately to ear; body color brown; male dewlap reddish-orange with dark spots.

SIMILAR SPECIES — *Anolis tenorioensis* is a *small brown* highland anole that may be distinguished from geographically proximal similar species as follows: *A. humilis*, *A. cupreus*, *A. laeviventris*, and *A. sericeus* have keeled ventral scales (smooth in *A. tenorioensis*); *A. kemptoni*, found to the east of the range of *A. tenorioensis*, has a bicolor orange and pink male dewlap (reddish-orange with dark spots in *A. tenorioensis*); lowland forms *A. limifrons*, *A. biscutiger*, and *A. polylepis* have longer hindlimbs (adpressed toe to beyond eye; to ear in *A. tenorioensis*). *Anolis monteverde* and *A. altae* are very similar to *A. tenorioensis* but are allopatric to it, found at high elevations to the southeast of the range of *A. tenorioensis*. Both species differ from *A. tenorioensis* in lacking dark spots on the male dewlap. *Anolis arenal* is similar to *A. tenorioensis* but is a lowland form (known only from below 600 m) with a differently colored male dewlap (unspotted, dark centrally; orange-red distally in *A. arenal*).

RANGE — Known only from the type locality.

NATURAL HISTORY — The type description states that specimens were collected at night sleeping on low vegetation. I have found this form sleeping on twigs and vines from 1.5 to 3.5 meters up in edge and low canopy-covered habitat.

— ANOLIS TEQUENDAMA — PLATE 1.214, 2.138

DESCRIPTION — Moreno-Arias, R. A., M. A. Méndez-Galeano, I. Beltrán, M. Vargas-Ramírez. 2023. Revealing the anole diversity in the highlands of Northern Andes: New and resurrected species of the *Anolis heterodermus* species group. *Vertebrate Zoology* 73:161–188.
TYPE SPECIMEN — ICN-R 4548.
TYPE LOCALITY — "Vereda Sabaneta, San Francisco municipality, Cundinamarca department, Colombia (4.891173°N −74.289925°W, 2850 m asl)."
MORPHOLOGY — Body length to 66 M, 68 F; ventral scales smooth; middorsal scales form a crest of triangular plates (see Lazell 1969: fig. 1); interparietal larger than tympanum; scales across the snout between the second canthals 3–5; toe lamellae 18–21; suboculars and supralabials in contact; supraorbital semicircles in contact; supraocular scales abruptly enlarged; hindlimbs short, adpressed toe to ear or posterior to ear; body color brown or green with white, gray, blue, and/or greenish-yellow, with light stripe from supralabial area back to shoulder; dewlap brown with light scales in male, yellow-orange with light scales and dark streaks in female; heterogeneous lateral scutellation.
SIMILAR SPECIES — *Anolis tequendama* is a distinctive highland *twig* anole. The heterogeneous lateral scalation, wherein large smooth oval scales are separated by granular interstitial scales, is differentiating in *A. tequendama*, especially in combination with its short hindlimbs and tail and large smooth head scales. Among potential eastern Andean anoles that share these traits, *A. vanzolinii* and *A. inderenae* are larger than *A. tequendama* (to >100 mm SVL; to 68 mm in *A. tequendama*) and possess greater numbers of toe lamellae (23–28; 18–21 in *A. tequendama*); *A. heterodermus*, *A. quimbaya*, and *A. richteri* differ from *A. tequendama* in male and female dewlap color (*A. heterodermus*: purple, red, or black in males and females; *A. quimbaya*: orange in male, green in females; *A. richteri*: cream to yellow in males and females, sometimes with red distally; *A. tequendama*: brown with light scales in male, yellow with dark streaks in female).
RANGE — High elevations along the eastern edge of the Southern Magdalena River Valley.

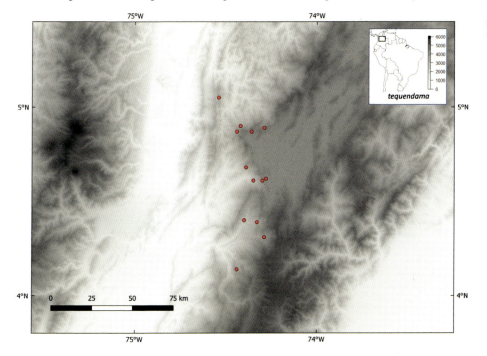

SPECIES ACCOUNTS

NATURAL HISTORY — The type description (Moreno-Arias et al. 2023) states, "This species preferentially uses small branches and narrow surfaces such as twigs and exhibits very slow movements." See Dunn (1944; as *Phenacosaurus heterodermus*); Osorno and Osorno (1946; as *P. heterodermus*); Méndez-Galeano and Calderón -Espinosa (2020; as *A. heterodermus*).

— *ANOLIS TETARII* — PLATE 1.215

DESCRIPTION — Barros, T. R., E. E. Williams, A. Viloria. 1996. The genus *Phenacosaurus* (Squamata: Iguania) in western Venezuela: *Phenacosaurus tetarii*, new species, *Phenacosaurus euskalerriari*, new species, and *Phenacosaurus nicefori* Dunn, 1944. Breviora 504:3–16.
TYPE SPECIMEN — MBLUZ 215.
TYPE LOCALITY — "The roads that lead to the Páramo del Tetari, Sierra de Perijá, Estado Zulia, Venezuela (10 06 34 N, 72 53 00 W), 2,790 m elevation."
MORPHOLOGY — Body length to 86 M, 70 F; ventral scales smooth; middorsal scales form an uneven crest of triangular plates; interparietal larger than tympanum; scales across the snout between the second canthals 4; toe lamellae 23; suboculars and supralabials in contact; scales separating supraorbital semicircles 0–1; body dark brown dorsally, blotched, yellow laterally; male dewlap yellow; lateral scales heterogeneous.
SIMILAR SPECIES — *Anolis tetarii* is a highland *twig* anole that may be sympatric with *A. fuscoauratus* and/or *A. planiceps*, and is amply distinct from both in its large smooth head scales (scales across the snout at second canthals 8–14 in *A. fuscoauratus* and *A. planiceps*; 4 in *A. tetarii*). *Anolis euskalerriari*, found southeast of the range of *A. tetarii*, differs from *A. tetarii* in its homogeneous lateral scales (heterogeneous in *A. tetarii*) and pink male dewlap (yellow in *A. tetarii*).
RANGE — From 2400 to 3050 meters in the northern Perijá mountains, northwestern Venezuela and northeastern Colombia.

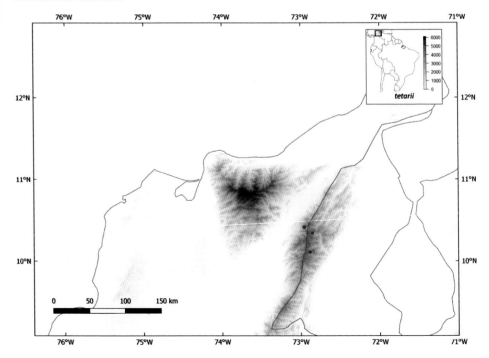

NATURAL HISTORY — Only information is from the type description: specimens were collected in the transition between cloud forest and páramo, on a shrub and "spongy lichens and dead leaves."

— *ANOLIS TIGRINUS* — PLATE 1.216

DESCRIPTION — Peters, W.C.H. 1863. Über einige neue Arten der Saurier-Gattung *Anolis*. *Monatsberichte der Königlichen Preussischen Akademie der Wissenschaften zu Berlin*. 1863 (März).
TYPE SPECIMEN — ZMB 4686.
TYPE LOCALITY — "Chili." Subsequent authors have interpreted this location as the country Chile, which obviously is in error.
MORPHOLOGY — Body length to 57 M, 52 F; ventral scales smooth; middorsal scales enlarged in 0–2 rows; interparietal larger than tympanum; scales across the snout between the second canthals 5–8; toe lamellae 16–22; suboculars and supralabials in contact; scales separating supraorbital semicircles 0–1; supraocular scales gradually enlarged; hindlimbs short, adpressed toe to posterior to ear; body color lichenous green, gray, and brown; male dewlap light tan, darker tan distally; female dewlap light tan with black blotches.
SIMILAR SPECIES — *Anolis tigrinus* is a *twig* anole that may be sympatric with *A. planiceps*, *A. fuscoauratus*, or *A. squamulatus* (Ugueto et al. 2009). *Anolis squamulatus* is much larger than *A. tigrinus* (SVL > 100 mm; to 57 mm in *A. tigrinus*) and green (gray to lichenous in *A. tigrinus*). *Anolis planiceps* has keeled ventral scales (smooth in *A. tigrinus*). *Anolis fuscoauratus* has smaller head scales than *A. tigrinus* (8–14 scales across the snout; 5–8 in *A. tigrinus*).
RANGE — From 950 to 2100 meters in northern Venezuela. An individual assigned to *Anolis tigrinus* was collected at 400 meters elevation on the island of Tobago, approximately 280 km northeast of other known localities for *A. tigrinus* (Anton et al. 2015; point not included on range map). This individual may represent a spatially and ecologically unusual introduction—high-elevation mainland

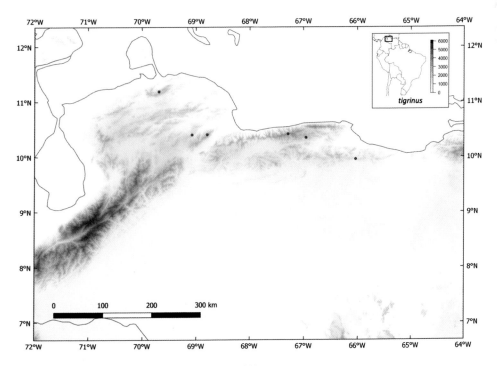

SPECIES ACCOUNTS

"twig" anoles such as *A. tigrinus* are otherwise not known to establish naturalized populations (see Latella et al. 2011)--or an undescribed species.

NATURAL HISTORY — In forest and forest edge, possibly in disturbed areas; active diurnally on twigs, branches, or leaves above 2 meters above ground; sleeps on twigs above 3 meters above ground. See Test et al. (1966); Ugueto et al. (2009).

— *ANOLIS TOLIMENSIS* — PLATE 1.217, 2.92

DESCRIPTION — Werner, F. 1916. Bemerkungen über einige niedere Wirbeltiere der Anden von Kolumbien mit Beschreibungen neuer Arten. *Zoologischer Anzeiger* 47:303–304.
TYPE SPECIMEN — ZMH R01800
TYPE LOCALITY — "Coñon del Tolima." This locality probably is Combeima Canyon, west of Ibague in Tolima Department, Colombia. Werner's description paper states that his "Coñon" specimens were taken at 1700 meters in the central Cordillera. Combeima Canyon reaches 1700 meters near the town of Villa Restrepo (lat 4.53 lon −75.31). This site seems a reasonable guess at an appropriate type locality for *Anolis tolimensis*.
MORPHOLOGY — Body length to 55 M, 52 F; ventral scales smooth to weakly keeled; middorsal scales enlarged in 0–2 rows; interparietal larger than tympanum; scales across the snout between the second canthals 8–13; toe lamellae 14–17; suboculars and supralabials in contact; scales separating supraorbital semicircles 1–3; supraocular scales abruptly or gradually enlarged; adpressed toe to eye or between ear and eye; body color brown; male dewlap pink posteriorly, orange anteriorly; female dewlap absent.
SIMILAR SPECIES — *Anolis tolimensis* is a *small brown* highland anole. Similar Andean "fuscoauratid" anoles are best distinguished by geography and male dewlap color: The male dewlap of *A. urraoi* is

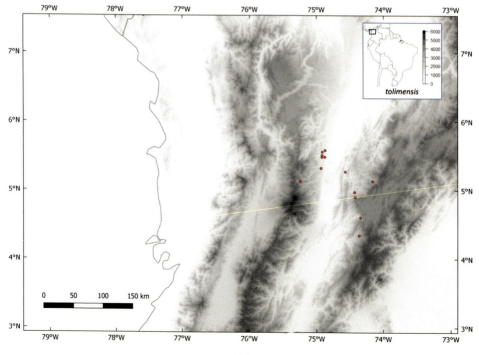

pale pink posteriorly, orange anteriorly; *A. urraoi* is found in the northern Cordillera Occidental. The male dewlap of *A. antonii* is red-orange to orange-red anteriorly, pale pink posteriorly; *A. antonii* is found in the southern Cordillera Occidental. The male dewlap of *A. fuscoauratus* usually is pale pink; *A. fuscoauratus* is found on the Amazonian slope of the Cordillera Oriental. The male dewlap of *A. mariarum* is red-orange anteriorly, yellow posteriorly; *A. mariarum* is found in the northern Cordillera Central. The male dewlap of *A. tolimensis* is solid pinkish red; *A. tolimensis* is found in the Cordillera Central in upper regions of the southern Magdalena River valley. *Anolis auratus* and *A. tropidogaster* may be distinguished from *A. tolimensis* by their keeled ventral scales (smooth to weakly keeled in *A. tolimensis*). *Anolis sulcifrons* has larger head scales than *A. tolimensis* (supraorbital semicircles in contact; separated by scales in *A. tolimensis*).
RANGE — From 1100 to 1650 meters along the Magdalena River valley, Colombia.
NATURAL HISTORY — Common in forest, edge, and disturbed habitats; sleeps on vegetation, usually twigs, above 1 meter above ground. See Ardila-Marín et al. (2008a, b); Ríos-Orjuela et al. (2019).
COMMENT — The distributional limits of the Andean *fuscoauratus*-like anoles are not clear (Grisales-Martínez et al. 2017; Espitia Sanabria 2023). I tentatively consider *fuscoauratus*-like anoles along the Magdalena River valley in Caldas, Cundimaca, and Tolima Departments to be *Anolis tolimensis*. See Espitia Sanabria (2023) for issues of species boundary and geographic range in *A. tolimensis*, and Comment under *A. antonii*.

— *ANOLIS TOWNSENDI* — PLATE 1.218, 2.57

DESCRIPTION — Stejneger, L. 1900. Description of two new lizards of the genus *Anolis* from Cocos and Malpelo Islands. *Bulletin of the Museum of Comparative Zoology* 36:163–4.
TYPE SPECIMEN — USNM 22106.

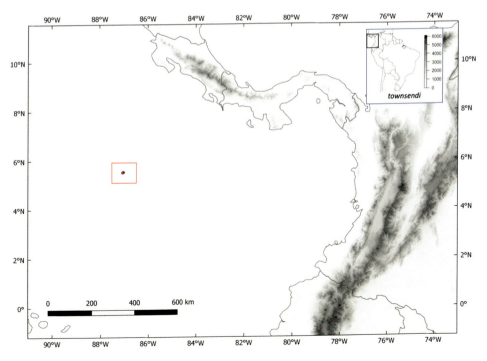

SPECIES ACCOUNTS

TYPE LOCALITY — "Cocos Island, Pacific Ocean, off Costa Rica, Central America." Cocos Island is located at lat 5.53 lon −87.06, 0–850 m.
MORPHOLOGY — Body length to 57 M, 44 F; ventral scales keeled; middorsal scales enlarged in 0–2 rows; interparietal equal to or larger than tympanum; scales across the snout between the second canthals 9–13; toe lamellae 15–19; suboculars and supralabials in contact; scales separating supraorbital semicircles 2–3; supraocular scales gradually enlarged; hindlimbs long, adpressed toe to between eye and tip of snout; body color brown, often with prominent light lateral stripe; male dewlap yellow with orange centrally.
SIMILAR SPECIES — *Anolis townsendi* is the only anole found on Cocos Island.
RANGE — Throughout Cocos Island, Pacific Ocean, Costa Rica.
NATURAL HISTORY — In all habitats on Cocos Island, on all visible perches and perch heights. See Carpenter (1965); Jenssen and Rothblum (1977); Savage (2002); Phillips et al. (2019).

— *ANOLIS TRACHYDERMA* — PLATE 1.219, 2.98

DESCRIPTION — Cope, E. D. 1876. Report on the reptiles brought by Professor James Orton from the middle and upper Amazon and western Peru. *Journal of the Academy of Natural Sciences of Philadelphia* 8:168.
TYPE SPECIMEN — ANSP 11363. Malnate's (1971) list of ANSP type material erroneously lists ANSP 11368.
TYPE LOCALITY — "Nauta." That is, Nauta, Loreto Province, Peru (lat −4.51 lon −73.58, 100 m).
MORPHOLOGY — Body length to 61 M, 57 F; ventral scales smooth to weakly keeled; middorsal scales enlarged in 0–2 rows; interparietal larger or smaller than tympanum; scales across the snout between the second canthals 10–20; toe lamellae 13–20; suboculars and supralabials in contact or separated by a row of scales; scales separating supraorbital semicircles 1–4; supraocular scales gradually enlarged;

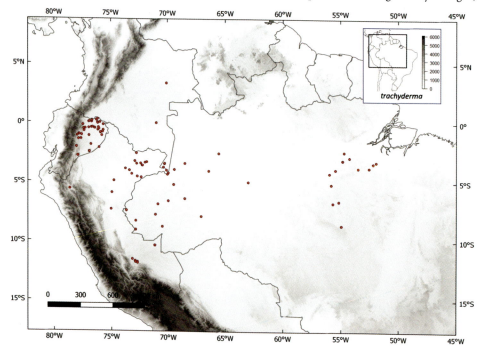

adpressed toe to eye or to between eye and snout or past snout; body color brown, often with light lateral stripe; male dewlap small, reaching approximately to axillae, red-orange proximally, yellow distally; female dewlap absent.

SIMILAR SPECIES — *Anolis trachyderma* is a *small brown* anole that is a member of the complement of anoles usually found at lowland Amazonian sites. Among these species, *A. transversalis* differs from *A. trachyerma* in body size (adults generally over 70 mm SVL in *A. transversalis*; to 61 mm in *A. trachyderma*) and eye color (blue in *A. transversalis*; brown in *A. trachyderma*); *A. punctatus* differ from *A. trachyderma* in body size (to 89 mm SVL) and green body color; *A. chrysolepis*, *A. bombiceps*, *A. tandai*, *A. planiceps*, and *A. scypheus* have strongly keeled ventral scales (smooth or barely keeled in *A. trachyderma*); *A. ortonii* and *A. fuscoauratus* differ in hindlimb length (adpressed toe reaches to ear in *A. ortonii*, *A. fuscoauratus*; past eye in *A. trachyderma*). The combination of brown body color and small male dewlap (reaching to forelimbs) will separate *A. trachyderma* from rare Amazonian species *A. caquetae*, *A. dissimilis*, and *A. phyllorhinus*.

RANGE — From 50 to 1000 meters in Amazonian Brazil, Ecuador, Colombia, and Peru.

NATURAL HISTORY — Common in forest; active diurnally on ground or low vegetation, usually below 1 meter above ground; sleeps on twigs or leaves below 1 meter above ground. See Duellman (1978); Ávila Pires (1995); Vitt and Zani (1996b); Vitt et al. (1999); Vitt (2000); Vitt et al. (2002); Pinto-Aguirre (2014); Narváez (2017); Moreno-Arias et al. (2020); Torres-Carvajal et al. (2019); Arteaga et al. (2023); Pinto and Torres-Carvajal (2023).

— *ANOLIS TRANSVERSALIS* — PLATE 1.220, 2.109

DESCRIPTION — Duméril, M. C., A. A. Duméril. 1851. *Catalogue méthodique de la collection des reptiles*. Muséum National d'Histoire Naturelle, Gide et Baudry, Paris. 57–58.

TYPE SPECIMEN — MNHN 2449.

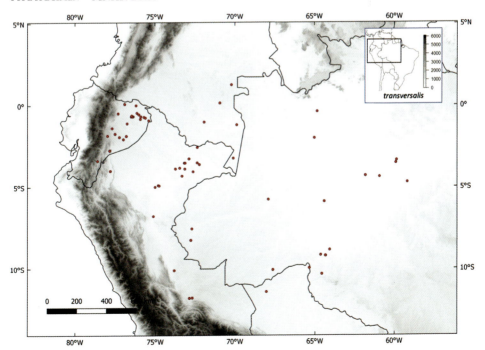

SPECIES ACCOUNTS

TYPE LOCALITY — "Amérique Méridionale." That is, South America. Williams and Vanzolini (1966) restricted the type locality to Sarayacu, Peru, based on Guichenot (1855). The town of Sarayacu along the Río Ucayali in Peru is located at lat −6.79 lon −75.11, 140 m.

MORPHOLOGY — Body length to 84 M, 80 F; ventral scales smooth; middorsal scales enlarged in 0–2 rows; interparietal smaller than tympanum; scales across the snout between the second canthals 4–9; toe lamellae 20–27; suboculars and supralabials in contact; scales separating supraorbital semicircles 0–1; adpressed toe to ear, occasionally to eye; body color green to brown, with transverse bands or ocelli; tail banded; male dewlap orange-yellow; female dewlap black with light scale rows; iris blue.

SIMILAR SPECIES — *Anolis transversalis* is a *large brown* or *green* lowland Amazonian anole unlikely to be confused with most potentially sympatric species due to its blue eyes and large size.

RANGE — From 50 to 950 meters in Amazonian Colombia, Ecuador, Peru, and Brazil.

NATURAL HISTORY — Rare in forest; active diurnally on tree trunks; sleeps high, usually above 1.5 meters above ground, on twigs, leaves, or vines. See Duellman (1978); Ávila-Pires (1995); Vitt et al. (2003a); Macedo et al. (2008); Pinto-Aguirre (2014); Narváez (2017); Moreno-Arias et al. (2020); Torres-Carvajal et al. (2019); Arteaga et al. (2023); Pinto and Torres-Carvajal (2023).

— *ANOLIS TRINITATIS* — PLATE 1.221, 2.163

DESCRIPTION — Reinhardt, J., C. F. Lutken. 1862. Bidrag tii det vestindiske Öriges og navnligen tii de dansk-vestindiske Oers Herpetologie. *Videnskabelige Meddelelser fra den Naturhistoriske Forening i Kjöbenhavn* (10–18): 153–291.

TYPE SPECIMEN — ZMUC-R37145.

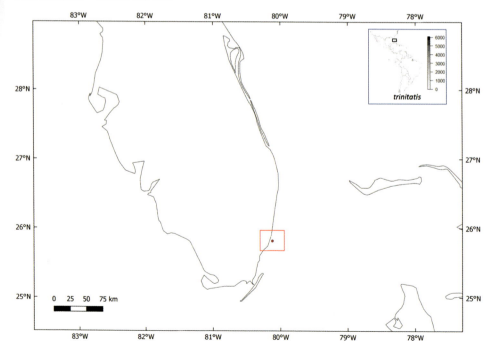

308

TYPE LOCALITY — "Trinidad." Lazell (1972:77) states: "Although this species does occur in Trinidad, that island is not within its natural range; it is an introduction. In addition, it is not very common there, and may be losing ground to the likewise introduced *Anolis aeneus* ... I feel it is mandatory to revise the type locality to Kingstown, St. Vincent; this is the principal town and port within the natural range of this species, and it is abundant there." If the holotype specimen was collected from a breeding population on Trinidad, I suggest the appropriate type locality is Trinidad (lat 10.4, lon −61.3 0–940 m), regardless of how that specimen's status is viewed.

MORPHOLOGY — Body length to 74 M, 57 F; ventral scales smooth; middorsal scales enlarged in 0–2 rows; interparietal larger than tympanum; scales across the snout between the second canthals 5–10; toe lamellae 25–27; suboculars and supralabials in contact; supraorbital semicircles usually in contact, occasionally separated by a row of scales; supraocular scales gradually enlarged; hindlimbs short, adpressed toe approximately to ear; body color bright green with blue, some yellow along the mouth, females duller and with a light lateral stripe; male dewlap yellow, sometimes with blue; female dewlap small, "not distinctively colored" (Lazell 1972).

SIMILAR SPECIES — *Anolis trinitatis* is a *large green* anole. In Florida, most other green anole species are either much larger (maximum SVL > 100 mm in *A. equestris, A. garmani*; to 74 mm in *A. trinitatis*) or have keeled ventral scales (*A. carolinensis*; smooth in *A. trinitatis*). *Anolis callainus* is green and has smooth ventral scales but lacks the vivid body coloration of *A. trinitatis* and differs in male dewlap color (gray anteriorly, black posteriorly; yellow, sometimes with blue in *A. trinitatis*).

RANGE — Krysko et al. (2011) document 2 vouchered specimens and 11 captured individuals from the Fontainebleau Hotel in Miami, Florida, suggesting a breeding population. The species is native to St. Vincent in the Lesser Antilles. My group visited the Fontainebleau locality in 2021 and failed to find any *Anolis trinitatis*.

NATURAL HISTORY — Not studied in Florida. On St. Vincent found in all habitats, but "rarely perch above ten feet above ground" (Lazell 1972).

— *ANOLIS TRIUMPHALIS* — PLATE 1.222, 2.81

DESCRIPTION — Nicholson, K. E., G. Köhler. 2014. A new species of the genus *Norops* from Darién, Panama, with comments on *N. sulcifrons* (Cope 1899) (Reptilia, Squamata, Dactyloidae). *Zootaxa* 3895:225–237.

TYPE SPECIMEN — SMF 98033.

TYPE LOCALITY — "Filo del Tallo, on main road (via Puerto Kimba) adjacent to the park (Filo del Tallo), 8.450981°N, 78.00002°W, 128 m elevation, Darién, Panama."

MORPHOLOGY — Body length to 55 M (unknown in females); ventral scales smooth; middorsal scales enlarged in 0–2 rows; interparietal larger than tympanum; scales across the snout between the second canthals 8–11; toe lamellae 14–16; suboculars and supralabials in contact; supraorbital semicircles in contact; hindlimbs short, adpressed toe to ear; body color grayish-white, lichenous; male dewlap red with yellow along scale rows and distally.

SIMILAR SPECIES — *Anolis triumphalis* is a *twig* or *small* gray to brown anole with smooth ventral scales. Potentially sympatric species *A. apletophallus* and *A. maculiventris* have smaller head scales than *A. triumphalis* (1–5 scales separating supraorbital semicircles; supraorbital semicircles in contact in *A. triumphalis*) and longer hindlimbs (adpressed toe reaches to beyond eye; to ear in *A. triumphalis*). *Anolis elcopeensis* is smaller than *A. triumphalis* (SVL to 45 mm; to 55 mm in *A. triumphalis*) and has smaller head scales (1–3 scales separating supraorbital semicircles). *Anolis sulcifrons* and *A. pentaprion*, found east and west of the range of *A. triumphalis*, respectively, are very similar to *A. triumphalis*. These species differ from *A. triumphalis* in male dewlap color (*A. sulcifrons*: red with black blotches; *A. pentaprion*: reddish-purple; *A. triumphalis*: red with yellow along scale rows).

RANGE — From 0 to 150 m in eastern Panama to northern Pacific Ecuador and southwestern Caribbean Colombia.

SPECIES ACCOUNTS

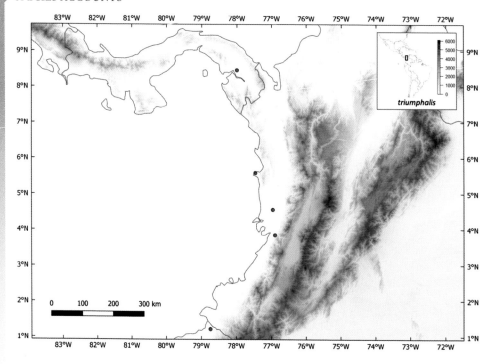

NATURAL HISTORY — The type specimen was collected in a disturbed area. The only confirmed specimen I have observed was sleeping on a twig near the ground (given the ecology of other *pentaprion*-like anoles—active on tree trunks and branches, sleeping high on large leaves—I suspect this behavior is aberrant).

— *ANOLIS TROPIDOGASTER* — PLATE 1.223, 2.104

DESCRIPTION — Hallowell, E. 1856. Notes on reptiles in the collection of the Academy of Natural Sciences of Philadelphia. *Proceedings of the Academy of Natural Sciences of Philadelphia* 8:224–225.
TYPE SPECIMEN — ANSP 7618.
TYPE LOCALITY — "New Grenada." That is, northwestern South America or Panama.
MORPHOLOGY — Body length to 55 M, 54 F; ventral scales keeled; middorsal scales gradually enlarged in several rows; interparietal larger or smaller than tympanum; scales across the snout between the second canthals 10–18; toe lamellae 13–17; suboculars and supralabials in contact or separated by rows of scales; scales separating supraorbital semicircles 2–4; supraocular scales gradually enlarged; hindlimbs long, adpressed toe to between eye and tip of snout; body color brown; tail banded; male dewlap yellow and reddish-orange to red; female dewlap usually absent.
SIMILAR SPECIES — *Anolis tropidogaster* is a widespread *small brown* lowland anole with keeled ventral scales that differs from potentially sympatric small brown anoles with keeled ventrals as follows: *Anolis humilis*, *A. auratus*, and *A. notopholis* have abruptly enlarged rows of keeled middorsal scales (middorsals weakly enlarged in *A. tropidogaster*); *A. graniliceps* has smooth to weakly keeled ventral scales (strongly keeled in *A. tropidogaster*) and a small male dewlap (extending to forelimbs; well on to chest in *A. tropidogaster*). Amazonian forms *A. scypheus*, *A. planiceps*, and *A. bombiceps* are larger than *A. tropidogaster* (at least 65 mm SVL in adults; to 55 mm in *A. tropidogaster*),

310

Anolis tropidolepis

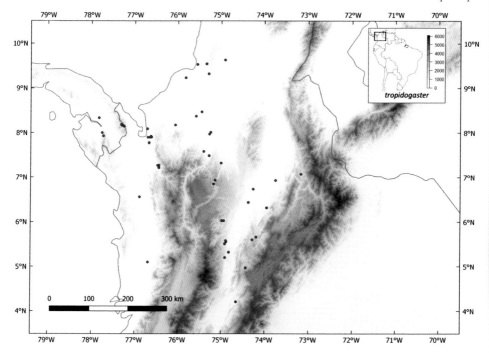

with some blue or red in adult male and female dewlaps (male dewlap yellow and reddish-orange to red, female dewlap usually absent in *A. tropidogaster*). *Anolis gracilipes* has a tiny interparietal, smaller than tympanum (approximately equal to tympanum in *A. tropidogaster*) and a solid orange male dewlap. Geographically proximal brown semiaquatic anoles differ from *A. tropidogaster* in possessing abruptly enlarged middorsal scales (*A. rivalis*, *A. macrolepis*), smooth ventral scales (*A. maculigula*), and/or larger body size (SVL to 72 mm in *A. poecilopus*). *Anolis vittigerus* differs from *A. tropidogaster* in its distinctive small male and female dewlaps (red [male] or blue or white [female] with a dark central spot). *Anolis gaigei* is externally indistinguishable from *A. tropidogaster* but possesses a very different hemipenes (bilobed in *A. gaigei*; elongated and unilobed in *A. tropidogaster*; see Köhler et al. 2012).

RANGE — From 0 to 1100 meters in eastern Panama west to northern Pacific and southwestern Caribbean Colombia and the Magdalena River valley.

NATURAL HISTORY — Common in forest and edge habitat; active diurnally on low vegetation; sleeps on twigs, leaves, or vines up to 1.5 meters above ground. See Medina-Rangel (2011, 2013); Carvajal-Cogollo and Urbina-Cordona (2015); Moreno-Arias et al. (2020).

COMMENT — Most ecological studies that refer to *Anolis tropidogaster* are of *A. gaigei* (see Köhler et al. 2012). The range and status of *A. tropidogaster* should be evaluated relative to the taxonomic status of *A. gaigei* (see comment under *A. gaigei*).

— *ANOLIS TROPIDOLEPIS* — PLATE 1.224, 2.51

DESCRIPTION — Boulenger, G. A. 1885. *Catalogue of the lizards in the British Museum (Natural History)*, 2nd ed., vol. 2. 53.
TYPE SPECIMEN — BMNH 85.3.24.11. (According to Dunn [1930].)

SPECIES ACCOUNTS

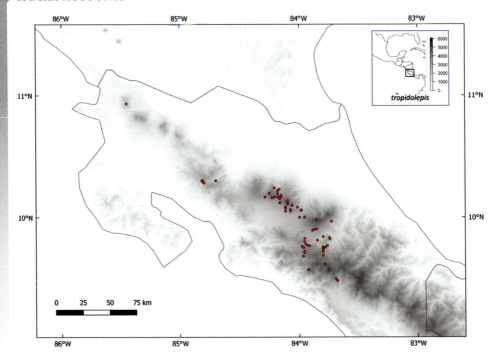

TYPE LOCALITY — "Irazu." Presumably Volcán Irazu, Cartago Province, Costa Rica. Savage (1974) stated that Boulenger's "Irazu" specimens were likely from the area between the city of Cartago and the volcano. That is, in the region of lat 9.9 lon −83.9.

MORPHOLOGY — Body length to 59 M, 58 F; ventral scales keeled; middorsal scales gradually enlarged in 2 to several rows; interparietal smaller than tympanum; scales across the snout between the second canthals 9–15; toe lamellae 12–15; suboculars and supralabials in contact or separated by a row of scales; scales separating supraorbital semicircles 2–6; supraocular scales gradually enlarged; hindlimbs long, adpressed toe to between eye and tip of snout; body color brown, frequently with middorsal diamonds, lines radiating from the eye, a dark interorbital bar, and banded tail; male dewlap red; males have a bulging hemipenial area.

SIMILAR SPECIES — *Anolis tropidolepis* is a *small brown* highland anole distinguishable from most potentially sympatric small brown anoles by its long hindlimbs (adpressed toe to anterior to eye; posterior to eye in *A. salvini, A. kemptoni, A. fungosus, A. altae, A. arenal, A. laeviventris, A. monteverde, A. tenorioensis*). *Anolis datzorum* has larger head scales than *A. tropidolepis* (7–9 scales across the snout between the second canthals; 9–15 in *A. tropidolepis*) and a solid orange dewlap in males and females (red in *A. tropidolepis*). Other geographically proximal small brown anoles (e.g., *A. polylepis, A. limifrons, A. biscutiger, A. auratus, A. marsupialis, A. cupreus, A. humilis*) are found at lower elevations than *A. tropidolepis*. *Anolis pachypus* is a very similar highland anole found east of the range of *A. tropidolepis* that is distinguished from *A. tropidolepis* by its yellow with reddish-orange male dewlap (solid red in *A. tropidolepis*).

RANGE — From 1100 to 2650 meters from central to northern Costa Rica.

NATURAL HISTORY — Common in forest, edge, and disturbed habitats characterized by fog and rain; active diurnally on the ground or low vegetation; sleeps on vegetation, often shrubs, usually below 1 meter above ground. See Echelle et al. (1971a); Fitch (1972, 1973a, b; 1975); Andrews (1976); Berkum (1986); Pounds (1988); Savage (2002); Kaiser and Kaiser (2021; as *Anolis leditzigorum*).

COMMENT — I consider *Anolis alocomyos* (type locality: "along the road between Santa María de Dota and Copey, 9.65696°N, 83.94831°W, elev. 1,720 m asl, Provincia de San José, Costa Rica") and *A. leditzigorum* (type locality: "4 km [on road] N Santa Elena, 10.34156°N, 84.80506°W, elev. 1,560 m asl, Provincia de Puntarenas, Costa Rica") to be junior synonyms of *A. tropidolepis* (see appendix 1 for justification). For those who wish to recognize these species, *A. alocomyos* was described from the southern end of the range of *A. tropidolepis*, in the region of Dota Canton, San Jose Province, Costa Rica. *Anolis leditzigorum* was reported from the northern distribution of *A. tropidolepis*, ranging from eastern Alajuela Province to the northern border of Costa Rica.

— ANOLIS TROPIDONOTUS — PLATE 1.225, 2.41

DESCRIPTION — Peters, W.C.H. 1863. Über einige neue Arten der Saurier-Gattung *Anolis*. Monatsberichte der Königlichen Preussischen Akademie der Wissenschaften zu Berlin.
TYPE SPECIMEN — ZMB 382.
TYPE LOCALITY — Unknown. Frequently listed as "Huanusco," presumably due to the mention of this locale in the type description, or "Huatusco," due to Smith and Taylor's (1950) correction of Huanusco to Huatusco, Veracruz (lat 19.15 lon −96.97, 1300 m).
MORPHOLOGY — Body length to 55 M, 53 F; ventral scales keeled; middorsal scales greatly enlarged in several rows; interparietal usually smaller than tympanum; scales across the snout between the second canthals 6–12; toe lamellae 14–17; suboculars and supralabials in contact or separated by a row of scales; scales separating supraorbital semicircles 0–3; supraocular scales gradually enlarged; hindlimbs long, appressed toe to between eye and tip of snout or past snout; body color brown, frequently with light middorsal diamonds; male dewlap red-orange with yellow edge and darker hue centrally; female dewlap absent; puncturelike axillary pocket.

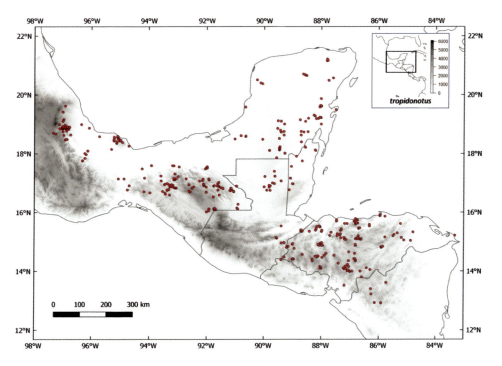

SPECIES ACCOUNTS

SIMILAR SPECIES — *Anolis tropidonotus* is a widespread *small brown* anole. Its puncturelike axillary pocket, terrestrial ecology, greatly enlarged keeled middorsal scales, and strongly keeled ventral scales will distinguish *A. tropidonotus* from all geographically proximal congeners except *A. compressicauda*, *A. humilis*, and *A. uniformis*. *Anolis compressicauda* differs from *A. tropidonotus* in its blue eyes and pink male dewlap (brown eyes, male dewlap red-orange with yellow edge in *A. tropidonotus*). *Anolis uniformis*, which frequently is sympatric with *A. tropidonotus*, is smaller than *A. tropidonotus* (SVL to 41 mm; to 53 mm in *A. tropidonotus*) and has equally sized scales anterior and posterior to the ear (anterior scales > posterior scales in *A. tropidonotus*) and a pink male dewlap. *Anolis humilis*, found south of *A. tropidonotus*, differs from *A. tropidonotus* in its equally sized scales around the ear and in male dewlap color (red with yellow edge).

RANGE — From 0 to 1850 meters from Mexico to Nicaragua.

NATURAL HISTORY — Common in forest and open areas; active diurnally on ground or low vegetation; often observed on tree buttresses or low on tree trunks; sleeps in leaf litter. See Jackson (1973); Henderson and Fitch (1975); Lee (1996); D'Cruze (2005); D'Cruze and Stafford (2006); Luja et al. (2008); McCranie and Köhler (2015); Flores-Villela (2019).

COMMENT — I consider *A. wampuensis* (type locality: "the confluence of Rios Aner and Wampu, 15° 039 N, 85° 079 W, 110 m elevation, Departamento de Olancho, Honduras"), *A. mccraniei* (type locality: "Municipalidad de Gualaco: Montaña de Jacaleapa, headwaters of Río del Oro, E slope Cerro de Bañaderos, 15.083288° N, 86.208250° W, elev. 1,180 m asl, Departamento de Olancho, Honduras"), *A. wilsoni* (Parque Nacional Pico Bonito, Estación Forestal CURLA, 15.70167°N, 86.84667°W, elev. 170 m asl, Departamento de Atlántida, Honduras), and *A. spilorhipis* ("Cerro Ombligo, 1280 m") to be junior synonyms of *A. tropidonotus* (see appendix 1 for justification). For those who wish to recognize *A. wampuensis*, it is the form of *A. tropidonotus* found along the Wampu River near its confluence with the Rio Aner, Honduras. Specimens assigned to *Anolis mccraniei* are found in most of Honduras and neighboring Nicaragua and Guatemala. *Anolis wilsoni* was described from northern Honduras. *Anolis spilorhipis* was delimited as the version of *A. tropidonotus* in Chiapas, Mexico (see Köhler and Townsend 2016).

— *ANOLIS UNIFORMIS* — PLATE 1.226, 2.40

DESCRIPTION — Cope, E. D. 1885. A contribution to the herpetology of Mexico. *Proceedings of the American Philosophical Society held at Philadelphia for promoting useful knowledge* 22: 392–393.

TYPE SPECIMEN — Syntypes: MCZ 10933; USNM 6774 (n=6), 24734–48, 24750, 24859.

TYPE LOCALITY — "Many specimens from Guatemala ... and one from Yucatan." According to Stuart (1955) and Smith and Taylor (1950) USNM 24859 is from Yucatan and the other syntypes are from Guatemala. Smith and Taylor (1950) restricted the type locality to 2 miles north of Santa Teresa, El Petén, Guatemala (lat 16.65 lon −90.13, 130 m).

MORPHOLOGY — Body length to 40 M, 41 F; ventral scales keeled; middorsal scales greatly enlarged in several rows; interparietal smaller than tympanum; scales across the snout between the second canthals 6–9; toe lamellae 12–16; suboculars and supralabials in contact or separated by a row of scales; scales separating supraorbital semicircles 1–3; supraocular scales gradually enlarged; adpressed toe to between eye and tip of snout; body color brown; male dewlap pink with dark central blotch; female dewlap absent; puncturelike axillary pocket.

SIMILAR SPECIES — *Anolis uniformis* is a widespread *small brown* anole found near the ground. Its puncturelike axillary pocket, greatly enlarged keeled middorsal scales, and strongly keeled ventral scales will distinguish *A. uniformis* from all geographically proximal congeners except *A. tropidonotus*, *A. compressicauda*, and *A. humilis*. *Anolis compressicauda* differs from *A. uniformis* in its blue eyes and solid pink dewlap (brown eyes, male dewlap pink with dark central blotch in *A. uniformis*). *Anolis tropidonotus*, which frequently is sympatric with *A. uniformis*, is larger than *A. uniformis* (SVL to 55 mm; to 41 mm in *A. uniformis*) and displays scales anterior to the ear larger than those posterior to it (scales around the ear roughly equal in *A. uniformis*) and a red-orange to yellow male dewlap. *Anolis humilis* differs from *A. uniformis* in male dewlap color (red with yellow distal edge).

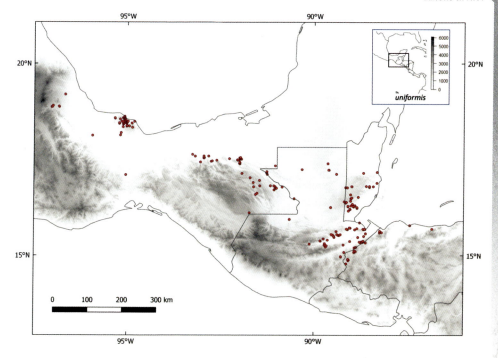

RANGE — From 0 to 1500 meters from Mexico to northern Honduras.
NATURAL HISTORY — Common in forest; diurnally usually in shade and terrestrial, but may inhabit low vegetation such as tree buttresses; McCranie and Köhler (2015:188) and I have found *A. uniformis* to sleep in leaf litter, but clearly it uses low vegetation as a sleep substrate in at least some instances (Cabrera-Guzmán and Reynoso 2010). See Echelle et al. (1978); Campbell et al. (1989); Eifler (1995); Villareal-Benítez (1997); Villareal-Benítez and Heras-Lara (1997); D'Cruze (2005); D'Cruze and Stafford (2006); Urbina-Cardona et al. (2006); Cabrera-Guzmán and Reynoso (2010); Cabrera-Guzmán and Garrido-Olvera (2014); McCranie and Köhler (2015); Russildi et al. (2016); Cardona and Reynoso (2017).

— *ANOLIS URRAOI* — PLATE 1.227

DESCRIPTION — Grisales-Martínez, F. A., J. A. Velasco, W. Bolivar, E. E. Williams, J. M. Daza. 2017. The taxonomic and phylogenetic status of some poorly known *Anolis* species from the Andes of Colombia with the description of a nomen nudum taxon. *Zootaxa* 4303:213–230.
TYPE SPECIMEN — MHUA-R 12733.
TYPE LOCALITY — "Department of Antioquia, Urrao municipality, paddock on the banks of Río Penderisco: 6.314253, −76.138528, 1822 m."
MORPHOLOGY — Body length to 54 (males and females about equal-sized according to Grisales-Martínez et al. 2017); ventral scales weakly keeled; middorsal scales enlarged in 0–2 rows; interparietal equal to or smaller than tympanum; scales across the snout between the second canthals 10–18; toe lamellae 13–16; suboculars and supralabials in contact; scales separating supraorbital semicircles 3–4; supraocular scales gradually enlarged; addpressed toe to between eye and tip of snout; body color brown; male dewlap pale peach posteriorly, orange anteriorly.

SPECIES ACCOUNTS

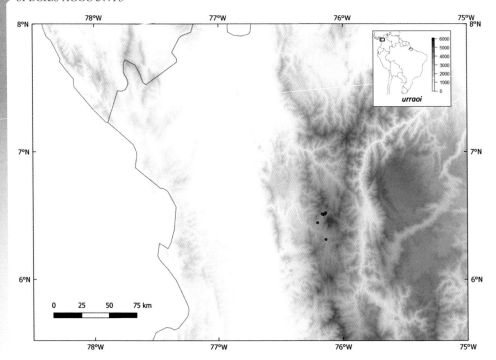

SIMILAR SPECIES — *Anolis urraoi* is a *small brown* highland anole. Similar Colombian Andean anoles are best distinguished by male dewlap and locality: The male dewlap of *A. tolimensis* is solid pinkish-red; *A. tolimensis* is found in the Cordillera Central in upper regions of the southern Magdalena River valley. The male dewlap of *A. antonii* is red-orange to orange-red anteriorly, pale pink posteriorly; *A. antonii* is found in the southern Cordillera Occidental. The male dewlap of *A. fuscoauratus* usually is pale pink in Colombia; *A. fuscoauratus* is found on the Amazonian slope of the Cordillera Oriental and in the northern Magdalena River valley. The male dewlap of *Anolis mariarum* is red-orange anteriorly, yellow posteriorly; *A. mariarum* is found in the northern Cordillera Central. The male dewlap of *Anolis urraoi* is pale pink posteriorly, orange anteriorly; *A. urraoi* is found in the northern Cordillera Occidental.
RANGE — From 1700 to 2250 meters in the northwestern Andes of Colombia.
NATURAL HISTORY — Common in open and disturbed areas on vegetation and human-made objects up to 3 meters above ground. Sleeps on leaves, often ferns, or twigs from 0.5 to 4 meters above ground.
COMMENT — The distributional and taxonomic limits of *A. urraoi* are not clear. See Comment under *A. antonii*.

— *ANOLIS USTUS* — PLATE 1.228

DESCRIPTION — Boulenger, G. A. 1885. *Catalogue of the lizards in the British Museum (Natural History)*, vol. 2. 73.
TYPE SPECIMEN — Syntypes BMNH 1946.8.5.60-61.
TYPE LOCALITY — "Belize."
MORPHOLOGY — Body length to 45 M, 49 F; ventral scales keeled; middorsal scales enlarged in 0–2 rows or gradually enlarged in several rows; interparietal larger than tympanum; scales across the

Anolis utilensis

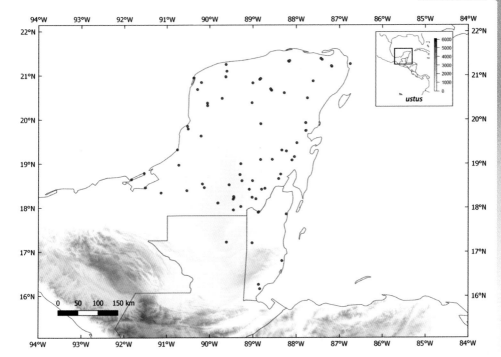

snout between the second canthals 5–9; toe lamellae 13–17; suboculars and supralabials in contact; scales separating supraorbital semicircles 0–2; supraocular scales gradually enlarged; hindlimbs short, adpressed toe to ear; body color grayish-brown; male and female dewlaps small, extending posteriorly only to arms, orange-yellow with blue central spot.

SIMILAR SPECIES — *Anolis ustus* is a *small brown* lowland anole. The combination of short hindlimbs (adpressed toe reaches to ear) and keeled ventral scales will separate *A. ustus* from potentially sympatric small anole species in Yucatán. Also, all anole species except *A. sericeus* and *A. ustus* lack the distinctive male dewlap colors of *A. ustus* (orange with a blue central blotch). *Anolis ustus* differs from *A. sericeus* in its small dewlaps in males and females (i.e., barely extending past the axillae; male dewlap large, extending to chest, female dewlap very small or absent in *A. sericeus*).

RANGE — From 0 to 600 meters in Belize and in Yucatán, Campeche, and Quintana Roo States, Mexico.

NATURAL HISTORY — Very common in forest, edge, and disturbed habitats; active diurnally on vegetation up to 4 meters above ground; sleeps on twigs, leaves, or vines at all visible heights. See Lee (1980; as *A. sericeus*); Lee (1996; as *A. sericeus*); Lara-Tufiño et al. (2016).

— *ANOLIS UTILENSIS* — PLATE 1.229, 2.78

DESCRIPTION — Köhler, G. 1996a. A new species of the *Norops pentaprion* group from Isla de Utila, Honduras. *Senckenbergiana biologica* 75 (1/2):23–31.

TYPE SPECIMEN — SMF 77051.

TYPE LOCALITY — "Honduras, Islas de la Bahia, Isla de Utila, 2 km NNE of the town Utila." I.e., lat 16.11, lon −88.89, elevation 10 m.

MORPHOLOGY — Body length to 61 M, 54 F; ventral scales smooth; middorsal scales enlarged in 0–2 rows or gradually enlarged in several rows; interparietal larger than tympanum; scales across the

SPECIES ACCOUNTS

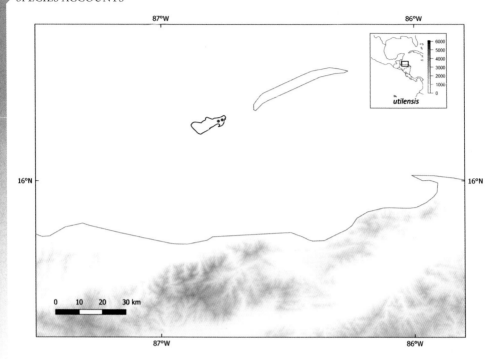

snout between the second canthals 6–8; toe lamellae 18–20; suboculars and supralabials in contact; supraorbital semicircles in contact; supraocular scales gradually enlarged; hindlimbs short, appressed toe to ear or posterior to ear; body color mostly gray with some brown, white, and black; male and female dewlaps reddish-pink.

SIMILAR SPECIES — *Anolis utilensis* is a *small brown* or *twig* anole, amply distinct from the other four anole species known from Utila. It is the only anole species on Utila with smooth ventrals, and its white-lichenous coloration and pink male and female dewlaps also are distinctive (pink dewlap also present in male *A. allisoni*; *A. bicaorum* is basically brown but may appear white under some conditions).

RANGE — Probably islandwide on Utila.

NATURAL HISTORY — Common diurnally from 1.5 to 6 meters above ground on trunks and branches of large trees in and outside of mangroves; sleeps high (often 4+ meters), frequently on leaves of large trees. See Powell (2003); Gutsche et al. (2004; as *Norops utilensis*); Gutsche (2005; as *Norops utilensis*); Hallman and Huy (2012); McCranie and Köhler (2015; as *Norops utilensis*); Brown et al. (2017; as *Norops utilensis*).

COMMENT — The status of *Anolis utilensis* relative to *A. beckeri* should be investigated.

— *ANOLIS VANZOLINII* — PLATE 1.230, 2.139

DESCRIPTION — Williams, E. E., G. Orcés-V., J. C. Matheus, R. Bleiweiss. 1996. A new giant phenacosaur from Ecuador. *Breviora* 505:1–32.

TYPE SPECIMEN — MECN 0309.

TYPE LOCALITY — "Ecuador: S La Alegria, at an elevation of 2,360 m, ca. 14 km by road from La Bonita (77 37 42 W, 0 27 30 N), Provincia de Sucumbios (formerly the northwest part of Provincia de Napo)."

Anolis vaupesianus

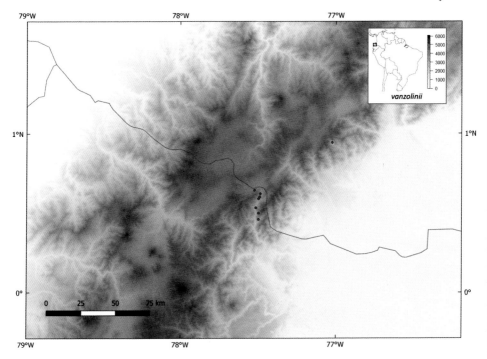

MORPHOLOGY — Body length to 108 M, 111 F; ventral scales smooth; middorsal scales form a crest of triangular plates; interparietal larger than tympanum; scales across the snout between the second canthals 3–4; toe lamellae 24–28; suboculars and supralabials in contact; scales separating supraorbital semicircles 0–1; supraocular scales abruptly or gradually enlarged; hindlimbs short, adpressed toe to ear or posterior to ear; body color brown with yellow, white, and tan; tail banded; male and female dewlaps yellow; heterogeneous lateral scutellation.

SIMILAR SPECIES — *Anolis vanzolinii* is a large, distinctive highland *twig* anole. The heterogeneous lateral scutellation, wherein large smooth oval scales are separated by granular interstitial scales, is distinctive in *A. vanzolinii*, especially in combination with its short hindlimbs (adpressed toe reaches to ear or posterior to ear) and very large smooth head scales (3–4 scales across the snout). Potentially sympatric species *A. quimbaya* shares these traits with *A. vanzolinii* but is smaller (SVL to 76 mm, to 104 mm in *A. vanzolinii*) and has fewer toe lamellae (20–24; 24–28 in *A. vanzolinii*).

RANGE — From 1950 to 2700 meters on the Amazonian Andean slope near the Ecuador-Colombia border.

NATURAL HISTORY — Rare in edge habitat and tall trees of disturbed areas; active diurnally on narrow vegetation; sleeps on twigs or leaves above 1 meter above ground. See Torres-Carvajal et al. (2019); Arteaga et al. (2023).

— *ANOLIS VAUPESIANUS* — PLATE 1.231

DESCRIPTION — Williams, E. E. 1982. Three new species of the *Anolis punctatus* complex from Amazonian and inter-Andean Colombia, with comments on the eastern members of the *punctatus* species group. *Breviora* 467:2–9.

TYPE SPECIMEN — MCZ 156309.

SPECIES ACCOUNTS

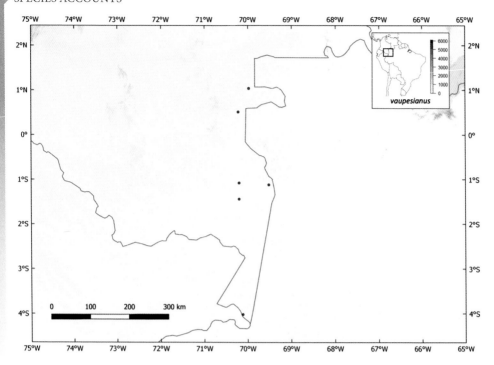

TYPE LOCALITY — "Timbo, a Cubeo village ca. 1 N 70 W on the Rio Vaupes, E of Mitu, Vaupes, Colombia." Timbo is located at lat 1.172 lon −70.009, 170 m.

MORPHOLOGY — Body length to 82 M, 75 F; ventral scales smooth to weakly keeled; middorsal scales enlarged in 0–2 rows; interparietal larger than tympanum; scales across the snout between the second canthals 8–11; toe lamellae 22–26; suboculars and supralabials in contact; scales separating supraorbital semicircles 0–2; supraocular scales gradually enlarged; adpressed toe to between eye and tip of snout; body color green, presumably changeable to brown; male dewlap dark purple to black.

SIMILAR SPECIES — *Anolis vaupesianus* is a *large green* lowland Amazonian anole unlikely to be confused with most potentially sympatric species due to size, dorsal color, or ventral scales. *Anolis bombiceps*, *A. scypheus*, *A. planiceps*, *A. tandai*, *A. chrysolepis* are predominantly brown with strongly keeled ventral scales (ventrals smooth to weakly keeled in *A. vaupesianus*). *Anolis fuscoauratus* and *A. ortonii* are smaller (SVL < 60 mm; to 82 mm in *A. vaupesianus*) and brown or gray with differently colored male dewlaps (*A. fuscoauratus*: pink; *A. ortonii*: red with yellow; *A. vaupesianus*: dark purple to black). *Anolis transversalis* is predominantly brown with blue eyes (eyes brown in *A. vaupesianus*). *Anolis caquetae*, known from a single specimen from one eastern Colombian locality, may be smaller than *A. vaupesianus* (SVL in mature male holotype 58 mm) and is distinctive in its large interparietal scale (Williams 1974b). *Anolis punctatus* differs from *A. vaupesianus* by its orange male dewlap.

RANGE — From 50 to 200 meters in southeast Colombia.

NATURAL HISTORY — Only information is from the type description: at least one individual was caught in a village, indicating tolerance for edge or disturbed habitat. Specimens were "associated with trees" (Williams 1982).

COMMENT — As described (I have not examined specimens of this form), this species appears to be a version of *Anolis punctatus* with a dark rather than orange-yellow dewlap. The name *vaupesianus* is likely to be a junior synonym of *A. caquetae* (D. L. Mahler, personal communication 2024; see Williams 1982:33–34).

— *ANOLIS VENTRIMACULATUS* — PLATE 1.232, 2.123

DESCRIPTION — Boulenger, G. A. 1911. Descriptions of new reptiles from the Andes of South America, preserved in the British Museum. *Annals and Magazine of Natural History* 8:20–21.
TYPE SPECIMEN — Lectotype BMNH 1946.8.13.5 designated by Williams and Duellman (1984).
TYPE LOCALITY — "Rio San Juan, Choco, S.W. Colombia." The San Juan is a major river in Chocó that flows 380 km from the western Andes to the Pacific about 50 km northwest of Buenaventura. Which is to say, the type locality of *Anolis ventrimaculatus* was only vaguely recorded. Williams and Duellman (1984:263) suggested Pueblo Rico, Risaralda, Colombia (5.22 lat −76.03 lon, 1560 m) as a likely type locality but cautioned, "Boulenger's careful avoidance of precision leaves the question open."
MORPHOLOGY — Body length to 80 M, 62 F; ventral scales smooth; middorsal scales enlarged in 0–2 rows; interparietal smaller than tympanum; scales across the snout between the second canthals 11–21; toe lamellae 16–22; suboculars and supralabials in contact or separated by a row of scales; scales separating supraorbital semicircles 2–5; supraocular scales subequal; hindlimbs long, adpressed toe to anterior to tip of snout; body color brown usually with some green, with bands and light spots, often with a dark shoulder ocellus; tail banded; male dewlap dirty orange; female dewlap absent.
SIMILAR SPECIES — *Anolis ventrimaculatus* is a *large brown* or *green* highland anole. Its combination of long hindlimbs and small head scales will separate *Anolis ventrimaculatus* from many geographically proximal large anoles. *Anolis gemmosus* is smaller than *A. ventrimaculatus* (SVL to 66 mm; to 80 mm in *A. ventrimaculatus*) and has shorter hindlimbs (adpressed toe reaches to between eye and tip of snout; to past snout in *A. ventrimaculatus*). *Anolis purpurescens* has broad toepads (narrow in *A. ventrimaculatus*) and shorter hindlimbs than *A. ventrimaculatus* (adpressed toe reaches to eye). *Anolis antioquiae* has a well-developed dewlap in females (dewlap absent in female *A. ventrimaculatus*) and

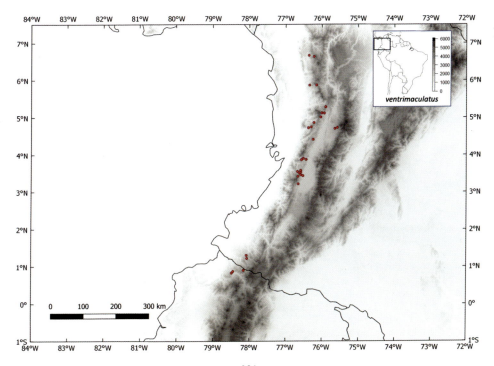

SPECIES ACCOUNTS

usually lacks an interparietal scale (interparietal usually present in *A. ventrimaculatus*). *Anolis dracula* has male and female dewlaps that differ in color from the male dewlap of *A. ventrimaculatus* (dark brown with red blotches and blue blotches in males and females; dewlap absent in females, dirty orange in males of *A. ventrimaculatus*). *Anolis danieli*, *A. huilae*, *A. propinquus*, and *A. apollinaris* have broad toepads and larger head scales than *A. ventrimaculatus* (8–12 scales across the snout; 11–21 in *A. ventrimaculatus*). *Anolis megalopithecus* differs from *A. ventrimaculatus* in body color (brown with red; green and brown in *A. ventrimaculatus*) and male dewlap color (red with black markings). *Anolis eulaemus* is larger than *A. ventrimaculatus* (SVL to 101 mm) and differs in body color (brown, often lacking green) and dewlap color (brown with black flecks proximally). *Anolis maculigula* is larger than *A. ventrimaculatus* (SVL to 107 mm), has shorter hindlimbs (adpressed toe reaches to between eye and tip of snout) and a dark blue and pink male dewlap, and is invariably associated with streams. *Anolis fraseri*, *A. frenatus*, and *A. princeps* are larger than *A. ventrimaculatus* (adults > 100 mm SVL), have broad toepads, and differ in dewlap presence and color (white or pale yellow, present in both sexes in these species). *Anolis mirus* has fewer toe lamellae than *A. ventrimaculatus* (12–15; 16–22 in *A. ventrimaculatus*).

RANGE — From 850 to 2200 meters, Pacific central Colombia to northern Pacific Ecuador.

NATURAL HISTORY — Common to very common in forest and edge habitat; active diurnally on understory vegetation; sleeps mainly on leaves, especially ferns, usually low (<1 meter above ground); apparently has been found at a few low-elevation sites but appears restricted to above ~1700 meters throughout most of its range. See Kattan (1984); Molina-Zuluaga and Gutiérrez-Cárdenas (2007); García González (2011); Barragán-Contreras and Calderón-Espinosa (2013); Narváez (2017); Torres-Carvajal et al. (2019); Arteaga et al. (2023).

— *ANOLIS VICARIUS* — PLATE 1.233

DESCRIPTION — Williams, E. E., 1986. *Anolis vicarius*, new species, related to *A. granuliceps*. Caldasia 15:451–459.

TYPE SPECIMEN — ICN 5916.

TYPE LOCALITY — "Heights of the Cordillera Occidental between the towns of Frontino and Dabeiba, rio Amparrado, on the trail between Pegadorcito and Amparrado, Department of Antioquia, Colombia." A point on a trail near the Amparradó River near Amparradó Carmen, at Williams's cited elevation (800 m), is located at lat 6.92 lon −76.38.

MORPHOLOGY — Body length to 47 M (females unknown); ventral scales smooth; middorsal scales enlarged in 0–2 rows; interparietal smaller than tympanum; scales across the snout between the second canthals 17; toe lamellae 16; suboculars and supralabials in contact; scales separating supraorbital semicircles 4; supraocular scales gradually enlarged; body color brown with black anterior blotching; male dewlap yellow.

SIMILAR SPECIES — *Anolis vicarius* is a *small brown* lowland anole currently formally known from one collected specimen. Its distinctive traits are its tiny head scales (17 scales across the snout) and black blotching around the nape. *Anolis gracilipes*, *A. auratus*, *A. notopholis*, *A. binotatus*, *A. vittigerus*, *A. poecilopus*, and *A. rivalis* apparently are larger than *A. vicarius* (SVL to >60 mm in these species, except to 54 mm in *A. auratus*; 47 mm in *A. vicarius*) and have strongly keeled ventral scales (smooth in *A. vicarius*). *Anolis granuliceps* has a smaller male dewlap than *A. vicarius*, extending barely to the axillae, and faint ventral keeling. *Anolis maculiventris* and *A. apletophallus* tend to have larger head scales than *A. vicarius* (9–16 scales across the snout). *Anolis apletophallus* is nearly indistinguishable from *A. vicarius* as described, but may differ in its lack of black blotching at the nape (present in *A. vicarius*).

RANGE — From 0 to at least 800 m and almost certainly higher, in northern Pacific Colombia.

NATURAL HISTORY — Unknown.

COMMENT — Although publicly known from a single museum specimen, other preserved museum specimens are extant (D. L. Mahler personal communication 2023), and individuals assignable to this

Anolis villai

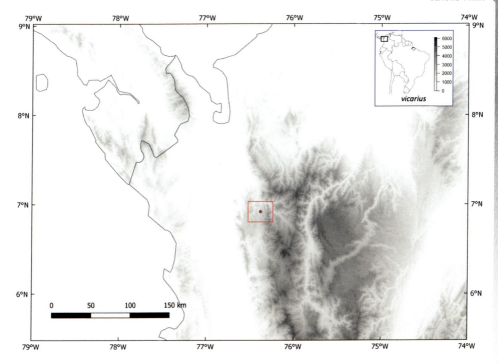

form based on Williams's description have been photographed recently (Calderón-Espinosa 2023; iNaturalist).

— *ANOLIS VILLAI* — PLATE 1.234; MAP overleaf

DESCRIPTION — Fitch, H. S., R. W. Henderson. 1976b. A new anole (Reptilia: Iguanidae) from Great Corn Island, Caribbean Nicaragua. *Contributions in Biology and Geology*, Milwaukee Public Museum 9:1–8.
TYPE SPECIMEN — KU 159646.
TYPE LOCALITY — "Great Corn Island." Great Corn Island is off the eastern coast of Nicaragua at lat 12.1667 lon –83.0333, 0–113 m.
MORPHOLOGY — Body length to 55 M, 38 F; ventral scales keeled; middorsal scales gradually enlarged in several rows; interparietal equal to or larger than tympanum; scales across the snout between the second canthals 8–10; toe lamellae 9–15; suboculars and supralabials in contact or separated by a row of scales; scales separating supraorbital semicircles 1–4; supraocular scales gradually enlarged; addpressed toe to between eye and tip of snout; body color brown, with light lateral stripe; male dewlap pale red distally, dull orange proximally.
SIMILAR SPECIES — *Anolis villai* is a *small brown* anole that shares Great Corn Island with *A. sericeus*. *Anolis sericeus* differs from *A. villai* in its short hindlimbs (addpressed toe reaches to ear; to between eye and tip of snout in *A. villai*) and distinctive male dewlap (orange with a blue central spot; red and orange in *A. villai*).
RANGE — Great Corn Island.
NATURAL HISTORY — In forest, shaded areas; active diurnally on stems and tree trunks approximately 1 to 2 meters above ground; see Sunyer et al. (2013).

SPECIES ACCOUNTS

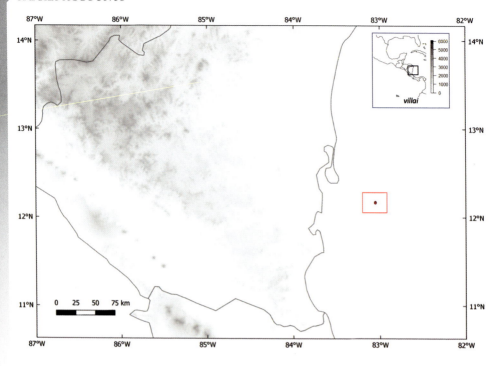

— *ANOLIS VITTIGERUS* — FIGURE 3.20; PLATE 1.235

DESCRIPTION — Cope, E. D. 1862. Contributions to Neotropical saurology. *Proceedings of the Academy of Natural Sciences of Philadelphia* 14:179–180.

TYPE SPECIMEN — Cope (1862) lists "Smithsonian (No. 4332)." This specimen is lost according to Köhler (2001).

TYPE LOCALITY — "Truando region, New Granada." Possibly lat 7.43 lon –77.11, elevation 0. See comments above on the Truando region in the discussion of the type locality for *Anolis poecilopus*.

MORPHOLOGY — Body length to 60 M, 70 F; ventral scales keeled; middorsal scales gradually enlarged in several rows; interparietal larger or smaller than tympanum; scales across the snout between the second canthals 6–13; toe lamellae 16–20; suboculars and supralabials in contact or separated by rows of scales; scales separating supraorbital semicircles 0–2; supraocular scales gradually enlarged; addressed toe to eye or just beyond; body color brown, usually with a light lateral line, dorsal markings, and a dark interorbital bar; male dewlap red with dark central spot; female dewlap pale bluish with dark central spot.

SIMILAR SPECIES — *Anolis vittigerus* is a widespread *large brown* lowland anole. The simplest distinguishing traits for this species are its unique male and female dewlaps (small, to axillae or just beyond, in both sexes; male: red with dark central blotch, female: pale blue with dark central blotch). *Anolis vittigerus* is geographically replaced by *A. lemurinus* in western Panama. Dewlaps in *A. lemurinus* are similar to those in *A. vittigerus* but lack the dark central spot, and the female dewlap may be pale blue, reddish, or white.

RANGE — From 0 to 1250 meters from central Panama to Ecuador.

NATURAL HISTORY — Common in forest, forest edge, and disturbed areas; active diurnally on tree trunks or other vegetation; sleeps on vegetation, often leaves, 1 to 3 meters above ground. See Rand and Myers (1990); Rengifo et al. (2014, 2015; as *A. lyra*); Carvajal-Cogollo and Urbina-Cardona

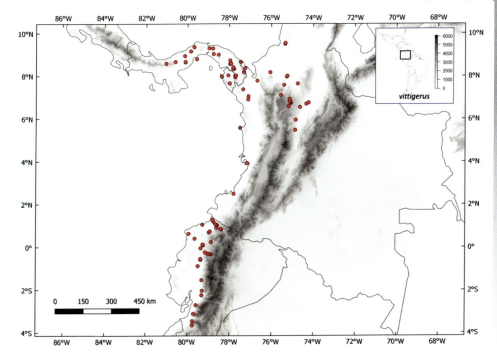

(2015); Narváez (2017; as *A. lyra*); Viteri (2015); Rengifo et al. (2019, 2021; as *A. lyra*); Batista et al. (2020); Pinto-Erazo et al. (2020; as *A. lyra*); Moreno-Arias et al. (2020).
COMMENT — I consider *Anolis lyra* (type locality: "Ecuador, Pichincha, Finca Victoria, 37 km SE of Santo Domingo") to be a junior synonym of *A. vittigerus* (see appendix 1 for justification). For those who wish to recognize *A. lyra*, it is the form of *A. vittigerus* found from central Colombia south into Ecuador.

— *ANOLIS WERMUTHI* — PLATE 1.236

DESCRIPTION — Köhler, G., M. Obermeier. 1998. A new species of anole of the *Norops crassulus* group from central Nicaragua (Reptilia: Sauria: Iguanidae). *Senckenbergiana biologica* 77:127–137.
TYPE SPECIMEN — SMF 77323.
TYPE LOCALITY — "Road from Matagalpa to Jinotega at km 146 (13° 01.995' N, 85° 55.848' W), 1400 m elevation, Departamento Jinotega, Nicaragua."
MORPHOLOGY — Body length to 51 M, 56 F; ventral scales keeled; middorsal scales greatly enlarged in several rows; interparietal larger or smaller than tympanum; scales across snout between the second canthals 5–8; toe lamellae probably about 16 (specimens not examined for this trait); suboculars and supralabials in contact; scales separating supraorbital semicircles 1–2; supraocular scales gradually enlarged; addressed toe to eye; body color brown; male dewlap orange-red; female dewlap orange.
SIMILAR SPECIES — *Anolis wermuthi* is a *small brown* highland anole. Geographically proximal similar species differ from *A. wermuthi* in their short hindlimbs (addressed toe reaches to ear in *A. laeviventris*, *A. sericeus*, *A. beckeri*, *A. pentaprion*; to eye in *A. wermuthi*). *Anolis humilis* and *A. tropidonotus* differ from *A. wermuthi* in the presence of a puncturelike axillary pocket (absent in *A. wermuthi*). *Anolis cupreus*, *A. limifrons*, and *A. lemurinus* differ from *A. wermuthi* in their lack of abruptly

SPECIES ACCOUNTS

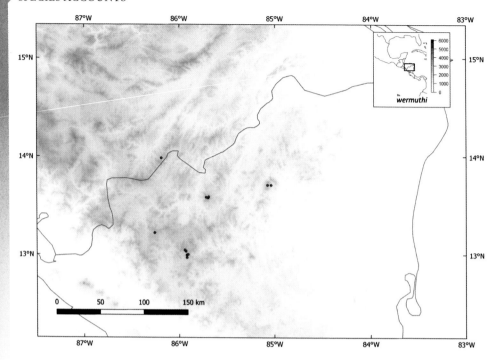

enlarged keeled middorsal scales (abruptly enlarged middorsals in *A. wermuthi*). *Anolis oxylophus* is larger than *A. wermuthi* (SVL to 76 mm; to 56 mm in *A. wermuthi*) and semiaquatic. All of these species except *A. laeviventris* tend to be found at lower elevations than *A. wermuthi* and have homogeneous flank scales (flank scales heterogenous mix of small and large scales in *A. wermuthi*). *Anolis caceresae*, *A. sminthus*, and *A. heteropholidotus* are similar highland forms found northwest of the range of *A. wermuthi*; these species differ from *A. wermuthi* in possessing greater numbers of abruptly enlarged middorsal scales (at least 10; 6–8 in *A. wermuthi*).

RANGE — From 1000 to 1850 meters from central Nicaragua to eastern Honduras.

NATURAL HISTORY — In forest; active diurnally on ground or vegetation up to 0.5 meters above ground; sleeps on twigs or leaves up to 1.5 meters above ground. See Sunyer (2009).

— *ANOLIS WILLIAMSMITTERMEIERORUM* — PLATE 1.237, 2.149

DESCRIPTION — Poe, S., C Yañez-Miranda. 2007. A new species of phenacosaur *Anolis* from Peru. *Herpetologica* 63:219–223.

TYPE SPECIMEN — MZUNAP 02.000181.

TYPE LOCALITY — "Venceremos, approximately 94 km west of Rioja, Department of San Martin, Peru (between old km markers 390–1, near new km marker 380), S 05 40.405 W 77 45.310, 1739 m."

MORPHOLOGY — Body length to 66 M, 60 F; ventral scales smooth; middorsal scales enlarged in 0–2 rows; interparietal larger than tympanum; scales across the snout between the second canthals 3–6; toe lamellae 17–19; suboculars and supralabials in contact; supraorbital semicircles in contact; supraocular scales abruptly or gradually enlarged; hindlimbs short, appressed toe to posterior to ear; body color mix of gray, white, green, and brown, with faint banding; tail banded; male dewlap pale peach, darker peach at edge; female dewlap black and white.

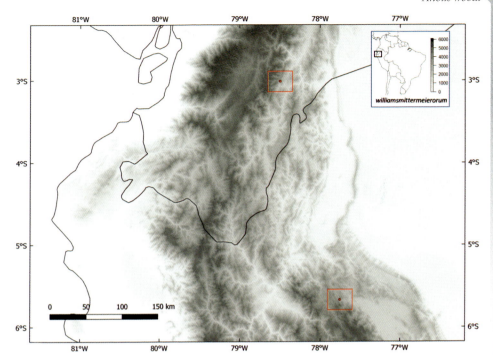

SIMILAR SPECIES — *Anolis williamsmittermeierorum* is a highland *twig* anole. It is commonly found sympatrically with *A. soinii*, which is larger than *A. williamsmittermeierorum* (SVL to 82 mm; to 66 mm in *A. williamsmittermeierorum*) and green (lichenous in *A. williamsmittermeierorum*), and *A. fuscoauratus*, which is smaller than *A. williamsmittermeierorum* (SVL to 53 mm) and has smaller head scales (8–14 scales across the snout; 3–6 in *A. williamsmiettermeierorum*). *Anolis podocarpus* is larger than *A. williamsmittermeierorum* (SVL to 96 mm) and has smaller head scales (14–20 scales across the snout). *Anolis peruensis* is similar to *A. williamsmittermeierorum* and parapatric or sympatric with it but differs in dewlap colors (yellow in male, black in female; pale peach in male, white with black blotches in female *A. williamsmittermeierorum*). *Anolis laevis* differs from *A. williamsmittermeierorum* in its strongly protruding snout overlapping the lower jaw (snout not protruding in *A. williamsmittermeierorum*) and crested middorsum (dorsum not crested in *A. williamsmittermeierorum*).
RANGE — Known only from the vicinity of the type locality and another locality in southern Ecuador. The Ecuador locality may represent an undescribed species. See Torres-Carvajal et al. (2019); Arteaga et al. (2023).
NATURAL HISTORY — Common in edge habitat of wet forest; sleeps on twigs above 1.5 meters above ground.

— *ANOLIS WOODI* — PLATE 1.238, 2.61

DESCRIPTION — Dunn, E. R. 1940. New and noteworthy herpetological material from Panamá. *Proceedings of the Academy of Natural Sciences of Philadelphia* 92:110–111.
TYPE SPECIMEN — AMNH 62647.
TYPE LOCALITY — "El Volcán, Chiriqui, Panama." El Volcán is located at 8.779, −82.644, 1400 m.
MORPHOLOGY — Body length to 92 M, 78 F; ventral scales keeled; middorsal scales enlarged in 0–2

SPECIES ACCOUNTS

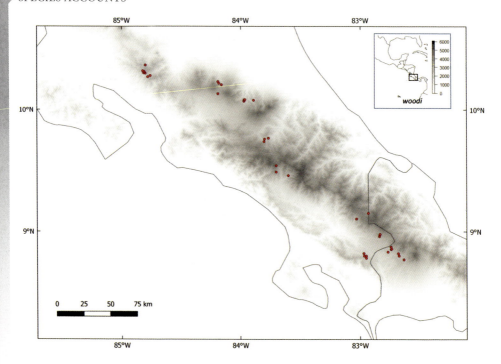

rows; interparietal smaller than tympanum; scales across the snout between the second canthals 9–14; toe lamellae 15–18; suboculars and supralabials in contact or separated by rows of scales; scales separating supraorbital semicircles 1–3; supraocular scales gradually enlarged; hindlimbs long, appressed toe to between eye and tip of snout; body color brown; male dewlap orange but changeable in color to olive (Savage 2002); iris blue.

SIMILAR SPECIES — *Anolis woodi* is a *large brown* or *green* highland anole with blue eyes. Within the range of *A. woodi*, only *A. woodi*, *A. microtus*, *A. ginaelisae*, and *A. aquaticus* may have blue eyes. *Anolis microtus* and *A. ginaelisae* are giant species (over 100 mm SVL; to 92 mm in *A. woodi*) with larger head scales than *A. woodi* (5–9 scales across the snout; 9–14 in *A. woodi*). *Anolis aquaticus*, *A. robinsoni*, and *A. riparius* are semiaquatic species, invariably found within about one meter of a stream, with different male dewlaps than *A. woodi* (red with yellow streaks in *A. aquaticus*, *A. riparius*; dark brown in *A. robinsoni*; orange varying to olive in *A. woodi*).

RANGE — From 900 to 1800 meters from western Panama through Costa Rica.

NATURAL HISTORY — Rare in forest; active diurnally on tree trunks and branches up to 6 meters above ground; sleeps on twigs or leaves. See Fitch (1975); Fitch et al. (1976; as *A. attenuatus*); Pounds (1988); Savage (2002).

— *ANOLIS YOROENSIS* — PLATE 1.239, 2.53

DESCRIPTION — McCranie, J. R., K. E. Nicholson, G. Köhler. A new species of *Norops* (Squamata: Polychrotidae) from northwestern Honduras. *Amphibia-Reptilia* 22:465–473.
TYPE SPECIMEN — USNM 541012.
TYPE LOCALITY — "2.5 airline km NNE La Fortuna, 15° 26' N, 87° 18' W, 1600 m elevation, Cordillera Nombre de Dios, Departamento de Yoro, Honduras."

Anolis yoroensis

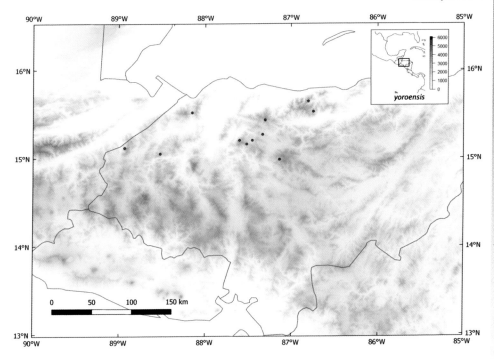

MORPHOLOGY — Body length to 45 M, 47 F; ventral scales keeled; middorsal scales enlarged in 0–2 rows; interparietal larger or smaller than tympanum; scales across the snout between the second canthals 5–14; toe lamellae 13–15; suboculars and supralabials in contact or separated by a row of scales; scales between supraorbital semicircles 1–4; supraocular scales gradually enlarged; adpressed toe to eye; body color brown, often with two parallel lateral stripes; male dewlap peach-orange.

SIMILAR SPECIES — *Anolis yoroensis* is a *small brown* highland anole. Among geographically proximal similar species, *A. cusuco*, *A. kreutzi*, *A. laeviventris*, *A. beckeri*, *A. sericeus*, and *A. ustus* differ from *A. yoroensis* in having shorter hindlimbs (adpressed toe reaches to ear; to eye in *A. yoroensis*); *A. carpenteri*, *A. rodriguezii*, *A. limifrons*, and *A. zeus* differ from *A. yoroensis* in their smooth ventral scales (keeled in *A. yoroensis*). *Anolis cupreus*, *A. purpurgularis*, and *A. pijolense* are larger than *A. yoroensis* (maximum SVL > 55 mm; to 47 mm in *A. yoroensis*) and differ in male dewlap color (*A. cupreus*: orange-brown proximally, pale pink-peach distally; *A. purpurgularis*: pinkish-purple; *A. pijolense*: pink with purple centrally; *A. yoroensis*: peach-orange). *Anolis amplisquamosus*, *A. heteropholidotus*, *A. muralla*, *A. morazani*, *A. rubribarbaris*, *A. crassulus*, *A. caceresae*, *A. sminthus*, *A. hetropholidotus*, *A. tropidonotus*, *A. humilis*, and *A. uniformis* differ from *A. yoroensis* in their abruptly enlarged keeled middorsal scales (middorsals uniform in *A. yoroensis*). *Anolis ocelloscapularis* shares similar body proportions and keeling and an orange male dewlap with *A. yoroensis*, and overlaps in most scale counts. *Anolis ocelloscapularis* has an ocellated shoulder spot (absent in *A. yoroensis*) and suboculars and supralabials in contact (may be separated by a row of scales in *A. yoroensis*).

RANGE — From 650 to 1600 meters in northwestern Honduras.

NATURAL HISTORY — Common in forest and edge habitat; active diurnally on vegetation up to 3 meters, also on ground; sleeps on twigs or leaves, usually below 1.5 meters above ground. See Townsend and Wilson (2008); McCranie and Köhler (2015).

SPECIES ACCOUNTS

— *ANOLIS ZEUS* — PLATE 1.240, 2.27

DESCRIPTION — Köhler, G., J. R. McCranie. 2001. Two new species of anoles from northern Honduras. *Senckenbergiana biologica* 81:236–240.
TYPE SPECIMEN — SMF 77196.
TYPE LOCALITY — "Liberia, Parque Nacional Pico Bonito, 90 m elevation, Departamento de Atlántida, Honduras." McCranie and Castañeda (2005) state that Liberia, Atlántida, is located at lat 15.702 lon −86.866.
MORPHOLOGY — Body length to 43 M, 44 F; ventral scales smooth; middorsal scales enlarged in 0–2 rows; interparietal larger or smaller than tympanum; scales across the snout between the second canthals 7–13; toe lamellae 16–17; suboculars and supralabials in contact; scales separating supraorbital semicircles 1–3; supraocular scales gradually enlarged; hindlimbs long, adpressed toe to between eye and tip of snout; body color brown; male dewlap white with or without basal yellow blotch.
SIMILAR SPECIES — *Anolis zeus* is a *small brown* lowland anole, essentially the northernmost version of *A. limifrons*. Its combination of smooth ventral scales and long hindlimbs (adpressed toe reaches to between eye and tip of snout) will differentiate *A. zeus* from most potentially sympatric small brown anoles. *Anolis rodriguezii* differs from *A. zeus* in its orange male dewlap (white with or without basal yellow spot in *A. zeus*). *Anolis limifrons*, found south of the range of *A. zeus*, and *A. zeus* are separable only based on geography (they appear to be molecularly distinct [Hofmann et al. 2019]).
RANGE — From 0 to 900 meters in northwestern Honduras.
NATURAL HISTORY — Common in forest, edge, and disturbed habitat; active diurnally on low vegetation; sleeps on twigs, leaves, or vines at all visible vertical levels. See McCranie and Köhler (2015).

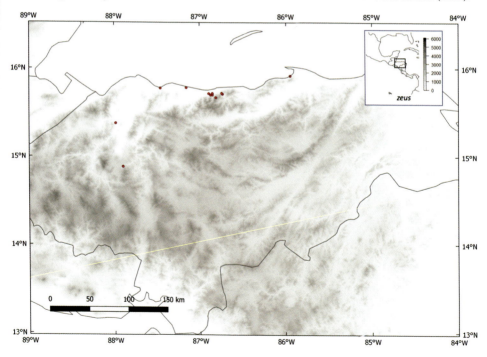

BODIES – PLATE SET 2

2.1 *A. liogaster* p.190

2.2 *A. dunni* p.138

2.3 *A. omiltemanus* p.232

2.4 *A. gadovii* p.154

2.5 *A. taylori* p.298

2.6 *A. alvarezdeltoroi* p.71

2.7 *A. milleri* p.216

2.8 *A. parvicirculatus* p.242

2.9 *A. cymbops* p.129

2.10 *A. hobartsmithi* p.167

PLATE SET 2 – BODIES

2.11 *A. naufragus* p.223

2.12 *A. schiedii* p.286

2.13 *A. megapholidotus* p.210

2.14 *A. rubiginosus* p.278

2.15 *A. subocularis* p.295

2.16 *A. nebuloides* p.225

2.17 *A. boulengerianus* p.99

2.18 *A. quercorum* p.269

2.19 *A. nebulosus* p.226

2.20 *A. microlepidotus* p.214

BODIES – PLATE SET 2

2.21 *A. compressicauda* p.118

2.22 *A. duellmani* p.137 ABM

2.23 *A. sericeus* p.289

2.24 *A. anisolepis* p.75

2.25 *A. macrinii* p.195 RA

2.26 *A. limifrons* p.187 TK

2.27 *A. zeus* p.330 TK

2.28 *A. apletophallus* p.81

2.29 *A. biscutiger* p.96 TK

2.30 *A. rodriguezii* p.277

PLATE SET 2 – *BODIES*

2.31 *A. elcopeensis* p.139

2.32 *A. gruuo* p.163

2.33 *A. altae* p.70

2.34 *A. kemptoni* p.178

2.35 *A. pseudokemptoni* p.260

2.36 *A. fortunensis* p.148

2.37 *A. monteverde* p.218

2.38 *A. arenal* p.85

2.39 *A. tenorioensis* p.299

2.40 *A. uniformis* p.314

BODIES – PLATE SET 2

2.41   *A. tropidonotus* p.313

2.42   *A. humilis* p.169

2.43   *A. amplisquamosus* p.72 TK

2.44   *A. laeviventris* p.182

2.45   *A. cusuco* p.127 TK

2.46   *A. datzorum* p.131

2.47   *A. sminthus* p.291

2.48   *A. benedikti* p.91

2.49   *A. magnaphallus* p.201

2.50   *A. pachypus* p.239

335

PLATE SET 2 – BODIES

2.51 *A. tropidolepis* p.311

2.52 *A. ocelloscapularis* p.231 TK

2.53 *A. yoroensis* p.328

2.54 *A. polylepis* p.256

2.55 *A. cupreus* p.124

2.56 *A. dollfusianus* p.134

2.57 *A. townsendi* p.305 OK

2.58 *A. lemurinus* p.186 TK

2.59 *A. bicaorum* p.92 TK

2.60 *A. serranoi* p.290

BODIES – PLATE SET 2

2.61   *A. woodi*   p.327

2.62   *A. capito*   p.107

2.63   *A. johnmeyeri*   p.176

2.64   *A. carpenteri*   p.112

2.65   *A. kunayalae*   p.180

2.66   *A. ibanezi*   p.171

2.67   *A. purpurescens*   p.265

2.68   *A. maia*   p.202

2.69   *A. petersii*   p.247   TK

2.70   *A. insignis*   p.174

PLATE SET 2 – BODIES

2.71 *A. savagei* p.285 RS

2.72 *A. brooksi* p.102

2.73 *A. kathydayae* p.176

2.74 *A. ginaelisae* p.159

2.75 *A. microtus* p.215 MR

2.76 *A. fungosus* p.152

2.77 *A. beckeri* p.89

2.78 *A. utilensis* p.317 TK

2.79 *A. cristifer* p.123

2.80 *A. pentaprion* p.243

BODIES – PLATE SET 2

2.81 *A. triumphalis* p.309 TK

2.82 *A. salvini* p.283

2.83 *A. lionotus* p.191

2.84 *A. oxylophus* p.238 CA

2.85 *A. aquaticus* p.83

2.86 *A. barkeri* p.88

2.87 *A. purpuronectes* p.267

2.88 *A. fuscoauratus* p.153

2.89 *A. maculiventris* p.200 TK

2.90 *A. antonii* p.80

PLATE SET 2 – BODIES

2.91 *A. mariarum* p.204

2.92 *A. tolimensis* p.304

2.93 *A. graniliceps* p.162

2.94 *A. bombiceps* p.98

2.95 *A. scypheus* p.287

2.96 *A. tandai* p.297 IP

2.97 *A. ortonii* p.235

2.98 *A. trachyderma* p.306

2.99 *A. gracilipes* p.161 TK

2.100 *A. peraccae* p.244

BODIES – PLATE SET 2

2.101 *A. binotatus* p.93 OT

2.102 *A. festae* p.146

2.103 *A. fasciatus* p.144

2.104 *A. tropidogaster* p.310

2.105 *A. gaigei* p.155

2.106 *A. notopholis* p.230

2.107 *A. auratus* p.86

2.108 *A. sulcifrons* p.296

2.109 *A. transversalis* p.307 OT

2.110 *A. gemmosus* p.158

PLATE SET 2 – *BODIES*

2.111 *A. otongae* p.237 OT

2.112 *A. poei* p.255 OT

2.113 *A. aequatorialis* p.67 TK

2.114 *A. dracula* p.136 OT

2.115 *A. mirus* p.217

2.116 *A. parilis* p.240 OT

2.117 *A. nemonteae* p.228 SR

2.118 *A. anoriensis* p.77

2.119 *A. eulaemus* p.141

2.120 *A. fitchi* p.147 TK

BODIES – PLATE SET 2

2.121 *A. megalopithecus* p.209

2.122 *A. podocarpus* p.253 SR

2.123 *A. ventrimaculatus* p.321

2.124 *A. antioquiae* p.78 EA

2.125 *A. calimae* p.104

2.126 *A. chloris* p.115 OT

2.127 *A. dissimilis* p.132 MA

2.128 *A. hyacinthogularis* p.170 OT

2.129 *A. parvauritus* p.241

2.130 *A. jacare* p.175 EA

PLATE SET 2 – BODIES

2.131 *A. frenatus* p.150

2.132 *A. latifrons* p.185

2.133 *A. princeps* p.257 FA

2.134 *A. danieli* p.130

2.135 *A. huilae* p.168 DB

2.136 *A. boettgeri* p.97 EL

2.137 *A. heterodermus* p.164

2.138 *A. tequendama* p.301

2.139 *A. vanzolinii* p.318

2.140 *A. neglectus* p.227 IP

BODIES – PLATE SET 2

2.141 *A. neblininus* p.224 IP

2.142 *A. euskalerriari* p.142 TB

2.143 *A. inderenae* p.173 RM

2.144 *A. lamari* p.183

2.145 *A. menta* p.211

2.146 *A. santamartae* p.284 JPG

2.147 *A. orcesi* p.234 TK

2.148 *A. lososi* p.192 SR

2.149 *A. williamsmittermeierorum* p.326

2.150 *A. peruensis* p.245

PLATE SET 2 – *BODIES*

2.151 *A. laevis* p.181 PV

2.152 *A. poecilopus* p.254

2.153 *A. macrolepis* p.196

2.154 *A. lynchi* p.194

2.155 *A. maculigula* p.199 EA

2.156 *A. aeneus* p.66 RSa

2.157 *A. lineatus* p.189 AO

2.158 *A. annectens* p.76 LR

2.159 *A. extremus* p.143 IL

2.160 *A. marmoratus* p.205

BODIES – PLATE SET 2

2.161 *A. carolinensis* p.111

2.162 *A. callainus* p.105 RG

2.163 *A. trinitatis* p.308 RP

2.164 *A. garmani* p.157 IL

2.165 *A. equestris* p.140

2.166 *A. cybotes* p.128

2.167 *A. sagrei* p.282

2.168 *A. cristatellus* p.121 TK

2.169 *A. distichus* p.133 JJ

2.170 *A. allisoni* p.69

PLATE SET 2 – *BODIES*

2.171 *A. baleatus* p.87

APPENDIX 1

EVALUATIONS OF VALIDITY FOR SOME SPECIES OF *ANOLIS*

Below I give justifications for the sample of *Anolis* species recognized in this book. I focus especially on cases where my species list conflicts with the Reptile Database as of this writing (Uetz et al. 2021), because this excellent resource has become the de facto authority for reptile species inferences. Some of these are cases where a published synonymy seems to have escaped public appreciation (e.g., *Anolis nigrolineatus*), or where I am aware of current unpublished work that will bear on a species inference. But some species discussed below are cases where, in my view, the publication review process should have headed off recognition of that species based on evidence presented in the description paper.

These treatments should not be construed as comments on the general quality of work from relevant authors. The workers involved generally made inferences based on their best interpretations of data in front of them, which is to say they acted according to appropriate scientific standards. In all cases discussed below, significant advances in understanding of anole intra- and interspecies variation were made in the cited publications. Here I dispute some of the resulting species inferences, in some cases based on new data but in many cases based on re-examination of presented data. These are practical decisions intended to help fieldworkers attempting to identify species and comparative biologists seeking units for evolutionary analysis. Quite simply, it does not seem useful, or evolutionarily correct, to recognize parapatrically distributed "species" that cannot be distinguished based on morphology or DNA.

ANOLIS DAMULUS, A. GIBBICEPS, A. IMPETIGOSUS

These species are occasionally recognized in species lists. I agree with previous authors that *Anolis damulus*, *A. impetigosus*, and *A. gibbiceps* are best considered *nomina dubia* (Nicholson et al. 2012; Rivas et al. 2012).

ANOLIS LEMNISCATUS, A. BITECTUS

Work in preparation by Fernando Ayala-Varela has established the synonymy of these names with *A. binotatus*.

ANOLIS BOCOURTII

I previously have recognized this form as valid (e.g., Poe et al. 2017a), but our unpublished molecular data suggest to me that this species is best considered a synonym of *A. fuscoauratus*, as it is usually treated.

ANOLIS MICROLEPIS

I have examined the holotype of *Anolis microlepis* (UIMNH 37971), live topotypical specimens of this form, and individuals assignable to *A. rodriguezii* from throughout much of its commonly recognized range (I have not examined the holotype of *A. rodriguezii*). The type specimen and topotypical specimens of *A. microlepis* are indistinguishable from *A. rodriguezii* as commonly recognized.

ANOLIS OSA

Anolis osa is still sometimes recognized as valid, even though this purported species is not monophyletic relative to *A. polylepis* according to mitochondrial DNA (Köhler et al. 2012: fig. 2), interbreeds freely with *A. polylepis* (Köhler et al. 2012b: 4), and differs from *A. polylepis* only in minor, continuously varying details of hemipenial morphology (Köhler et al. 2012b: fig. 1; Köhler et al. 2010). For these reasons, I consider the name *A. osa* to be a junior synonym of *A. polylepis*.

APPENDIX 1

ANOLIS CRYPTOLIMIFRONS

This species is externally indistinguishable from *A. limifrons* but reported to differ from it in hemipenial structure. Individuals assignable to *Anolis cryptolimifrons* may be parapatric or syntopic with *A. limifrons* (Kohler and Sunyer 2008; most collections of these forms do not have hemipenes everted so specific judgment is ambiguous). Species boundaries between these putative species according to hemipenes, the sole trait potentially to differentiate them, are contradicted by molecular data. Based on phylogenetic analysis of published 16S ribosomal RNA sequences, some individuals identified as *A. cryptolimifrons* based on hemipenes are estimated to share a more recent evolutionary history with individuals identified as *A. limifrons* relative to other individuals identified as *A. cryptolimifrons* (Köhler et al. 2014c: 276; Poe unpublished analyses of Genbank data). Given this molecular evidence (and noting the limited survey of hemipenes within *A. limifrons*), it seems most prudent at this time not to recognize *A. cryptolimifrons* as valid.

ANOLIS IBAGUE

Anolis ibague was described by Williams (1975) and diagnosed from the (at that time) poorly known species *A. sulcifrons* and similar (i.e., *pentaprion*-like) anole species based on its unusual head scales. In particular, Williams (1975: 5) stated that "The characters of *A. ibague* that are extreme are the great size of the interparietal, of certain of the supraorbital scales, and of the sublabials." My group has visited the type locality of *A. ibague* and collected individuals of *A. sulcifrons* there that are indistinguishable from specimens of *A. sulcifrons* we collected elsewhere in the Magdalena River valley, and which also display the listed diagnostic traits of *A. ibague*. That is, we observed numerous individuals of *A. sulcifrons* at Ibague and elsewhere with the interparietal in contact with the supraorbital semicircles, large supraorbital semicircles, and large sublabials. Williams's (1975) table 1 suggests he was unaware of the considerable variation in traits that occurs within *A. sulcifrons*, as he listed point estimates for traits of *A. sulcifrons* (e.g., scales between interparietal and supraorbital semicircles) that are known to vary in all species of *Anolis* (Williams [1975] did not present a list of specimens examined). Thus, although I have not examined the type specimen of *A. ibague*, I tentatively consider the name *A. ibague* to be a junior synonym of *A. sulcifrons*.

ANOLIS NIGROLINEATUS

Williams (1965) described *Anolis nigrolineatus* without reference to *A. festae*, but later (Williams 1982: 29) stated "*A. nigrolineatus* may be a strict synonym of *A. festae* Peracca (syntypes examined)." Future work may determine southern *A. festae* (i.e., *A. nigrolineatus*) to be distinct from *A. festae* proper, but my observations of northern and southern *A. festae* in life suggest to me that Williams's (1982) later view is correct and *A. nigrolineatus* should be treated as a synonym of *A. festae*.

ANOLIS DELTAE

Williams (1974b) described *Anolis deltae* from a single specimen from the Orinoco delta of Venezuela. The characteristics of this specimen are distinct from all mainland South American *Anolis*, but accord well with southern Lesser Antillean species of the *roquet* group of *Anolis*. In particular, the *deltae* holotype's combination of supraorbital semicircles in extensive contact with each other and with the interparietal scale, lack of postcloacal scales in males, and presence of tail crest is found only in *roquet* group *Anolis* from the southern Lesser Antilles. Williams was unaware of the now well-documented "invasions" of Venezuela by *A. extremus* and *A. aeneus* of the *roquet* group (Peters & Donoso-Barros 1970; Rivas et al. 2012). Although I have not examined the holotype of *A. deltae*, I recommend that the name *A. deltae* be treated as a junior synonym of a *roquet*-group southern Antillean anole, either *A. aeneus* or *A. extremus*, based on Williams's (1974b) description. There are no data in Williams's description that distinguish the *A. deltae* type specimen from these species.

ANOLIS QUAGGULUS

Anolis quaggulus (Cope 1885) was diagnosed from *A. humilis* based on hemipenial structure and modal tendencies in some external traits (Köhler et al. 2006). Recent detailed molecular work has shown type-locality *A. quaggulus* to be phylogenetically nested within *A. humilis* (Phillips et al. 2015; i.e., *A. humilis* as currently recognized is paraphyletic), and that purported hemipenial differences between these species do not hold up to scrutiny. That is, for example, individuals with *quaggulus* hemipenes were recorded from within the range of Phillips et al.'s (2015) "*N. humilis* 3"—a population of *A. humilis* that is phylogenetically separated from type-population *A. quaggulus* (compare range maps in Phillips et al. [2015], Köhler et al. [2006]). Phillips et al. (2015) found significant molecular structure within *A. humilis/quaggulus*, and I concur with those authors that the name *quaggulus* is likely to survive the completion of pertinent taxonomic work. For now, however, it seems practical to treat *A. quaggulus* as a junior synonym within *A. humilis*. There are no morphological traits with which to distinguish the type population of *A. quaggulus* from surrounding *A. humilis*, geographic species boundaries for a molecularly diagnosed *A. quaggulus* are unclear (e.g., should Phillips et al.'s "*N. humilis* 5" be included in *A. quaggulus*?), and no suggested taxonomy exists for the other distinct lineages within currently recognized *A. humilis* (see Phillips et al. [2015] for additional discussion).

ANOLIS LYRA

Poe et al. (2009) recognized *Anolis lyra* as distinct from *A. vittigerus* based on analysis of external morphological traits. These purported species are identical in male dewlap color pattern (red with a dark central spot), and hypothesized external scale differences have not held up upon recent fieldwork (pers. obs.). Molecular work has found an acceptable species-level divergence between type locality versions of these forms (9.8% p-distances for gene COI; data from Poe et al. 2017a). But the continuous distribution of *A. vittigerus* and *A. lyra* from Panama to Ecuador suggests that such divergences may represent simple isolation by distance within a single species. Pending multigene analysis with broad geographic sampling, it seems most prudent to recognize a single species, *A. vittigerus*, for the *lemurinus*-like anole distributed along the Pacific versant from central Panama to Ecuador. As with several species discussed here, it is likely that the newer name, *A. lyra*, will survive further molecular scrutiny, but likely in some altered geographic form. Until then, there seems little point in recognizing a species that is identifiable purely by geography.

ANOLIS LEDITZIGORUM, A. ALOCOMYOS

Anolis leditzigorum and *A. alocomyos* (Köhler et al. 2014c) were described and compared to *A. tropidolepis*. These three Costa Rican species were diagnosed relative to each other based on 16S DNA, hemipenial structure, and dewlap coloration; according to the describing authors the forms are indistinguishable according to scalation and proportion (Köhler et al. 2014c). Regarding the dewlap: "The dewlap of adult males of *N. leditzigorum* is purplish red, whereas that of *N. tropidolepis* is orange red." (Köhler et al. 2014c: 271). But figure 12c of Köhler et al. shows an individual assigned to *A. leditzigorum* with an orange-red dewlap. *Anolis alocomyos* is reported to be "differentiated further from *N. benedikti, N. pachypus*, and *N. pseudopachypus* by the presence of a uniform purplish red male dewlap in life." However, figure 6B shows an individual assigned to *A. alocomyos* with an orange-red dewlap. The male dewlaps of figures 1B (*A. tropidolepis*), 6B (*A. leditzigorum*), and 12C (*A. alocomyos*) appear identical. Thus, according to evidence presented by the describing authors, male dewlap coloration is not a diagnostic trait for any of these species relative to each other.

Regarding the hemipenes: The hemipenes of *Anolis tropidolepis* and *A. leditzigorum* appear indistinguishable in Köhler et al.'s (2014c) figure 3. The hemipenes of *A. alocomyos* (fig. 3e, but see fig. 8g also) appears to possess longer lobes than those of *A. tropidolepis/leditzigorum*. This trait was listed as the only diagnostic trait for *A. alocomyos* relative to *A. tropidolepis/leditzigorum*. However, one of the three listed male individuals of *A. alocomyos* was reported to possess "the mitochondrial (i.e., maternal) genotype of *Norops* [*Anolis*] *alocomyos* while exhibiting hemipenial phenotype of the geographically neighboring *N. tropidolepis*." (Köhler et al. 2014c: 276). That is, of the three individuals assigned to this species that by virtue of their maleness allows diagnosis of this species (assuming hemipenes were everted), one (1/3 of the sample) lacks the single diagnostic trait of this species relative to *A. tropidolepis*. The authors interpreted the combination of "*alocomyos* DNA" and "*tropidolepis*/

APPENDIX 1

leditzigorum hemipenes" to indicate hybridization between *A. alocomyos* and *A. tropidolepis*. Given the limited sampling of purported *A. alocomyos*, a more parsimonious interpretation is simply that there is some intraspecific variation in the degree of development of the hemipenial lobes. The number of specimens examined with everted hemipenes was not listed in Köhler et al. (2014c).

Regarding the 16S results: There clearly is some mitochondrial structure within the species previously referred to as *Anolis tropidolepis* (Köhler et al. 2014c: fig. 2). Köhler et al. assigned names to some of these clades. The degrees of divergence (i.e., the p-distances) among these putative species are compatible with either conspecificity or separate species status for these forms (cited species with low divergences, such as *A. monteverde* and *A. altae*, may actually be conspecific). In such cases, one hopes for additional independent evidence—e.g., morphology, nuclear genes, chromosomes, allozymes—to corroborate these lineages (e.g., Jockusch et al 2012: *Batrachoseps*, Glor and Laport 2012: *Anolis distichus*; Nantarat et al. 2014: land snails). But as noted above, no additional traits have been found that unambiguously track the apparent mitochondrial lineages.

It is possible that the mitochondrial lineages recognized as species by Köhler et al. (2014c) within *Anolis tropidolepis* are recently diverged species; this is the conclusion of these authors (Köhler et al. 2014c: 260, 275). But there are numerous cases where mitochondrial results alone have been shown to be unreliable in species inferences (e.g., Ballard et al. 2004). The shallow mitochondrial splits, the geographical contiguity of the "species," and, especially, the lack of diagnostic traits that corroborate the mitochondrial lineages suggest that the most realistic interpretation of these patterns is mitochondrial structure and normal intraspecific variation in hemipenes, male dewlap coloration, DNA, and external morphology in a single species, *A. tropidolepis*. I adopt this interpretation of a single species cautiously and pragmatically: cautiously because the mitochondrial population structure discovered by Köhler et al. (2014c) in 16S is potentially suggestive of some evolutionary differentiation, and additional DNA work on this complex is warranted; pragmatically because there are no known diagnostic morphological or molecular traits with which workers in the field could distinguish these putative species. Alternatively to recognizing two species, *A. pachypus* and *A. tropidolepis*, it would be reasonable to recognize a "*tropidolepis* species complex" in acknowledgment of the considerable variation discovered by Köhler et al. (2014c).

ANOLIS CARLLIEBI, A. SACAMECATENSIS

Anolis carlliebi and *A. sacamecatensis* (Köhler, Trejo Pérez, Petersen, & Méndez de la Cruz 2014) were diagnosed relative to Mexican endemic *A. quercorum* according to genetic distance, hemipenial morphology, and "the ranges and average values of several morphometric and pholidotic characters." However, the reported genetic distances (4.8%–7.1% for 16S) are compatible with either conspecific or separate species status for these species, and the hemipenes (reported as unilobed in *carlliebi*, "slightly to distinctly bilobed" in *A. quercorum* and *A. sacamecatensis*) appear very similar in the presented photos (compare Köhler et al. 2014: figs. 57, 64, 70). Hemipenes are known to evolve rapidly (Klaczko et al. 2015), and individuals from at least some populations recorded to display different hemipenes are able to interbreed (e.g., *A. polylepis* and *A. "osa"*; Köhler et al. 2012). Köhler et al. (2014) reported hemipenial morphology for only a single male of each of *A. carlliebi* and *A. sacamecatensis*.

Morphological differences of *Anolis carlliebi* and *A. sacamecatensis* relative to *A. quercorum* are stated to be "most obvious in (1) number of middorsal scales in one head length; (2) ratio shank length/HL; and (3) subdigital lamellae on Phalanges II–IV of Toe IV." However, according to table 9 of Köhler et al. (2014), the variation in these species in these traits is nearly completely enveloped by variation in *A. quercorum*: middorsals: *quercorum* (22–38), *carlliebi* (26–40), *sacamecatensis* (24–34); shank length: *quercorum* (0.82–0.96), *carlliebi* (0.92–1.03), *sacamecatensis* (0.81–0.93); lamellae: *quercorum* (19–26), *carlliebi* (20–24), *sacamecatensis* (19–22).

Anolis quercorum, *A. carlliebi*, and *A. sacamecatensis* display identical dewlaps (Köhler et al. 2014: figs. 52a, 58a, 65a), and their hypothesized ranges are not separated by geographic barriers. Although there may be some mitochondrial structure within what has been recognized as *A. quercorum* (Köhler et al. 2014: fig. 51; but a test for isolation by distance is warranted), the lack of diagnostic traits correlated with recovered mitochondrial clades suggests caution in recognizing these clades as separate species. For now, the most prudent course seems to be to consider these three species as a single species, *A. quercorum*.

ANOLIS NIETOI

Anolis nietoi (Köhler et al. 2014) was diagnosed from *A. nebuloides* and *A. megapholidotus* based on phylogenetic results and "the ranges and average values of several morphometric and pholidotic characters." However, the reported phylogenetic results are poorly supported (52% bootstrap value for the hypothesized *A. nietoi* clade in COI; also see differing clade relationships between COI and 16S), and there are no traits listed that distinguish *A. nietoi* from *A. megapholidotus*. In particular, the species differences that are reported to be "most obvious in (1) number of middorsal scales in one head length; and (2) number of subdigital lamellae on Phalanges II–IV of Toe IV" show near complete overlap in these traits in *A. nietoi* and *A. megapholidotus* (middorsals, *nietoi*: 16–24, *megapholidotus*: 14–20; lamellae, *nietoi*: 20–24, *megapholidotus*: 17–23). Given the lack of diagnostic traits separating *A. nietoi* and *A. megapholidotus* and the questionable DNA support, it seems most prudent not to recognize *A. nietoi* as a separate species from *A. megapholidotus* at this time.

ANOLIS STEVEPOEI

Anolis stevepoei (Köhler, Trejo Pérez, Petersen, & Méndez de la Cruz, 2014) was diagnosed from *A. nebuloides* and *A. megapholidotus* by "(1) a rather gradual transition over 3 to 4 scales between the enlarged dorsal scales and the lateral body scales (vs. a more abrupt transition over only 1 or 2 scales); (2) the edges of the field of enlarged dorsal scale rows forming an undulating line due to single enlarged scales or groups thereof outside the main field (vs. the edges of this field well defined and in a more or less straight line) (3) and by having a pink to purple male dewlap (vs. more reddish)." Variation in middorsal scale rows in *nebuloides-megapholidotus Anolis* is evident but subtle, and continuously varying between populations. For example, Köhler et al.'s (2014) figure 49b (*A. nebuloides*) and figure 31a (*A. stevepoei*) do not appear distinguishable based on the listed diagnostic middorsal traits. As for the purported dewlap difference, the range of variation depicted by Köhler et al. (2014) for *A. nebuloides* (figures 45a–d) encompasses the dewlap variation shown for *A. megapholidotus* (fig. 41a; which, incidentally, appears redder than my photo of an *A. megapholidotus* dewlap from its type locality) and *A. stevepoei*. Compare, for example, Köhler et al.'s (2014) figure 45a (*A. nebuloides*) to their figures 26a–b (*A. stevepoei*). As there appears to be no way to distinguish *A. stevepoei* and *A. nebuloides* in the field, and these species are similar according to mitochondrial DNA, it seems most prudent to consider *A. stevepoei* to be a junior synonym of *A. nebuloides*.

ANOLIS ZAPOTECORUM

Anolis zapotecorum (Köhler, Trejo Pérez, Petersen & Méndez de la Cruz, 2014) was diagnosed from *A. nebuloides* and *A. megapholidotus* by "(1) a rather gradual transition over 3 to 4 scales between the enlarged dorsal scales and lateral body scales (vs. a more abrupt transition over only 1 or 2 scales); (2) the edges of the field of enlarged dorsal scale rows forming an undulating line due to single enlarged scales or groups thereof outside the main field (vs. the edges of this field forming well defined and in a more or less straight line) (3) and by having a pink male dewlap (vs. more reddish)." As in *A. stevepoei*, these traits do not actually distinguish *A. zapotecorum* from *A. nebuloides* according to photos presented by Köhler et al. (2014; dewlap: compare figure 33b, *A. zapotecorum*, to figure 45a, *A. nebuloides*; middorsals: compare figure 38a, *A. zapotecorum*, to figure 49b, *A. nebuloides*). As there is no corroborating diagnostic support for the recovered mitochondrial clades in this group (Köhler et al., 2014: fig. 19), it seems most realistic to recognize a pair of traditionally acknowledged and geographically coherent, but mitochondrially variable, species, *A. nebuloides* (including *A. zapotecorum*, *A. stevepoei*) and *A. megapholidotus* (including *A. nietoi*).

ANOLIS WAMPUENSIS, A. SPILORHIPIS, A. MCCRANIEI, A. WILSONI

Köhler et al. (2016) divided *Anolis tropidonotus* into four species including *A. spilorhipis* (previously described as a subspecies of *A. tropidonotus*), *A. mccraniei*, and *A. wilsoni*. Unmentioned by these

APPENDIX 1

authors, *A. wampuensis* (McCranie and Köhler 2001) was described previously as indistinguishable from *A. tropidonotus* but for its smaller body size, ecology, and lack of a dark streak on the dewlap. *Anolis wampuensis* was not included in Köhler et al.'s (2016) review of *A. tropidonotus*, even though one of their new *tropidonotus*-like species (*A. mccraniei*) is found within 20 km of localities for *A. wampuensis*, and *A. wampuensis* is acknowledged to be a vicariant form of *A. tropidonotus* (McCranie and Köhler 2001). McCranie and Köhler (2016: 203) stated with reference to *A. tropidonotus* that "Were it not for the extreme differences in habitat between the two nominal forms (see below), we would consider them [i.e., *A. wampuensis* and *A. tropidonotus*] conspecific." I concur that the stated morphological differences between *A. tropidonotus* and *A. wampuensis* are not compelling evidence of separate species status (the body size difference is based on seven male specimens of *A. wampuensis*, and *A. tropidonotus* [sensu lato] is variable in the presence of a dark blotch or streak on the dewlap according to Álvarez del Toro and Smith [1956]; see also McCranie and Köhler [2016]: fig. 72b). The diagnostic habitat difference was described by McCranie and Köhler (2016: 203) as *A. wampuensis* occurring in "undisturbed broadleaf rainforest," a habitat in which "*Norops tropidonotus* has also never been found." However, those authors noted finding *A. wampuensis* in a disturbed area ("the edge of a cornfield"), and obviously *A. tropidonotus* occurred in "undisturbed" areas before humans arrived. I have found *A. tropidonotus* in deep forest and disturbed areas, but I have not visited the Wampu river area and so I cannot assess whether that forest is substantially different from forested areas where I have found *A. tropidonotus*. But even granting the habitat difference, it seems unlikely that a small radius of eastern localities harbors a distinct species that is identical to a geographically surrounding species but for its existence in remnant habitat. *Anolis wampuensis* may eventually be the valid name for some eastern populations of *A. tropidonotus/mccraniei*. For now, however, it seems most prudent to consider *A. wampuensis* to be a junior synonym of *A. tropidonotus*.

The remaining *tropidonotus*-complex species also are in need of scrutiny. These species were identified based on phylogenetic analysis of 16S DNA data, with corroboration attempted using morphological traits. In particular, the species were reported to differ in hemipenes, dewlap color pattern, and some scale counts.

The genetic distances between purported species (minimum of 2.9%–3.2% in 16S DNA) would be among the lowest species differences in *Anolis*; they are compatible with conspecificity. The presented phylogenetic tree (Köhler et al. 2016: fig. 1) depicts an explosive speciation event whereby four species result—there are zero-length branches for a polytomy of three species and an unsupported branch showing *A. wilsoni* as sister to these three. The simultaneous radiation of four species would be interesting and remarkable if true. However, a more parsimonious interpretation is that the four clades simply show normal intraspecific variation in a mitochondrial gene within a single species. The use of single mitochondrial genes in species diagnoses has been criticized (e.g., Ballard and Whitlock 2004), but certainly such evidence is telling if it can be corroborated with independent evidence such as that from morphology or additional genes. Unfortunately, the additional evidence in Köhler et al. (2016) is equivocal at best.

The hemipenes represent a strong possibility of corroborating evidence, as the unilobed organ of *Anolis tropidonotus* (Köhler et al. 2016: fig. 2b) appears distinctly different from the bilobed anatomies of the other three species. However, Köhler et al. (2016) do not present a list of localities from which hemipenes were examined, so it is impossible to know the geographic distribution of this trait, its consistency within populations or hypothesized species, or its condition at apparent areas of parapatry for these species (*A. mccraniei* and *A. wilsoni* appear nearly sympatric in the presented range map [Köhler et al. 2016: fig. 9], and *A. tropidonotus* is presented as geographically bracketing *A. spilorhipis*). Neighboring *Anolis* populations with very different hemipenes may interbreed freely (e.g., *A. polylepis* and the Osa peninsula population of *A. polylepis* that was recognized as *A. osa*; Köhler et al. 2012). The dewlap variation is similarly unhelpful, as there is no mention in Köhler et al. (2016) of the observed geographic distribution of purported differences, and intraspecific variation in the supposedly diagnostic dewlap trait (the presence and structure of a dark blotch/streak) is known to be common in *A. tropidonotus* (e.g., Álvarez del Toro and Smith 1956; pers. obs.). Other mentioned diagnostic traits are acknowledged to overlap in these species (see Köhler et al. 2016: table 2).

Kohler et al. (2016) and others (e.g., Álvarez del Toro and Smith 1956; Phillips et al. 2015) have documented interesting molecular and morphological variation within *Anolis tropidonotus* sensu lato, and it would be surprising if such a widespread form did not harbor multiple species. However, based on the above arguments, I currently recognize all of these forms as a composite *A. tropidonotus* complex. I do not doubt that some or all of the names *A. spilorhipis*, *A. wilsoni*, *A. mccraniei*, and *A. wampuensis* may survive upon more geographically detailed analyses. However, I do not see value in recognizing the current taxonomy and distribution. The molecular and hemipenes sampling is too thin, the species boundaries are too fuzzy, the mitochondrial divergences too shallow, and the external diagnoses (which workers must use to identify which form they have in hand) do not hold up.

APPENDIX 2

SPECIES LISTS OF *ANOLIS* BY COUNTRY, REGION, AND GESTALT

UNITED STATES

Small brown: *Anolis cristatellus, A. cybotes, A. distichus, A. sagrei*
Large brown: *Anolis trinitatis*
Small *or* large green: *Anolis allisoni, A. carolinensis, A. callainus*
Giant: *Anolis equestris, A. garmani.*

MEXICO
PACIFIC, WEST OF ISTHMUS OF TÉHUANTÉPEC
Small brown: *Anolis boulengerianus, A. brianjuliani, A. dunni, A. immaculogularis, A. liogaster, A. megapholidotus, A. microlepidotus, A. milleri, A. nebuloides, A. nebulosus, A. omiltemanus, A. peucephilus, A. quercorum, A. rubiginosus, A. sericeus, A. subocularis*
Large brown: *Anolis dunni, A. gadovii, A. taylori*
Large green: *Anolis macrinii*

PACIFIC, EAST OF ISTHMUS OF TÉHUANTÉPEC
Small brown: *Anolis anisolepis, A. campbelli, A. compressicauda, A. crassulus, A. cuprinus, A. dollfusianus, A. hobartsmithi, A. laeviventris, A. matudai, A. parvicirculatus, A. pygmaeus, A. sericeus*
Large brown: *Anolis alvarezdeltoroi, A. cristifer, A. serranoi*
Giant: *Anolis petersii*
Twig: *Anolis cristifer*
Semiaquatic: *Anolis barkeri, A. purpuronectes*

CARIBBEAN VERSANT INCLUDING YUCATÁN
Small brown: *Anolis beckeri, A. campbelli, A. cristatellus, A. cymbops, A. duellmani, A. hobartsmithi, A. laeviventris, A. milleri, A. naufragus, A. rodriguezii, A. sagrei, A. schiedii, A. sericeus, A. tropidonotus, A. uniformis, A. ustus*
Large brown: *Anolis capito, A. lemurinus*
Small green: *Anolis carolinensis*
Large green: *Anolis biporcatus, A. carolinensis*
Giant: *Anolis petersii*
Twig: *Anolis beckeri*
Semiaquatic: *Anolis barkeri, A. purpuronectes*

BELIZE

Small brown: *Anolis beckeri, A. rodriguezii, A. sagrei, A. tropidonotus, A. uniformis, A. ustus*
Large brown: *Anolis capito, A. lemurinus*
Small green: *Anolis allisoni*
Large green: *Anolis allisoni, A. biporcatus*
Twig: *Anolis beckeri*

SPECIES LISTS OF *ANOLIS* BY COUNTRY, REGION, AND GESTALT

GUATEMALA

Small brown: *Anolis beckeri, A. campbelli, A. cobanensis, A. crassulus, A. dollfusianus, A. heteropholidotus, A. laeviventris, A. macrophallus, A. rodriguezii, A. sagrei, A. sericeus, A. tropidonotus, A. uniformis*
Large brown: *Anolis capito, A. cristifer, A. lemurinus*
Large green: *Anolis biporcatus*
Giant: *Anolis petersii*
Twig: *Anolis beckeri, A. cristifer*

HONDURAS

MAINLAND

Small brown: *Anolis amplisquamosus, A. beckeri, A. caceresae, A. cupreus, A. cusuco, A. heteropholidotus, A. humilis, A. kreutzi, A. laeviventris, A. morazani, A. muralla, A. ocelloscapularis, A. pijolense, A. purpurgularis, A. rodriguezii, A. rubribarbaris, A. sagrei, A. sericeus, A. sminthus, A. tropidonotus, A. uniformis, A. wermuthi, A. yoroensis, A. zeus*
Large brown: *Anolis capito, A. lemurinus, A. johnmeyeri*
Large green: *Anolis biporcatus*
Giant: *Anolis loveridgei, A. petersii*
Twig: *Anolis beckeri*

UTILA

Small brown: *Anolis utilensis, A. sagrei, A. sericeus*
Large brown: *Anolis bicaorum*
Small *or* large green: *Anolis allisoni*
Twig: *Anolis utilensis*

ROATAN

Small brown: *Anolis sagrei*
Large brown: *Anolis roatanensis*
Small *or* large green: *Anolis allisoni*

GUANAJA

Small *or* large green: *Anolis allisoni*
Small brown: *Anolis sagrei*

NICARAGUA

MAINLAND

Small brown: *Anolis beckeri, A. cupreus, A. humilis, A. laeviventris, A. pentaprion, A. sericeus, A. tropidonotus, A. wermuthi*
Large brown: *Anolis capito, A. lemurinus*
Large green: *Anolis biporcatus*
Twig: *Anolis beckeri, A. pentaprion*

GREAT CORN ISLAND

Small brown: *Anolis sericeus, A. villai*

APPENDIX 2

EL SALVADOR

Small brown: *Anolis crassulus, A. heteropholidotus, A. laeviventris, A. macrophallus, A. tropidonotus*
Large brown: *Anolis serranoi*

COSTA RICA

PACIFIC VERSANT

Small brown: *Anolis altae, A. arenal, A. auratus, A. benedikti, A. biscutiger, A. cupreus, A. datzorum, A. humilis, A. kemptoni, A. laeviventris, A. limifrons, A. marsupialis, A. monteverde, A. pachypus, A. polylepis, A. salvini, A. sericeus, A. tenorioensis, A. tropidolepis*
Large brown: *Anolis capito, A. charlesmyersi, A. lemurinus, A. woodi*
Small green: *Anolis carpenteri, A. datzorum*
Large green: *Anolis biporcatus, A. woodi*
Giant: *Anolis frenatus, A. insignis, A. microtus, A. savagei*
Twig: *Anolis charlesmyersi, A. fungosus, A. salvini*
Semiaquatic: *Anolis aquaticus, A. oxylophus, A. riparius, A. robinsoni*

CARIBBEAN VERSANT

Small brown: *Anolis altae, A. arenal, A. benedikti, A. cristatellus, A. cupreus, A. humilis, A. laeviventris, A. limifrons, A. monteverde, A. pachypus, A. sagrei, A. sericeus, A. tenorioensis, A. tropidolepis*
Large brown: *Anolis capito, A. lemurinus, A. woodi*
Small green: *Anolis carpenteri*
Large green: *Anolis biporcatus, A. ibanezi, A. woodi*
Giant: *Anolis frenatus, A. insignis, A. microtus*
Twig: *Anolis fungosus, A. pentaprion*
Semiaquatic: *Anolis aquaticus, A. oxylophus*

COCOS ISLAND

Small brown: *Anolis townsendi*

PANAMA

WEST OF CANAL

Small brown: *Anolis auratus, A. benedikti, A. biscutiger, A. charlesmyersi, A. datzorum, A. elcopeensis, A. fortunensis, A. gaigei, A. gruuo, A. humilis, A. kemptoni, A. limifrons, A. magnaphallus, A. pachypus, A. pentaprion, A. polylepis, A. pseudokemptoni, A. pseudopachypus, A. salvini*
Large brown: *Anolis capito, A. lemurinus, A. vittigerus, A. woodi*
Small green: *Anolis carpenteri, A. datzorum*
Large green: *Anolis biporcatus, A. ibanezi, A. kunayalae*
Giant: *Anolis brooksi, A. casildae, A. frenatus, A. ginaelisae, A. kathydayae, A. kunayalae, A. microtus*
Twig: *Anolis charlesmyersi, A. fungosus, A. pentaprion, A. salvini*
Semiaquatic: *Anolis aquaticus, A. lionotus, A. oxylophus*

EAST OF CANAL

Small brown: *Anolis apletophallus, A. auratus, A. elcopeensis, A. gaigei, A. humilis, A. maculiventris[?], A. pentaprion, A. triumphalis, A. tropidogaster*
Large brown: *Anolis capito, A. vittigerus*

SPECIES LISTS OF *ANOLIS* BY COUNTRY, REGION, AND GESTALT

Small green: *Anolis chloris*
Large green: *Anolis biporcatus, A. ibanezi, A. kunayalae, A. maia, A. purpurescens*
Giant: *Anolis brooksi, A. frenatus, A. kunayalae, A. latifrons*
Twig: *Anolis pentaprion, A. triumphalis*
Semiaquatic: *Anolis poecilopus*

COLOMBIA
WESTERN ANDES TO PACIFIC
Small brown: *Anolis anchicayae, A. antonii, A. apletophallus, A. auratus, A. binotatus*[?], *A. calimae, A. festae, A. gracilipes, A. granuliceps, A. maculiventris, A. mariarum, A. notopholis, A. peraccae, A. triumphalis, A. urraoi, A. vicarius*
Large brown: *Anolis anoriensis, A. dracula, A. eulaemus, A. megalopithecus, A. ventrimaculatus, Anolis vittigerus*
Small green: *Anolis calimae, A. chloris, A. festae, A. gemmosus, A. peraccae*
Large green: *Anolis antioquiae, A. biporcatus, A. dracula, A. parvauritus, A. purpurescens, A. ventrimaculatus*
Giant: *Anolis danieli, A. dracula, A. eulaemus, A. frenatus, A. latifrons, A. maculigula, A. megalopithecus, A. mirus, A. princeps, A. propinquus*
Twig: *Anolis calimae, A. quimbaya, A. triumphalis, A. vanzolinii*
Semiaquatic: *Anolis lynchi, A. macrolepis, A. maculigula, A. poecilopus, A. rivalis*

MAGDALENA RIVER VALLEY
Small brown: *Anolis antonii, A. auratus, A. fuscoauratus, A. granuliceps, A. sulcifrons, A. tolimensis*
Large brown: *Anolis anoriensis, A. vittigerus*
Large green: *Anolis apollinaris, A. limon*
Giant: *Anolis apollinaris, A. frenatus*
Semiaquatic: *Anolis poecilopus*

AMAZONIA TO EASTERN SLOPE OF ANDES
Small brown: *Anolis auratus, A. fuscoauratus, A. lamari, a. ortonii, A. ruizii, A. trachyderma*
Large brown: *Anolis bombiceps, A. fitchi, A. scypheus, A. transversalis*
Small green: *Anolis lamari, A. ruizii*
Large green: *Anolis caquetae, A. huilae, A. punctatus, A. vaupesianus*
Twig: *Anolis heterodermus, A. inderenae, A. lamari, A. nicefori, A. quimbaya, A. richteri, A. tequendama, A. ruizii*

NORTH OF ANDES, INCLUDING SANTA MARTA MOUNTAINS
Small brown: *Anolis auratus, A. gaigei, A. menta, A. santamartae, A. solitarius, A. tropidogaster*
Large brown: *Anolis jacare, A. onca*
Small green: *Anolis menta, A. santamartae, A. solitarius*
Large green: *Anolis apollinaris, A. biporcatus, A. jacare*
Giant: *Anolis apollinaris, A. frenatus*
Twig: *Anolis euskalerriari, A. menta, A. santamartae, A. solitarius, A. tetarii*

GORGONA ISLAND
Small brown: *Anolis medemi*
Blue: *Anolis gorgonae*
Large green: *Anolis parvauritus, A. purpurescens*
Giant: *Anolis princeps*

APPENDIX 2

MALPELO ISLAND
Giant *or* large brown: *Anolis agassizi*

ECUADOR
WESTERN ANDES TO PACIFIC
Small brown: *Anolis binotatus, A. gracilipes, A. festae, A. granuliceps, A. maculiventris, A. peraccae, A. sagrei, A. triumphalis*
Large brown: *Anolis aequatorialis, A. dracula, A. ventrimaculatus, A. vittigerus*
Small green: *Anolis chloris, A. festae, A. gemmosus, A. otongae, A. poei, A. proboscis*
Large green: *Anolis aequatorialis, A. dracula, A. fasciatus, A. parilis, A. parvauritus, A. purpurescens*
Giant: *Anolis fraseri, A. nemonteae, A. princeps*
Twig: *Anolis quimbaya, A. proboscis, A. triumphalis*
Semiaquatic: *Anolis lynchi*

EASTERN ANDES TO AMAZONIA
Small brown: *Anolis fuscoauratus, A. hyacinthogularis, A. ortonii, A. trachyderma*
Large brown: *Anolis bombiceps, A. fitchi, A. podocarpus, A. scypheus, A. transversalis*
Small green: *Anolis hyacinthogularis*
Large green: *Anolis punctatus, A. soinii*
Giant: *Anolis podocarpus*
Twig: *Anolis lososi, A. orcesi, A. vanzolinii, A. williamsmittermeierorum*

PERU
EASTERN ANDES TO AMAZONIA
Small brown: *Anolis fuscoauratus, A. ortonii, A. tandai, A. trachyderma*
Large brown: *Anolis bombiceps, A. scypheus, A. tandai, A. transversalis*
Small green: *Anolis dissimilis*
Large green: *Anolis boettgeri, A. cuscoensis, A. punctatus, A. soinii*
Twig: *Anolis laevis, A. peruensis, A. williamsmittermeierorum*

BOLIVIA
Small brown: *Anolis fuscoauratus, A. meridionalis, A. ortonii, A. tandai*[?]
Large brown: *Anolis tandai*[?], *A. transversalis*
Large green: *Anolis punctatus*

PARAGUAY
Small brown: *Anolis meridionalis*

BRAZIL
Small brown: *Anolis auratus, A. brasiliensis, A. chrysolepis, A. fuscoauratus, A. meridionalis, A. ortonii, A. tandai, A. trachyderma*
Large brown: *Anolis bombiceps, A. chrysolepis, A. planiceps, A. scypheus, A. transversalis*

Small green: *Anolis carolinensis, A. dissimilis*
Large green: *Anolis carolinensis, A. phyllorhinus, A. punctatus*
Twig: *Anolis nasofrontalis, A. neglectus, A. pseudotigrinus*

FRENCH GUIANA

Small brown: *Anolis chrysolepis, A. fuscoauratus, A. ortonii*
Large brown: *Anolis chrysolepis*
Large green: *Anolis marmoratus, A. punctatus*

SURINAME

Small brown: *Anolis chrysolepis, A. cybotes, A. fuscoauratus, A. lineatus, A. ortonii*
Large brown: *Anolis chrysolepis, A. planiceps*
Small green: *Anolis callainus*
Large green: *Anolis marmoratus*[?], *A. punctatus*
Giant: *Anolis baleatus*

GUYANA

Small brown: *Anolis chrysolepis, A. fuscoauratus, A. ortonii*
Large brown: *Anolis aeneus, A. chrysolepis, A. planiceps*
Large green: *Anolis aeneus, A. punctatus*

VENEZUELA

Small brown: *Anolis anatoloros, A. auratus, A. fuscoauratus, A. ortonii*
Large brown: *Anolis annectens, A. jacare, A. onca, A. transversalis*
Large green: *Anolis biporcatus, A. extremus, A. jacare, A. punctatus, A. squamulatus*
Twig: *Anolis anatoloros, A. bellipeniculus, A. carlostoddi, A. euskalerriari, A. neblininus, A. nicefori, A. tetarii, A. tigrinus*

NOTES

INTRODUCTION

1. *Dactyloa* includes a radiation of island forms, the Southern Lesser Antillean "*roquet* series," in addition to its mainland radiation. We (Poe et al. 2015a) gave the name *Continenteloa* to the main mainland radiation within *Dactyloa*. However, the *Continenteloa* clade (see note [3]) was poorly supported and its relationship to the *roquet* series clade is ambiguous (compare, e.g., trees in Prates et al. [2020] and Poe et al. [2017a]). Given the uncertain monophyly of *Continenteloa* and the strong support and relative familiarity of the name *Dactyloa*, I will refer to continenteloan anoles as dactyloan here.
2. A clade is a set of species that includes an evolutionary ancestral population and all its descendants.
3. In addition to draconuran and dactyloan/continenteloan anoles, there are several anole species from island clades that have "invaded" mainland sites and established breeding populations. Also note that some exclusively island forms such as *Anolis agassizi* and *A. gorgonae* have deep evolutionary roots on the mainland. The coverage of this book includes both island forms that have "invaded" mainland sites and island forms associated with mainland evolutionary radiations.
4. Two species are sympatric if individuals of both species live in the same area.
5. Keep in mind that these are trends, not absolute rules. Exceptions include draconuran anole *Anolis petersii*, which is large and has well-developed dewlaps in males and females, and dactyloan anole *A. chloris*, which is small and lacks a dewlap in females.
6. For example, Pacific South American *Anolis ventrimaculatus* is a large species that tends to occur low on vegetation, thus violating the "crown" part of the Caribbean "crown-giant" ecomorph. But there are many exceptions to the distinctness of mainland and island faunas. For example, strikingly similar "twig" species with large head scales, short limbs and tails, and cryptic coloration and behavior are found both in the eastern Andes (e.g., *A. orcesi*) and on each of the Greater Antilles (e.g., *A. insolitus*).
7. Assemblages are sets of similar species coexisting in nature; that is, in sympatry.
8. Among these: Ávila-Pires (1995) for Amazonia; Lee (1996) for Yucatan; McCranie and Köhler (2015) for Honduras; Savage (2002) for Costa Rica; Moreno-Arias et al. (2021) for Colombia; the Bioweb site of the Pontificia Universidad Católica of Ecuador (Torres-Carvajal et al. 2019).
9. Officially (Wiley 1981): "a single lineage of ancestor-descendant populations of organisms which maintains its identity from other such lineages [in space and time] and which has its own evolutionary tendencies and historical fate."
10. A weakness of most species "concepts" is that they are operational, focusing on measurable characteristics such as ability to interbreed or morphological similarity. Any operational species definition will fail in some situations (e.g., most species can interbreed with some other species under certain circumstances). Therefore, it is critical that a species concept be truly conceptual, describing a theory of species in nature, rather than operational, describing one or more criteria for species recognition. See Frost and Kluge (1994) and de Queiroz (1998) for further discussion of this distinction.
11. What value is counted as "substantial" varies by gene and geographic sampling.
12. For example, individuals of *Anolis tropidogaster* and *A. gaigei* are externally indistinguishable and geographically mixed, but their hemipenes are so strikingly different that I treat both these forms as valid.
13. Cryptic species are those that are not easily delimited by humans.
14. A type locality is the area from which a species was originally described.
15. For example, there is no trait that consistently distinguishes purportedly parapatric[24] forms *Anolis vittigerus* and *A. lyra* (see appendix 1).
16. Geographically disjunct naturalized populations of anoles occur in several countries outside of this range, including for example Taiwan and Germany.

INTRODUCTION

17 The Köppen-Geiger system categorizes regions according to temperature and precipitation patterns. It was developed by German climatologists Vladmir Köppen and Rudolf Geiger in the late 19th and early 20th centuries.
18 Vicariance refers to geologic events such as rising sea levels isolating islands or movement of tectonic plates creating mountain ranges that divide and separate species.
19 Ranking of countries with the most anole species: 1) Colombia (74), 2) Cuba (62), 3) Mexico (52), 4) Panama (46), 5) Ecuador (41).
20 For example, the Jubones River basin in southwestern Ecuador separates similar species *Anolis fraseri* and *A. nemonteae*.
21 See chapter 3 for a description of "twig" and other anole gestalts.
22 Some treatments consider the SNSM to be a part of the Andes. Although geologically warranted (Oppenheim 1952), this grouping downplays the faunal distinctiveness of SNSM.
23 Endemic species are restricted to a particular region.
24 Parapatric species abut in range; that is, one species "replaces" another in a neighboring geographic area.
25 Williams (1970) labeled those anole species lacking sympatric congeners as "solitary."
26 Additional anole species are sure to be found in this underexplored area.
27 Often *Anolis bombiceps* and one of *A. chrysolepis, A. scypheus, A. tandai, A. brasiliensis*.
28 This geographic pattern of Pacific and Caribbean lowland presence to the south, switching to Caribbean restriction around Nicaragua, is partially or completely shared by several anole species as well as other organisms. Note that this range rather neatly maps to the Köppen-Geiger Tropical Rainforest and Tropical Monsoon areas of this region (fig. 0.3).
29 The high elevations of the Serranía del Darién, which reaches 1875 meters at Cerro Tacaruna, are relatively unexplored and likely to harbor undescribed anole species.
30 "We" meaning Julian Velasco.
31 Current consensus is that final isthmus closure occurred about 2.8 mya (O'Dea et al. 2016), but some estimates are as old as 15 mya (Montes et al. 2015).
32 Although it would not be surprising if some apparent Caribbean and Pacific conspecific populations actually constitute respective cryptic species.
33 See Savage (2002: chapter 2) for an excellent herpetologically inclined treatment of the environment of Costa Rica.
34 These generalities of range depend on taxonomic decisions. For example, *Anolis wellbornae* as recognized by Köhler and Vesely (2010) is present only in the northwest of Nicaragua.
35 *Anolis allisoni* is widespread and well studied on Cuba, but in Central America it is restricted to its type locality, the Bay Islands of Honduras. In North America this form is found in southwestern Florida.
36 As with many widespread highland forms, *Anolis laeviventris* likely is composed of multiple species.
37 This region of Pacific Mexico contains the most boring anole assemblages. All listed species except *Anolis macrinii* are small brown nondescript anoles.
38 Linnaeus's student Anders Sparrman's (1784) Lesser Antillean *Anolis* ("*Lacerta*") *bimaculatus* is the earliest anole species name that is still valid today.
39 "Alpha" taxonomy refers to taxonomy related to individual species. The term is used to distinguish species taxonomy from other forms of taxonomy such as that involving groups of species.
40 Duméril: National Museum of Natural History, France; Peters: Berlin Zoological Museum; Boulenger: British Museum. Cope mainly worked independently, but many of his important anole specimens were deposited in the Philadelphia Academy of Natural Sciences and the Smithsonian National Museum of Natural History.
41 Nearly all of the actual field procurement of lizards during this period likely was performed by local people in service of visiting "collectors."

NOTES

42 Several species description papers from this era cite the collector in the title. For example, Cope's (1876) publication describing *Anolis laevis*, *A. bombiceps*, *A. trachyderma*, and *A. bocourtii* is titled "Report on the reptiles brought by Professor James Orton from the middle and upper Amazon, and western Peru."
43 Smith was a prodigious herpetologist and author of the most emphatic announcement of novelty ever to title a species description: "Two new lizards, one new, of the genus *Anolis* from Mexico" (Smith 1968).
44 Williams is more famous for his Caribbean anole work, but his alpha-taxonomic contributions were focused in South America. Thirty of 43 (!) anole species described by Williams are South American.
45 Williams collaborated with many accomplished South American herpetologists, including Fernando Castro, Stephen Ayala, and Paolo Vanzolini.
46 It was also around this time that Caribbean anoles began to become a model system in ecology and comparative biology. The seminal Caribbean anole "ecomorph" paper, by Bruce Collette, was published in 1961.
47 A widely circulated unpublished chapter on Colombian anoles by Fernando Castro, Ernest Williams, and Stephen Ayala was an important early reference for anole biologists working on mainland forms. See Moreno-Arias et al. (2021) for an excellent update of this work.
48 Note to users of GBIF data for large-scale comparative analyses: As of this writing, the anole data therein is too flawed to be useful. Almost every species listing includes erroneous localities, some of them egregiously so (e.g., a South American form mistakenly reported from Cuba). Nevertheless, GBIF is a fantastic tool without which the maps of this book and a diversity of valuable research would not be possible.
49 For a model example of deducing the type locality for a particular species see Cisneros-Heredia's (2017) careful work on *Anolis aequatorialis*. Savage's (1974) treatment for Costa Rica is a thorough example of deducing type localities for all herps of an entire area on limited information.
50 Before these guidelines were adopted, early versions of this book included some "Similar Species" sections that were pages long.
51 Very common anoles are known colloquially as *trash anoles*.
52 Although obviously one would not expect detectability and ecology to be independent.
53 To give a concrete example: I do not know if populations of *Anolis fungosus* are composed of more individuals than populations of *A. kemptoni*. I do know that I have found thousands of individuals of *A. kemptoni* and six *A. fungosus*.
54 The amount of time it would take to summarize all ecological information for every mainland anole species places that goal well outside of the scope of this book.
55 These references are limited to natural history information in mainland environments and are not comprehensive for the biology of a given species. Note that I do not include island studies for those species found on both Caribbean islands and the mainland. Note further that this approach excludes a great many classes of information on species, including, for example, phylogenetic studies, molecular genetic explanations, parasite lists, morphometric studies, physiological studies, species lists for a locality, and range extensions. I also generally have not listed popular field guides that do not cite published literature.

CHAPTER 1.
BIOLOGY OF MAINLAND ANOLES

1 Not surprisingly, these alternative research foci are a result of the predilections of particular researchers and their academic lineages. In the case of the above examples, Ernest Williams and Jonathan Losos in the Caribbean and Laurie Vitt on the mainland have been instrumental.
2 Color change and expanded toepads are not unique to *Anolis* among lizards. True chameleons are famous for their color-changing ability, and several gecko lineages have evolved versions of toepads.

3   The terrestrial species *Anolis onca* and *A. annectens* lack toepads. Other terrestrial forms like *A. humilis* display extremely narrow toepads.
4   See Poe (1998), Velasco (2008), Castañeda and de Queiroz (2013) for additional documentation of osteological variation.
5   This trait signals the ability for caudal autotomy, i.e., the ability to easily sacrifice the tail to a predator.
6   Etheridge's (1959) "beta" condition of caudal vertebrae, wherein transverse processes extend anteriorly from each anterior vertebra, is one of the more consistent and useful morphological systematic traits in anoles. Its presence evolutionarily delimits the *Norops* clade, a group of anoles present on Jamaica (*Placopsis* clade), Cuba (*Trachypilus* clade), and the American mainland (*Draconura* clade).
7   See discussion of the use and variation in the dewlap elsewhere in this chapter.
8   The set of hyoid bones has been co-opted for many uses over evolution, including, for example, tongue projection in chameleons and support for the larynx in humans.
9   Occasionally two eggs are present (e.g., Smith et al. 1972), presumably due to egg retention.
10 David Crews and collaborators have been instrumental in elucidating the anatomical, neural, physiological, and behavioral correlates of reproduction in *Anolis carolinensis* (e.g., Crews 1980).
11 Both these references are volumes in Carl Gans's Biology of the Reptilia series. Although now dated, these volumes continue to be useful starting points for gaining general reptile knowledge, especially of anatomy.
12 And note that the few anole lifespan studies have been on small species. Larger species would be expected to live longer.
13 *Cladistics* was the name given to a kind of early quantitative analysis of evolutionary history that evolved into the approaches we use to infer phylogeny today.
14 These are rough and informal definitions. See, e.g., Sanford et al. (2002) for an extended discussion of these terms.
15 Quantitative methods that explicitly account for evolutionary relationship are usually assigned the overgeneral term "comparative methods." See Felsenstein (1985) for a description of the problem these methods are designed to address.
16 Note that a finding of difference or similarity between mainland and island forms by itself is not explanatory (a standard definition of an "island" is a body of land surrounded by water. But all bodies of land are surrounded by water). Some other aspect associated with these geographic regions (area size, predation rates, climate variability, topography, etc.) must be invoked in order for conclusions to be biologically meaningful. Commonly, mainland environments are thought to include more predators and competitors than island environments.
17 For alternative conclusions based on analyses incorporating fewer anole species and different assumptions, see Pinto et al. (2008), Feiner et al. (2021), Patton et al. (2021).
18 Niche breadth refers to the set of resources used and environments occupied by an individual, species, or group of species. See Carscadden et al. (2020).
19 Modularity in this case refers to the morphological integration of body parts. For example, an association between foot length and limb length may prompt analytical consideration of these parts together as a single module.
20 See endnote 7 of the Introduction regarding assemblages.
21 At 72 species, Patton et al. (2021) currently is the most taxonomically extensive comparative ecological study of anoles. Nicholson et al.'s (2012) phylogenetic mapping of "ecomodes" was an admirable and interesting first pass at a comprehensive treatment of anole ecology. The weaknesses of the Nicholson et al. analysis are its adoption of the Caribbean ecomorphs as a cladewide ecological framework and its ecological categorization of many species that had not yet been studied ecologically. The weakness of the Patton et al. (2021) study is that, even given this work's impressive sampling and quantitative rigor, it is hard to have confidence in cladewide conclusions when only ~20% of the clade has been sampled. Ecological data is difficult and time-consuming to collect. Thus, despite the progress of studies like Patton et al. (2021) and the

NOTES

anole clade's deserved notoriety as a model for understanding ecomorphology, taxonomically comprehensive studies of anole ecology are a distant goal.

22 That is, the dimorphism evolved before the solitary geography.
23 Reinforcement processes allow species to maintain their evolutionary separation from other species.
24 The term "genomic" generally refers to studies that incorporate whole-genome information. The first four anole genomes to be assembled were from mainland species: *Anolis carolinensis*, *A. frenatus*, *A. auratus*, *A. apletophallus*.
25 The one anole species believed possibly to have gone extinct during human history is *Anolis roosevelti*, native to islands east of Puerto Rico. Curiously, apparent habitat for this species still is present, although reduced, on at least three of the four islands where presence of the species has been documented. I have taken seven trips to attempt to find this lizard, visiting each of its purported localities (Culebra, Vieques, Tortola, St. John) and additional close-by Caribbean islands. The species may be extinct, or maybe it is just uncommon and/or hard to find. I have not yet given up on this species being found.
26 The IUCN is an administrative body that endeavors, among other things, to categorize the level of extinction imperilment for species.
27 As of this writing, the IUCN lists ratings for several species that are properly interpreted as synonyms of other species, e.g., *Anolis simmonsi*, which is a name for *A. nebuloides* (see Nieto-Montes de Oca et al. 2013).
28 It is an unending source of frustration that land development proceeds unfettered in most places where we do fieldwork to collect anoles, but obtaining government permission to preserve some small number of individuals for scientific study is a task of Augean difficulty.
29 Alternatively, for example, habitat degradation could be caused by an asteroid impact or (more gradually) climate change.
30 Except in the case of *Anolis sagrei*, which survives fine in parking lots.
31 We humans are not good enough at finding anoles to significantly impact populations, and "invasive" anoles tend to be limited to human-created habitats.
32 The fact of human causation in global warming has now been established, repeatedly and unequivocally, in published studies.
33 The additional step, almost never acknowledged but critical to conservation and subject of an entire field of philosophy called Environmental Ethics, is to assess why and whether species and populations have value and should be preserved. We will leave this issue aside for now.
34 *Anolis sagrei* has dispersed to no fewer than 12 countries during recorded human history, and in each case you will find conservationist hand-wringing over the potential "negative impact" of this species. Yet no species has been shown to have gone extinct due to *A. sagrei*. In fact, in almost all cases, an effect of *A. sagrei* on "native" fauna has not been detected. So keep in mind that evidence for the "impact" part of the "negative impact" statement is flimsy, and the "negative" part of this statement is contingent on a debatable ethical view that many conservationists mistakenly treat as self-evident.
35 Note that unless additional steps of ethical reasoning are taken (e.g., change is defined as "bad"; or "natural," "native," or "ancestral" behaviors are defined as "good"), the interpretation of these effects of *Anolis sagrei* as negative for the species *A. carolinensis* is not logically warranted. That is, why would it be "bad" for individuals of a species to perch differently or evolve larger toepads?
36 Transformative in the sense of Sarkar (2005).
37 Any claimed human benefit of anoles (examples: mosquito control, toepads as a model for adhesion, marker of climate change) could be fulfilled by other species or human efforts (examples, respectively: lots of animals eat mosquitos, and chemicals can kill mosquitos; geckos have more complex and impressive toepads than anoles, and directed non-animal research might be more efficient anyway; all species have some degree of temperature sensitivity and so may act as markers of climate change, and thermometers exist).

CHAPTER 2.
COLLECTING: HOW TO FIND AND CATCH MAINLAND ANOLES

1. The securing is as important as the finding. A herper who exults in seeing a herp she has failed to capture may be chided for applying "birder's rules," in recognition that the goal of bird enthusiasts is to *see* birds, with percase little consideration of the uncertainty of identification that necessarily accompanies the failure to procure an observed specimen in hand.
2. Perhaps contrary to popular belief, birds are members of Reptilia.
3. For example, a careful reader might detect some antipathy in Carl Linnaeus's 1750 description of herps: "Most amphibia [= reptiles and amphibians] are abhorrent because of their cold body, pale colour, cartilaginous skeleton, filthy skin, fierce aspect, calculating eye, offensive smell, harsh voice, squalid habitation, and terrible venom." But otherwise he was cool with them.
4. VertNet lists 1126 specimens of *Anolis polylepis* and 8 specimens of *A. insignis* from Costa Rica. Which species do you think would be harder to find during a trip to that country?
5. Hiking boots and running shoes function as giant ladders to guide stinging ants to your legs.
6. Remsen's focus is birds, but the principles are applicable to lizards as well (see, e.g., Stebbins [1966]).
7. The approach described here formerly was called "noosing." Herpetologist Earyn McGee pioneered the replacement of the term "noose" with a less offensive descriptor such as "lasso," a movement that thankfully has been taken up by many University Animal Care and use committees. I concur wholeheartedly with the replacement of "noose," a term that carries an unavoidable pejorative racial history. However, I respectfully suggest the term "snare" to be more appropriate than "lasso." According to merriam-webster.com, a "lasso" is a "rope or long thong of leather with a noose used especially for catching horses and cattle." Setting aside the horse/cattle link and the rope/thong requirement, the problem is the association of this term specifically and solely with a length of line, with no reference to the pole that is an integral part of lizard noosing/lassoing/snaring, nor to the creativity with which lizard nooses are occasionally fashioned (see note 9). Merriam-webster.com identifies a snare as "a contrivance often consisting of a noose for entangling birds or mammals." I think our lizard nooses/lassos/snares are more accurately described as contrivances—some thing *incorporating* a line formed as a loop that is designed for securing lizards—rather than as just a line ("rope," "thong") formed as a loop.
8. SpiderWire is a favorite of mine.
9. I have watched local people construct an entire functioning snare out of a long blade of grass. It is humbling to be outsnared by somebody who is snaring with a blade of grass.
10. Imagine how you would feel if a 150-foot-tall animal locked eyes with you and then came directly toward you in a threatening way. You'd run in the other direction, right? Now imagine that animal is walking toward you but its gaze is focused on something else, just beyond your position. You might hunker down and hope it hasn't seen you, especially if that strategy has worked before and you have a brain the size of a pea. I'm reasonably sure those scenarios both describe scenes from the classic TV show *Land of the Lost* and analogize the anole's cognitive threat-evaluation process when confronted with an approaching herpetologist.
11. In fact, overly rapid loop closure can have dire consequences. I have observed an anole decapitation performed by an overzealous lizard snarer.
12. See, for example, Andrews (1971), Talbot (1979), Lister and Aguayo (1992).
13. Williams and Rivero (Williams et al. 1965) note that Thomas—a legendary anole finder—collected "more than 30 individuals" of *Anolis occultus*, in addition to the one specimen those authors found. Thomas (1965) stated that "all but two" of the individuals of *A. occultus* that he had collected were found at night. I've caught dozens of *A. occultus*, all of them at night.
14. Likewise, an anole can drop and evade a clumsy herpetologist who has disturbed its perch.
15. The superiority of white over yellow light for anole herping may be a matter of personal preference. To complicate matters: When anole herping I wear glasses or contact lenses with a faint yellow tint to increase detection of contrast.

NOTES

16 Unhygienic, yes, but this actually is our standard sequence for securing anoles: Find an anole using a light in hand; transfer light to mouth while continuing light direction toward anole; grab anole using both hands.
17 Although the listed demerits of flashlights entail my regular use of bike lights, my current backup light is a 262,000-candela compact flashlight with a throw of 1024 meters. The weight and girth of this light, as well as the discrete central spot of its beam (see above), render nightlong use impractical. But its extreme throw provides situational utility for distant observations. And it comes with a rifle mount.
18 We have employed several poles for anole work, but our fallback pole for length, stiffness, portability, value, and reliability is the 20-foot B'n'M Black Widow retractable panfish pole.
19 Colloquially, the practice of knocking an anole down off its perch and catching it is called a "shake and bake." The term originated with Erik Hulebak in 2003 and quickly gained favor over its competing term, the "smackem yakem."
20 Most exceptions to this rule, in my experience, occur in the Caribbean. *Anolis conspersus* of Grand Cayman, for example, is highly light-sensitive, sometimes dropping from a leaf at night before capture is possible, and then bolting for underbrush upon reaching the ground.
21 I have inadvertently kicked up numerous *Anolis tropidonotus* and *A. uniformis* from leaf litter while herping for anoles at night, which suggests to me that these species sleep in these microhabitats.
22 Curiously, *Anolis distichus* appears highly abundant at night at several localities in the Caribbean.
23 As Roberto Langstroth advised me regarding *Anolis meridionalis*, "They are where you find them."
24 See, e.g., Pianka (1994).
25 It seems that searching these places would work well for particular anole species in special situations.
26 Williams and Schwartz and their associates herped the Caribbean intensively in the 1960s, 1970s, and 1980s, resulting in classic works such as Schwartz and Henderson's (1991) guide to Caribbean herps and Williams's (1976b) landmark summary of anole taxonomy.
27 Harvard's Museum of Comparative Zoology, where Williams famously was curator, houses 3541 specimens of *Anolis cybotes* and seven of *A. sheplani* as of this writing.
28 I'm hoping for a Star Trek–like tricorder that is keyed to particular DNA sequences so specific anole species can be detected. I'd also like a dog voice translator and an invisibility ring.
29 The phenomenon of greater abundance and diversity of plants and animals at the margins of habitat types has been formalized in ecology as the *edge effect* (Odum 1971).
30 Strongly human-affected areas with island faunas are perhaps the ultimate places for anole abundance. Rates of anole collection on Disney World hotel grounds may exceed multiple individuals per minute.
31 Depending on one's motivational makeup, caffeine, sugar, or rum also may help.

CHAPTER 3.
HOW TO IDENTIFY MAINLAND ANOLES

1 For example, in their cautiously worded introduction to the description of *Anolis phyllorhinus* Myers and Carvalho (1945) stated, "We are fully aware of the difficult nature of the genus *Anolis*." More famously among anolologists, L. C. Stuart (1955), in his treatment of Guatemalan anoles, stated, "I have learned something of variation in the genus, but the more specimens I have examined the more convinced I have become that the rumor as to my knowledge is without substantial foundation."
2 Females often are more variable than males, males have distinctive dewlaps, and juveniles may be differently patterned than adults (see "Variation within Species" in this chapter).
3 This dewlap rule is foolproof on the mainland but not in the Caribbean. Cuba is home to two species of anole that lack a dewlap in males: *Anolis bartschi* and *A. vermiculatus*.

CHAPTER 3. HOW TO IDENTIFY MAINLAND ANOLES

4   Excepting those cases where a new, undescribed species has been discovered.
5   This approach is how I figured out the case of *Anolis utowanae* (Poe 2014), a name associated with a single, long-preserved juvenile female specimen.
6   Some demerits of dichotomous keys: they require correct scoring of all pertinent couplets for accurate identification; uncertainty in scoring characters is not accommodated; couplets must be followed in a specified order, even if scoring a particular character is difficult, tedious, or ambiguous (due to, e.g., specimen damage); gender-specific traits are not straightforward to incorporate; a single species call, rather than multiple possible species assignments, is the necessary final result; species are not ranked, or grouped, according to likelihood given some known traits. See Williams et al. (1995) for additional discussion.
7   Users of both the Williams and Köhler systems should be aware of broad agreement but (sometimes subtle) differences in character definitions. Köhler's (2014) treatment is especially useful for its excellent character photographs.
8   Keeled ventrals frequently are arranged in strongly overlapping rows; smooth ventrals often are tightly juxtaposed.
9   An exception is the dorsal head scales of some species, for which only anterior or posterior scales may be keeled.
10 This is not to say there are no trends in patterns of keeling. All dorsal body surfaces tend to display similar degrees of keeling, and the combination of keeled ventral and smooth dorsal torso scales is rare.
11 At the time of this writing, the *Anolekey* is mounted on the internet at stevenpoe.net [password: EEW]. I harbor no illusions that this key will still be found at this electronic location for the duration of whatever lifetime this book may enjoy. However, I expect that the key will continue to be available, perhaps as an "app" or in some other format that is findable using standard internet resources.
12 This judgment is due to experience with graduate students in my lab, UNM herpetology classes, and other field companions, as well as email solicitations to me for identifications, and my corrections of iNaturalist identifications (working under the screen name "YoThisRayRay").
13 These counts were made without regard for sample size, so should not be considered precise estimates. Also, comparing maximum sizes is not the best way to get at sexual size dimorphism; comparing mean lengths among sexually mature individuals of each sex is preferred. The approach here is employed because it is quick and allows inclusion of even poorly sampled species in counts.
14 This underscores the value of the male dewlap, which infrequently varies within species, as the trait that is most identificatory.
15 This discussion has been restricted to morphology, as that is most used for identification. But of course all traits—ecological, physiological, molecular—display some intraspecific variation, and the degree of that variation varies between traits and species. Intraspecific variation in ecological traits is described in the species accounts in chapter 4.
16 A storied event in anole lore is the encounter at Harvard's Museum of Comparative Zoology between celebrated evolutionary biologist Ernst Mayr and ultimate anole expert Ernest Williams upon Mayr's return from fieldwork in the West Indies. As chronicled by anole biologist Greg Mayer (2010), Mayr is said to have inquired of Williams the identifications of the three species of anole on St. Croix, an island known to harbor just the single anole species *Anolis acutus*. As legend has it, Williams drolly replied, "The males, the females, and the juveniles."

REFERENCES

Aguirre, J. 2014. Ecología de una comunidad de lagartijas del género *Anolis* en el Parque Nacional Yasuní. Thesis, Pontificia Universidad Católica del Ecuador.
Alexander, N. J., and P. F. Parakkal. 1969. Formation of α- and β-type keratin in lizard epidermis during the molting cycle. *Zeitschrift für Zellforschung und Mikroskopische Anatomie* 101:72–87.
Alföldi, J., F. Palma, M. Grabherr, C. Williams, L. Kong, E. Mauceli, P. Russell, et al. 2011. The genome of the green anole lizard and a comparative analysis with birds and mammals. *Nature* 477:587–591.
Alibardi, L. 2016. Sauropsids cornification is based on corneous beta-proteins, a special type of keratin-associated corneous proteins of the epidermis. *Journal of Experimental Zoology Part B: Molecular and Developmental Evolution* 326:338–351.
Alvarez del Toro, M. 1982. *Los reptiles de Chiapas*, tercera edición. Tuxtla Gutierrez, Chiapas: Instituto de Historia Natural.
Alvarez del Toro, M., and H. M. Smith. 1956. Notulae herpetologicae chiapasiae I. *Herpetologica* 12:3–17.
Amador, L., F. Ayala-Varela, A. E. Nárvaez, K. Cruz, and O. Torres-Carvajal. 2017. First record of the invasive Brown Anole, *Anolis sagrei* Duméril & Bibron, 1837 (Squamata: Iguanidae: Dactyloinae) in South America. *Check List* 13:1–6.
Amaral, A. D. 1933. Estudos sobre lacertílios neotrópicos. I. Novos gêneros e espécies de lagartos do Brasil. *Memórias do Instituto Butantan* 7:51–74.
Amorim, D. M., and R. W. Ávila. 2019. Infection patterns of helminths in *Norops brasiliensis* (Squamata, Dactyloidae) from a humid forest, Northeastern Brazil and their relation with body mass, sex, host size, and season. *Helminthologia* 56:168–174.
Anderson, C. G. 2024. The descent of *Anolis*: assemblages, convergence, and ecomorphological evolution (Doctoral dissertation, University of New Mexico).
Andrade-C., M. G., and J. D. Lynch. 2007. *Los tipos nomenclaturales depositados en la colección zoológica del Instituto de Ciencias Naturales*. Instituto de Ciencias Naturales, Facultad de Ciencias, Universidad Nacional.
Andrews, R. M. 1971. Structural habitat and time budget of a tropical *Anolis* lizard. *Ecology* 52:262–270.
———. 1976. Growth rate in island and mainland anoline lizards. *Copeia* 1976:477–482.
———. 1979a. Evolution of life histories: a comparison of *Anolis* lizards from matched island and mainland habitats. *Breviora* 454:1–51.
———. 1979b. Reproductive effort of female *Anolis limifrons* (Sauria: Iguanidae). *Copeia* 1979:620–626.
———. 1982. Spatial variation in egg mortality of the lizard *Anolis limifrons*. *Herpetologica* 38:165–171.
———. 1983. *Norops polylepis*. In *Costa Rican Natural History*, edited by D. H. Janzen. University of Chicago Press.
———. 1985. Mate choice by females of the lizard, *Anolis carolinensis*. *Journal of Herpetology* 19:284–289.
———. 1991. Population stability of a tropical lizard. *Ecology* 72:1204–1217.
Andrews, R. M., and J. D. Nichols. 1990. Temporal and spatial variation in survival rates of the tropical lizard *Anolis limifrons*. *Oikos* 57:215–221.
Andrews, R. M., and A. S. Rand. 1974. Reproductive effort in anoline lizards. *Ecology* 55:1317–1327.
Andrews, R. M., and A. S. Rand. 1982. Seasonal breeding and long-term population fluctuations in the lizard *Anolis limifrons*. In *The Ecology of a Tropical Forest: Seasonal Rhythms and Long-Term Changes*, edited by E. G. Leigh, A. S. Rand, and D. M. Windsor. Smithsonian Institution Press.
Andrews, R. M., and A. S. Rand. 1983. Limited dispersal of juvenile *Anolis limifrons*. *Copeia* 1983:429–434.
Andrews, R. M., and O. J. Sexton. 1981. Water relations of the eggs of *Anolis auratus* and *Anolis limifrons*. *Ecology* 62:556–562.

REFERENCES

Andrews, R. M., and J. A. Stamps. 1994. Temporal variation in sexual size dimorphism of *Anolis limifrons* in Panama. *Copeia* 1994:613–622.

Andrews, R. M., A. S. Rand, and S. Guerrero. 1983. Seasonal and spatial variation in the annual cycle of a tropical lizard. In *Advances in Herpetology and Evolutionary Biology: Essays in Honor of Ernest E. Williams*, edited by A.G.J. Rhodin and K. Miyata. Museum of Comparative Zoology, Harvard University.

Andrews, R. M., S. J. Stahl, and P. F. Nicoletto. 1989. Intra-population variation in age of sexual maturity of the tropical lizard *Anolis limifrons* in Panama. *Copeia* 1989:751–753.

Anton, T., R. Auguste, A. Braswell, J. C. Murphy, and M. G. Rutherford. 2015. An addition to the herpetofauna of Tobago, *Anolis tigrinus*. https://herpetologytt.blogspot.com/2015/10/an-addition-to-herpetofauna-of-tobago.html.

Anzai, W., A. Omura, A. C. Diaz, M. Kawata, and H. Endo. 2014. Functional morphology and comparative anatomy of appendicular musculature in Cuban *Anolis* lizards with different locomotor habits. *Zoological Science* 31:454–463.

Araújo-Nieto, J. V. 2017. Estrutura da Taxocenose de Lagartos de um Fragmento de Floresta Atlântica Setentrional. Dissertation, Universidade Federal de Pernambuco.

Araujo, C., B. Nascimento, F. Maffei, G. Moya, D. Corrêa, and S. Santos. 2014. Lizards from Estação Ecológica de Santa Bárbara, a remnant of Cerrado in the state of São Paulo, Brazil. *Check List* 10:1038–1043.

Ardila-Marín, D. A., D. G. Gaitán-Reyes, and E. J. Hernández-Ruz. 2008a. Biología reproductiva de una población de *Anolis tolimensis* (Sauria: Iguanidae) en los andes Colombianos. *Caldasia* 30:151–159.

Ardila-Marín, D. A., E. J. Hernández-Ruz, and D. G. Gaitán-Reyes. 2008b. Ecología de *Anolis tolimensis* (Sauria, Iguanidae) en la cordillera Oriental de Colombia. *Herpetotropicos* 4:71–78.

Arguedas, V. P., and M. D. Barquero. 2009. Structural habitat use by the many-scaled Anole, *Anolis polylepis*, Squamata: Polychrotidae. *Acta Herpetologica* 4 (2):135–142.

Armstead, J. V., F. Ayala-Varela, O. Torres-Carvajal, M. J. Ryan, and S. Poe. 2017. Systematics and ecology of *Anolis biporcatus* (Squamata: Iguanidae). *Salamandra* 53:285–293.

Armstrong, J. A., H. J. Gamble, and F. Goldby. 1953. Observations on the olfactory apparatus and the telencephalon of *Anolis*, a microsmatic lizard. *Journal of Anatomy* 87:288–288.

Arosemena, F. A., and R. D. Ibáñez. 1993. Una especie nueva de *Anolis* (Squamata: Iguanidae) del grupo *fuscoauratus* de Fortuna, Panamá. *Revista de Biología Tropical* 41:267–272.

Arosemana, F. A., and R. D. Ibáñez. 1994. Un saurio nuevo del género *Anolis* (Reptilia: Polychridae), grupo *fuscoauratus*, de las Tierras Altas de Chiriquí, Panamá. *Caribbean Journal of Science* 30:222–227.

Arosemena, F. A., R. D. Ibáñez, and F. de Sousa. 1991. Una especie nueva de *Anolis* (Squamata: Iguanidae) del grupo *latifrons* de Fortuna, Panamá. *Revista de Biología Tropical* 39:255–262.

Arteaga, A., L. Bustamante, J. Vieira, and J. M. Guayasamin. 2023. *Reptiles of Ecuador*. https://www.reptilesofecuador.com/.

Arteaga, A., L. Bustamante, and J. M. Guayasamin. 2013. *The Amphibians and Reptiles of Mindo: Life in the Cloudforest*. Quito, Ecuador: Universidad Tecnológica Indoamérica.

Autumn, K., M. Sitti, Y. A. Liang, A. M. Peattie, W. R. Hansen, S. Sponberg, T. W. Kenny, et al. 2002. Evidence for van der Waals adhesion in gecko setae. *Proceedings of the National Academy of Sciences* 99:12252–12256.

Ávila-Pires, T.C.S. 1995. Lizards of Brazilian Amazonia (Reptilia: Squamata). *Zoologische Verhandelingen* 299:1–706.

Avilés-Rodríguez, K. J., and J. J. Kolbe. 2019. Escape in the city: urbanization alters the escape behavior of *Anolis* lizards. *Urban Ecosystems* 22:733–742.

Ayala, S. C. 1986. Saurios de Colombia: lista actualizada, y distribución de ejemplares colombianos en los museos. *Caldasia* 15:555–575.

Ayala, S. C., D. M. Harris, and E. E. Williams. 1983. New or problematic *Anolis* from Colombia I: *Anolis calimae*, new species, from the cloud forest of western Colombia. *Breviora* 475:1–11.

REFERENCES

Ayala, S. C., D. M. Harris, and E. E. Williams. 1984. *Anolis menta*, Sp. n. (Sauria, Iguanidae), a new *tigrinus* group anole from the west side of the Santa Marta Mountains, Columbia. *Papeís Avulsos de Zoologia, Museum of Zoology, University of São Paulo* 35:135–145.
Ayala, S. C., and J. L. Spain. 1975. Annual oogenesis in the lizard *Anolis auratus* determined by a blood smear technique. *Copeia* 1975:138–141.
Ayala, S. C., and E. E. Williams. 1988. New or problematic *Anolis* from Colombia, VI: two fuscoauratid anoles from the Pacific lowlands, *A. maculiventris* Boulenger, 1898 and *A. medemi*, a new species from Gorgona Island. *Breviora* 490:1–16.
Ayala-Varela, F., and O. Torres-Carvajal. 2010. A new species of dactyloid anole (Iguanidae, Polychrotinae, *Anolis*) from the southeastern slopes of the Andes of Ecuador. *ZooKeys* 53:59–73.
Ayala-Varela, F. P., and J. A. Velasco. 2010. A new species of dactyloid anole (Squamata: Iguanidae) from the western Andes of Ecuador. *Zootaxa* 2577:46–56.
Ayala-Varela, F., S. Poe, A. Carvajal-Campos, L. Gray, J. Davis, and A. Almendáriz. 2011. *Anolis soinii* Poe and Yañez-Miranda, 2008 (Squamata: Iguanidae: Polychrotinae): distribution extension, first records for Ecuador and notes on geographic variation. *Check List* 7:629.
Ayala-Varela, F. P., D. Troya-Rodríguez, X. Talero-Rodríguez, and O. Torres-Carvajal. 2014. A new Andean anole species of the *Dactyloa* clade (Squamata: Iguanidae) from western Ecuador. *Amphibian & Reptile Conservation* [Special Section] 8 (1): 8–24.
Ayala-Varela, F., S. Valverde, S. Poe, A. E. Narváez, M. H. Yánez-Muñoz, and O. Torres-Carvajal. 2021. A new giant anole (Squamata: Iguanidae: Dactyloinae) from southwestern Ecuador. *Zootaxa* 4991:295–317.
Badillo-Saldaña, L. M., C. I. Beteta-Hernández, A. Ramírez-Bautista, J. D. Lara-Tufiño, and R. P. López. 2016. First records of nocturnal activity in two diurnal anole species (Squamata: Dactyloidae) from Mexico. *Mesoamerican Herpetology* 3:715–718.
Baeckens, S., T. Driessens, and R. Damme. 2016. Intersexual chemo-sensation in a "visually-oriented" lizard, *Anolis sagrei*. *PeerJ* 4:e1874–e1874.
Ballard, J.W.O., and M. C. Whitlock. 2004. The incomplete natural history of mitochondria. *Molecular Ecology* 13:729–744.
Ballinger, R. E., K. R. Marion, and O. J. Sexton. 1970. Thermal ecology of the lizard, *Anolis limifrons* with comparative notes on three additional Panamanian anoles. *Ecology* 51:246–254.
Barbour, T. 1905. The vertebrata of Gorgona island, Colombia. Reptilia; Amphibia. *Bulletin of the Museum of Comparative Zoology* 46:98–102.
———. 1923. Notes on reptiles and amphibians from Panama. *Occasional Papers of the University of Michigan* 129:1–16.
———. 1928. Reptiles from the Bay Islands. *Proceedings of the New England Zoölogical Club* 12:100–102.
———. 1930. The Anoles I. The forms known to occur on the Neotropical Islands. *Bulletin of the Museum of Comparative Zoology* 70:103–144
———. 1932. New anoles. *Proceedings of the New England Zoölogical Club* 11:100–101.
———. 1934. *Reptiles and Amphibians: Their Habits and Adaptations*. Houghton Mifflin.
Barnett, J., L. A. Rueda-Solano, J. L. Pérez-González, and S. Poe. 2022. Rediscovery of *Anolis lamari* Williams, 1992: morphological variation and nocturnal ecology (Squamata: Dactyloidae). *Herpetology Notes* 15:329–334.
Barquero, M. D., and V. Arguedas. 2009. Structural habitat use by the Many-scaled Anole, *Anolis polylepis* (Squamata: Polychrotidae). *Acta Herpetologica* 4:135–142.
Barquero, M. D., and F. Bolaños. 2018. Morphological and ecological variation of a tropical anoline lizard: are agonistic interactions shaping ecomorphological relationships? *Biological Journal of the Linnean Society* 124:350–362.
Barragán-Contreras, L. A., and M. L. Calderón-Espinosa. 2013. What do *Anolis* eat?: evaluation of sexual dimorphism and geographic variation in the diet of *Anolis ventrimaculatus* (Squamata: Dactyloidae) in Colombia. *Actualidades Biológicas* 35:199–208.

Barrio-Amorós, C. L. 2006. *Anfibios y Reptiles de Rancho Grande. Parque Nacional Henri Pittier, Venezuela*. Serie Informes 2, Técnicos-Fundación AndígenA.

———. 2013. La Herpetofauna del Mundo Perdido, Venezuela. *Río Verde* 12:69–78.

Barrio-Amorós, C. L., C. Brewer-Carías, and O. Fuentes-Ramos. 2011. Aproximación preliminar a la herpetocenosis de un bosque pluvial en la sección occidental de la Sierra de Lema. Guayana Venezolana. *Revista de Ecología Latinoamericana* 16:1–46.

Barros, T. R., L. F. Esqueda, A. Mijares-Urrutia, E. La Marca, and K. E. Nicholson. 2007. The anoline "lost link" rediscovered: variation and distribution of *Anolis annectens* Williams 1974 (Squamata, Polychrotidae). *Tropical Zoology* 20:41–53.

Barros, T. R., E. E. Williams, and A. Viloria. 1996. The genus *Phenacosaurus* (Squamata: Iguania) in western Venezuela: *Phenacosaurus tetarii*, new species, *Phenacosaurus euskalerriari*, new species, and *Phenacosaurus nicefori* Dunn. *Breviora* 504:1–30.

Baruch, E. M., M. A. Manger, and J. L. Stynoski. 2016. Ground anoles (*Anolis humilis*) discriminate between aposematic and cryptic model insects. *Journal of Herpetology* 50:245–248.

Batista, A., K. Mebert, M. Miranda, O. Garces, R. Fuentes, and M. Ponce. 2020. Endemism on a threatened sky island: new and rare species of herpetofauna from Cerro Chucantí, eastern Panama. *Amphibian and Reptile Conservation* 14:27–46.

Batista, A., M. Vesely, K. Mebert, S. Lotzkat, and G. Koehler. 2015. A new species of *Dactyloa* from eastern Panama, with comments on other *Dactyloa* species present in the region. *Zootaxa* 4039:057–084.

Battles, A. C., and J. J. Kolbe. 2019. Miami heat: urban heat islands influence the thermal suitability of habitats for ectotherms. *Global Change Biology* 25:562–576.

Battles, A. C., D. J. Irschick, and J. J. Kolbe. 2019. Do structural habitat modifications associated with urbanization influence locomotor performance and limb kinematics in *Anolis* lizards? *Biological Journal of the Linnean Society* 127:100–112.

Battles, A. C., M. Moniz, and J. J. Kolbe. 2018. Living in the big city: preference for broad substrates results in niche expansion for urban *Anolis* lizards. *Urban Ecosystems* 21:1087–1095.

Beck, H. E., N. E. Zimmermann, T. R. McVicar, N. Vergopolan, A. Berg, and E. F. Wood. 2018. Present and future Köppen-Geiger climate classification maps at 1-km resolution. *Scientific Data* 5:1–12.

Beebe, W. 1944. Field notes on the lizards of Kartabo, British Guiana, and Caripito, Venezuela. Part 2: Iguanidae. *Zoologica: Scientific Contributions of the New York Zoological Society* 29:195–216.

Bejarano-Bonilla, D. A., and M. H. Bernal-Bautista. 2019. Patrón de actividad diaria y de temperaturas ambientales y microambientales en una población de la lagartija endémica colombiana *Anolis huilae* (Squamata, Dactyloidae). *Revista de la Academia Colombiana de Ciencias Exactas, Físicas y Naturales* 43:38–43.

Bels, V. L. 1990. The mechanism of dewlap extension in *Anolis carolinensis* (Reptilia: Iguanidae) with histological analysis of the hyoid apparatus. *Journal of Morphology* 206:225–244.

Beltrán, I., and L. A. Barragán-Contreras. 2019. Male courtship display in two populations of *Anolis heterodermus* (Squamata: Dactyloidea) from the Eastern Cordillera of Colombia. *Herpetology Notes* 12:881–884.

Berkum, F. H. 1986. Evolutionary patterns of the thermal sensitivity of sprint speed in *Anolis* lizards. *Evolution* 40:594–604.

Berthold, A. A. 1846. *Über verschiedene neue oder seltene Reptilien aus Neu-Granada und Crustaceen aus China*. Dieterich.

Beuchat, C. A. 1986. Phylogenetic distribution of the urinary bladder in lizards. *Copeia* 1986:512–517.

Bienentreu, J.-F., G. Köhler, A. Hertz, and S. Lotzkat. 2013. Distribution extension for *Anolis salvini* Boulenger, 1885 (Reptilia: Squamata: Dactyloidae), in western Panama. *Check List* 9:169–174.

Birt, R., R. Powell, and B. Greene. 2001. Natural history of *Anolis barkeri*: a semiaquatic lizard from southern México. *Journal of Herpetology* 35:161–166.

REFERENCES

Boada Viteri, E. Á. 2015. Ecología de una comunidad de lagartijas del género *Anolis* (Iguanidae: Dactyloinae) de un bosque pie-montano del Ecuador Occidenta. Thesis, Pontificia Universidad Católica del Ecuador.

Bock, B. C., A. M. Ortega, A. M. Zapata, and V. P. Páez. 2009. Microgeographic body size variation in a high elevation Andean anole (*Anolis mariarum*; Squamata, Polychrotidae). *Revista de Biología Tropical* 57:1253–1262.

Bock, B. C., A. M. Zapata, and V. P. Páez. 2010. Survivorship rates of adult *Anolis mariarum* (Squamata: Polychrotidae) in two populations with differing mean and asymptotic body sizes. *Papéis Avulsos de Zoologia* 50:43–50.

Bocourt, M. F. 1873. Recherches zoologiques pour servir a L'histoire de la faune de Amerique Centrale et du Mexique. Mission Scientifique au Mexique et dans Amerique Centrale. Recherches Zoologiques, part 3, section 1. In *Etudes sur les reptiles*, edited by A. Duméril, M.-F. Bocourt, and F. Mocquard. Paris : Imprimerie Imperiale.

Boettger, O. 1885. Liste von reptilien und batrachiern aus Paraguay. *Zeitschrift für Naturwissenschaften* 58:213–248.

Boulenger, G. A. 1881. Description of a new species of *Anolis* from Yucatan. *Proceedings of the Zoological Society of London* 49:921–922.

———. 1885. Catalogue of the Lizards in the British Museum (Natural History), vol. 2.: Iguanidae, Xenosauridae, Zonuridae, Anguidae, Anniellidae, Helodermatidae, Varanidae, Xantusiidae, Teiidae, Amphisbaenidae. Trustees of the British Museum, London.

———. 1896. Descriptions of new reptiles and batrachians from Colombia. *Annals and Magazine of Natural History* 17:16–21.

———. 1898. An account of the reptiles and batrachians collected by Mr. W.F.H. Rosenberg in western Ecuador. *Proceedings of the Zoological Society of London* 66:107–128.

———. 1902. Descriptions of new batrachians and reptiles from north-western Ecuador. *Annals and Magazine of Natural History* 9:51–57.

———. 1905. Descriptions of new reptiles discovered in Mexico by Dr. H. Gadow, F.R.S. *Proceedings of the Zoological Society of London* 1905:245–247.

———. 1908. Descriptions of new batrachians and reptiles discovered by Mr. M. G. Palmer in southwestern Colombia. *Journal of Natural History* 2:515–522.

———. 1912. Descriptions of new reptiles from the Andes of South America, preserved in the British Museum. *Annals and Magazine of Natural History* 10:420–424.

———. 1919. Descriptions of two new lizards and a new frog from the Andes of Colombia. *Proceedings of the Zoological Society of London* 89:79–81.

Boyd, A., M. Ogle, G. R. Smith, J. Lemos-Espinal, and C. Dibble. 2007. *Anolis nebulosus* (Clouded Anole). Diet. *Herpetological Review* 38:75.

Brach, V. 1976. Habits and food of *Anolis equestris* in Florida. *Copeia* 1976:187–189.

Brandon, R. A., R. G. Altig, and E. H. Albert. 1966. *Anolis barkeri* in Chiapas, Mexico. *Herpetologica* 22:156–157.

Bravo, C. 2019. Evaporative water loss of some habitat-restricted Mexican lizard species. *Herpetological Conservation and Biology* 14:51–66.

Brown, T. W., D. F. Maryon, M. P. Van den Burg, and G. Lonsdale. 2017. Distribution and natural history notes on *Norops bicaorum* (Squamata: Dactyloidae) endemic to Isla de Utila, Honduras. *Mesoamerican Herpetology* 4:493–497.

Buckley, L. B., and W. Jetz. 2007. Insularity and the determinants of lizard population density. *Ecology Letters* 10:481–489.

Cabrera-Guzmán, E. and L. Garrido-Olvera. 2014. Helminth parasites of the lesser scaly anole, *Anolis uniformis* (Squamata: Dactyloidae), from Los Tuxtlas, southern Mexico: evidence of diet and habitat use. *South American Journal of Herpetology* 9:183–189.

Cabrera-Guzmán, E., and V. H. Reynoso. 2010. Use of sleeping perches by the lizard *Anolis uniformis* (Squamata: Polychrotidae) in the fragmented tropical rainforest at Los Tuxtlas, Mexico. *Revista Mexicana de Biodiversidad* 81:921–924.

REFERENCES

Calderón-Espinosa, M. L. 2023. *Anolis* de Colombia. https://anoliscolombia.wixsite.com/website.

Calderón-Espinosa, M. L., and L. A. Barragán-Contreras. 2014. Geographic body size and shape variation in a mainland *Anolis* (Squamata: Dactyloidae) from northwestern South America (Colombia). *Acta Biológica Colombiana* 19:167–174.

Calderón-Espinosa, M. L., and A. B. Forero. 2011. Morphological diversification in solitary endemic anoles: *Anolis concolor* and *Anolis pinchoti* from San Andrés and Providence islands, Colombia. *South American Journal of Herpetology* 6:205–210.

Calsbeek, R., C. Bonneaud, S. Prabhu, N. Manoukis, and T. B. Smith. 2007. Multiple paternity and sperm storage lead to increased genetic diversity in *Anolis* lizards. *Evolutionary Ecology Research* 9:495–503.

Camp, C. L. 1923. Classification of the lizards. *Bulletin of the American Museum of Natural History* 48:290–480.

Campbell, H. W. 1973. Ecological observations on *Anolis lionotus* and *Anolis poecilopus* (Reptilia, Sauria) in Panama. *American Museum Novitates* 2510:1–30.

Campbell, J. A. 1989. Rediscovery of a rare Mexican lizard, *Norops milleri* (Sauria: Iguanidae). *The Southwestern Naturalist* 34:560–561.

Campbell, J. A., D. R. Formanowicz Jr., and P. B. Medley. 1989a. The reproductive cycle of *Norops uniformis* (Sauria: Iguanidae) in Veracruz, Mexico. *Biotropica* 21:237–243.

Campbell, J. A., D. M. Hillis, and W. W. Lamar. 1989b. A new lizard of the genus *Norops* (Sauria: Iguanidae) from the cloud forest of Hidalgo, México. *Herpetologica* 45:232–242.

Campbell, T. S. 2000. Analyses of the effects of an exotic lizard (*Anolis sagrei*) on a native lizard (*Anolis carolinensis*) in Florida, using islands as experimental units. Dissertation, University of Tennessee.

Campbell-Staton, S. C., S. V. Edwards, and J. B. Losos. 2016. Climate-mediated adaptation after mainland colonization of an ancestrally subtropical island lizard, *Anolis carolinensis*. *Journal of Evolutionary Biology* 29:2168–2180.

Campos, I. H. 2016. Autoecologia de *Norops fuscoauratus* (Squamata, Dactyloidae) na Mata do CIMNC, domínio de Floresta Atlântica, Pernambuco, Brasil. Master's thesis, Universidade Federal de Pernambuco.

Campos, I. H., C. N. Oliveira, J. V. Araújo-Neto, S. V. Brito, M. C. Guarnieri, and S. C. Ribeiro. Helminth fauna of *Norops fuscoauratus* (D'Orbigny, 1837) (Squamata: Dactyloidae) in the Atlantic Forest, northeastern Brazil. 2022. *Brazilian Journal of Biology* 82:241819.

Camposano, B. J., K. L. Krysko, K. M. Enge, E. M. Donlan, and M. Granatosky. 2008. The knight anole (*Anolis equestris*) in Florida. *Iguana* 15:213–219.

Canseco-Márquez, L., and M. G. Gutiérrez-Mayén. 2010. *Anfibios y Reptiles del Valle de Tehuacán-Cuicatlán*. Comisión Nacional para el Conocimiento y Uso de la Biodiversidad, Fundación para la Reserva de la Biosfera Cuicatlán AC. Benemérita Universidad Autónoma de Puebla Mexico.

Cantwell, L. R., and T. G. Forrest. 2013. Response of *Anolis sagrei* to acoustic calls from predatory and nonpredatory birds. *Journal of Herpetology* 47:293–298.

Carpenter, C. C. 1965. The display of the Cocos Island anole. *Herpetologica* 21:256–260.

Carrier, D. R. 1987. Lung ventilation during walking and running in four species of lizards. *Experimental Biology* 47:33–42.

Carscadden, K.A., Emery, N.C., Arnillas, C.A., Cadotte, M.W., Afkhami, M.E., Gravel, D., Livingstone, S.W. and Wiens, J.J., 2020. Niche breadth: causes and consequences for ecology, evolution, and conservation. *The Quarterly Review of Biology*, 95 (3):179–214.

Carvajal-Cogollo, J. E., and N. Urbina-Cardona. 2015. Ecological grouping and edge effects in tropical dry forest: reptile-microenvironment relationships. *Biodiversity and Conservation* 24:1109–1130.

Cassel, M., C. Strüssmann, and A. Ferreira. 2012. Histological evidence of reproductive activity in lizards from the APM Manso, Chapada dos Guimarães, Mato Grosso State, Brazil. *Acta Scientiarum. Biological Sciences* 34:327–334.

REFERENCES

Cassin, J. 1860. Catalogue of birds collected during a survey of a route for a ship canal across the Isthmus of Darien, by order of the government of the United States, made by Lieut. N. Michler, of the US Topographical Engineers, with notes and descriptions of new species. *Proceedings of the Academy of Natural Sciences of Philadelphia* 12:132–144.

Castañeda, M. R., and K. de Queiroz. 2011. Phylogenetic relationships of the *Dactyloa* clade of *Anolis* lizards based on nuclear and mitochondrial DNA sequence data. *Molecular Phylogenetics and Evolution* 61:784–800.

———. 2013. Phylogeny of the *Dactyloa* clade of *Anolis* lizards: new insights from combining morphological and molecular data. *Bulletin of the Museum of Comparative Zoology* 160:345–398.

Castro Arango, J. A. 2017. Evaluación del comportamiento territorial de *Anolis antonii* en dos hábitats con diferente grado de perturbación. Thesis, Universidad del Tolima.

Castro-Herrera, F. 1988. Niche structure of an anole community in a tropical rain forest within the Choco region of Colombia. Dissertation, University of North Texas.

Castro-Herrera, F., A. Valencia, and D. Villaquirán. 2012. *Diversidad de anfibios y reptiles del parque nacional natural Isla Gorgona*. Cali, Colombia: Feriva Impresores S. A.

Chaves, G., M. J. Ryan, F. Bolaños, C. Márquez, G. Köhler, and S. Poe. 2023. Two new species of semiaquatic *Anolis* (Squamata: Dactyloidae) from Costa Rica. *Zootaxa* 5319:249–262.

Chun, W. 2010. Miscellaneous notes on some rare and unusual anoles. Anolis *Newsletter* VI: 14–22. Museum of Comparative Zoology, Harvard University.

Cisneros, F. J. 1880. *Memoria sobre la construcción de un ferro-carril de Puerto Berrío a Barbosa (Estado de Antioquia)*. Imprenta y Libreria de N. Ponce de Leon, New York.

Cisneros-Heredia, D. F. 2017. The type localities of *Anolis aequatorialis* Werner, 1894 (Sauria: Iguania: Dactyloidae) and *Pristimantis appendiculatus* (Werner, 1894) (Amphibia: Anura: Craugastoridae). *Zootaxa* 4216:190–196.

Clark Jr., D. R. 1973. Respuestas a la temperatura de tres lagartijas (*Anolis*) costarricenses. *Caribbean Journal of Science* 13:199–206.

Clause, J., and T. Brown. 2017. Additional information on the natural history observations in *Anolis cusuco* (Squamata: Dactyloidae). *Mesoamerican Herpetology* 4:215–218.

Cochran, D. M. 1931. A new lizard (*Anolis pinchoti*) from Old Providence Island. *Journal of the Washington Academy of Sciences* 21:354–355.

Collette, B. B. 1961. Correlations between ecology and morphology in anoline lizards from Havana, Cuba, and southern Florida. *Bulletin of the Museum of Comparative Zoology* 125:135–162.

Colli, G. R., R. P. Bastos, and A. F. Araujo. 2002. The character and dynamics of the Cerrado Herpetofauna. In *The Cerrados of Brazil: Ecology and Natural History of a Neotropical Savanna*, edited by P. S. Oliveira and R. J. Marquis. Columbia University Press.

Conner, J., and D. Crews. 1980. Sperm transfer and storage in the lizard, *Anolis carolinensis*. *Journal of Morphology* 163:331–348.

Cope, E. D. 1861. Notes and descriptions of anoles. *Proceedings of the Academy of Natural Sciences of Philadelphia* 13:208–215.

———. 1862. Contributions to Neotropical saurology. *Proceedings of the Academy of Natural Sciences of Philadelphia* 14:176–594.

———. 1864. Contributions to the herpetology of tropical America. *Proceedings of the Academy of Natural Sciences of Philadelphia* 16:166–176.

———. 1868. An examination of the Reptilia and Batrachia obtained by the Orton Expedition to Equador and the Upper Amazon, with notes on other species. *Proceedings of the Academy of Natural Sciences of Philadelphia* 20:96–140.

———. 1871. Ninth contribution to the herpetology of tropical America. *Proceedings of the Academy of Natural Sciences of Philadelphia* 23:200–224.

———. 1875. On the Batrachia and Reptilia of Costa Rica: with notes on the herpetology and ichthyology of Nicaragua and Peru. *Journal of the Academy of Natural Sciences of Philadelphia*, series 2, 8:93–154.

———. 1876. Report on the reptiles brought by Professor James Orton from the middle and upper Amazon and western Peru. *Journal of the Academy of Natural Science of Philadelphia*, ser. 2, 8:159–183.

———. 1885. A contribution to the herpetology of Mexico. *Proceedings of the American Philosophical Society* 22:379–404.

———. 1892. The osteology of the Lacertilia. *Proceedings of the American Philosophical Society* 30:185–221.

———. 1899a. Contributions to the herpetology of New Granada and Argentina, with descriptions of new forms. *Philadelphia Museums Scientific Bulletin* 1:1–22.

———. 1899b. On a collection of Batrachia and Reptilia from New Granada. *Philadelphia Museums Scientific Bulletin* 1:3–19.

Corn, M. J. 1971. Upper thermal limits and thermal preferenda for three sympatric species of *Anolis*. *Journal of Herpetology* 5:17–21.

———. 1981. Ecological separation of *Anolis* lizards in a Costa Rican rain forest. Dissertation, University of Florida.

Corn, M. J., and P. L. Dalby. 1973. Systematics of the anoles of San Andrés and Providencia Islands, Colombia. *Journal of Herpetology* 7:63–74.

Cosgrove, J. J., D. H. Beermann, W. A. House, B. D. Toddes, and E. S. Dierenfeld. 2002. Whole-body nutrient composition of various ages of captive-bred bearded dragons (*Pogona vitticeps*) and adult wild anoles (*Anolis carolinensis*). *Zoo Biology* 21:489–97.

Costa, B. M., D. L. Pantoja, M. C. M. Vianna, and G. R. Colli. 2013. Direct and short-term effects of fire on lizard assemblages from a Neotropical savanna hotspot. *Journal of Herpetology* 47:502–510.

Costa, G. C., D. O. Mesquita, and G. R. Colli. 2008. The effect of pitfall trapping on lizard diets. *The Herpetological Journal* 18:45–48.

Cox, C. L., S. Alexander, B. Casement, et al. 2020. Ectoparasite extinction in simplified lizard assemblages during experimental island invasion. *Biology Letters* 16:20200474.

Crandell, K. E., A. Herrel, M. Sasa, J. B. Losos, and K. Autumn. 2014. Stick or grip? Co-evolution of adhesive toepads and claws in *Anolis* lizards. *Zoology* 117:363–369.

Crews, D. 1980. Interrelationships among ecological, behavioral, and neuroendocrine processes in the reproductive cycle of *Anolis carolinensis* and other reptiles. *Advances in the Study of Behavior* 11:1–74.

Crump, M. L. 1971. Quantitative analysis of the ecological distribution of a tropical herpetofauna. *Occasional Papers of the Museum of Natural History, University of Kansas* 3:1–62.

Culbertson, K. A., and N. C. Herrmann. 2019. Asymmetric interference competition and niche partitioning between native and invasive *Anolis* lizards. *Oecologia* 190:811–820.

Cunha, O. R. da. 1981. Lacertílios da Amazônia VII – Lagartos da região norte do Território Federal de Roraima, Brasil. (Lacertilia; Gekkonidae, Iguanidae, Scincidae e Teiidae). *Boletim do Museu Paraense Emílio Goeldi. Nova série Zoologia*, Belém, n. 107.

Curio, E., and H. Möbius. 1978. Versuche zum Nachweis eines Riechvermögens von *Anolis* l. *lineatopus* (Rept., Iguanidae). *Zeitschrift für Tierpsychologie* 47:281–292.

Curlis, J. D., R. W. Davis, E. Zetkulic, and C. L. Cox. 2017. Condition dependence of shared traits differs between sympatric *Anolis* lizards. *Journal of Experimental Zoology Part A: Ecological and Integrative Physiology* 327:110–118.

Cuvier, G. 1817. *Les reptiles, les poissons, les molusques et les annèlides* (vol. 2). Paris: Chez Deterville.

da Silva, F. M., A. C. Menks, A.L.C. Prudente, J.C.L. Costa, A.E.M. Travassos, and U. Galatti. 2011. Squamate Reptiles from municipality of Barcarena and surroundings, state of Pará, north of Brazil. *Check List* 7:220–226.

Dalrymple, G. H. 1980. Comments on the density and diet of a giant anole *Anolis equestris*. *Journal of Herpetology* 14:412–415.

Daltry, J. C. 2009. *The Status and Management of Saint Lucia's Forest Reptiles and Amphibians*. Technical Report to the National Forest Demarcation and Bio-Physical Resource Inventory Project.

REFERENCES

D'Angiolella, A. B., J. Klaczko, M. T. Rodrigues, and T.C.S. Ávila-Pires. 2016. Hemipenial morphology and diversity in South American anoles (Squamata: Dactyloidae). *Canadian Journal of Zoology* 94:251–256.

Dappen, N. 2003. Microhabitat use and escape behavior of male, female and juvenile *Norops oxylophus* (Polychrotidae). *Tropical Ecology and Conservation* (Monteverde Institute) 401.

Daudin, F. M. 1802. *Histoire Naturelle, génerale et particulière des reptiles*, ouvrage faisant suite, a l'histoire naturelle, générale et particuli & egravere composée par Leclerc de Buffon, et redigée par CS SONNINI.

Davis, W. B. 1954. Three new anoles from Mexico. *Herpetologica* 10:1–6.

Davis, W. B., and J. R. Dixon. 1961. Reptiles (exclusive of snakes) from the Chilpancingo region, Mexico. *Proceedings of the Biological Society of Washington* 74:37–56.

D'Cruze, N. C. 2005. Natural history observations of sympatric *Norops* (Beta *Anolis*) in a subtropical mainland community. *Herpetological Bulletin* 91:10–18.

D'Cruze, N. C., and P. J. Stafford. 2006. Resource partitioning of sympatric *Norops* (Beta *Anolis*) in a subtropical mainland community. *The Herpetological Journal* 16:273–280.

de Oliveira, D. P., S. M. Souza, L. Frazão, A. P. D. Almeida, and T. Hrbek. 2014. Lizards from central Jatapú River, Amazonas, Brazil. *Check List* 10:46–53.

de Queiroz, K. 1998. The general lineage concept of species, species criteria, and the process of speciation: a conceptual unification and terminological recommendations. In *Endless Forms: Species and Speciation*, edited by D. J. Howard and S. H. Berlocher. Oxford University Press.

Delaney, D. M., and D. A. Warner. 2017. Effects of age- and sex-specific density on behaviour and survival in a territorial lizard (*Anolis sagrei*). *Animal Behaviour* 129:31–41.

Diele-Viegas, L. M., F. P. Werneck, and C.F.D. Rocha. 2019. Climate change effects on population dynamics of three species of Amazonian lizards. *Comparative Biochemistry and Physiology Part A: Molecular & Integrative Physiology* 236:110530.

Dixon, J. R., and P. Soini. 1975. The reptiles of the upper Amazon basin, Peru. I. Lizards and amphisbaenians. *Milwaukee Public Museum Contributions to Biology and Geology* 4:1–58.

Dixon, J. R., and P. Soini. 1986. *The Reptiles of the Upper Amazon Basin, Iquitos Region, Peru*. Part 1, Lizards and amphisbaenians. Milwaukee Public Museum.

Doan, T. M. 1996. Basking behavior of two *Anolis* lizards in south Florida. *Florida Scientist* 59:16.

Dor, A., J. Valle-Mora, S. E. Rodríguez-Rodríguez, and P. Liedo. 2014. Predation of *Anastrepha ludens* (Diptera: Tephritidae) by *Norops serranoi* (Reptilia: Polychrotidae): functional response and evasion ability. *Environmental Entomology* 43:706–715.

D'Orbigny, A. 1847. *Voyage dans l'Amérique Méridionale ... exécuté pendant les années 1826–1833*. (1, Reptiles): 1–12, pis. 1–6,13–15.

Duellman, W. E. 1978. *The Biology of an Equatorial Herpetofauna in Amazonian Ecuador*. Miscellaneous publication no. 65. University of Kansas.

———.1979. *The South American Herpetofauna: Its Origin, Evolution, and Dispersal*. Monograph of the Museum of Natural History, the University of Kansas 7:1–485.

Duméril, A.M.C., and G. Bibron. 1837. *Erpétologie générale ou histoire naturelle complète des reptiles*, vol. 4: *Contenant l'histoire de quarante-six genres et de cent quarante-six espèces de la famille des iguaniens, de l'ordre des sauriens*. Paris: Librairie encyclopédique de Roret.

———. 1841. *Erpétologie générale ou histoire naturelle complète des reptiles*, vol. 8: *comprenant l'histoire génerale des batraciens, et la description des cinquante-deux genres et des cent soixante-trois espèces des deux premiers sous-ordres: les péroméles ...* Imp. de Fain et Thunot.

Duméril, A.M.C., and A.H.A. Duméril. 1851. *Catalogue méthodique de la collection des reptiles*. Gide et Baudry.

Duméril, A.M.C., M. Bocourt, and F. Mocquard. 1870. *Mission Scientifique au Mexique et dans l'Amérique Centrale. Recherches Zoologiques, Troisième Partie, Première Section: Études sur les Reptiles*. Paris : Imprimerie Impériale.

Dunn, E. R. 1930. Notes on Central American *Anolis*. *Proceedings of the New England Zoölogical Club* 12:15–24.

———. 1940. New and noteworthy herpetological material from Panama. *Proceedings of the Academy of Natural Sciences of Philadelphia* 92:105–122
———. 1944. The lizard genus *Phenacosaurus*. *Caldasia* 3:57–62.
Dunn, E. R., and J. T. Emlen. 1932. Reptiles and amphibians from Honduras. *Proceedings of the Academy of Natural Sciences of Philadelphia* 84:21–32.
Dunn, E. R., and L. H. Saxe Jr. 1950. Results of the Catherwood-Chaplin West Indies Expedition, 1948. Part V. Amphibians and reptiles of San Andrés and Providencia. *Proceedings of the Academy of Natural Sciences of Philadelphia* 102:141–165.
Dunn, E. R., and L. C. Stuart. 1951. Comments on some recent restrictions of type localities of certain South and Central American amphibians and reptiles. *Copeia* 1951:55–61.
Echelle, A. A., A. F. Echelle, and H. S. Fitch. 1971a. A comparative analysis of aggressive display in nine species of Costa Rican *Anolis*. *Herpetologica* 27:271–288
———. 1971b. A new anole from Costa Rica. *Herpetologica* 27:354–362.
Echelle, A. F., A. A. Echelle, and H. S. Fitch. 1978. Behavioral evidence for species status of *Anolis uniformis* (Cope). *Herpetologica* 34:205–207.
Edwards, J. R., and S. P. Lailvaux. 2012. Display behavior and habitat use in single and mixed populations of *Anolis carolinensis* and *Anolis sagrei* lizards. *Ethology* 118:494–502.
———. 2013. Do interspecific interactions between females drive shifts in habitat use? a test using the lizards *Anolis carolinensis* and *A. sagrei*. *Biological Journal of the Linnean Society* 110:843–851.
Eifler, D. A. 1995. *Anolis uniformis*: Feeding behavior. *Herpetological Review* 26:204.
Eifler, D. A., and M. A. Eifler. 2010. Use of habitat by the semiaquatic lizard, *Norops aquaticus*. *The Southwestern Naturalist* 55:466–469.
Elzen, P., and K. Schuchmann. 1980. Notes on *Anolis notopholis* Boulenger, 1896. *Bonner zoologische Beiträge* 31:319–322.
Escondida, L. 2001. Natural history of *Anolis barkeri*: a semiaquatic lizard from southern México. *Journal of Herpetology* 35:161–166.
Espitia Sanabria, D. E. 2023. Filogenia y taxonomía de tres especies de *Anolis* (Squamata: Anolidae) presentes en los Andes colombianos. Thesis, University of Caldas, Colombia.
Etheridge, R. E. 1959. The relationships of the anoles (Reptilia: Sauria: Iguanidae): an interpretation based on skeletal morphology. Ph.D. Dissertation, University of Michigan.
Evans, L. T. 1938. Cuban field studies on territoriality of the lizard *Anolis sagrei*. *Journal of Comparative Psychology* 25:97–125.
Fairchild, G. B., and C. O. Handley Jr. 1966. Gazetteer of collecting localities in Panama. In *Ectoparasites of Panama*, edited by R. L. Wenzel and V. J. Tipton. Chicago: Field Museum of Natural History.
Faria, A. S., A. P. Lima, and W. E. Magnusson. 2004. The effects of fire on behaviour and relative abundance of three lizard species in an Amazonian savanna. *Journal of Tropical Ecology* 20:591–594.
Faria, A. S., M. Menin, and I. L. Kaefer. 2019. Riparian zone as a main determinant of the structure of lizard assemblages in upland Amazonian forests. *Austral Ecology* 44:850–858.
Felsenstein, J. 1985. Phylogenies and the comparative method. *The American Naturalist* 125:1–15.
Fetters, T. L., and J. W. McGlothlin. 2017. Life histories and invasions: accelerated laying rate and incubation time in an invasive lizard, *Anolis sagrei*. *Biological Journal of the Linnean Society* 122:635–642.
Fisher, S. R., L.A.D. Pinto, and R. N. Fisher. 2020. Establishment of brown anoles (*Anolis sagrei*) across a southern California county and potential interactions with a native lizard species. *PeerJ* 8:e8937.
Fitch, H. S. 1968. Temperature and behavior of some equatorial lizards. *Herpetologica* 24:35–38.
———. 1970. Reproductive cycles of lizards and snakes. *Miscellaneous Publications of the Museum of Natural History, University of Kansas.* 52:1–247.
———. 1972. Ecology of *Anolis tropidolepis* in Costa Rican cloud forest. *Herpetologica* 28:10–21.
———. 1973a. A field study of Costa Rican lizards. *The University of Kansas Science Bulletin* 50:39–126.

REFERENCES

———. 1973b. Population structure and survivorship in some Costa Rican lizards. *Occasional Papers of the Museum of Natural History, University of Kansas* 18:1–41.

———. 1973c. Observations on the population ecology of the Central American iguanid lizard *Anolis cupreus*. *Caribbean Journal of Science* 13:215–228.

———. 1975. Sympatry and interrelationships in Costa Rican anoles. *Occasional Papers of the Museum of Natural History, University of Kansas* 40:1–60.

———. 1978. Two new anoles (Reptilia: Iguanidae) from Oaxaca with comments on other Mexican species. *Contributions in Biology and Geology, Milwaukee Public Museum* 20:1–15.

Fitch, H. S., A. A. Echelle, and A. F. Echelle. 1972. Variation in the Central American iguanid lizard *Anolis cupreus* with the description of a new subspecies. *Occasional Papers of the Museum of Natural History, University of Kansas* 8:1–20.

———. 1976. Field observations on rare or little known mainland anoles. *The University of Kansas Science Bulletin* 51:91–128.

Fitch, H. S., and R. W. Henderson. 1973. A new anole (Reptilia: Iguanidae) from southern Veracruz, Mexico. *Journal of Herpetology* 7:125–128.

———. 1976a. A field study of the rock anoles (Reptilia, Lacertilia, Iguanidae) of southern Mexico. *Journal of Herpetology* 10:303–311.

———. 1976b. A new anole (Reptilia: Iguanidae) from Great Corn Island, Caribbean Nicaragua. *Milwaukee Public Museum Contributions in Biology and Geology* 9:1–8.

Fitch, H. S., and R. A. Seigel. 1984. Ecological and taxonomic notes on Nicaraguan anoles. *Milwaukee Public Museum Contributions in Biology and Geology* 57:1–13.

Fleishman, L. J. 1988a. Sensory and environmental influences on display form in *Anolis auratus*, a grass anole from Panama. *Behavioral Ecology and Sociobiology* 22:309–316.

———. 1988b. Sensory influences on physical design of a visual display. *Animal Behaviour* 36:1420–1424.

———. 1988c. The social behavior of *Anolis auratus*, a grass anole from Panama. *Journal of Herpetology* 22:13–23.

———. 1992. The influence of the sensory system and the environment on motion patterns in the visual displays of anoline lizards and other vertebrates. *The American Naturalist* 139:S36–S61.

Fleishman, L. J., E. R. Loew, and M. Leal. 1993. Ultraviolet vision in lizards. *Nature* 365:397–397.

Fleming, T. H., and R. S. Hooker. 1975. *Anolis cupreus*: the response of a lizard to tropical seasonality. *Ecology* 56:1243–1261.

Flores-Villela, O. A. 2019. Dietary analysis of three species of the genus *Anolis* (Sauria: Dactyloidae) in "Los Tuxtlas," Veracruz, Mexico. *Revista Latinoamericana De Herpetología*, 2 (1):26–30.

Flores-Villela, O., and A. Muñoz-Alonso. 1990. *Anolis omiltemanus* Davis: Abaniquillo Amarillo. *Catalogue of American Amphibians and Reptiles* 490:1–2.

———. 1993. Anfibios y reptiles. In *Historia Natural del Parque Ecológico Estatal Omiltemi. Chilpancingo, Guerrero, México*, edited by I. Luna-Vega and J. Llorente-Bousquets. México, DF: CONABIO-UNAM.

Flores-Villela, O. A., F. M. Quijano, and G. G. Porter. 1995. *Recopilación de claves para la determinación de anfibios y reptiles de México*. Universidad Nacional Autónoma de México, Facultad de Ciencias.

Flot, J.-F. 2015. Species delimitation's coming of age. *Systematic Biology* 64:897–899.

Font, E., and L. C. Rome. 1990. Functional morphology of dewlap extension in the lizard *Anolis equestris* (Iguanidae). *Journal of Morphology* 206:245–258.

Frank, H., and J. Flanders. 2016. *Anolis aquaticus* (*Norops aquaticus*) (Water Anole) sleep site fidelity. *Herpetological Review* 47:131–132.

Freitas, M., D. Coutinho Machado, N. Venâncio, D. França, and D. Verissimo. 2013. First record for Brazil of the Odd Anole lizard, *Anolis dissimilis* Williams, 1965 (Squamata: Polychrotidae) with notes on coloration. *Herpetology Notes* 6:383–385.

Frost, D. R., and A. G. Kluge. 1994. A consideration of epistemology in systematic biology, with special reference to species. *Cladistics* 10:259–294.

REFERENCES

Gabe, M., and H. Girons. 1967. Données histologiques sur le tégument et les glandes épidermoides céphaliques des lépidosauriens. *Cells Tissues Organs* 67:571–594.

Gainsbury, A. M., and G. R. Colli. 2014. Effects of abandoned eucalyptus plantations on lizard communities in the Brazilian Cerrado. *Biodiversity and Conservation* 23:3155–3170.

Gallego-Carmona, C. A., C. F. Castro, K. A. Torres, and J.S.F. Rodríguez. 2012. Relación uso de hábitat y ectoparasitismo en una población de *Anolis antonii* (Dactyloidae) en Llanitos, Tolima-Colombia. *The Biologist* 10:Extra 2.

Gallego-Carmona, C. A., J. A. Castro-Arango, and M. H. Bernal-Bautista. 2016. Effect of habitat disturbance on the body condition index of the Colombian endemic lizard *Anolis antonii* (Squamata: Dactyloidae). *South American Journal of Herpetology* 11:183–187.

Gamble, T., A. J. Geneva, R. E. Glor, and D. Zarkower. 2014. *Anolis* sex chromosomes are derived from a single ancestral pair. *Evolution* 68:1027–1041.

García, A. 2008. The use of habitat and time by lizards in a tropical deciduous forest in western Mexico. *Studies on Neotropical Fauna and Environment* 43:107–115.

García González, M. X. 2011. Estructura poblacional, uso nocturno de hábitat y dieta de *Anolis ventrimaculatus* (Squamata: polychrotidae) en un bosque de niebla sobre la cordillera occidental de Colombia. BSc thesis, Universidad del Valle.

Garman, S. 1887. On West Indian reptiles (Iguanidae). *Bulletin of the Essex Institute* 19:25–50.

Garner, A. M., M. C. Wilson, A. P. Russell, A. Dhinojwala, and P. H. Niewiarowski. 2019. Going out on a limb: how investigation of the anoline adhesive system can enhance our understanding of fibrillar adhesion. *Integrative and Comparative Biology* 59:61–69.

Gasnier, T. R., W. E. Magnusson, and A. P. Lima. 1994. Foraging activity and diet of four sympatric lizard species in a tropical rainforest. *Journal of Herpetology* 28:187–192.

Gerber, G. P., and A. C. Echternacht. 2000. Evidence for asymmetrical intraguild predation between native and introduced *Anolis* lizards. *Oecologia* 124:599–607.

Giery, S. T., N. P. Lemoine, C. M. Hammerschlag-Peyer, R. N. Abbey-Lee, and C. A. Layman. 2013. Bidirectional trophic linkages couple canopy and understorey food webs. *Functional Ecology* 27:1436–1441.

Giery, S. T., E. Vezzani, S. Zona, and J. T. Stroud. 2017. Frugivory and seed dispersal by the invasive knight anole (*Anolis equestris*) in Florida, USA. *Food Webs* 11:13–16.

Giraldelli, G. R. 2007. Estrutura de comunidades de lagartos ao longo de um gradiente de vegetação em uma área de Cerrado em Coxim. Dissertation, Universidade Federal de Mato Grosso do Sul.

Glor, R. E., M. E. Gifford, A. Larson, J. B. Losos, L. R. Schettino, A.R.C. Lara, and T. R. Jackman. 2004. Partial island submergence and speciation in an adaptive radiation: a multilocus analysis of the Cuban green anoles. *Proceedings of the Royal Society of London. Series B* 271:2257–2265.

Godman, F.D.C., and O. Salvin. 1901. *Biologia Centrali-Americana: InsectA. Lepidoptera-Rhopalocera*. London: R. H. Porter.

González, L. A., J. Velásquez, H. Ferrer, and A. Prieto Arcas. 2007. Hábitos alimentarios del lagarto *Anolis onca* O'Shaughnessy, 1875 (Sauria: Polychrotidae) en una zona xerofítica de la laguna de Bocaripo, península de Araya, estado Sucre, Venezuela. *Acta Biologica Venezuelica* 27:25–35.

González, S., R. Dirzo, and R. C. Vogt. 1997. *Historia Natural de Los Tuxtlas*. CONABIO, Instituto de Ecología-UNAM, Instituto de Biología-UNAM, Mexico City, Mexico.

Goodman, D. 1971. Differential selection of immobile prey among terrestrial and riparian lizards. *American Midland Naturalist* 86:217–219.

Gordon, R. E. 1956. The biology and biodemography of *Anolis carolinensis carolinensis*, Voigt. Dissertation, Tulane University.

Gorman, G. C. 1973. The chromosomes of the Reptilia: a cytotaxonomic interpretation. In *Cytotaxonomy and Vertebrate Evolution*, edited by B. Chiarelli and E. Capanna. London: Academic Press.

Gorman, G. C., D. Buth, M. Soulé, and S. Y. Yang. 1983. The relationships of the Puerto Rican *Anolis*. Electrophoretic and karyotypic studies. In *Advances in Herpetology and Evolutionary Biology*, edited by A.G.J. Rhodin and K. Miyata. Museum of Comparative Zoology, Harvard University.

REFERENCES

Gray, J. E. 1840. Catalogue of the species of reptiles collected in Cuba by W. S. MacLeay, Esq., with some notes of their habits extracted from his MS. *Annals and Magazine of Natural History* 5:108–115.

Gray, L. N., A. J. Barley, D. M. Hillis, C. J. Pavón-Vázquez, S. Poe, and B. A. White. 2020. Does breeding season variation affect evolution of a sexual signaling trait in a tropical lizard clade? *Ecology and Evolution* 10:3738–3746.

Gray, L. N., A. J. Barley, S. Poe, R. C. Thomson, A. Nieto-Montes de Oca, and I. J. Wang. 2019. Phylogeography of a widespread lizard complex reflects patterns of both geographic and ecological isolation. *Molecular Ecology* 28:644–657.

Gray, L. N., R. Meza-Lázaro, S. Poe, and A. Nieto-Montes de Oca. 2016. A new species of semiaquatic *Anolis* (Squamata: Dactyloidae) from Oaxaca and Veracruz, Mexico. *The Herpetological Journal* 26:253–262.

Greenberg, B., and G. K. Noble. 1944. Social behavior of the American chameleon (*Anolis carolinensis* Voigt). *Physiological Zoology* 17:392–439.

Greenberg, N. 1977. A neuroethological study of display behavior in the lizard *Anolis carolinensis* (Reptilia, Lacertilia, Iguanidae). *American Zoologist* 17:191–201.

———. 1982. A forebrain atlas and stereotaxic technique for the lizard, *Anolis carolinensis*. *Journal of Morphology* 174:217–236.

Grisales-Martínez, F. A., J. A. Velasco, W. Bolivar, E. E. Williams, and J. M. Daza. 2017. The taxonomic and phylogenetic status of some poorly known *Anolis* species from the Andes of Colombia with the description of a nomen nudum taxon. *Zootaxa* 4303:213–230.

Grote, M. 2003. Modificación de la estructura de las comunidades de Saurios por cambios antrópicos en un bosque seco de la Isla de Providencia. Thesis, Pontificia Universidad Javeriana, Bogotá D.C., Colombia.

Guarnizo, C. E., F. P. Werneck, L. G. Giugliano, M. G. Santos, J. Fenker, L. Sousa, A. B. D'Angiolella, et al. 2016. Cryptic lineages and diversification of an endemic anole lizard (Squamata, Dactyloidae) of the Cerrado hotspot. *Molecular Phylogenetics and Evolution* 94:279–289.

Guichenot, A. 1855. Reptiles. In *Animaux Nouveaux ou Rares Recueillis pendant l'Expédition dans les Parties Centrales de l'Amérique du Sud, de Rio de Janeiro a Lima, et de Lima au Para; Exécutée par Ordre du Gouvernement Francais pendant les années 1843 a 1847, sous la direction du Comte Francis de Castelnau*. Paris: P. Bertrand, Libraire-Éditeur.

Günther, A. 1859. Second list of cold-blooded vertebrata collected by Mr. Fraser in the Andes of western Ecuador. *Proceedings of the Zoological Society of London* 1859:402–420.

Gutsche, A. 2005. The world's most endangered anole? *Iguana* 12:240–243.

Gutsche, A., J. R. McCranie, and K. E. Nicholson. 2004. Field observations on a nesting site of *Norops utilensis* Köhler 1996 (Reptilia: Squamata) with comments about its conservation status. *Salamandra* 40:297–302.

Guyer, C. 1986a. Seasonal patterns of reproduction of *Norops humilis* (Sauria: Iguanidae) in Costa Rica. *Revista de Biología Tropical* 34:247–251.

———. 1986b. The role of food in regulating population density in a tropical mainland anole, *Norops humilis*. Dissertation, University of Miami.

———. 1988a. Food supplementation in a tropical mainland anole, *Norops humilis*: demographic effects. *Ecology* 69:350–361.

———. 1988b. Food supplementation in a tropical mainland anole, *Norops humilis*: effects on individuals. *Ecology* 69:362–369.

Guyer, C., and M. A. Donnelly. 2004. *Amphibians and Reptiles of La Selva, Costa Rica, and the Caribbean Slope: A Comprehensive Guide*. University of California Press.

Guyer, C., and J. M. Savage. 1986. Cladistic relationships among anoles (Sauria: Iguanidae). *Systematic Zoology* 35:509–531.

Hall, J. M., and D. A. Warner. 2017. Body size and reproduction of a non-native lizard are enhanced in an urban environment. *Biological Journal of the Linnean Society* 122:860–871.

Hallmen, M., and A. Huy. 2012. *Anolis utilensis*. Habitat. *Herpetological Review* 43:642–643.

REFERENCES

Hallowell, E. 1856. Notes on the reptiles in the collection of the Academy of Natural Sciences of Philadelphia. *Proceedings of the Academy of Natural Sciences of Philadelphia* 8:221–238.

———. 1861. Report upon the Reptilia of the North Pacific exploring expedition, under command of Capt. John Rogers, U.S.N. *Proceedings of the Academy of Natural Sciences of Philadelphia* 12:481–482.

Hamilton, M. 2021. Walter Garbe photographs from Brazil. https://americanindian.si.edu/collections-search/edan-record/ead_collection%3Asova-nmai-ac-137.

Hamlett, G. W. 1952. Notes on breeding and reproduction in the lizard *Anolis carolinensis*. *Copeia* 1952:183–185.

Hedges, S. B., and K. L. Burnell. 1990. The Jamaican radiation of *Anolis* (Sauria: Iguanidae): an analysis of relationships and biogeography using sequential electrophoresis. *Caribbean Journal of Science* 26:31–44.

Henderson, R. W. 1982. Trophic relationships and foraging strategies of some New World tree snakes (*Leptophis, Oxybelis, Uromacer*). *Amphibia-Reptilia* 3:71–80.

Henderson, R. W., and H. S. Fitch. 1975. A comparative study of the structural and climatic habitats of *Anolis sericeus* (Reptilia: Iguanidae) and its syntopic congeners at four localities in southern Mexico. *Herpetologica* 31:459–471.

Henle, K., and A. Ehrl. 1991. Zur reptilienfauna Perus nebst Beschreibung eines neuen *Anolis* (Iguanidae) und zweier neuer Schlangen (Colubridae). *Bonner Zoologische Beiträge* 42:143–180.

Heras-Lara, L., and Villareal-Benítez, J. L. 1997. Historia natural de especies *Anolis sericeus*. In *Historia Natural de Los Tuxtlas*, edited by E. González-Soriano, R. Dirzo, and R. C. Vogt. Universidad Autonoma de Mexico, CONABIO, Mexico.

Hernández-Ruz, E. J., O. V. Castaño-Mora, G. Cárdenas-Arévalo, and P. A. Galvis-Peñuela. 2001. Caracterización preliminar de la "comunidad" de reptiles de un sector de la Serranía del Perijá, Colombia. *Caldasia* 23:475–489.

Hernández-Salinas, U., A. Ramírez-Bautista, and R. Cruz-Elizalde. 2016. Variation in feeding habits of the arboreal lizard *Anolis nebulosus* (Squamata: Dactyloidae) from island and mainland populations in Mexican Pacific. *Copeia* 104:831–837.

Hernández-Salinas, U., A. Ramírez-Bautista, R. Cruz-Elizalde, S. Meiri, and C. Berriozabal-Islas. 2019. Ecology of the growth of *Anolis nebulosus* (Squamata: Dactyloidae) in a seasonal tropical environment in the Chamela region. Jalisco, Mexico. *Ecology and Evolution* 9:2061–2071.

Hernández-Salinas, U., A. Ramírez-Bautista, and V. Mata-Silva. 2014. Species richness of squamate reptiles from two islands in the Mexican Pacific. *Check List* 10:1264–1269.

Herrel, A., B. Vanhooydonck, R. Joachim, and D. J. Irschick. 2004. Frugivory in polychrotid lizards: effects of body size. *Oecologia* 140:160–168.

Herrel, A., B. Vanhooydonck, J. Porck, and D. J. Irschick. 2008. Anatomical basis of differences in locomotor behavior in *Anolis* lizards: a comparison between two ecomorphs. *Bulletin of the Museum of Comparative Zoology* 159:213–238.

Herrera, F. C. 1978. Saurios en la zona de estudios biológicos de Providencia, Anorí, Antioquia. *Actualidades Biológicas* 7:37–41.

Herrera, N., V. Henríquez, and A. M. Rivera. 2005. Contribuciones al conocimiento de la herpetofauna de El Salvador. *Mesoamericana* 9:1–6.

Herrmann, N. C. 2017. Substrate availability and selectivity contribute to microhabitat specialization in two Central American semiaquatic anoles. *Breviora* 555:1–13.

Hertz, P. E. 1974. Thermal passivity of a tropical forest lizard, *Anolis polylepis*. *Journal of Herpetology* 8:323–327.

Hertz, P. E., Y. Arima, A. Harrison, R. B. Huey, J. B. Losos, and R. E. Glor. 2013. Asynchronous evolution of physiology and morphology in *Anolis* lizards. *Evolution* 67:2101–2113.

Hicks, R. A., and R. L. Trivers. 1983. The social behavior of *Anolis valencienni*. In *Advances in Herpetology and Evolutionary Biology: Essays in Honor of Ernest E. Williams*, edited by A.G.J. Rhodin and K. Miyata. Museum of Comparative Zoology, Harvard University.

Hilje, B., G. Chaves, J. Klank, F. Timmerman, J. Feltham, S. Gillingwater, T. Piraino, et al. 2020. Amphibians and reptiles of the Tirimbina Biological Reserve: a baseline for conservation, research

REFERENCES

and environmental education in a lowland tropical wet forest in Costa Rica. *Check List* 16:1633–1655.

Hillis, D. M. 1996. The Hillis lab at the University of Texas at Austin. *Herpetological Review*.

Hofmann, E. P., and J. H. Townsend. 2017. Origins and biogeography of the *Anolis crassulus* subgroup (Squamata: Dactyloidae) in the highlands of nuclear Central America. *BMC Evolutionary Biology* 17:1–14.

Hofmann, E. P., and J. H. Townsend. 2018. A cryptic new species of anole (Squamata: Dactyloidae) from the Lenca Highlands of Honduras, previously referred to as *Norops crassulus* (Cope, 1864). *Annals of Carnegie Museum* 85:91–111.

Hofmann, E. P., K. E. Nicholson, I. R. Luque-Montes, et al. 2019. Cryptic diversity, but to what extent? Discordance between single-locus species delimitation methods within mainland anoles (Squamata: Dactyloidae) of northern central America. *Frontiers in Genetics* 10:11.

Hoogmoed, M. S. 1973. Notes on the herpetofauna of Surinam. IV. The lizards and amphisbaenians of Surinam. *Biogeographica* 4: i–ix + 1–419.

———. 1979. The Herpetofauna of the Guianan region. In *The South American Herpetofauna: Its Origin, Evolution, and Dispersal*, edited by W. E. Duellman. Museum of Natural History, University of Kansas Monographs 7:241–279.

———. 1980. Introduced species of reptiles in Surinam. Notes on the herpetofauna of Surinam VIII. *Amphibia-Reptilia* 1:277–285.

Hover, E. L., and T. A. Jenssen. 1976. Descriptive analysis and social correlates of agnostic displays of *Anolis limifrons* (Sauria, Iguanidae). *Behaviour* 58:173–191.

Hoyos-Hoyos, J. M., P. Isaacs-Cubides, N. Devia, D. M. Galindo-Uribe, and A. R. Acosta-Galvis. 2012. An approach to the ecology of the herpetofauna in agroecosystems of the Colombian coffee zone. *South American Journal of Herpetology* 7:25–34.

Huey, R. B. 1982. Temperature, physiology, and the ecology of reptiles. In *Biology of the Reptilia*, vol. 12: Physiology, edited by C. Gans and F. H. Pough. London: Academic Press.

Huey, R. B., E. R. Pianka, and T. W. Schoener. 1982. *Lizard Ecology: Studies of a Model Organism*. Harvard University Press.

Huie, J. M., I. Prates, R. C. Bell, and K. de Queiroz. 2021. Convergent patterns of adaptive radiation between island and mainland *Anolis* lizards. *Biological Journal of the Linnean Society* 134:85–110.

Huleback, E. P. 2008. Evolution of the *Anolis limifrons* complex. Master's thesis, University of New Mexico.

Huleback, E. P., S. Poe, R. Ibáñez, and E. E. Williams. 2007. A striking new species of *Anolis* lizard (Squamata, Iguania) from Panama. *Phyllomedusa* 6:5–10.

Hutchins, E. D., G. J. Markov, W. L. Eckalbar, et al. 2014. Transcriptomic analysis of tail regeneration in the lizard *Anolis carolinensis* reveals activation of conserved vertebrate developmental and repair mechanisms. *PloS One* 9:e105004.

Ingram, T., A. Harrison, D. L. Mahler, et al. 2016. Comparative tests of the role of dewlap size in *Anolis* lizard speciation. *Proceedings of the Royal Society B: Biological Sciences* 283:2199.

Irschick, D. J., B. Vanhooydonck, A. Herrel, and J.A.Y. Meyers. 2005a. Intraspecific correlations among morphology, performance and habitat use within a green anole lizard (*Anolis carolinensis*) population. *Biological Journal of the Linnean Society* 85:211–221.

Irschick, D. J., E. Carlisle, J. Elstrott, et al. 2005b. A comparison of habitat use, morphology, clinging performance and escape behaviour among two divergent green anole lizard (*Anolis carolinensis*) populations. *Biological Journal of the Linnean Society* 85:223–234.

Irschick, D. J., L. J. Vitt, P. A. Zani, and J. B. Losos. 1997. A comparison of evolutionary radiations in mainland and Caribbean *Anolis* lizards. *Ecology* 78:2191–2203.

Jackman, T. R., A. Larson, K. de Queiroz, and J. B. Losos. 1999. Phylogenetic relationships and tempo of early diversification in *Anolis* lizards. *Systematic Biology* 48:254–285.

Jackson, J. F. 1973. Notes on the population biology of *Anolis tropidonotus* in a Honduran highland pine forest. *Journal of Herpetology* 7:309–311.

REFERENCES

Jensen, B., A.F.M. Moorman, and T. Wang. 2014. Structure and function of the hearts of lizards and snakes. *Biological Reviews* 89:302–336.

Jenssen, T. A. 1969. Ethoecology and display analysis of *Anolis nebulosus* (Sauria, Iguanidae). Dissertation, University of Oklahoma.

———. 1970a. Female response to filmed displays of *Anolis nebulosus* (Sauria, Iguanidae). *Animal Behaviour* 18:640–647.

———. 1970b. The ethoecology of *Anolis nebulosus* (Sauria, Iguanidae). *Journal of Herpetology* 4:1–38.

———. 1971. Display analysis of *Anolis nebulosus* (Sauria, Iguanidae). *Copeia* 1971:197–209.

———. 1975. Display repertoire of a male *Phenacosaurus heterodermus* (Sauria: Iguanidae). *Herpetologica* 31:48–55.

Jenssen, T. A., and E. L. Hover. 1976. Display analysis of the signature display of *Anolis limifrons* (Sauria: Iguanidae). *Behaviour* 57:227–240.

Jenssen, T. A., and S. C. Nunez. 1998. Spatial and breeding relationships of the lizard, *Anolis carolinensis*: Evidence of intrasexual selection. *Behaviour* 135:981–1003.

Jenssen, T. A., and L. M. Rothblum. 1977. Display repertoire analysis of *Anolis townsendi* (Sauria: Iguanidae) from Cocos Island. *Copeia* 1977:103–109.

Jenssen, T. A., J. D. Congdon, R. U. Fischer, R. Estes, D. Kling, and S. Edmands. 1995a. Morphological characteristics of the lizard *Anolis carolinensis* from South Carolina. *Herpetologica* 51:401–411.

Jenssen, T. A., J. Congdon, R. Fischer, R. Estes, D. Kling, S. Edmands, and H. Berna. 1996. Behavioural, thermal, and metabolic characteristics of a wintering lizard (*Anolis carolinensis*) from South Carolina. *Functional Ecology* 10 (2):201–209.

Jenssen, T. A., N. Greenberg, and K. A. Hovde. 1995b. Behavioral profile of free-ranging male lizards, *Anolis carolinensis*, across breeding and post-breeding seasons. *Herpetological Monographs* 9:41–62.

Jenssen, T. A., K. A. Hovde, and K. G. Taney. 1998. Size-related habitat use by nonbreeding *Anolis carolinensis* lizards. *Copeia* 1998:774–779.

Jiménez, R. R., and J. A. Rodríguez-Rodríguez. 2015. The relationship between perch type and aggressive behavior in the lizard *Norops polylepis* (Squamata: Dactyloidae). *Phyllomedusa* 14:43–51.

Jockusch, E. L., I. Martínez-Solano, R. W. Hansen, and D. B. Wake. 2012. Morphological and molecular diversification of slender salamanders (Caudata: Plethodontidae: *Batrachoseps*) in the southern Sierra Nevada of California with descriptions of two new species. *Zootaxa* 3190:1–30.

Johnson, J. D., C. A. Ely, and R. G. Webb. 1976. Biogeographical and taxonomic notes on some herpetozoa from the northern highlands of Chiapas, Mexico. *Transactions of the Kansas Academy of Science* 79:131–139.

Johnson, J. D., V. Mata-Silva, and L. D. Wilson. 2015. A conservation reassessment of the Central American herpetofauna based on the EVS measure. *Amphibian & Reptile Conservation* 9:1–94.

Johnson, M. A., M. V. Lopez, T. K. Whittle, B. K. Kircher, A. K. Dill, D. Varghese, and J. Wade. 2014. The evolution of copulation frequency and the mechanisms of reproduction in male *Anolis* lizards. *Current Zoology* 60:768–777.

Kahrl, A. F., and R. M. Cox. 2017. Consistent differences in sperm morphology and testis size between native and introduced populations of three *Anolis* lizard species. *Journal of Herpetology* 51:532–537.

Kahrl, A. F., M. A. Johnson, and R. M. Cox. 2019. Rapid evolution of testis size relative to sperm morphology suggests that post-copulatory selection targets sperm number in *Anolis* lizards. *Journal of Evolutionary Biology* 32:302–309.

Kaiser, C. M., and H. Kaiser. 2021. Coming home to roost: comments on individual sleep-site fidelity, sleep-site choice, and sleeping positions in *Anolis* (*Norops*) *leditzigorum* (Köhler et al., 2014) in Costa Rica (Squamata: Dactyloidae). *Herpetology Notes* 14:375–378.

Kamath, A., and J. B. Losos. 2018. Estimating encounter rates as the first step of sexual selection in the lizard *Anolis sagrei*. *Proceedings of the Royal Society B: Biological Sciences* 285:20172244.

REFERENCES

Kästle, W. 1963. Zur ethologie des Grasanolis (*Norops auratus*) (Daudin). *Zeitschrift für Tierpsychologie* 20:16–33.

———. 1965. Kästle, W. (1965). Zur Ethologie des Anden-Anolis *Phenacosaurus richteri*. *Zeitschrift für Tierpsychologie* 22:751–769.

Kattan, G. 1984. Sleeping perch selection in the lizard *Anolis ventrimaculatus*. *Biotropica* 16:328–329.

Kawamura, S., and S. Yokoyama. 1998. Functional characterization of visual and nonvisual pigments of American chameleon (*Anolis carolinensis*). *Vision Research* 38:37–44.

Kelly-Hernández, A., V. Vásquez-Cruz, N. M. Cerón-de la Luz, E. León-López, U. O. García-Vázquez, and L. Canseco-Márquez. 2018. Historia natural y nuevos registros de distribución de *Anolis schiedii* (Squamata: Dactyloidae) una lagartija endémica de Veracruz. *Revista Latinoamericana de Herpetología* 1:44–46.

Kennedy, J. 1965. Observations on the distribution and ecology of Barker's anole, *Anolis barkeri* Schmidt (Iguanidae). *Zoologica* 50:41–43.

Kiester, A. R. 1979. Conspecifics as cues: a mechanism for habitat selection in the Panamanian grass anole (*Anolis auratus*). *Behavioral Ecology and Sociobiology* 5:323–330.

Kiester, A. R., G. C. Gorman, and D. C. Arroyo. 1975. Habitat selection behavior of three species of *Anolis* lizards. *Ecology* 56:220–225.

Klaczko, J., C. A. Gilman, and D. J. Irschick. 2017. Hemipenis shape and hindlimb size are highly correlated in *Anolis* lizards. *Biological Journal of the Linnean Society* 122:627–634.

Klaczko, J., T. Ingram, and J. Losos. 2015. Genitals evolve faster than other traits in *Anolis* lizards. *Journal of Zoology* 295:44–48.

Köhler, G. 1996a. A new species of anole of the *Norops pentaprion* group from Isla de Utila, Honduras (Reptilia: Sauria: Iguanidae). *Senckenbergiana biologica* 75:23–32.

———. 1996b. Additions to the known herpetofauna of Isla de Utila (Islas de la Bahia, Honduras) with the description of a new species of the genus *Norops* (Reptilia: Sauria: Iguanidae. *Senckenbergiana biologica* 76:19–28.

———. 1999. Eine neue Saumfingerart der Gattung *Norops* von der Pazifikseite des nördlichen Mittelamerika. *Salamandra*, Rheinbach 35:37–52.

———. 2001. Type material and use of the name *Anolis bourgeaei* Bocourt (Sauria: Polychrotidae). *Copeia* 2001:274–275.

———. 2003. *Reptiles of Central America*. Offenbach, Germany: Herpeton.

———. Köhler, G., 2007. Assessing the status of *Anolis salvini* Boulenger 1885 and *A. bouvierii* Bocourt 1873 based on the primary types. *Senckenbergiana biologica* 87:1–6.

———. 2008. *Reptiles of Central America*, 2nd ed. Offenbach: Herpeton.

———. 2009. New species of *Anolis* formerly referred to as *Anolis altae* from Monteverde, Costa Rica (Squamata: Polychrotidae). *Journal of Herpetology* 43:11–20.

———. 2010. A revision of the Central American species related to *Anolis pentaprion* with the resurrection of *A. beckeri* and the description of a new species (Squamata: Polychrotidae). *Zootaxa* 2354:1–18.

———. 2011a. A new species of anole related to *Anolis altae* from Volcán Tenorio, Costa Rica (Reptilia, Squamata, Polychrotidae). *Zootaxa* 3120:29–42.

———. 2011b. A new species of *Anolis* (Squamata: Iguania: Dactyloidae) formerly referred to as *A. pachypus* from the Cordillera de Talamanca of western Panama and adjacent Costa Rica. *Zootaxa* 3125:1–21.

———. 2014. Characters of external morphology used in *Anolis* taxonomy—definition of terms, advice on usage, and illustrated examples. *Zootaxa* 3774:201–257.

Köhler, G., and M. Acevedo. 2004. The anoles (genus *Norops*) of Guatemala. I. The species of the Pacific versant below 1500 m elevation. *Salamandra* 40:113–140.

Köhler, G., and A. Bauer. 2001. *Dactyloa biporcata* Wiegmann, 1834 (curently *Anolis biporcatus*) and *Anolis petersii* Bocourt, 1873 (Reptilia, Sauria): proposed conservation of the specific names and designation of a neotype for *A. biporcatus*. *Bulletin of Zoological Nomenclature* 58:122–125.

REFERENCES

Köhler, G., and S. Hedges. 2020. A replacement name for the Hispaniolan anole formerly referred to as *Anolis chlorocyanus* Duméril & Bibron, 1837. *Caribbean Herpetology* 70:1–3.

Köhler, G., and S. B. Hedges. 2016. A revision of the green anoles of Hispaniola with description of eight new species (Reptilia, Squamata, Dactyloidae). *Novitates Caribaea* 1–135.

Köhler, G., and J. Kreutz. 1999. *Norops macrophallus* (Werner, 1917), a valid species of anole from Guatemala and El Salvador (Squamata: Sauria: Iguanidae). *Herpetozoa* 12:57–65.

Köhler, G., and J. R. McCranie. 2001. Two new species of anoles from northern Honduras (Reptilia, Squamata, Polychrotidae). *Senckenbergiana biologica* 81:235–246.

Köhler, G., and M. Obermeier. 1998. A new species of anole of the *Norops crassulus* group from central Nicaragua (Reptilia: Sauria: Iguanidae). *Senckenbergiana biologica* 77:127–138.

Köhler, G., and E. N. Smith. 2008. A new species of anole of the *Norops schiedei* group from western Guatemala (Squamata: Polychrotidae). *Herpetologica* 64:216–223.

Köhler, G., and J. Sunyer. 2008. Two new species of anoles formerly referred to as *Anolis limifrons* (Squamata: Polychrotidae). *Herpetologica* 64:92–108.

Köhler, G., and J. Vargas. 2019. A new species of anole from Parque Nacional Volcán Arenal, Costa Rica (Reptilia, Squamata, Dactyloidae: *Norops*). *Zootaxa* 4608:261–278.

Köhler, G., and M. Vesely. 2010. A revision of the *Anolis sericeus* complex with the resurrection of *A. wellbornae* and the description of a new species (Squamata: Polychrotidae). *Herpetologica* 66:207–228.

Köhler, G., A. Batista, M. Vesely, M. Ponce, A. Carrizo, and S. Lotzkat. 2012a. Evidence for the recognition of two species of *Anolis* formerly referred to as *A. tropidogaster* (Squamata: Dactyloidae). *Zootaxa* 3348:1–23.

Köhler, G., D. M. Dehling, and J. Koehler. 2010. Cryptic species and hybridization in the *Anolis polylepis* complex, with the description of a new species from the Osa Peninsula, Costa Rica (Squamata: Polychrotidae). *Zootaxa* 2718:23–38.

Köhler, G., R. Gómez Trejo Pérez, M. Garcia-Pareja, C. B. P. Petersen, and F. R. Mendez Cruz. 2013a. A contribution to the knowledge of *Anolis macrinii* Smith, 1968 (Reptilia: Squamata: Dactyloidae). *Breviora* 537:1–14.

Köhler, G., R. G. T. Pérez, C. B. P. Petersen, and F. Méndez de la Cruz. 2014a. A revision of the Mexican *Anolis* (Reptilia, Squamata, Dactyloidae) from the Pacific versant west of the Isthmus de Tehuantepec in the states of Oaxaca, Guerrero, and Puebla, with the description of six new species. *Zootaxa* 3862:1–210.

Köhler, G., R. Gómez Trejo Pérez, C. B. P. Petersen, and F. R. Méndez de la Cruz. 2014b. A new species of pine anole from the Sierra Madre del Sur in Oaxaca, Mexico (Reptilia, Squamata, Dactyloidae: *Anolis*). *Zootaxa* 3753:453–468.

Köhler, G., R. Gómez Trejo Pérez, M. G. Pareja, C. B. P. Petersen, and F. R. Méndez-de la Cruz. 2013b. Notes on *Anolis omiltemanus* Davis, 1954 (Reptilia: Squamata: Dactyloidae). *Herpetology Notes* 6:401–412.

Köhler, J., M. Hahn, and G. Köhler. 2012b. Divergent evolution of hemipenial morphology in two cryptic species of mainland anoles related to *Anolis polylepis*. *Salamandra* 48:1–11.

Köhler, G., J. R. McCranie, and L. D. Wilson. 1999. Two new species of anoles of the *Norops crassulus* group from Honduras (Reptilia: Sauria: Polychrotidae). *Amphibia-Reptilia* 20:279–298.

Köhler, G., J. R. McCranie, and L. D. Wilson. 2001. A new species of anole from western Honduras (Squamata: Polychrotidae). *Herpetologica* 247–255.

Köhler, G., J. R. McCranie, K. E. Nicholson, and J. Kreutz. 2003. Geographic variation in hemipenial morphology in *Norops humilis* (Peters 1863), and the systematic status of *Norops quaggulus* (Cope 1885)(Reptilia, Squamata, Polychrotidae). *Senckenbergiana biologica* 82:213–222.

Köhler, G., C. B. P. Petersen, and F. R. Méndez de La Cruz. 2019. A new species of anole from the Sierra Madre del Sur in Guerrero, Mexico (Reptilia, Squamata, Dactyloidae: *Norops*). *Vertebrate Zoology* 69:145–160.

REFERENCES

Köhler, J., S. Poe, M. J. Ryan, and G. Köhler. 2015. *Anolis marsupialis* Taylor 1956, a valid species from southern Pacific Costa Rica (Reptilia, Squamata, Dactyloidae). *Zootaxa* 3915:111–122.

Köhler, G., M. Ponce, J. Sunyer, and A. Batista. 2007. Four new species of anoles (genus *Anolis*) from the Serranía de Tabasará, west-central Panama (Squamata: Polychrotidae). *Herpetologica* 63:375–391.

Köhler, G., J. H. Townsend, and C.B.P. Petersen. 2016. A taxonomic revision of the *Norops tropidonotus* complex (Squamata, Dactyloidae), with the resurrection of *N. spilorhipis* (Álvarez del Toro and Smith, 1956) and the description of two new species. *Mesoamerican Herpetology* 3:8–41.

Köhler, G., J. Vargas, and S. Lotzkat. 2014c. Two new species of the *Norops pachypus* complex (Squamata, Dactyloidae) from Costa Rica. *Mesoamerican Herpetology* 1:254–280.

Kolbe, J. J., A. C. Battles, and K. J. Avilés-Rodríguez. 2016. City slickers: poor performance does not deter *Anolis* lizards from using artificial substrates in human-modified habitats. *Functional Ecology* 30:1418–1429.

Kolbe, J. J., N. Gilbert, J. T. Stroud, and Z. A. Chejanovski. 2021. An experimental analysis of perch diameter and substrate preferences of *Anolis* lizards from natural forest and urban habitats. *Journal of Herpetology* 55:215–221.

Kolbe, J. J., P. S. VanMiddlesworth, N. Losin, N. Dappen, and J. B. Losos. 2012. Climatic niche shift predicts thermal trait response in one but not both introductions of the Puerto Rican lizard *Anolis cristatellus* to Miami, Florida, USA. *Ecology and Evolution* 2:1503–1516.

Krysko, K. L., J. P. Burgess, M. R. Rochford, et al. 2011. Verified non-indigenous amphibians and reptiles in Florida from 1863 through 2010: outlining the invasion process and identifying invasion pathways and stages. *Zootaxa* 3028:1–64.

Lailvaux, S. P., A. Herrel, B. VanHooydonck, J. J. Meyers, and D. J. Irschick. 2004. Performance capacity, fighting tactics and the evolution of life–stage male morphs in the green anole lizard (*Anolis carolinensis*). *Proceedings of the Royal Society of London, Series B* 271:2501–2508.

Landauro, C. Z., and V. Morales. 2007. Ensamblaje ecológico en las lagartijas arbóreas (Squamata, Polychrotidae, *Anolis*) en la Amazonía sur del Perú. *Biotempo* 7:55–60.

Langstroth, R. P. 2006. Notas sobre *Anolis meridionalis* Boettger, 1885 (Squamata: Iguania: Polychrotidae) en Bolivia y comentarios sobre *Anolis steinbachi*. *Kempffiana* 2:154–172.

Lara-Resendiz, R. A., Y. Ramírez-Enríquez, I. Valle-Jiménez, J. H. Valdez-Villavicencio, F. R. Méndez-de la Cruz, and P. Galina-Tessaro. 2017. River rocks as sleeping perches for *Norops oxylophus* and *Basiliscus plumifrons* in the Cordillera de Talamanca, Costa Rica. *Mesoamerican Herpetology* 4:418–422.

Lara-Tufiño, J. D., A. Nieto-Montes de Oca, A. Ramírez-Bautista, and L. N. Gray. 2016. Resurrection of *Anolis ustus* Cope, 1864 from synonymy with *Anolis sericeus* Hallowell, 1856 (Squamata, Dactyloidae). *ZooKeys* 619:147–162.

Latella, I. M., S. Poe, and J. Tomasz Giermakowski. 2011. Traits associated with naturalization in *Anolis* lizards: comparison of morphological, distributional, anthropogenic, and phylogenetic models. *Biological Invasions* 13:845–856.

Lattanzio, M. 2009. Escape tactic plasticity of two sympatric *Norops* (Beta *Anolis*) species in Northeast Costa Rica. *Amphibia-Reptilia* 30:1–6.

Lazell, J. D. 1969. The genus *Phenacosaurus* (Sauria: Iguanidae). *Breviora* 325:1–24.

———. 1972. The anoles (Sauria, Iguanidae) of the lesser Antilles. *Bulletin of the Museum of Comparative Zoology* 143:1–115.

Leal, M. 1999. Honest signalling during prey–predator interactions in the lizard *Anolis cristatellus*. *Animal Behaviour* 58:521–526.

Lee, J. C. 1980. Variation and systematics of the *Anolis sericeus* complex (Sauria: Iguanidae). *Copeia* 1980:310–320.

———. 1996. *Amphibians and Reptiles of the Yucatan Peninsula*. Cornell University Press.

Lee, J. C., D. Clayton, S. Eisenstein, and I. Perez. 1989. The reproductive cycle of *Anolis sagrei* in southern Florida. *Copeia* 1989:930–937.

REFERENCES

Licht, P., and R. E. Jones. 1967. Effects of exogenous prolactin on reproduction and growth in adult males of the lizard *Anolis carolinensis*. *General and Comparative Endocrinology* 8:228–244.

Linnaeus, C. 1758. *Systema Naturae*. Stockholm: Laurentii Salvii.

Lister, B. C., and A. G. Aguayo. 1992. Seasonality, predation, and the behaviour of a tropical mainland anole. *Journal of Animal Ecology* 61:717–733.

Llano-Mejía, J., Á. M. Cortés-Gómez, and F. Castro-Herrera. 2010. Lista de anfibios y reptiles del departamento del Tolima, Colombia. *Biota Colombiana* 11:89–106.

Logan, M. L., S. G. Fernandez, and R. Calsbeek. 2015. Abiotic constraints on the activity of tropical lizards. *Functional Ecology* 29:694–700.

Logan, M. L., R. K. Huynh, R. A. Precious, and R. G. Calsbeek. 2013. The impact of climate change measured at relevant spatial scales: new hope for tropical lizards. *Global Change Biology* 19:3093–3102.

Logan, M. L., C. E. Montgomery, S. M. Boback, R. N. Reed, and J. A. Campbell. 2012. Divergence in morphology, but not habitat use, despite low genetic differentiation among insular populations of the lizard *Anolis lemurinus* in Honduras. *Journal of Tropical Ecology* 28:215–222.

Logan, M. L., L. K. Neel, D. J. Nicholson, et al. 2021. Sex-specific microhabitat use is associated with sex-biased thermal physiology in *Anolis* lizards. *Journal of Experimental Biology* 224:jeb235697.

Lombo, J. G. 1989. Caracterización de los patrones de conducta agresiva territorial del lagarto de la sabana de Bogotá *Phenacosaurus heterodermus* (Sauria: Iguanidae). *Caldasia* 76:112–118.

López-González, C. A., and A. González-Romero. 1997. The lizard community from Cozumel Island, Quintana Roo, Mexico. *Acta Zoológica Mexicana* (nueva serie) 72:27–38.

López-Herrera, D. F., M. León-Yusti, S. C. Guevara-Molina, and F. Vargas-Salinas. 2016. Reptiles in biological corridors and roadkills in Barbas-Bremen, Quindío, Colombia. *Revista de la Academia Colombiana de Ciencias Exactas, Físicas y Naturales* 40:484–493.

López-Victoria, M. 2006. Los lagartos de Malpelo (Colombia): aspectos sobre su ecología y amenazas. *Caldasia* 28:129–134.

López-Victoria, M., P. A. Herrón, and J. C. Botello. 2011. Notes on the ecology of the lizards from Malpelo Island, Colombia. *Boletín de Investigaciones Marinas y Costeras-INVEMAR* 40:79–89.

Losos, J. B. 2009. *Lizards in an Evolutionary Tree*. University of California Press.

Losos, J. B., R. M. Andrews, O. J. Sexton, and A. L. Schuler. 1991. Behavior, ecology, and locomotor performance of the giant anole, *Anolis frenatus*. *Caribbean Journal of Science* 27:173–179.

Losos, J. B., T. R. Jackman, A. Larson, K. de Queiroz, and L. Rodríguez-Schettino. 1998. Contingency and determinism in replicated adaptive radiations of island lizards. *Science* 279:2115–2118.

Losos, J. B., M. L. Woolley, D. L. Mahler, et al. 2012. Notes on the natural history of the little-known Ecuadorian horned anole, *Anolis proboscis*. *Breviora* 531:1–17.

Lotzkat, S., J.-F. Bienentreu, A. Hertz, and G. Koehler. 2011. A new species of *Anolis* (Squamata: Iguania: Dactyloidae) formerly referred to as *A. pachypus* from the Cordillera de Talamanca of western Panama and adjacent Costa Rica. *Zootaxa* 3125:1–21.

Lotzkat, S., A. Hertz, J.-F. Bienentreu, and G. Koehler. 2013. Distribution and variation of the giant alpha anoles (Squamata: Dactyloidae) of the genus *Dactyloa* in the highlands of western Panama, with the description of a new species formerly referred to as *D. microtus*. *Zootaxa* 3626:1–54.

Lotzkat, S., J. J. Köhler, A. Hertz, and G. Köhler. 2010. Morphology and colouration of male *Anolis datzorum* (Squamata: Polychrotidae). *Salamandra* 46:48–52.

Lotzkat, S., L. Stadler, A. Batista, A. Hertz, M. Ponce, N. Hamad, and G. Köhler. 2012. Distribution extension for *Anolis gruuo* Köhler, Ponce, Sunyer and Batista, 2007 (Reptilia: Squamata: Dactyloidae) in the Comarca Ngöbe-Buglé of western Panama, and first records from Veraguas province. *Check List* 8:620–625.

Lovern, M. B., M. M. Holmes, and J. Wade. 2004. The green anole (*Anolis carolinensis*): a reptilian model for laboratory studies of reproductive morphology and behavior. *Ilar Journal* 45:54–64.

Luja, V. H., S. Herrando-Pérez, D. González-Solís, and L. Luiselli. 2008. Secondary rain forests are not havens for reptile species in tropical Mexico. *Biotropica* 40:747–757.

REFERENCES

Luna-Reyes, R., C. Cundapí-Pérez, P. E. Pérez-López, A. López-Villafuerte, M. A. Rodríguez-Reyes, and J. A. Luna-Sánchez. 2017. Riqueza y diversidad de anfibios y reptiles en Nuevo San Juan Chamula y Veinte Casas, Reserva de la Biosfera Selva El Ocote. In *Vulnerabilidad social y biológica ante el cambio climático en la Reserva de la Biósfera Selva El Ocote*, edited by L. Ruiz-Montoya, G. Alvarez-Gordillo, N. Ramírez-Marcial, and B. Cruz-Salazar. Chiapas: El Colegia de la Frontera Sur (ECOSUR).

Luna-Reyes, R., R. Vidal-López, E., Hernández-García, and H. Montesino-Castillejos. 2012. Anfibios y reptiles de la región marina prioritaria Corredor Puerto Madero, Chiapas, México. In *Recursos acuáticos costeros del sureste*, edited by A. J. Sánchez, X. Chiappa-Carrara, and R. Brito-Pérez. Mérida: RECORECOS/UNAM.

Macedo, L. C., P. S. Bernarde, and A. S. Abe. 2008. Lagartos (Squamata: Lacertilia) em áreas de floresta e de pastagem em Espigão do Oeste, Rondônia, sudoeste da Amazônia, Brasil. *Biota Neotropica* 8:133–139.

Macip-Ríos, R., and A. Muñoz-Alonso. 2008. Diversidad de lagartijas en cafetales y bosque primario en el Soconusco chiapaneco. *Revista mexicana de biodiversidad* 79:185–195.

Macri, S., Y. Savriama, I. Khan, and N. Di-Poi. 2019. Comparative analysis of squamate brains unveils multi-level variation in cerebellar architecture associated with locomotor specialization. *Nature Communications* 10:1–16.

Macrini, T. E., D. J. Irschick, and J. B. Losos. 2003. Ecomorphological differences in toepad characteristics between mainland and island anoles. *Journal of Herpetology* 37:52–58.

Magnusson, W. E. 1993. Body temperatures of field-active Amazonian savanna lizards. *Journal of Herpetology* 27:53–58.

Magnusson, W. E., and E. V. da Silva. 1993. Relative effects of size, season and species on the diets of some Amazonian savanna lizards. *Journal of Herpetology* 27:380–385.

Magnusson, W. E., L. J. de Paiva, R. M. da Rocha, C. R. Franke, L. A. Kasper, and A. P. Lima. 1985. The correlates of foraging mode in a community of Brazilian lizards. *Herpetologica* 41:324–332.

Magnusson, W. E., A. P. Lima, A. S. Faria, R. L. Victoria, and L. A. Martinelli. 2001. Size and carbon acquisition in lizards from Amazonian savanna: evidence from isotope analysis. *Ecology* 82:1772–1780.

Malnate, E. V. 1971. A catalog of primary types in the herpetological collections of the Academy of Natural Sciences, Philadelphia (ANSP). *Proceedings of the Academy of Natural Sciences of Philadelphia* 123:345–375.

Manley, G. A. 2002. Evolution of structure and function of the hearing organ of lizards. *Journal of Neurobiology* 53:202–211.

Márquez, C. M., and L. D. Márquez. 2009. Reproductive biology in the wild and in captivity of *Anolis aquaticus* (Sauria: Polychrotidae) in Costa Rica. *Boletín Técnico, Serie Zoológica* 8.

Márquez, C., L. Márquez, S. Rea, and J. Márquez. 2009. Demografía de la población de *Anolis aquaticus* (Sauria Polychrotidae) de la quebrada La Palma, Puriscal, Costa Rica. *Revista Ecuatoriana de Medicina y Ciencias Biológicas: REMCB* 30:62–77.

Márquez, C., J. M. Mora, F. Bolaños, and S. Rea. 2005. Aspectos de la biología poblacional en el campo de *Anolis aquaticus*, Sauria: Polychridae en Costa Rica. *Ecología Aplicada* 4:59–69.

Márquez-Baltán, C. 1994. Historia natural de *Anolis aquaticus* Taylor 1956 (Sauria: Polychridae) en la quebrada la Palma, Puriscal, San Jose, Costa Rica. Thesis, University of Costa Rica.

Martins, M. 1991. The lizards of Balbina, Central Amazonia, Brazil: a qualitative analysis of resource utilization. *Studies on Neotropical Fauna and Environment* 26:179–190.

Mayer, G. 2010. Ernst Mayr, Ernest E. Williams, and the nondimensional species concept. In Mahler, D. L., A. Herrel, J. B. Losos. *Anolis NewsLetter VI*:144. Harvard University Press.

Mautz, W. J. 1981. Use of cave resources by a lizard community. In *Woodcock Ecology and Management*, Wildlife Research Report 13: Papers from the Seventh Woodcock Symposium Held at the Pennsylvania State University, University Park, Pennsylvania, 28–30 October 1980. US Department of the Interior, Fish and Wildlife Service.

REFERENCES

McCranie, J. R., and F. E. Castañeda. 2005. The herpetofauna of Parque Nacional Pico Bonito, Honduras. *Phyllomedusa* 4:3–16.

McCranie, J. R., and G. Kohler. 2001. A new species of anole from eastern Honduras related to *Norops tropidonotus* (Reptilia, Squamata, Polychrotidae). *Senckenbergiana biologica* 81:227–234.

McCranie, J. R., and G. Köhler. 2015. The anoles (Reptilia: Squamata: Dactyloidae: *Anolis: Norops*) of Honduras: systematics, distribution, and conservation. *Bulletin of the Museum of Comparative Zoology* 161:1–280.

McCranie, J. R., G. A. Cruz, and P. A. Holm. 1993a. A new species of cloud forest lizard of the *Norops schiedei* group (Sauria: Polychrotidae) from northern Honduras. *Journal of Herpetology* 4:386–392.

McCranie, J. R., G. Köhler, and L. D. Wilson. 1999. Two new species of anoles of the *Norops crassulus* group from Honduras (Reptilia: Sauria: Polychrotidae). *Amphibia-Reptilia* 20:279–298.

McCranie, J. R., G. Köhler, and L. D. Wilson. 2000. Two new species of anoles from northwestern Honduras related to *Norops laeviventris* Wiegmann 1834 (Reptilia, Squamata, Polychrotidae). *Senckenbergiana biologica* 80:213–224.

McCranie, J. R., K. Nicholson, and G. Köhler. 2001. A new species of *Norops* (Squamata: Polychrotidae) from northwestern Honduras. *Amphibia-Reptilia* 22:465–473.

McCranie, J. R., L. D. Wilson, and K. L. Williams. 1992. A new species of anole of the *Norops crassulus* group (Sauria: Polychrotidae) from northwestern Honduras. *Caribbean Journal of Science* 28:208–215.

McCranie, J. R., L. D. Wilson, and K. L. Williams. 1993b. Another new species of lizard of the *Norops schiedei* group (Sauria: Polychrotidae) from northern Honduras. *Journal of Herpetology* 27:393–399.

Medina, M., J. B. Fernández, P. Charruau, F. M. Cruz, and N. Ibargüengoytía. 2016. Vulnerability to climate change of *Anolis allisoni* in the mangrove habitats of Banco Chinchorro Islands, Mexico. *Journal of Thermal Biology* 58:8–14.

Medina-Rangel, G. F. 2011. Diversidad alfa y beta de la comunidad de reptiles en el complejo cenagoso de Zapatosa, Colombia. *Revista de Biología Tropical* 59:935–968.

———. 2013. Seasonal change in the use of resources of the reptilian assemblages in the Zapatosa's wetland complex, departamento del Cesar (Colombia). *Caldasia* 35:103–122.

Medina-Rangel, G. F., and G. Cárdenas-Árevalo. 2015. Relaciones espaciales y alimenticias del ensamblaje de reptiles del complejo cenagoso de Zapatosa, departamento del Cesar (Colombia). *Papéis Avulsos de Zoologia* 55:143–165.

Méndez-Galeano, M. A., and M. L. Calderón-Espinosa. 2017. Thermoregulation in the Andean lizard *Anolis heterodermus* (Squamata: Dactyloidae) at high elevation in the eastern Cordillera of Colombia. *IheringiA. Série Zoologia* 107:e2017018.

Méndez-Galeano, M. A., R. F. Paternina-Cruz, and M. L. Calderón-Espinosa. 2020. The highest kingdom of *Anolis*: thermal biology of the Andean lizard *Anolis heterodermus* (Squamata: Dactyloidae) over an elevational gradient in the Eastern Cordillera of Colombia. *Journal of Thermal Biology* 89:102498.

Merrem, B. 1820. *Versuch eines Systems der Amphibien: Tentamen systematis amphibiorum*. Marburg: Johann Christian Krieger.

Mertens, R. 1952. Die Amphibien und Reptilien von El Salvador, auf Grund der Reisen von R. Mertens und A. Zilch. *Abhandlungen der Senckenbergischen Naturforschenden Gesellschaft* 487:1–120.

Meshaka Jr., W. E. 1993. Hurricane Andrew and the colonization of five invading species in south Florida. *Florida Scientist* 56:193–201.

———. 1999a. The herpetofauna of the Doc Thomas house in South Miami, Florida. *Florida Field Naturalist* 27:121–123.

———. 1999b. The herpetofauna of the Kampong. *Florida Scientist* 27:153–157.

———. 2010. Summer anoles, some are not: differences in the activity patterns of *Anolis equestris* and *A. sagrei* in south Florida. Anolis Newsletter VI:145–151.

Meshaka Jr., W. E., and K. Rice. 2005. The Knight Anole: ecology of a successful colonizing species in extreme southern mainland Florida. In *Amphibians and Reptiles: Status and Conservation in Florida*, edited by W. E. Meshaka and K. J. Babbitt. Krieger Publishing Company, Malabar, Florida.

REFERENCES

Meshaka Jr., W. E., H. T. Smith, J. W. Gibbons, T. Jackson, M. Mandica, and K. A. Boler. 2008. An exotic herpetofaunal bioblitz survey at a state park in southern Florida. *Herpetology* 26:14–16.

Mesquita, D. O., G. R. Colli, F. G. França, and L. J. Vitt. 2006. Ecology of a Cerrado lizard assemblage in the Jalapão region of Brazil. *Copeia* 2006:460–471.

Mesquita, D. O., G. R. Colli, and L. J. Vitt. 2007. Ecological release in lizard assemblages of neotropical savannas. *Oecologia* 153:185–195.

Mesquita, D. O., G. C. Costa, A. S. Figueredo, et al. 2015. The autecology of *Anolis brasiliensis* (Squamata, Dactyloidae) in a Neotropical savanna. *The Herpetological Journal* 25:233–244.

Meyer, J. R. 1968. Distribution and variation of the Mexican lizard, *Anolis barkeri* Schmidt (Iguanidae), with redescription of the species. *Copeia* 1968:89–95.

Mitchell, B. J. 1989. Resources, group behavior, and infant development in white-faced capuchin monkeys, *Cebus capucinus*. Dissertation, University of California, Berkeley, CA.

Miyata, K. I. 1983. Notes on *Phenacosaurus heterodermus* in the Sabana de Bogotá, Colombia. *Journal of Herpetology* 17:102–105.

———. 1985. A new *Anolis* of the *lionotus* group from northwestern Ecuador and southwestern Colombia (Sauria: Iguanidae). *Breviora* 481:1–13.

———. 2013. Studies on the ecology and population biology of little known Ecuadorian Anoles. *Bulletin of the Museum of Comparative Zoology* 161:45–78.

Molina Zuluaga, C., and D. A. Gutiérrez Cárdenas. 2007. Uso nocturno de perchas en dos especies de *Anolis* (Squamata: Polychrotidae) en un bosque Andino de Colombia. *Papéis Avulsos de Zoologia* 47:273–281.

Monagan Jr., I. V., J. R. Morris, A. R. Davis Rabosky, I. Perfecto, and J. Vandermeer. 2017. *Anolis* lizards as biocontrol agents in mainland and island agroecosystems. *Ecology and Evolution* 7:2193–2203.

Monroe Jr., B. L. 1965. A distributional survey of the birds of Honduras. Dissertation, Louisiana State University and Agricultural & Mechanical College.

Montes, C., A. Cardona, C. Jaramillo, et al. 2015. Middle Miocene closure of the Central American seaway. *Science* 348:226–229.

Montgomery, C. E., E.J.G. Rodriquez, H. L. Ross, and K. R. Lips. 2011. Communal nesting in the anoline lizard *Norops lionotus* (Polychrotidae) in central Panama. *The Southwestern Naturalist* 56:83–88.

Moreno, L.E.R., J. T. Rengifo Mosquera, and J. M. Robledo. 2007. Comunidad de reptiles presente en el sotobosque de la selva pluvial central del departamento del Chocó. *Revista Institucional Universidad Tecnológica del Chocó* 26:23–36.

Moreno-Arias, R. A., and M. L. Calderón-Espinosa. 2016. Patterns of morphological diversification of mainland *Anolis* lizards from northwestern South America. *Zoological Journal of the Linnean Society* 176:632–647.

Moreno-Arias, R. A., and J. N. Urbina-Cardona. 2013. Population dynamics of the Andean Lizard *Anolis heterodermus*: fast-slow demographic strategies in fragmented scrubland landscapes. *Biotropica* 45:253–261.

Moreno-Arias, R. A., P. Bloor, and M. L. Calderón-Espinosa. 2020. Evolution of ecological structure of anole communities in tropical rain forests from north-western South America. *Zoological Journal of the Linnean Society* 190:298–313.

Moreno-Arias, R. A., G. F. Medina-Rangel, J. E. Carvajal-Cogollo, and O. V. Castaño-Mora. 2009. Herpetofauna de la Serranía de Perijá. In *Colombia Diversidad Biótica VIII: Media y Baja Montaña de la Serranía de Perijá*, edited by J. Rangel. Bogotá: Instituto de Ciencias Naturales – Universidad Nacional de Colombia – CORPOCESAR.

Moreno-Arias, R. A., M. A. Méndez-Galeano, I. Beltrán, and M. Vargas-Ramírez. 2023. Revealing anole diversity in the highlands of the Northern Andes: new and resurrected species of the *Anolis heterodermus* species group. *Vertebrate Zoology* 73:161–188.

Moritz, C., and R. Agudo. 2013. The future of species under climate change: resilience or decline? *Science* 341:504–508.

REFERENCES

Mosauer, W. 1936. The re-discovery of *Anolis gadovii*. *Herpetologica* 1:61–63.

Mothes, C. C., J. T. Stroud, S. L. Clements, and C. A. Searcy. 2019. Evaluating ecological niche model accuracy in predicting biotic invasions using south Florida's exotic lizard community. *Journal of Biogeography* 46:432–441.

Motte, M., and P. Cacciali. 2009. Descripción de un neotipo para *Anolis meridionalis* Boettger, 1885 (Sauria: Polychrotidae). *Cuadernos de Herpetología* 23:19–24.

Motte, M., K. Núñez, P. C. Sosa, F. Brusquetti, N. J. Scott, and A. L. Aquino. 2009. Categorización del estado de conservación de los anfibios y reptiles de Paraguay. *Cuadernos de Herpetología* 23:5–18.

Muñoz, A., A. Horváth, R. Percino, R. Ramírez, R. Macip, R. Martínez, M. Moreno, et al. 2002. Evaluación de la diversidad de vertebrados terrestres en cafetales de la Reserva de la Biosfera El Triunfo. Informe final. Chiapas: El Colegio de la Frontera Sur (ECOSUR-IDSMAC).

Muñoz, M. M., K. E. Crandell, S. C. Campbell-Staton, et al. 2015. Multiple paths to aquatic specialisation in four species of Central American *Anolis* lizards. *Journal of Natural History* 49:1717–1730.

Muñoz, M. M., A. Herrel, M. Sasa-Marín, and J. Losos. 2009. How similar are aquatic *Anolis* lizards: a detailed ecological and behavioral analysis of two Costa Rican species (*A. oxylophus* and *A. aquaticus*). *Integrative and Comparative Biology* 49:E121.

Munoz-Alonso, L. A. 1988. Estudio herpetofaunístico del Parque Ecológico Estatal de Omiltemi, Mpio. de Chilpancingo, Guerrero. PhD Thesis, Universidad Nacional Autónoma de Mexico.

Muñoz-Alonso, L. A., J. Nieblas-Camacho, M. A. Chau-Cortez, et al. 2017. Diversidad de anfibios y reptiles en la Reserva de la Biosfera Selva El Ocote: su vulnerabilidad ante la fragmentación y el cambio climático. In *Vulnerabilidad social y biológica ante el cambio climático en la Reserva de la Biosfera Selva El Ocote*, edited by L. Ruiz-Montoya, G.D.C.Á. Gordillo, N. Ramírez-Marcial, and B. Cruz-Salazar. Chiapas: El Colegio de la Frontera Sur (ECOSUR).

Muñoz-Nolasco, F. J., D. M. Arenas-Moreno, R. Santos-Bibiano, A. Bautista-del Moral, F. J. Gandarilla-Aizpuro, D. Brindis-Badillo, and F. Méndez–de la Cruz. 2019. Evaporative water loss of some habitat-restricted Mexican lizard species. *Herpetological Conservation and Biology* 14:51–66.

Muñoz-Nolasco, F. J., D. Cruz-Sáenz, O. J. Rodríguez-Ruvalcaba, and I. E. Terrones-Ferreiro. 2015. Notes on the herpetofauna of western Mexico 12: herpetofauna of a temperate forest in Mazamitla, southeastern Jalisco, Mexico. *Bulletin of the Chicago Herpetological Society* 50:45–50.

Myers, C. W. 1971. Central American lizards related to *Anolis pentaprion*: two new species from the Cordillera de Talamanca. *American Museum Novitates* 2471:1–40.

———. 1974. The systematics of Rhadinaea (Colubridae), a genus of New World snakes. La sistemática de Rhadinaea (Colubridae), un género de serpientes del Nuevo Mundo. *Bulletin of the American Museum of Natural History* 153:1–262.

Myers, C. W., and M. A. Donnelly. 1997. A tepui herpetofauna on a granitic mountain (Tamacuari) in the borderland between Venezuela and Brazil: report from the Phipps Tapirapecó Expedition. *American Museum Novitates* 3213:1–71.

Myers, C. W., and M. A. Donnelly. 2008. The summit herpetofauna of Auyantepui, Venezuela: report from the Robert G. Goelet American Museum–Terramar Expedition. *Bulletin of the American Museum of Natural History* 2008:1–147.

Myers, C. W., M. A. Donnelly, and R. G. Goelet. 1996. A new herpetofauna from Cerro Yaví, Venezuela: first results of the Robert G. Goelet American Museum–Terramar Expedition to the northwestern tepuis. *American Museum Novitates* 3172:1–56.

Myers, C. W., E. E. Williams, and R. W. McDiarmid. 1993. A new anoline lizard (*Phenacosaurus*) from the highland of Cerro de la Neblina, Southern Venezuela. *American Museum Novitates* 3070:1–15.

Myers, G. S., and A. Leitão de Carvalho. 1945. A strange leaf-nosed lizard of the genus *Anolis* from Amazonia. *Boletim Do Museu Nacional Zoologia* 43:1–44.

REFERENCES

Myers, M. F. 1997. Morphometry of the limb skeleton of *Anolis garmani*: practical and theoretical implications. Master's thesis, University of Calgary.

Nantarat, N., P. Tongkerd, C. Sutcharit, C. M. Wade, F. Naggs, and S. Panha. 2014. Phylogenetic relationships of the operculate land snail genus *Cyclophorus* Montfort, 1810 in Thailand. *Molecular Phylogenetics and Evolution* 70:99–111.

Narváez, A. E. 2017. Processes influencing community structure in *Anolis* lizards (Dactyloidae) from Ecuador. Dissertation, La Trobe University.

Narváez, A. E., T. Ghia, M. M. Moretta-Urdiales, and N. M. Moreira. 2020. Feeding habits of *Anolis sagrei*, an introduced species, in urban ecosystems of Guayas Province. *Urban Ecosystems* 23:1371–1376.

Narváez, A. E., A. Marmol, and A. Argoti. 2019. Blow fly infestation on *Anolis parvauritus*: Notes of the effects of myasis on lizard's behaviour. *Herpetology Notes* 12:847–852.

Neel, L. K., M. L. Logan, D. J. Nicholson, et al. 2021. Habitat structure mediates vulnerability to climate change through its effects on thermoregulatory behavior. *Biotropica* 53:1121–1133.

Nicholson, K. E., and G. Köhler. 2014. A new species of the genus *Norops* from Darién. *Zootaxa* 3895:225–237.

Nicholson, K. E., and P. M. Richards. 2011. Home-range size and overlap within an introduced population of the Cuban Knight Anole, *Anolis equestris* (Squamata: Iguanidae). *Phyllomedusa* 10:65–73.

Nicholson, K. E., B. I. Crother, C. Guyer, and J. M. Savage. 2012. It is time for a new classification of anoles (Squamata: Dactyloidae). *Zootaxa* 3477:1–108.

Nicholson, K. E., R. E. Glor, J. J. Kolbe, A. Larson, S. B. Hedges, and J. B. Losos. 2005. Mainland colonization by island lizards. *Journal of Biogeography* 32:929–938.

Nicholson, K. E., R. Ibañez, C. A. Jaramillo, and K. R. Lips. 2001. Morphological variation in the tropical anole, *Anolis casildae* (Squamata: Polychrotidae). *Revista de Biología Tropical* 49:709–714.

Nicholson, K. E., J. McCranie, and G. Köhler. 2001. A new species of *Norops* (Squamata: Polychrotidae) from northwestern Honduras. *Amphibia-Reptilia* 22:465–473.

Nieto-Montes de Oca, A. 1994a. A taxonomic review of the *Anolis schiedii* group (Squamata: Polychrotidae). PhD Thesis, University of Kansas.

———. 1994b. Rediscovery and redescription of *Anolis schiedii* (Wiegmann) (Squamata: Polychridae) from central Veracruz. *Herpetologica* 50:325–335.

———. 1995. Key to the species of the *Anolis schiedii* group south and east of the Isthmus of Tehuantepec. In *Recopilación de Claves para la Determinación de Anfibios y Reptiles de México*, vol. 10, edited by O. A. Flores Villella, F. Mendoza Quijano, and G. Gonzalez Porter. Publicaciones Especiales de la Museo Zoologia Facultad de Ciencias, UNAM.

———. 1996. A new species of *Anolis* (Squamata: Polychrotidae) from Chiapas, México. *Journal of Herpetology* 30:19–27.

———. 2001. The systematics of *Anolis hobartsmithi* (Squamata: Polychrotidae), another species of the *Anolis schiedii* group from Chiapas, Mexico. *Mesoamerican Herpetology: Systematics, Zoogeography, and Conservation*. Centennial Museum, University of Texas at El Paso, Special Publication 1:1–200.

Nieto-Montes de Oca, A., S. Poe, S. Scarpetta, L. Gray, and C. S. Lieb. 2013. Synonyms for some species of Mexican anoles (Squamata: Dactyloidae). *Zootaxa* 3637:484–492.

Nogueira, C., G. R. Colli, and M. Martins. 2009. Local richness and distribution of the lizard fauna in natural habitat mosaics of the Brazilian Cerrado. *Austral Ecology* 34:83–96.

Nogueira, C., P. H. Valdujo, and F.G.R. França. 2005. Habitat variation and lizard diversity in a cerrado area of central Brazil. *Studies on Neotropical Fauna and Environment* 40:105–112.

Northcutt, R. G. 2002. Understanding vertebrate brain evolution. *Integrative and Comparative Biology* 42:743–756.

Norval, G., W.-F. Hsiao, S.-C. Huang, and C. K. Chen. 2009. The diet of an introduced lizard species, the brown anole (*Anolis sagrei*), in Chiayi county, Taiwan. *Russian Journal of Herpetology* 17:131–138.

REFERENCES

Norval, G., J.-J. Mao, H.-P. Chu, and L.-C. Chen. 2002. A new record of an introduced species, the brown anole (*Anolis sagrei* Duméril & Bibron 1837). *Zoological Studies-Taipei* 41:332–336.

Nunez, S. C., and T. A. Jenssen. 1998. Spatial and breeding relationships of the lizard, *Anolis carolinensis*: evidence of intrasexual selection. *Behaviour* 135:981–1003.

Nunez, S. C., T. A. Jenssen, and K. Ersland. 1997. Female activity profile of a polygynous lizard (*Anolis carolinensis*): evidence of intersexual asymmetry. *Behaviour* 134:205–223.

Nyffeler, M., G. B. Edwards, and K. L. Krysko. 2017. A vertebrate-eating jumping spider (Araneae: Salticidae) from Florida, USA. *The Journal of Arachnology* 45:238–241.

O'Bryant, E. L., and J. Wade. 2001. Sexual dimorphism in neuromuscular junction size on a muscle used in courtship by green anole lizards. *Journal of Neurobiology* 50:24–30.

O'Dea, A., H. A. Lessios, A. G. Coates, et al. 2016. Formation of the Isthmus of Panama. *Science Advances* 2:e1600883–e1600883.

Odendaal, F. J., M. D. Rausher, B. Benrey, and J. Nunez-Farfan. 1987. Predation by *Anolis* lizards on *Battus philenor* raises questions about butterfly mimicry systems. *Journal of the Lepidopterists' Society* 41:141–144.

Odum, E. P. 1971. *Fundamentals of Ecology*. 3d ed. W. B. Saunders Company.

Oliveira, J. A., and L.J.C.L. Moraes. 2021. Mating behavior of *Anolis punctatus* (Squamata: Dactyloidae) in the Brazilian Amazonia. *Phyllomedusa* 20:185–190.

Oliveira, J.C.F., T. M. de Castro, D. Vrcibradic, M. C. Drago, and I. Prates. 2018. A second Caribbean anole lizard species introduced to Brazil. *Herpetology Notes* 11:761–764.

Oppenheim, V. 1952. The structure of Colombia. *American Geophysical Union Transactions* 33:739–748.

Orrell, K., and T. Jenssen. 2003. Heterosexual signaling by the lizard *Anolis carolinensis*, with intersexual comparisons across contexts. *Behaviour* 140:603–634.

Orton, J. 1876. *The Andes and the Amazon*. Harper and Brothers.

Osgood, W. H. 1912. Mammals from western Venezuela and eastern Colombia. *Field Museum of Natural History Zoological Publications* 10:33–66.

O'Shaughnessy, A.W.E. 1875. List and revision of the species of Anolidæ in the British-Museum collection, with descriptions of new species. *Annals and Magazine of the Museum of Natural History* 15:270–281.

Osorno-Mesa, H., and E. Osorno-Mesa. 1946. Anotaciones sobre lagartos del género *Phenacosaurus*. *Caldasia* 17:123–130.

Oswandel, J. J. 2010. *Notes of the Mexican War, 1846–1848*. University of Tennessee Press.

Paemelaere, E. A., C. Guyer, and F. S. Dobson. 2011. Survival of alternative dorsal-pattern morphs in females of the anole *Norops humilis*. *Herpetologica* 67:420–427.

Parfit, D. 2011. *On What Matters*. Oxford University Press.

Park, O. 1938. Studies in Nocturnal Ecology, VII. Preliminary observations on Panama rain forest animals. *Ecology* 19:208–223.

Parker, H. W. 1926. The reptiles and batrachians of Gorgona Island, Colombia. *Annals and Magazine of Natural History* 17:549–554.

Parmelee, J. R., and C. Guyer. 1995. Sexual differences in foraging behavior of an anoline lizard, *Norops humilis*. *Journal of Herpetology* 29:619–621.

Paterson, A. V. 1999. Effects of prey availability on perch height of female bark anoles, *Anolis distichus*. *Herpetologica* 66:242–247.

Pavón-Vázquez, C. J., I. Solano-Zavaleta, and L. N. Gray. 2014. Morphological variation and natural history of *Anolis duellmani* (Squamata: Dactyloidae). *Mesoamerican Herpetology* 1:146–153.

Paynter, R. A. 1982. *Ornithological Gazetteer of Venezuela*. Museum of Comparative Zoology, Harvard University.

———. 1993. *Ornithological Gazetteer of Ecuador*. Museum of Comparative Zoology, Harvard University.

REFERENCES

———. 1997. *Ornithological Gazetteer of Colombia*. Museum of Comparative Zoology, Harvard University.

Perdomo, J., and E. Marca. 2016. *Anolis jacare* (Little Andean Chameleon) reproduction. *Herpetological Review* 47:293.

Pérez-Higareda, G., H. M. Smith, and D. Chiszar. 1997. *Anolis pentaprion* (lichen anole). Frugivory and cannibalism. *Herpetological Review* 28:201–202.

Perez-Martinez, C. A., A. Kamath, A. Herrel, and J. B. Losos. 2021. The anoles of La Selva: niche partitioning and ecological morphology in a mainland community of *Anolis* lizards. *Breviora* 570:1–27.

Pérez-Rojas, D. A., D. Escamilla-Quitián, M. F. Estupiñan-Tibaduiza, and J. E. Carvajal-Cogollo. 2020. Annotated checklist of the amphibians and reptiles of the Santander highland, Colombia. *Check List* 16:611–620.

Perry, G. 1995. The evolutionary ecology of lizard foraging: a comparative study. Dissertation, University of Texas, Austin.

———. 1996. The evolution of sexual dimorphism in the lizard *Anolis polylepis* (Iguania): evidence from intraspecific variation in foraging behavior and diet. *Canadian Journal of Zoology* 74:1238–1245.

Peters, J. A., and R. Donoso-Barros. 1970. Catalogue of the Neotropical Squamata, part II: lizards and amphisbaenians. *Bulletin of the United States National Museum* 297. Smithsonian Institution Press.

Peters, J. A., and G. Orcés-V. 1956. A third leaf-nosed species of the lizard genus *Anolis* from South America. *Breviora* 62:1–8.

Peters, W.C.H. 1863. Derselbe machte eine Mittheilung über einige neue Arten der Saurier-Gattung *Anolis*. *Monatsberichte der Königlich Preussischen Akademie der Wissenschaften zu Berlin* 1863:135–149.

———. 1874. Über neue Saurier (*Spæriodactylus, Anolis, Phrynosoma, Tropidolepisma, Lygosoma, Ophioscincus*) aus Centralamerica, Mexico und Australien. *Monatsberichte der Koniglichen Preussischen Akademie der Wissenschaften zu Berlin* 1863:738–747.

Peterson, A. T., L. C. Marquez, J.L.C. Jiménez, et al. 2004. A preliminary biological survey of Cerro Piedra Larga. *Anales del Instituto de Biología, Universidad Nacional Autónoma de México Serie Zoología* 75:439–466.

Peterson, J. A. 1983. The evolution of the subdigital pad of *Anolis* 2. Comparisons among the iguanid genera related to the anolines and a view from outside the radiation. *Journal of Herpetology* 17:371–397.

———. 1984. The microstructure of the scale surface in iguanid lizards. *Journal of Herpetology* 18:437–467.

Phillips, J. G., S. E. Burton, M. M. Womack, E. Pulver, and K. E. Nicholson. 2019. Biogeography, systematics, and ecomorphology of Pacific island anoles. *Diversity* 11:141.

Phillips, J. G., J. Deitloff, C. Guyer, S. Huetteman, and K. E. Nicholson. 2015. Biogeography and evolution of a widespread Central American lizard species complex: *Norops humilis* (Squamata: Dactyloidae). *BMC Evolutionary Biology* 15:1–13.

Pianka, E. R. 1994. *The Lizard Man Speaks*. University of Texas Press.

Pianka, E. R., and L. J. Vitt. 2003. *Lizards: Windows to the Evolution of Diversity*. University of California Press.

Pinilla-Renteria, E., J. T. Rengifo Mosquera, and J. S. Londoño. 2015. Dimorfismo, uso de hábitat y dieta de *Anolis maculiventris* (Lacertilia: Dactyloidae), en bosque pluvial tropical del Chocó, Colombia. *Acta Biológica Colombiana* 20:89–100.

Pinto, J., and O. Torres-Carvajal. 2023. *Basic and Applied Herpetology* 37:107–114

Pinto Aguirre, J. A. 2014. Ecología de una comunidad de lagartijas del género *Anolis* en el Parque Nacional Yasuní. Thesis, Pontificia Universidad Católica del Ecuador.

Pinto-Erazo, M. A., M.L.C. Espinosa, G.F.M. Rangel, and M.Á.M. Galeano. 2020. Herpetofauna from two municipalities of southwestern Colombia. *Biota Colombiana* 21:41–57.

REFERENCES

Pisani, G. R. 1973. A guide to preservation techniques for amphibians and reptiles. *Herpetological Circulars* 1:1–22.

Poe, S. 1998. Skull characters and the cladistic relationships of the Hispaniolan dwarf twig *Anolis*. *Herpetological Monographs* 12:192–236.

———. 2004. Phylogeny of anoles. *Herpetological Monographs* 18:37–89.

———. 2014. The travels of Thomas Barbour on the ship *Utowana* in 1931 and the taxonomic status of *Anolis utowanae*. *Breviora* 538:1–9.

———. 2016. Review of the anoles (Reptilia: Squamata: Dactyloidae: *Anolis*: *Norops*) of Honduras. *The Quarterly Review of Biology* 91:227–228.

Poe, S. Manuscript. The AnoleKey: a computer program to identify *Anolis* (Squamata: Anolidae). http://www.stevenpoe.net.

Poe, S., and C. G. Anderson. 2019. The existence and evolution of morphotypes in *Anolis* lizards: coexistence patterns, not adaptive radiations, distinguish mainland and island faunas. *PeerJ* 6:e6040.

Poe, S., and B. Armijo. 2014. Lack of effect of herpetological collecting on the population structure of a community of *Anolis* (Squamata: Dactyloidae) in a disturbed habitat. *Herpetology Notes* 7:153–157.

Poe, S., and R. Ibáñez. 2007. A new species of *Anolis* lizard from the Cordillera de Talamanca of western Panama. *Journal of Herpetology* 41:263–270.

Poe, S., and M. J. Ryan. 2017. Description of two new species similar to *Anolis insignis* (Squamata: Iguanidae) and resurrection of *Anolis* (*Diaphoranolis*) *brooksi*. *Amphibian and Reptile Conservation* 11:1–16.

Poe, S., and C. Yañez-Miranda. 2007. A new species of phenacosaur *Anolis* from Peru. *Herpetologica* 63:219–223.

———. 2008. Another new species of green *Anolis* (Squamata: Iguania) from the eastern Andes of Peru. *Journal of Herpetology* 42:564–571.

Poe, S., F. Ayala-Varela, I. M. Latella, et al. 2012. Morphology, phylogeny, and behavior of *Anolis proboscis*. *Breviora* 530:1–11.

Poe, S., M. Cash, A. J. Aguilar Kirigin, and O. Torres-Carvajal. Manuscript. *Anolis fuscoauratus* in Bolivia and Peru: redescription from an inferred type locality, phylogenetics, dewlap color variation, and the status of *Anolis bocourtii* and *Anolis scapularis*.

Poe, S., I. M. Latella, M. J. Ryan, and E. W. Schaad. 2009a. A new species of *Anolis* lizard (Squamata, Iguania) from Panama. *Phyllomedusa* 8:81–87.

Poe, S., I. Latella, F. Ayala-Varela, C. Yañez-Miranda, and O. Torres-Carvajal. 2015a. A new species of phenacosaur *Anolis* (Squamata; Iguanidae) from Peru and a comprehensive phylogeny of Dactyloa-clade *Anolis* based on new DNA sequences and morphology. *Copeia* 103:639–650.

Poe, S., A. Nieto Montes de Oca, O. Torres-Carvajal, et al. 2017a. A phylogenetic, biogeographic, and taxonomic study of all extant species of *Anolis* (Squamata; Iguanidae). *Systematic Biology* 66:663–697.

Poe, S., S. Scarpetta, and E. W. Schaad. 2015b. A new species of *Anolis* (Squamata: Iguanidae) from Panama. *Amphibian and Reptile Conservation* 9:1–13.

Poe, S., J. Velasco, K. Miyata, and E. E. Williams. 2009b. Descriptions of two nomen nudum species of *Anolis* lizard from northwestern South America. *Breviora* 516:1–16.

Poe, S., C. Yañez-Miranda, and E. Lehr. 2008. Notes on variation in *Anolis boettgeri* Boulenger 1911, assessment of the status of *Anolis albimaculatus* Henle and Ehrl 1991, and description of a new species of *Anolis* (Squamata: Iguania) similar to *Anolis boettgeri*. *Journal of Herpetology* 42:251–259.

Ponce, M., and G. Köhler. 2008. Morphological variation in anoles related to *Anolis kemptoni* in Panama. *Salamandra* 44:65–84.

Poulin, B., G. Lefebvre, R. Ibáñez, C. Jaramillo, C. Hernández, and A. S. Rand. 2001. Avian predation upon lizards and frogs in a neotropical forest understorey. *Journal of Tropical Ecology* 17:21–40.

REFERENCES

Pounds, J. A. 1988. Ecomorphology, locomotion, and microhabitat structure: patterns in a tropical mainland *Anolis* community. *Ecological Monographs* 58:299–320.
Pounds, J. A., and M. P. Fogden. 2000. Appendix 8: Amphibians and Reptiles of Monteverde. In *Monteverde: Ecology and Conservation of a Tropical Cloud Forest*, edited by N. M. Nadkarni and N. T. Wheelwright. Oxford University Press.
Pounds, J. A., M. P. Fogden, and J. H. Campbell. 1999. Biological response to climate change on a tropical mountain. *Nature* 398:611–615.
Powell, R. 2003. Species profile: Utila's reptiles. *Iguana* 10:36–38.
Powell, R., and R. A. Birt. 2001. *Anolis barkeri*. *Catalogue of American Amphibians and Reptiles* 727:1–3.
Prado-Irwin, S. 2022. Phylogeography and signal evolution in a widespread Central American anole. Dissertation, Harvard University.
Prates, I. 2017. Climate-driven habitat shifts in South America: biogeography, historical demography, and population genomics of anole lizards. Dissertation, City University of New York.
Prates, I., P. R. Melo-Sampaio, L. de Oliveira Drummond, M. Teixeira Jr., M. T. Rodrigues, and A. C. Carnaval. 2017. Biogeographic links between southern Atlantic Forest and western South America: rediscovery, re-description, and phylogenetic relationships of two rare montane anole lizards from Brazil. *Molecular Phylogenetics and Evolution* 113:49–58.
Prates, I., P. R. Melo-Sampaio, K. de Queiroz, A. C. Carnaval, M. T. Rodrigues, and L. Oliveira Drummond. 2020. Discovery of a new species of *Anolis* lizards from Brazil and its implications for the historical biogeography of montane Atlantic Forest endemics. *Amphibia-Reptilia* 41:87–103.
Prates, I., A. Penna, M. T. Rodrigues, and A. C. Carnaval. 2018. Local adaptation in mainland anole lizards: integrating population history and genome-environment associations. *Ecology and Evolution* 8:11932–11944.
Prates, I., M. T. Rodrigues, P. R. Melo-Sampaio, and A. C. Carnaval. 2015. Phylogenetic relationships of Amazonian anole lizards (*Dactyloa*): taxonomic implications, new insights about phenotypic evolution and the timing of diversification. *Molecular Phylogenetics and Evolution* 82:258–268.
Pratt, C. M. 1948. The morphology of the ethmoidal region of *Sphenodon* and lizards. *Proceedings of the Zoological Society of London* 118:171–201.
Pruett, J. E., A. Fargevieille, and D. A. Warner. 2020. Temporal variation in maternal nest choice and its consequences for lizard embryos. *Behavioral Ecology* 31:902–910.
Putman, B. J., K. R. Azure, and L. Swierk. 2019. Dewlap size in male water anoles associates with consistent inter-individual variation in boldness. *Current Zoology* 65:189–195.
Quirola, D. R., A. Mármol, O. Torres-Carvajal, A. E. Narváez, F. Ayala-Varela, and I. T. Moore. 2017. Use of a rostral appendage during social interactions in the Ecuadorian *Anolis proboscis*. *Journal of Natural History* 51:1625–1638.
Ramírez-Bautista, A. 2002. *Anolis nebulosus* (Wiegmann 1834) Lagartija arborícola. In *Historia Natural de Chamela*, edited by F. A. Noguera, J. H. Vega Rivera, A. N. García Aldrete, and M. Quesada Avendaño. Coyoacan: Instituto de Biología, UNAM.
———. 2003. Some reproductive characteristics of a tropical arid lizard assemblage from Zapotitlán Salinas, Puebla, México. *Herpetological Review* 34:328–330.
Ramírez-Bautista, A., and M. Benabib. 2001. Perch height of the arboreal lizard *Anolis nebulosus* (Sauria: Polychrotidae) from a tropical dry forest of Mexico: effect of the reproductive season. *Copeia* 2001:187–193.
Ramírez-Bautista, A., and R. Cruz-Elizalde. 2013. Reptile community structure in two fragments of cloud forest of the Sierra Madre Oriental, Mexico. *Northwestern Journal of Zoology* 9:410–417.
Ramírez-Bautista, A., and L. J. Vitt. 1997. Reproduction in the lizard *Anolis nebulosus* (Polychrotidae) from the Pacific coast of Mexico. *Herpetologica* 53:423–431.
Ramírez-Bautista, A., U. Hernández-Salinas, and R. C. Elizalde. 2017. Patterns of temporal variation in growth rate from a mainland population of *Anolis nebulosus* (Squamata: Dactyloidae. Mexican Pacific Coast. *PeerJ Preprints* 5:e3092v3091.

REFERENCES

Ramírez-Bautista, A., L. Oliver López, and V. Mata-Silva. 2002. *Anolis quercorum* (Gray Anole). General Ecology. *Herpetological Review* 33:203–204.

Ramírez-Jaramillo, S. R. 2018. Microhábitats nocturnos en dos especies de *Anolis* (Iguania: Dactyloidae) al noroccidente de Pichincha. Ecuador. *Revista Biodiversidad Neotropical* 8:7–13.

Ramírez-Perilla, J., G. de Pérez, M. P. Ramírez Pinilla, and M. Vargas. 1991. Ciclo ovárico de *Phenacosaurus heterodermus* (Sauria: Iguanidae) con relación a niveles circulantes de lipoproteínas séricas y variación anual de lluvias. *Trianea* 4:513–526.

Ramírez-Pinilla, M. P., G. de Pérez, and J. Ramírez-Perilla. 1989. Histología del tracto reproductivo de la hembra del lagarto *Phenacosaurus heterodermus* (Reptilia: Sauria: Iguanidae). *Trianea* 3:93–103.

Rand, A. S., and S. S. Humphrey. 1968. Interspecific competition in the tropical rain forest: ecological distribution among lizards at Belém, Pará. *Proceedings of the United States National Museum* 125:1–17.

Rand, A. S., and C. W. Myers. 1990. The herpetofauna of Barro Colorado Island, Panama: an ecological summary. In *Four Neotropical Rainforests*, edited by A. H. Gentry. Yale University Press.

Rand, A. S., and P. J. Rand. 1967. Field notes on *Anolis lineatus* in Curaçao. *Studies on the Fauna of Curaçao and other Caribbean Islands* 24:112–117.

Rand, A. S., G. Gorman, and W. Rand. 1975. Natural history, behavior, and ecology of *Anolis agassizi*. *Smithsonian Contributions in Zoology* 176:27–38.

Rand, A. S., S. Guerrero, and R. M. Andrews. 1983. The ecological effects of malaria on populations of the lizard *Anolis limifrons* on Barro Colorado Island, Panama. In *Advances in Herpetology and Evolutionary Biology*, edited by A.G.J. Rhodin and K. Miyata. Museum of Comparative Zoology, Harvard University.

Ream, K., and K. Reider. 2013. *Anolis oxylophus* (stream anole), peccary wallows as novel habitat. *Herpetological Review* 44:313–314.

Reinhardt, J. T., and C. F. Lütken. 1862. Bidrag til det vestindiske Öriges og navnligen de dansk-vestindiske Öers Herpetologie. *Videnskabelige Meddelelser fra den naturhistoriske Forening i Kjöbenhavn* IV:153–291.

Remsen Jr, J. V. 1977. On taking field notes. *American Birds* 31(5):946–953.

Rengifo Mosquera, J. T., and L.E.R. Moreno. 2011. Reptiles del departamento del Chocó, Colombia. *Revista Biodiversidad Neotropical* 1:38–47.

Rengifo M., J. T., and L. E. Rentería-Moreno. 2012. Reptiles del departamento del Chocó, Colombia. *Revista Biodiversidad Neotropical* 1:38–47.

Rengifo M., J. T., F. Castro-Herrera, and F.J.P. Iraizos. 2015. Uso de hábitat y relaciones ecomorfológicas de un ensamble de *Anolis* (Lacertilia: Dactyloidae) en la región natural Chocoana, Colombia. *Acta Zoológica Mexicana* 31:159–172.

Rengifo M., J. T., F. Castro-Herrera, and F. J. Purroy. 2014. Diversidad de una comunidad de *Anolis* (Iguania: Dactyloidae) en la selva pluvial central, departamento del Chocó, Colombia. *Basic and Applied Herpetology* 28:51–63.

Rengifo M., J. T., F. Castro-Herrera, and M. Y. Rengifo Palacios. 2019. Importance of the genus *Anolis* (Lacertilia: Dactyloidae), as indicators of habitat status, in tropical rain forest of Chocó. *Revista Colombiana de Ciencia Animal RECIA* 11:67–79.

Rengifo P., M. Y., J. T. Rengifo M., and J. E. Serna. 2021. Diversidad de *Anolis* (Lacertilia: Dactyloidae) en bosque pluvial tropical, del Chocó-Colombia. *Revista Colombiana de Ciencia Animal RECIA* 13:27–36.

Reynolds, R. G., A. R. Puente-Rolón, A. L. Castle, M. Schoot, and A. J. Geneva. 2018. Herpetofauna of Cay Sal Bank, Bahamas and phylogenetic relationships of *Anolis fairchildi*, *Anolis sagrei*, and *Tropidophis curtus* from the region. *Breviora* 560:1–19.

Rheubert, J. L., D. S. Siegel, and S. E. Trauth. 2014. *Reproductive Biology and Phylogeny of Lizards and Tuatara*. CRC Press.

REFERENCES

Ribeiro-Júnior, M. A. 2015. Catalogue of distribution of lizards (Reptilia: Squamata) from the Brazilian Amazonia. I. Dactyloidae, Hoplocercidae, Iguanidae, Leiosauridae, Polychrotidae, Tropiduridae. *Zootaxa* 3983:1–110.

Ríos, E. E., C. F. Hurtado P., J. T. Rengifo M., and F. Castro-Herrera. 2011. Lagartos en comunidades naturales de dos localidades en la región del Chocó de Colombia. *Herpetotropicos* 5:85–92.

Ríos-Orjuela, J. C., J. S. Camacho-Bastidas, and A. Jerez. 2020. Appendicular morphology and locomotor performance of two morphotypes of continental anoles: *Anolis heterodermus* and *Anolis tolimensis*. *Journal of Anatomy* 236:252–273.

Ríos Rodas, L., M. R. Barragán Vázquez, M.A.T. Pérez, and D.I.T. Ramírez. 2017. Ampliación de distribución de *Anolis compressicauda* Smith & Kerster, 1955 (Squamata: Dactyloidae) en el estado de Tabasco, México. *Acta Zoológica Mexicana* Nueva Serie 33:120–122.

Rivas, G. A., C. R. Molina, G. N. Ugueto, T. R. Barros, C. L. Barrio-Amorós, and P. J. Kok. 2012. Reptiles of Venezuela: an updated and commented checklist. *Zootaxa* 3211:1–64.

Robinson, D. C. 1962. Notes on the lizard *Anolis barkeri* Schmidt. *Copeia* 1962:640–642.

Rodrigues, M. T., V. Xavier, G. Skuk, and D. Pavan. 2002. New specimens of *Anolis phyllorhinus* (Squamata, Polychrotidae): the first female of the species and of proboscid anoles. *Papéis Avulsos de Zoologia* 42:363–380.

Rodríguez Schettino, L. 1999. *Iguanid Lizards of Cuba*. University Press of Florida.

Rogowitz, G. L. 1996. Evaluation of thermal acclimation of metabolism in two eurythermal lizards, *Anolis cristatellus* and *A. sagrei*. *Journal of Thermal Biology* 21:11–14.

Rojas-Runjaic, F. J., M. C. Castellanos-Montero, I. Márquez, Y. Roos-Arzola, D. Rojas, D. Quihua, and E. E. Infante-Rivero. 2023. Confirmation of the presence of *Anolis gaigei* Ruthven, 1916 (Squamata, Anolidae) in Venezuela and new distribution records. *Memoria de la Fundación La Salle de Ciencias Naturales* 81 (191).

Rubio-Rocha, L. C., B. C. Bock, and V. P. Páez. 2011. Continuous reproduction under a bimodal precipitation regime in a high elevation anole (*Anolis mariarum*) from Antioquia, Colombia. *Caldasia* 33:91–104.

Ruby, D. E. 1984. Male breeding success and differential access to females in *Anolis carolinensis*. *Herpetologica* 40:272–280.

Rueda-Almonacid, J. 1989. Un nuevo y extraordinario saurio de color rojo (Iguanidae: *Anolis*) para la Cordillera Occidental de Colombia. *Trianea* 3:85–92.

Rueda-Almonacid, J., and J. Hernández-Camacho. 1988. *Phenacosaurus inderenae* (Sauria: Iguanidae), nueva especie gigante, proveniente de la Cordillera Oriental de Colombia. *Trianea* 2:339–350.

Russildi, G., V. Arroyo-Rodríguez, O. Hernández-Ordóñez, E. Pineda, and V. H. Reynoso. 2016. Species- and community-level responses to habitat spatial changes in fragmented rainforests: assessing compensatory dynamics in amphibians and reptiles. *Biodiversity and Conservation* 25:375–392.

Ruthven, A. G. 1916. Three new species of *Anolis* from the Santa Marta Mountains, Colombia. *Occasional Papers of the Museum of Zoology, University of Michigan* 32:1–8.

———. 1922. The amphibians and reptiles of the Sierra Nevada de Santa Marta, Colombia. *Miscellaneous Publications of the Museum of Zoology, University of Michigan* 8:1–69.

Ryan, M. J., and S. Poe. 2014. Seasonal shifts in relative density of the lizard *Anolis polylepis* (Squamata, Dactyloidae) in forest and riparian habitats. *Journal of Herpetology* 48:495–499.

Sabaj, M. H. 2020. Codes for natural history collections in ichthyology and herpetology. *Copeia* 108(3):593–669.

Salazar, J. C., M. R. Castañeda, G. A. Londoño, B. L. Bodensteiner, and M. M. Muñoz. 2019. Physiological evolution during adaptive radiation: a test of the island effect in *Anolis* lizards. *Evolution* 73:1241–1252.

Salzburg, M. A. 1984. *Anolis sagrei* and *Anolis cristatellus* in southern Florida: a case study in interspecific competition. *Ecology* 65:14–19.

Sanford, G. M., W. I. Lutterschmidt, and V. H. Hutchison. 2002. The comparative method revisited. *BioScience* 52:830–836.

REFERENCES

Santos, D. L., S. P. de Andrade, E. P. Victor-Jr, and W. Vaz-Silva. 2014. Amphibians and reptiles from southeastern Goiás, Central Brazil. *Check List* 10:131–148.

Sarkar, S. 2005. *Biodiversity and Environmental Philosophy: An Introduction*. Cambridge University Press.

Savage, J. M. 1973. Herpetological collections made by Dr. John F. Bransford, Assistant Surgeon, USN during the Nicaragua and Panama Canal surveys (1872–1885). *Journal of Herpetology* 7:35–38.

———. 1974. Type localities for species of amphibians and reptiles described from Costa Rica. *Revista de Biología Tropical* 22:71–122.

———. 2002. *The Amphibians and Reptiles of Costa Rica: A Herpetofauna Between Two Continents, Between Two Seas*. University of Chicago Press.

Savage, J. M., and J. J. Talbot. 1978. The giant anoline lizards of Costa Rica and western Panama. *Copeia* 1978:480–492.

Savit, A. Z. 2006. Reptiles of the Santa Lucía cloud forest, Ecuador. *Iguana* 13:94–103.

Scarpetta, S., L. Gray, A. Nieto Montes de Oca, et al. 2015. Morphology and ecology of the Mexican cave anole *Anolis alvarezdeltoroi*. *Mesoamerican Herpetology* 2:261–270.

Schlaepfer, M. A. 2006. Growth rates and body condition in *Norops polylepis* (Polychrotidae) vary with respect to sex but not mite load. *Biotropica* 38:414–418.

Schmidt, K. P. 1933. Amphibians and reptiles collected by the Smithsonian Biological Survey of the Panama Canal Zone. *Smithsonian Miscellaneous Collection* 89 (1):1–20.

———. 1936. New amphibians and reptiles from Honduras in the Museum of Comparative Zoology. *Proceedings of the Biological Society of Washington* 49:43–50.

———. 1939. A new lizard from Mexico with a note on the genus *Norops*. *Zoological Series of the Field Museum of Natural History* 24:7–10.

Schneider, C. J., J. B. Losos, and K. de Queiroz. 2001. Evolutionary relationships of the *Anolis bimaculatus* group from the northern Lesser Antilles. *Journal of Herpetology* 35:1–12.

Schwartz, A., and O. H. Garrido. 1972. The lizards of the *Anolis equestris* complex in Cuba. *Studies on the Fauna of Curaçao and other Caribbean Islands* 39:1–86.

Schwartz, A., and R. W. Henderson. 1991. *Amphibians and Reptiles of the West Indies: Descriptions, Distributions, and Natural History*. University Press of Florida.

Schwenk, K. 1985. Occurrence, distribution and functional significance of taste buds in lizards. *Copeia* 1985:91–101.

Scott Jr., N. J., D. E. Wilson, C. Jones, and R. M. Andrews. 1976. The choice of perch dimensions by lizards of the genus *Anolis* (Reptilia, Lacertilia, Iguanidae). *Journal of Herpetology* 10:75–84.

Serrano, J. M. 2018. La condición física del lagarto *Anolis cupreus* (Squamata: Dactyloidae) y su relación con la estructura del hábitat durante la estación seca. *Cuadernos de Herpetología* 32:119–121.

Sexton, O. J. 1967. Population changes in a tropical lizard *Anolis limifrons* on Barro Colorado Island, Panama Canal Zone. *Copeia* 1967:219–222.

Sexton, O. J., and H. F. Heatwole. 1968. An experimental investigation of habitat selection and water loss in some anoline lizards. *Ecology* 49:762–767.

Sexton, O. J., J. Bauman, and E. Ortleb. 1972. Seasonal food habits of *Anolis limifrons*. *Ecology* 53:182–186.

Sexton, O. J., H. F. Heatwole, and E. H. Meseth. 1963. Seasonal population changes in the lizard, *Anolis limifrons*, in Panama. *American Midland Naturalist* 69:482–491.

Sexton, O. J., E. P. Ortleb, L. M. Hathaway, R. E. Ballinger, and P. Licht. 1971. Reproductive cycles of three species of anoline lizards from the Isthmus of Panama. *Ecology* 52:201–215.

Siliceo-Cantero, H. H., and A. García. 2014. Differences in growth rate, body condition, habitat use and food availability between island and mainland lizard populations of *Anolis nebulosus* in Jalisco, Mexico. *Journal of Tropical Ecology* 30:493–501.

———. 2015. Activity and habitat use of an insular and a mainland populations of the lizard *Anolis nebulosus* (Squamata: Polychrotidae) in a seasonal environment. *Revista Mexicana de Biodiversidad* 86:406–411.

REFERENCES

Siliceo-Cantero, H. H., A. García, R. G. Reynolds, G. Pacheco, and B. C. Lister. 2016. Dimorphism and divergence in island and mainland Anoles. *Biological Journal of the Linnean Society* 118:852–872.

Siliceo-Cantero, H. H., J. J. Zúñiga-Vega, K. Renton, and A. García. 2017. Assessing the relative importance of intraspecific and interspecific interactions on the ecology of *Anolis nebulosus* lizards from an island vs. a mainland population. *Herpetological Conservation and Biology* 12:673–682.

Simpson, G. G. 1951. The species concept. *Evolution* 5:285–298.

Slevin, J. R. 1942. Notes on a collection of reptiles from Boquete, Panama: with the description of a new species of *Hydromorphus*. *Proceedings of the California Academy of Sciences* 23:463–480.

Smith, H. M. 1933. Notes on some Mexican lizards of the genus *Anolis*, with the description of a new species, *A. megapholidotus*. *Transactions of the Kansas Academy of Science* 36:315–320.

———. 1936. A new *Anolis* from Mexico. *Copeia* 1936:9.

———. 1956. A new anole (Reptilia: Squamata) from Chiapas, Mexico. *Herpetologica* 12:1–2.

———. 1964. A new *Anolis* from Oaxaca, Mexico. *Herpetologica* 20:31–33.

———. 1968a. A new pentaprionid anole (Reptilia: Lacertilia) from Pacific slopes of Mexico. *Transactions of the Kansas Academy of Science* 71:195–200.

———. 1968b. Two new lizards, one new, of the genus *Anolis* from Mexico. *Journal of Herpetology* 2:143–146.

Smith, H. M., and H. W. Kerster. 1955. New and noteworthy Mexican lizards of the genus *Anolis*. *Herpetologica* 11:193–201.

Smith, H. M., and D. R. Paulson. 1968. A new lizard of the *schiedi* group of *Anolis* from Mexico. *The Southwestern Naturalist* 13:365–368.

Smith, H. M., and R. A. Spieler. 1945. A new anole from Mexico. *Copeia* 1945:165–168.

Smith, H. M., and E. H. Taylor. 1950. An annotated checklist and key to the reptiles of Mexico exclusive of the snakes. *Bulletin of the United States National Museum* 199:1–253.

Smith, H. M., F. W. Burley, and T. H. Fritts. 1968. A new anisolepid *Anolis* (Reptilia: Lacertilia) from Mexico. *Journal of Herpetology* 2:147–151.

Smith, H. M., G. Sinelnik, J. D. Fawcett, and R. E. Jones. 1972. A survey of the chronology of ovulation in anoline lizard genera. *Transactions of the Kansas Academy of Science* 75:107–120.

Socci, A. M., M. A. Schlaepfer, and T. A. Gavin. 2005. The importance of soil moisture and leaf cover in a female lizard's (*Norops polylepis*) evaluation of potential oviposition sites. *Herpetologica* 61:233–240.

Souza, F. L., M. Uetanabaro, P. Landgref-Filho, L. Piatti and C. P. A. Prado. 2010. Herpetofauna, municipality of Porto Murtinho, Chaco region, state of Mato Grosso do Sul. Brazil. *Check List* 6:470–475.

Sparrman, A. 1784. *Lacerta bimaculata*, eine neue Eidechse aus Amerika. *Der Königlich-Schwedischen Akademie der Wissenschaften Neue Abhandlungen, aus der Naturlehre Haushaltungskunst und Mechanik, für das Jahr 1784* 3: 173–174.

Stamps, J. A. 1977. Social behavior and spacing patterns in lizards. In *Biology of the Reptilia*, vol. 7: *Ecology and Behavior*, edited by C. Gans and D. W. Tinkle. London: Academic Press.

———. 1983. The relationship between ontogenetic habitat shifts, competition and predator avoidance in a juvenile lizard (*Anolis aeneus*). *Behavioral Ecology and Sociobiology* 12:19–33.

———. 1999. Relationships between female density and sexual size dimorphism in samples of *Anolis sagrei*. *Copeia* 1999:760–765.

Stamps, J. A., J. B. Losos, and R. M. Andrews. 1997. A comparative study of population density and sexual size dimorphism in lizards. *The American Naturalist* 149:64–90.

Stapley, J., M. Garcia, and R. M. Andrews. 2015. Long-term data reveal a population decline of the tropical lizard *Anolis apletophallus*, and a negative affect of El Niño years on population growth rate. *PloS One* 10:e0115350.

Stapley, J., C. Wordley, and J. Slate. 2011. No evidence of genetic differentiation between anoles with different dewlap color patterns. *Journal of Heredity* 102:118–124.

Stebbins, R. C. 1966. *A Field Guide to Western Reptiles and Amphibians*. Houghton Mifflin, Boston. 279 pp.

Steffen, J. 2010. Perch height differences among female *Anolis polylepis* exhibiting dorsal pattern polymorphism. *Reptiles and Amphibians: Conservation and Natural History* 17:89–94.

Stehle, C. M., A. C. Battles, M. N. Sparks, and M. A. Johnson. 2017. Prey availability affects territory size, but not territorial display behavior, in green anole lizards. *Acta Oecologica* 84:41–47.

Stejneger, L. 1899. A new name for the great crested *Anolis* of Jamaica. *The American Naturalist* 33:601–602.

———. 1900. Description of two new lizards of the genus *Anolis* from Cocos and Malpelo Islands. *Bulletin of the Museum of Comparative Zoology* 36:161–163.

Stimie, M. 1964. The cranial anatomy of the iguanid *Anolis carolinensis*. Thesis, University of Stellenbosch.

Stroud, J. T. 2018. Using introduced species of *Anolis* lizards to test adaptive radiation theory. Dissertation, Florida International University.

Stuart, L. C. 1942. Comments on several species of *Anolis* from Guatemala, with descriptions of three new forms. *Occasional Papers of the Museum of Zoology, University of Michigan* 464:1–10.

———. 1955. A brief review of the Guatemalan lizards of the genus *Anolis*. *Miscellaneous Publications of the Museum of Zoology, University of Michigan* 91:1–31.

———. 1958. A study of the herpetofauna of the Uaxactun-Tikal area of northern El Peten, Guatemala. *Contributions from the Laboratory of Vertebrate Biology, University of Michigan* 75:1–30.

Stuart, Y. E., T. Campbell, P. Hohenlohe, R. G. Reynolds, L. Revell, and J. Losos. 2014. Rapid evolution of a native species following invasion by a congener. *Science* 346:463–466.

Suárez, J.E.C., and N. D. Gutiérrez. 2013. Perch use by *Anolis polylepis* Peters, 1874 (Polychrotidae) in a tropical humid forest at the Piro Biological Station, Costa Rica. *Herpetology Notes* 6:219–222.

Suárez-Varón, G., O. Suárez-Rodriguez, K. Gribbins, and O. Hernández-Gallegos. 2016. Reproductive and parental care notes for *Norops beckeri* (Boulanger, 1891) in northern Guatemala. *Ecological Monographs* 33:83–112.

Sunyer, J., K. E. Nicholson, J. G. Phillips, J. A. Gubler, and L. A. Obando. 2013. Lizards (Reptilia: Squamata) of the Corn Islands, Caribbean Nicaragua. *Check List* 9:1383–1390.

Sunyer Mac Lennan, J. 2009. Taxonomy, zoogeography, and conservation of the herpetofauna of Nicaragua. PhD Thesis, Frankfurt: Goethe University Frankfurt am Main.

Talavera, J. B., A. Carriere, L. Swierk, and B. J. Putman. 2021. Tail autotomy is associated with boldness in male but not female water anoles. *Behavioral Ecology and Sociobiology* 75:1–10.

Talbot, J. J. 1977. Habitat selection in two tropical anoline lizards. *Herpetologica* 33:114–123.

———. 1979. Time budget, niche overlap, inter-and intraspecific aggression in *Anolis humilis* and *A. limifrons* from Costa Rica. *Copeia* 1979:472–481.

Tamsitt, J. R., and D. Valdivieso. 1963. The herpetofauna of the Caribbean islands San Andrés and Providencia. *Revista de Biología Tropical* 11:131–139.

Taylor, E. H. 1956. A review of the lizards of Costa Rica. *The University of Kansas Science Bulletin* 38:1–322.

Test, F. H., H. Heatwole, and O. J. Sexton. 1966. Reptiles of Rancho Grande and vicinity: Estado Aragua, Venezuela. *Miscellaneous Publications of the Museum of Zoology, University of Michigan* 128:1–63.

Thawley, C. J., H. A. Moniz, A. J. Merritt, A. C. Battles, S. N. Michaelides, and J. J. Kolbe. 2019. Urbanization affects body size and parasitism but not thermal preferences in *Anolis* lizards. *Journal of Urban Ecology* 5:1–9.

Thomas, O. 2020. Predation on a Slender Anole (*Anolis fuscoauratus*) by a whip scorpion. *IRCF Reptiles and Amphibians* 26:253–254.

Thomas, R. 1965. A new anole (Sauria: Iguanidae) from Puerto Rico. Part II. Field observations of *Anolis occultus* Williams and Rivero. *Breviora* 231:10–18.

Thominot, A. 1887. Description de trois espèces nouvelles d'Anolis et d'un amphisbaenien. *Bulletin de la Societé Philomatique Paris*, ser. 7, 11 (4):182–190.

REFERENCES

Thompson, M. E., B. J. Halstead, and M. A. Donnelly. 2018. Thermal quality influences habitat use of two anole species. *Journal of Thermal Biology* 75:54–61.
Tiatragul, S., and D. A. Warner. 2019. Beating the heat: nest characteristics of anoles across suburban and forest habitats in South Miami. *Anolis Newsletter* VII:289–283.
Tiatragul, S., J. M. Hall, N. G. Pavlik, and D. A. Warner. 2019. Lizard nest environments differ between suburban and forest habitats. *Biological Journal of the Linnean Society* 126:392–403.
Tiatragul, S., A. Kurniawan, J. J. Kolbe, and D. A. Warner. 2017. Embryos of non-native anoles are robust to urban thermal environments. *Journal of Thermal Biology* 65:119–124.
Tinius, A. 2016. Geometric morphometric analysis of the breast-shoulder apparatus of Greater Antillean anole ecomorphs. Dissertation, University of Calgary, Alberta, Canada.
Tinius, A., A. P. Russell, H. A. Jamniczky, and J. S. Anderson. 2020. Ecomorphological associations of scapulocoracoid form in Greater Antillean *Anolis* lizards. *Annals of Anatomy-Anatomischer Anzeiger* 231:151527.
Tokarz, R. R. 1985. Body size as a factor determining dominance in staged agonistic encounters between male brown anoles (*Anolis sagrei*). *Animal Behaviour* 33:746–753.
———. 1998. Mating pattern in the lizard *Anolis sagrei*: implications for mate choice and sperm competition. *Herpetologica* 54:388–394
Tokarz, R. R., A. V. Paterson, and S. McMann. 2003. Laboratory and field test of the functional significance of the male's dewlap in the lizard *Anolis sagrei*. *Copeia* 2003:502–511.
Tollis, M., E. D. Hutchins, J. Stapley, et al. 2018. Comparative genomics reveals accelerated evolution in conserved pathways during the diversification of anole lizards. *Genome Biology and Evolution* 10:489–506.
Torres-Carvajal, O., F. Ayala-Varela, and A. Carvajal-Campos. 2010. *Anolis heterodermus* Duméril, 1851: distribution extension, first record for Ecuador and notes on color variation. *Check List* 6:189–190.
Torres-Carvajal, O., F. P. Ayala-Varela, S. E. Lobos, S. Poe, and A. E. Narváez. 2018. Two new Andean species of *Anolis* lizard (Iguanidae: Dactyloinae) from southern Ecuador, *Journal of Natural History* 52:13–16
Torres-Carvajal, O., G. Pazmiño-Otamendi, and D. Salazar-Valenzuela. 2019. Reptiles of Ecuador: a resource-rich portal, with a dynamic checklist and photographic guides. *Amphibian & Reptile Conservation* 13:209–229.
Townsend, J. H., and L. D. Wilson. 2008. *Guide to the Amphibians and Reptiles of Cusuco National Park, Honduras.* Salt Lake City: Bibliomania.
———. 2009. New species of cloud forest *Anolis* (Squamata: Polychrotidae) in the *crassulus* group from Parque Nacional Montaña de Yoro, Honduras. *Copeia* 2009:62–70.
Townsend, J. H., L. D. Wilson, B. L. Talley, D. C. Fraser, T. L. Plenderleith, and S. M. Hughes. 2006. Additions to the herpetofauna of Parque Nacional El Cusuco, Honduras. *Herpetological Bulletin* 96:29–39.
Troschel, F. H. 1848. Amphibien. In *Versuch einer Zusammenstellung der Fauna und Flora von Britisch-Guiana.* Part 3 of *Reisen in Britisch-Guiana in den Jahren 1840–1844. Im Auftrag Sr. Majestät des Konigs von Preussen ausgeführt*, by Richard Schomburgk. Leipzig: J. J. Weber.
Trowbridge, A. H. 1937. Ecological observations on amphibians and reptiles collected in southeastern Oklahoma during the summer of 1934. *American Midland Naturalist* 18:285–303.
Uetanabaro, M., F. L. Souza, P. L. Filho, A. F. Beda, and R. A. Brandão. 2007. Anfíbios e répteis do Parque Nacional da Serra da Bodoquena, Mato Grosso do Sul, Brasil. *Biota Neotropica* 7:279–289.
Ugueto, G. N., G. R. Fuenmayor, T. Barros, S. J. Sanchez-Pacheco, and J. E. Garcia-Perez. 2007. A revision of the Venezuelan Anoles I: a new *Anolis* species from the Andes of Venezuela with the redescription of *Anolis jacare* Boulenger 1903 (Reptilia: Polychrotidae) and the clarification of the status of *Anolis nigropunctatus* Williams 1974. *Zootaxa* 1501:1–30.
Ugueto, G. N., G. Rivas, T. Barros, and E. N. Smith. 2009. A revision of the Venezuelan anoles II: redescription of *Anolis squamulatus* Peters 1863 and *Anolis tigrinus* Peters 1863 (Reptilia: Polychrotidae). *Caribbean Journal of Science* 45:30–51.

REFERENCES

Urbina-Cardona, J. N., and V. H. Reynoso. 2017. Descripción y modelado del microhábitat de los anfibios y reptiles que habitan la selva alta perennifolia de Los Tuxtlas. In *Avances y Perspectivas en la Investigación de los Bosques Tropicales y sus Alrededores: la Región de Los Tuxtlas*, edited by V. H. Reynoso, R. I. Coates, and M. L. Vázquez Cruz. Instituto de Biología, Universidad Nacional Autónoma de México, Mexico City.

Urbina-Cardona, J. N., M. Olivares-Pérez, and V. H. Reynoso. 2006. Herpetofauna diversity and microenvironment correlates across a pasture–edge–interior ecotone in tropical rainforest fragments in the Los Tuxtlas Biosphere Reserve of Veracruz, Mexico. *Biological Conservation* 132:61–75.

Valdivieso, D., and J. R. Tamsitt. 1963. Records and observations on Colombian reptiles. *Herpetologica* 19:28–39.

Vance, T. 1991. Morphological variation and systematics of the green anole, *Anolis carolinensis* (Reptilia: Iguanidae). *Bulletin of the Maryland Herpetological Society* 27:43–89.

Vanegas-Guerrero, J., C. Fernández, W. Buitrago-González, and F. Vargas-Salinas. 2016. Urban Remnant Forests: are they important for herpetofaunal conservation in the Central Andes of Colombia? *Herpetological Review* 47:180–185.

Van-Silva, W., A. G. Guedes, P. L. de Azevedo-Silva, F. F. Gontijo, R. S. Barbosa, G. R. Aloísio, and F.C.G. de Oliveira. 2007. Herpetofauna, Espora hydroelectric power plant, state of Goiás, Brazil. *Check List* 3:338–345.

Vanzolini, P. E. 1972. Miscellaneous notes on the ecology of some Brazilian lizards (Sauria). *Papéis Avulsos de Zoologia* 26:83–115.

Vanzolini, P. E., and E. E. Williams. 1970. South American anoles: the geographic differentiation and evolution of the *Anolis chrysolepis* species group (Sauria, Iguanidae). *Arquivos de Zoologia* 19:125–298.

Vargas Salinas, F., and M. E. Bolaños-Lizalda. 1999. Anfibios y reptiles presentes en hábitats perturbados de selva lluviosa tropical en el bajo Anchicayá, Pacífico colombiano. *Revista de la Academia Colombiana de Ciencias Exactas, Físicas y Naturales* 23:499–511.

Vásquez-Cruz, V., A. Reynoso-Martínez, A. Fuentes-Moreno, and L. Canseco-Márquez. 2020. The distribution of Cuban Brown Anoles, *Anolis sagrei* (Squamata: Dactyloidae) in Mexico, with new records and comments on ecological interactions. *Reptiles & Amphibians* 27:29–35.

Velasco, J. A., and A. Herrel. 2007. Ecomorphology of *Anolis* lizards of the Chocó region in Colombia and comparisons with Greater Antillean ecomorphs. *Biological Journal of the Linnean Society* 92:29–39.

Velasco, J. A., and J. P. Hurtado-Gómez. 2014. A new green anole lizard of the "Dactyloa" clade (Squamata: Dactyloidae) from the Magdalena river valley of Colombia. *Zootaxa* 3785:201–216.

Velasco, J. A., S. Poe, C. González-Salazar, and O. Flores-Villela. 2019. Solitary ecology as a phenomenon extending beyond insular systems: exaptive evolution in *Anolis* lizards. *Biology Letters* 15:20190056.

Velasco, J. A., P.D.A. Gutiérrez-Cárdenas, and A. Quintero-Angel. 2010. A new species of *Anolis* of the *aequatorialis* group (Squamata: Iguania) from the central Andes of Colombia. *The Herpetological Journal* 20:231–236.

Velasco, J. A., E. Martinez-Meyer, O. Flores-Villela, A. Garcia, A. C. Algar, G. Köhler, and J. M. Daza. 2016. Climatic niche attributes and diversification in *Anolis* lizards. *Journal of Biogeography* 43:134–144.

Velasco, J. A., F. Villalobos, J. A. F. Diniz-Filho, S. Poe, and O. Flores-Villela. 2020. Macroecology and macroevolution of body size in *Anolis* lizards. *Ecography* 43:812–822.

Velásquez, J., L. A. González, and A. P. Arcas. 2011. Ecología térmica y patrón de actividad del lagarto *Anolis onca* (Squamata: Polychrotidae) en la Península de Araya, Venezuela. *Revista Multidisciplinaria del Consejo de Investigación de la Universidad de Oriente* 23:5–12.

Veludo, L. B. A. 2011. Ecologia de *Anolis meridionalis* (Squamata, Polychrotidae) no Cerrado brasileiro. Thesis, Universidade de Brasília.

Verwaijen, D., and R. Damme. 2008. Wide home ranges for widely foraging lizards. *Zoology* 111:37–47.

REFERENCES

Villarreal-Benítez, J. L. 1997. Historia natural del género *Anolis*. In *Historia Natural de Los Tuxtlas*, edited by E. González, R. Dirzo and R. Vogt. México, D.F.: Universidad Nacional Autónoma de México.

Villarreal-Benítez, J. L., and L. Heras-Lara. 1997. Historia natural de especies, *Anolis uniformis*. In *Historia Natural de Los Tuxtlas*, edited by E. González-Soriano, R. Dirzo, and R. C. Vogt. Universidad Nacional Autónoma de México.

Vitt, L. J. 1991. An introduction to the ecology of cerrado lizards. *Journal of Herpetology* 25:79–90.

———. 2000. Ecological consequences of body size in neonatal and small-bodied lizards in the neotropics. *Herpetological Monographs* 14:388–400.

Vitt, L. J., and J. P. Caldwell. 1993. Ecological observations on Cerrado lizards in Rondônia, Brazil. *Journal of Herpetology* 27:46–52.

Vitt, L. J., and C. M. de Carvalho. 1995. Niche partitioning in a tropical wet season: lizards in the lavrado area of northern Brazil. *Copeia* 1995:305–329.

Vitt, L. J., and P. A. Zani. 1996a. Ecology of the South American lizard *Norops chrysolepis* (Polychrotidae). *Copeia* 1996:56–68.

———. 1996b. Organization of a taxonomically diverse lizard assemblage in Amazonian Ecuador. *Canadian Journal of Zoology* 74:1313–1335.

———. 1998. Prey use among sympatric lizard species in lowland rain forest of Nicaragua. *Journal of Tropical Ecology* 14:537–559.

———. 2005. Ecology and reproduction of *Anolis capito* in rain forest of southeastern Nicaragua. *Journal of Herpetology* 39:36–42.

Vitt, L. J., T.C.S. Ávila-Pires, M. C. Espósito, S. S. Sartorius, and P. A. Zani. 2003a. Sharing Amazonian rain-forest trees: ecology of *Anolis punctatus* and *Anolis transversalis* (Squamata: Polychrotidae). *Journal of Herpetology* 37:276–285.

Vitt, L. J., T.C.S. Ávila-Pires, P. A. Zani, S. S. Sartorius, and M. C. Espósito. 2003b. Life above ground: ecology of *Anolis fuscoauratus* in the Amazon rain forest, and comparisons with its nearest relatives. *Canadian Journal of Zoology* 81:142–156.

Vitt, L. J., T. Cristina, S. Ávila-Pires, P. A. Zani, and M. C. Espósito. 2002. Life in shade: the ecology of *Anolis trachyderma* (Squamata: Polychrotidae) in Amazonian Ecuador and Brazil, with comparisons to ecologically similar anoles. *Copeia* 2002:275–286.

Vitt, L. J., S. S. Sartorius, T.C.S. Ávila-Pires, and M. C. Espósito. 2001. Life on the leaf litter: the ecology of *Anolis nitens tandai* in the Brazilian Amazon. *Copeia* 2001:401–412.

Vitt, L. J., D. B. Shepard, G. H. Vieira, J. P. Caldwell, G. R. Colli, and D. O. Mesquita. 2008. Ecology of *Anolis nitens brasiliensis* in Cerrado woodlands of Cantao. *Copeia* 2008:144–153.

Vitt, L. J., P. A. Zani, and R. D. Durtsche. 1995. Ecology of the lizard *Norops oxylophus* (Polychrotidae) in lowland forest of southeastern Nicaragua. *Canadian Journal of Zoology* 73:1918–1927.

Vitt, L. J., P. A. Zani, and M. C. Espósito. 1999. Historical ecology of Amazonian lizards: implications for community ecology. *Oikos* 87:286–294.

Voigt, F. S. 1832. *Das Thierreich, geordnet nach seiner Organisation. Als Grundlage der Naturgeschichte der Thiere und Einleitung in die vergleichenden Anatomie. Vom Baron von Cuvier … Nach der zweiten, vermehrten Ausgabe übersetzt und durch Zusätze erweitert von F. S. Voigt.* Vol. 2: Reptiles and Fishes. Leipzig: Brockhaus. [reptiles and amphibians: 1–179].

Wake, D. B., and J. D. Johnson. 1989. A new genus and species of plethodontid salamander from Chiapas, Mexico. *Contributions in Science, Museum of Natural History, Los Angeles County* 411:1–10.

Watling, J. I., J. H. Waddle, D. Kizirian, and M. A. Donnelly. 2005. Reproductive phenology of three lizard species in Costa Rica, with comments on seasonal reproduction of neotropical lizards. *Journal of Herpetology* 39:341–348.

Wegener, J. E., J. N. Pita-Aquino, J. Atutubo, A. Moreno, and J. J. Kolbe. 2019. Hybridization and rapid differentiation after secondary contact between the native green anole (*Anolis carolinensis*) and the introduced green anole (*Anolis porcatus*). *Ecology and Evolution* 9:4138–4148.

REFERENCES

Werner, F. 1894. Über einige Novitäten der herpetologischen Sammlung des Wiener zoolog. vergl. anatom. Instituts. *Zoologischer Anzeiger* 17:155–157.

———. 1916. Bemerkungen über einige niedere Wirbeltiere der Anden von Kolumbien mit Beschreibungen neuer Arten. *Zoologischer Anzeiger* 47:305–311.

———. 1917. Über einige neue Reptilien und einen neuen Frosch des Zoologischen Museums in Hamburg. *Mitteilungen aus dem Naturhistorischen Museum in Hamburg* 34:31–36.

Werner, Y. L. 1972. Temperature effects on inner-ear sensitivity in six species of iguanid lizards. *Journal of Herpetology* 6:147–177.

Wever, E. G. 1978. *The Reptile Ear*. Princeton University Press, Princeton, New Jersey.

White, B. A., S. R. Prado-Irwin, and L. N. Gray. 2019. Female signal variation in the *Anolis lemurinus* group. *Breviora* 564:1–10.

White, F. N. 1968. Functional anatomy of the heart of reptiles. *American Zoologist* 8:211–219.

Whitfield, S. M., K. E. Bell, T. Philippi, et al. 2007. Amphibian and reptile declines over 35 years at La Selva, Costa Rica. *Proceedings of the National Academy of Sciences* 104:8352–8356

Wiegmann, A.F.A. 1834. Herpetologia Mexicana, seu Descriptio Amphibiorum Novae Hispaniae, quae itineribus comitis de Sack, Ferdinandi Deppe et Chr. Guil. Schiede in Museum Zoologicum Berolinense pervenerunt. Pars prima, *Saurorum Species* amplectens, adiecto systematis saurorum prodromo, additisque multis in hune amphibiorum ordinem observationibus. Berlin: Sumptibus C. G. Lüderitz.

Wiley, E. O. 1978. The evolutionary species concept reconsidered. *Systematic Zoology* 27:17–26.

———. 1981. *Phylogenetics: The Theory and Practice of Phylogenetic Systematics*. Wiley & Sons, New York.

Willard, W. A. 1915. The cranial nerves of *Anolis carolinensis*. *Bulletin of the Museum of Comparative Zoology*, Harvard University 59: 17–116.

Williams, E. E. 1963. Studies on South American Anoles: description of *Anolis mirus*, new species, from Rio San Juan, Colombia, with comment on digital dilation and dewlap as generic and specific characters in the anoles. *Bulletin of the Museum of Comparative Zoology* 129:463–480.

———. 1965. South American *Anolis* (Sauria, Iguanidae): two new species of the *punctatus* group. *Breviora* 233:1–15.

———. 1966. South American *Anolis*: *Anolis biporcatus* and *Anolis fraseri* (Sauria, Iguanidae) compared. *Breviora* 239:1–14.

———. 1970. *Anolis jacare* Boulenger: A "solitary" anole from the Andes of Venezuela. *Breviora* 353:1–15.

———. 1974a. A case history in retrograde evolution; the onca lineage in anoline lizards: *Anolis annectens* new species, intermediate between the genera *Anolis* and *Tropidodactylus*. I. *Breviora* 421:1–21.

———. 1974b. South American *Anolis*: three new species related to *Anolis nigrolineatus* and *A. dissimilis*. *Breviora* 422:1–15.

———. 1975. South American *Anolis*: *Anolis parilis*, new species, near *A. mirus* Williams. *Breviora* 434:1–8.

———. 1976a. South American anoles: the species groups. *Papéis Avulsos de Zoologia* 29:249–268.

———. 1976b. West Indian anoles: a taxonomic and evolutionary summary. I. Introduction and a species list. *Breviora* 440:1–21.

———. 1982. Three new species of the *Anolis punctatus* complex from Amazonian and inter-Andean Colombia, with comments on the eastern members of the *punctatus* species group. *Breviora* 422:1–15.

———. 1983. Ecomorphs, faunas, island size, and diverse end points in island radiations of *Anolis*. In *Lizard Ecology: Studies of a Model Organism*, edited by R. B. Huey, E. Pianka, and T. W. Schoener. Harvard University Press.

———. 1984a. New or problematic *Anolis* from Colombia II: *Anolis propinquus*, another new species from the cloud forest of western Colombia. *Breviora* 477:1–7.

———. 1984b. New or problematic *Anolis* from Colombia. III. Two new semiaquatic anoles from Antioquia and Chocó, Colombia. *Breviora* 478:1–22.

REFERENCES

———. 1985. New or problematic *Anolis* from Colombia. IV: *Anolis antioquiae*, new species of the *Anolis eulaemus* subgroup from western Colombia. *Breviora* 482:1–9.

———. 1986. *Anolis vicarius*, new species related to *A. granuliceps*. *Caldasia* 15:452–459.

———. 1988. New or problematic *Anolis* from Colombia. V. *Anolis danieli*, a new species of the *latifrons* species group and a reassessment of *Anolis apollinaris* Boulenger. *Breviora* 489:1–25.

———. 1992. New or problematic *Anolis* from Colombia. VII. *Anolis lamari*, a new anole from the Cordillera Oriental of Colombia, with a discussion of tigrinus and punctatus species group boundaries. *Breviora* 495:1–24.

Williams, E. E., and W. E. Duellman. 1967. *Anolis chocorum*, a new *punctatus*-like anole from Darién, Panamá (Sauria, Iguanidae). *Breviora* 256:1–12.

———. 1984. *Anolis fitchi*, a new species of the *Anolis aequatorialis* group from Ecuador and Colombia. *University of Kansas Publications, Museum of Natural History* 10:257–266.

Williams, E. E., and J. A. Peterson. 1982. Convergent and alternative designs in the digital adhesive pads of scincid lizards. *Science* 215:1509–1511.

Williams, E. E., and P. E. Vanzolini. 1966. Studies on South American anoles. *Anolis transversalis* A. Duméril. *Papéis Avulsos de Zoologia* 19:197–204.

Williams, E. E., G. Orcés-V., J. C. Matheus, and R. Bleiweiss. 1996. A new giant phenacosaur from Ecuador. *Breviora* 505:1–32.

Williams, E. E., M. J. Praderio, S. Gorzula. 1996. A phenacosaur from Chimanta Tepui, Venezuela. *Breviora* 506:1–15.

Williams, E. E., H. Rand, A. S. Rand, and R. J. O'Hara. 1995. A computer approach to the comparison and identification of species in difficult taxonomic groups. *Breviora* 502:1–47.

Williams, E. E., J. A. Rivero, and R. Thomas. 1965. A new anole (Sauria: Iguanidae) from Puerto Rico. *Breviora* 231:1–9.

Wilson, L. D., and J. R. McCranie. 1982. A new cloud forest *Anolis* (Sauria: Iguanidae) of the *schiedei* group from Honduras. *Transactions of the Kansas Academy of Science* 85:133–141.

Wittorski, A., J. B. Losos, and A. Herrel. 2016. Proximate determinants of bite force in *Anolis* lizards. *Journal of Anatomy* 228:85–95.

Wolda, H. 1975. The ecosystem on Malpelo island. *Smithsonian Contributions to Zoology* 176:21–26.

Woolrich-Piña, G. A., G. R. Smith, J. A. Lemos-Espinal, and J. P. Ramírez-Silva. 2015a. Do gravid female *Anolis nebulosus* thermoregulate differently than males and non-gravid females? *Journal of Thermal Biology* 52:84–89.

———. 2015b. Observations on sexual dimorphism, sex ratio, and reproduction of *Anolis nebulosus* (Squamata: Dactyloidae) from Nayarit, Mexico. *Phyllomedusa* 14:67–71.

Wright, S. J. 1979. Competition between insectivorous lizards and birds in central Panama. *American Zoologist* 19:1145–1156.

Wylie, D. B., and C. I. Grunwald. 2016. First report of *Bothriechis schlegelii* (Serpentes: Viperidae: Crotalinae) from the state of Oaxaca, Mexico. *Mesoamerican Herpetology* 3:1066–1067.

Yánez-Muñoz, M. H. 2001. Aspectos ecológicos de *Dactyloa fitchi* Williams & Duellman (Sauria: Polychrotidae) en los bosques húmedos de La Sofía, Provincia de Sucumbíos. *Memorias de las XXV Jornadas Ecuatorianas de Biología*, Universidad Estatal de Guayaquil, Guayaquil, Ecuador.

Yánez-Muñoz, M. H., C. Reyes-Puig, J. P. Reyes-Puig, J. A. Velasco, F. Ayala-Varela, and O. Torres-Carvajal. 2018. A new cryptic species of *Anolis* lizard from northwestern South America (Iguanidae, Dactyloinae). *ZooKeys* 794:135–163.

Yánez-Muñoz, M. H., M. A. Urgilés, M. Altamirano-Benavides, and S. R. Cáceres. 2010. Redescripción de *Anolis proboscis* Peters & Orcés (Reptilia: Polychrotidae), con el descubrimiento de las hembras de la especie y comentarios sobre su distribución y taxonomía. *Avances* 2:B7–B15.

Yuasa, H. J., K. Mizuno, and H. J. Ball. 2015. Low efficiency IDO 2 enzymes are conserved in lower vertebrates, whereas higher efficiency IDO 1 enzymes are dispensable. *The FEBS Journal* 282:2735–2745.

Zimmerman, B., and M. Rodrigues. 1990. Frogs, snakes, and lizards of the INPA-WWF reserves near Manaus, Brazil. In *Four Neotropical Rainforests*, edited by A. Gentry. Yale University Press.

PHOTOGRAPHIC CREDITS

DEWLAP PLATES

ABBREVIATIONS FOR PHOTOGRAPHERS' NAMES

ABM = Adán Bautista-del Moral
AH = Anthony Herrel
AM = Aurelien Mirales
AMCZ = Photo courtesy of Archives of the Museum of Comparative Zoology
AN = Andrea Narváez
ANa = Alberto Nadal
ANi = Adrian Nieto
AO = Andrew Odum
CA = Chris Anderson
CBA = César Barrio Amorós
DB = David Bejarano
DG = Diego Gómez
DJ = Delmer Jonathan
DL = David Laurencio
DM = D. Luke Mahler
DQ = Diego Quirola
EA = Esteban Alzate
EL = Edgar Lehr
FG = Freddie Grisales
FO = Fabio Olmos
FOm = Feroze Omardeen
GK = Gunther Köhler
GM = Guido F. Medina-Rangel
HV = Hermes Vega
IL = Ian Latella
IP = Ivan Prates
JB = Joseph Barnett
JD = Juan Daza
JG = Josué Ramos Galdámez
JH = Juan Pablo Hurtado
JJ = Janson Jones
JL = Jose Daniel Lara-Trufiño
JPG = José Luis Pérez González
JS = John Sullivan
JV = Josh Vandermeulen
KP = Kyle Pote
MK = Michael Kielb
MR = Mason Ryan
MT = Miguel Trefaut Rodrigues
OK = Oliver Komar
OT = Omar Torres-Carvajal
PS = Paolo Sampaio
PV = Pablo Venegas
RA = Rodrigo Arrazola
RC = Rosario Castañeda
RF = Rob Foster
RG = Rich Glor
RL = Roberto Langstroth
RM = Rafael Moreno-Arias
RP = Robert Powell
RS = Rich Sajdak
SR = Santiago Ron
TB = Tito Barros
TK = Tom Kennedy
ZK = Zsombor Károlyi

BODY PLATES

ABBREVIATIONS FOR PHOTOGRAPHERS' NAMES

ABM = Adán Bautista-del Moral
AO = Andrew Odum
CA = Chris Anderson
DB = David Bejarano
EA = Esteban Alzate
EL = Edgar Lehr
FA = Fernando Ayala-Varela
IL = Ian Latella
IP = Ivan Prates
JJ = Janson Jones
JPG = José Luis Pérez González
LR = Luis Rodriguez
MA = Marco Antonio de Freitas
MR = Mason Ryan
OK = Oliver Komar
OT = Omar Torres-Carvajal
PV = Pablo Venegas
RA = Rafael Alejandro Calzada Arciniega
RG = Rich Glor
RM = Rafael Moreno-Arias
RP = Robert Powell
RS = Rick Stanley
RSa = Rich Sajdak
SR = Santiago Ron
TB = Tito Barros
TK = Tom Kennedy

INDEX

Note: **Bold** page numbers indicate Species Accounts. *Italic* page numbers indicate illustrations.

abundances, 21, 34
acronyms, museum, 14
alpha taxonomy, 10, 363n39, 364n44
Amazon river basin, 5–6
ANAM. *See* Autoridad Nacional de Ambiente de Panama
anatomy of anoles
   overview of, 17–21
   shared traits of, 1
   sources of data on, 13
Andes mountains, 5–6
Andrews, Robin, 12
AnoleKey, 36–37, 42–43, *43*, 369n11
Anolis aeneus, **66–67**
   body of, *346*
   dewlap of, *50*
   synonyms of, 350
Anolis aequatorialis, **67–68**
   body of, *342*
   collecting of, 33
   dewlap of, 45, *50*
Anolis agassizi, **68–69**
   dewlap of, *50*
   evolution of, 362n3
   range of, 6
Anolis allisoni, **69–70**
   body of, *347*
   dewlap of, *50*
   range of, 9, 363n35
Anolis alocomyos, 313, 351–52
Anolis altae, **70–71**
   body of, *334*
   dewlap of, *50*
   population declines in, 26
Anolis alvarezdeltoroi, **71–72**
   body of, *331*
   dewlap of, 46, *50*

range of, 9
sleeping behavior of, 32, *32*
Anolis amplisquamosus, **72–73**
   body of, *335*
   dewlap of, *50*
   range of, 9
Anolis anatoloros, **73–74**
   dewlap of, *50*
Anolis anchicayae, **74–75**
   dewlap of, *50*
   sexual dimorphism in, 47
Anolis anisolepis, **75–76**
   body of, *333*
   dewlap of, *50*
Anolis annectens, **76–77**
   anatomy of, 365n3
   body of, *346*
   dewlap of, *50*
Anolis anoriensis, **77–78**
   body of, *342*
   dewlap of, *50*
Anolis antioquiae, **78–80**
   body of, *343*
   dewlap of, *50*
   synonyms of, 210
Anolis antonii, **80–81**
   body of, *339*
   dewlap of, *50*
Anolis apletophallus, **81–82**
   body of, *333*
   dewlap of, *50*
   genome of, 366n24
   identification of, 15
   population declines in, 26
Anolis apollinaris, **82–83**
   dewlap of, *51*
   sexual dimorphism in, 47
Anolis aquaticus, **83–84**
   body of, *339*
   dewlap of, *51*
   diet of, 23
   sleeping behavior of, 32
Anolis arenal, **85**
   body of, *334*
   dewlap of, *51*

Anolis auratus, **86–87**
   abundances of, 21
   body of, *341*
   dewlap of, 45, *51*
   genome of, 366n24
   history of research on, 10
   natural history of, 16
Anolis baleatus, **87–88**
   body of, *348*
   dewlap of, *51*
Anolis barbatus, 30
Anolis barkeri, **88–89**
   body of, *339*
   dewlap of, *51*
   range of, 9
   sleeping behavior of, *32*
Anolis bartschi, 32, 368n3
Anolis beckeri, **89–90**
   body of, *338*
   dewlap of, *51*
   range of, 9
Anolis bellipeniculus, **90–91**
   dewlap of, *51*
   range of, 6
Anolis benedikti, **91–92**
   body of, *335*
   dewlap of, *51*
Anolis bicaorum, **92–93**
   body of, *336*
   dewlap of, *51*
   range of, 9
   vocalization in, 20
Anolis bimaculatus, 363n38
Anolis binotatus, **93–94**
   body of, *341*
   dewlap of, *51*
Anolis biporcatus, **95–96**
   dewlap of, 46, *51*
   head scales in identification of, 38
   identification of, 15
   intraindividual variation within, 48
   range of, 7, 9

INDEX

Anolis biscutiger, **96–97**
    body of, *333*
    dewlap of, *51*
Anolis bitectus, 94, 349
Anolis bocourtii, 154, 349
Anolis boettgeri, **97–98**
    body of, *344*
    dewlap of, *51*
Anolis bombiceps, **98–99**
    body of, *340*
    dewlap of, *51*
Anolis boulengerianus, **99–100**
    vs. *A. cuprinus*, 126
    body of, *332*
    dewlap of, *52*
    range of, *9*
Anolis brasiliensis, **100–101**
    dewlap of, *52*
Anolis brianjuliani, **101–2**
    dewlap of, *52*
Anolis brooksi, **102–3**
    body of, *338*
    collecting of, 30
    detectability of, 16
    dewlap of, 45, 46, *46*, *52*
    variation within, 46
Anolis caceresae, **103–4**
    dewlap of, *52*
Anolis calimae, **104–5**
    body of, *343*
    collecting of, 30
    dewlap of, *52*
Anolis callainus, **105–6**
    body of, *347*
    dewlap of, *52*
Anolis campbelli, **106–7**
    dewlap of, *52*
    range of, *9*
Anolis capito, **107–8**
    body of, *337*
    dewlap of, *52*
    range of, *7*, *7*, *9*, 363n28
Anolis caquetae, **109**
    dewlap of, *52*
    range of, *6*

Anolis carlliebi, 270, 352–53
Anolis carlostoddi, **110**
    dewlap of, *52*
    range of, *6*
Anolis carolinensis, **111–12**
    anatomy of, 17–21
    body of, *347*
    dewlap of, *52*
    effects of *A. sagrei* on, 26, 366n35
    genome of, 17, 25, 366n24
    history of research on, 10, 12, 17
    as model organism, 17
    range of, *9*, 17
    sexual dimorphism in, 45
Anolis carpenteri, **112–13**
    body of, *337*
    dewlap of, *52*
Anolis casildae, **113–14**
    dewlap of, *52*
Anolis charlesmyersi, **114–15**
    dewlap of, *52*
    head scales in identification of, *38*
    range of, *8*
Anolis chloris, **115–16**
    body of, *343*
    dewlap of, 45, *53*, 362n5
Anolis chlorocyanus, 105–6
Anolis chocorum, 266
Anolis chrysolepis, **116–17**
    dewlap of, *53*
    range of, *6*
Anolis cobanensis, **117–18**
    dewlap of, *53*
Anolis compressicauda, **118–19**
    body of, *333*
    dewlap of, *53*
    range of, *9*
Anolis concolor, **119–20**
    dewlap of, *53*
Anolis conspersus, 368n20
Anolis crassulus, **120–21**
    dewlap of, *53*, 76
    range of, 8–9, 14

Anolis cristatellus, **121–23**
    anatomy of, *18*
    body of, *347*
    dewlap of, *53*
    predation on, 23
Anolis cristifer, **123**
    body of, *338*
    dewlap of, *53*
    eye color in identification of, *41*
Anolis cryptolimifrons, 3, 188, 350
Anolis cupreus, **124–25**
    body of, *336*
    dewlap of, *53*
Anolis cuprinus, **125–26**
    dewlap of, *53*
    range of, *9*
Anolis cuscoensis, **126–27**
    dewlap of, *53*
    sexual dimorphism in, *47*
Anolis cusuco, **127–28**
    body of, *335*
    dewlap of, *53*
Anolis cybotes, **128–29**
    body of, *347*
    collecting of, 368-27
    dewlap of, *53*
Anolis cymbops, **129–30**
    body of, *331*
    dewlap of, *53*
    range of, *9*
Anolis damulus, 349
Anolis danieli, **130–31**
    body of, *344*
    dewlap of, *53*
Anolis datzorum, **131–32**
    body of, *335*
    dewlap of, *54*
Anolis deltae, 67, 144, 350
Anolis dissimilis, **132–33**
    body of, *343*
    dewlap of, *54*
    range of, *6*
Anolis distichus, **133–34**
    body of, *347*
    collecting of, 33, 368n22
    dewlap of, *54*

INDEX

Anolis dollfusianus, **134–35**
  body of, *336*
  dewlap of, *54*
Anolis dracula, **136–37**
  body of, *342*
  dewlap of, *54*
Anolis duellmani, **137–38**
  body of, *333*
  dewlap of, *54*
  range of, *9*
Anolis dunni, **138–39**
  body of, *331*
  dewlap of, *54*
  range of, *9*
Anolis elcopeensis, **139–40**
  body of, *334*
  dewlap of, *54*
  range of, *7*
Anolis equestris, **140–41**
  body of, *347*
  dewlap of, *54*
Anolis eulaemus, **141–42**
  body of, *342*
  dewlap of, *54*
Anolis euskalerriari, **142–43**
  body of, *345*
  dewlap of, *54*
Anolis extremus, **143–44**
  body of, *346*
  dewlap of, *54*
  synonyms of, *350*
Anolis fasciatus, **144–45**
  body of, *341*
  dewlap of, *54*
Anolis festae, **146–47**
  body of, *341*
  dewlap of, *54*
  synonyms of, *350*
Anolis fitchi, **147–48**
  body of, *342*
  dewlap of, *54*
  identification of, *42, 43*
Anolis fortunensis, **148–49**
  body of, *334*
  dewlap of, *55*

Anolis fraseri, **149–50**
  biogeographic barriers in, 363n20
  collecting of, 33
  dewlap of, *2, 55*
  range of, *6*, 363n20
Anolis frenatus, **150–52**
  body of, *344*
  dewlap of, *55*
  genome of, 366n24
  head scales in identification of, *38*
  home range size of, 22
  range of, *7, 8*
Anolis fungosus, **152–53**
  body of, *338*
  collecting of, 30
  detectability of, 364n53
  dewlap of, *55*
Anolis fuscoauratus, **153–54**
  body of, *339*
  dewlap of, 45, *45, 55*
  range of, *6*, 13
  synonyms of, 349
Anolis gadovii, **154–55**
  body of, *331*
  dewlap of, *55*
  range of, *9*
  sleeping behavior of, 32
Anolis gaigei, **155–56**
  body of, *341*
  as cryptic species, 362n12
  dewlap of, *55*
  identification of, 15
Anolis garmani, **157**
  body of, *347*
  dewlap of, *55*
Anolis gemmosus, **158–59**
  body of, *341*
  collecting of, 33
  dewlap of, *55*
Anolis gibbiceps, 349
Anolis ginaelisae, **159–60**
  body of, *338*
  dewlap of, *55*
  range of, *8*

Anolis gorgonae, **160–61**
  dewlap of, *55*
  evolution of, 362n3
  range of, *6*
Anolis gracilipes, **161–62**
  body of, *340*
  dewlap of, *55*
Anolis graniliceps, **162–63**
  body of, *340*
  dewlap of, *55*
Anolis gruuo, **163–64**
  body of, *334*
  dewlap of, *55*
Anolis haguei, 121
Anolis heterodermus, **164–65**
  body of, *344*
  dewlap of, *55*
  diet of, 23
  head scales in identification of, *39*
Anolis heteropholidotus, **166**
  dewlap of, *56*
Anolis hispaniolae, 129
Anolis hobartsmithi, **167–68**
  body of, *331*
  dewlap of, *56*, 76
  range of, *9*
Anolis huilae, **168–69**
  body of, *344*
  dewlap of, *56*
Anolis humilis, **169–70**
  anatomy of, 365n3
  body of, *335*
  collecting of, 32–33
  dewlap of, *56*
  history of research on, 12
  range of, *7*
  synonyms of, 350–51
Anolis hyacinthogularis, **170–71**
  body of, *343*
  dewlap of, *56*
Anolis ibague, 297, 350
Anolis ibanezi, **171–72**
  body of, *337*
  dewlap of, *56*
  range of, *7*

412

INDEX

Anolis immaculogularis, **172–73**
    dewlap of, *56*
    range of, *9*
Anolis impetigosus, 349
Anolis inderenae, **173–74**
    body of, *345*
    dewlap of, *56*
Anolis insignis, **174–75**
    body of, *337*
    dewlap of, *56*
    range of, *8*
Anolis intermedius, 183
Anolis jacare, **175–76**
    body of, *343*
    dewlap of, *56*
Anolis johnmeyeri, **176–77**
    body of, *337*
    dewlap of, *46, 46, 56*
Anolis kathydayae, **176–78**
    body of, *338*
    dewlap of, *56*
    range of, *8*
Anolis kemptoni, **178–79**
    body of, *334*
    detectability of, 364n53
    dewlap of, *56*
    range of, *14*
    sleeping behavior of, *31*
Anolis kreutzi, **179–80**
    dewlap of, *56*
Anolis kunayalae, **180–81**
    anatomy of, *18*
    body of, *337*
    dewlap of, *56*
    head scales in identification of, *38, 39*
    range of, *7*
    variation within, 46
Anolis laevis, **181–82**
    body of, *346*
    dewlap of, *57*
Anolis laeviventris, **182–83**
    body of, *335*
    dewlap of, *57*
    range of, *8*

Anolis lamari, **183–84**
    body of, *345*
    dewlap of, *57*
    identifying sex of, *36*
Anolis latifrons, **185–86**
    body of, *344*
    detectability of, 16
    dewlap of, *57*
Anolis leditzigorum, 313, 351–52
Anolis lemniscatus, 349
Anolis lemurinus, **186–87**
    body of, *336*
    dewlap of, *46, 57*
    range of, *9*
    sleeping behavior of, *31*
Anolis limifrons, **187–88**
    vs. *A. cryptolimifrons*, 3
    body of, *333*
    dewlap of, *57*
    history of research on, 12
    home range size of, 22
    number of specimens, 13
    range of, *7*
    synonyms of, 82, 97, 350
Anolis limon, **188–89**
    dewlap of, *57*
Anolis lineatus, **189–90**
    body of, *346*
    dewlap of, *57*
    history of research on, 10
Anolis liogaster, **190–91**
    body of, *331*
    dewlap of, *57*
    range of, *9*
Anolis lionotus, **191–92**
    body of, *339*
    dewlap of, *57*
    distribution of, 21
    range of, *7*
Anolis lososi, **192–93**
    body of, *345*
    dewlap of, *57*
Anolis loveridgei, **193–94**
    dewlap of, *57*

Anolis lynchi, **194–95**
    body of, *346*
    dewlap of, *57*
Anolis lyra, 325, 351, 362n15
Anolis macrinii, **195–96**
    body of, *333*
    dewlap of, *57*
    range of, *9*
Anolis macrolepis, **196–97**
    body of, *346*
    dewlap of, *57*
Anolis macrophallus, **197–98**
    dewlap of, *58*
    range of, *9*
Anolis maculigula, **199–200**
    body of, *346*
    dewlap of, *58*
    sleeping behavior of, 32
Anolis maculiventris, **200–201**
    body of, *339*
    dewlap of, *58*
    range of, *13*
Anolis magnaphallus, **201–2**
    body of, *335*
    dewlap of, *58*
    head scales in identification of, *38, 39*
Anolis maia, **202–3**
    body of, *337*
    dewlap of, *58*
Anolis mariarum, **204–5**
    body of, *340*
    dewlap of, *58*
Anolis marmoratus, **205–6**
    body of, *346*
    dewlap of, *58*
Anolis marsupialis, **206–7**
    dewlap of, *58*
Anolis matudai, **207–8**
    dewlap of, *58*
    range of, *9*
Anolis mccraniei, 314, 353–55
Anolis medemi, **208–9**
    dewlap of, *58*
    range of, *6*

INDEX

Anolis megalopithecus, **209–10**
  body of, *332, 343*
  dewlap of, *58*
  as synonym, 80
Anolis megapholidotus, **210–11**
  dewlap of, *58*
  range of, 9
  synonyms of, 353
Anolis menta, **211–12**
  body of, *345*
  dewlap of, *46, 58*
  range of, 6
Anolis meridionalis, **212–13**
  collecting of, 368n23
  dewlap of, *58*
Anolis microlepidotus, **214–15**
  body of, *332*
  dewlap of, *58*
  range of, 9
Anolis microlepis, 278, 349
Anolis microtus, **215–16**
  body of, *338*
  dewlap of, *59*
Anolis milleri, **216–17**
  body of, *331*
  dewlap of, *59*
  range of, 9
Anolis mirus, **217–18**
  body of, *342*
  dewlap of, *59*
Anolis monteverde, **218–19**
  body of, *334*
  dewlap of, *59*
  population declines in, 26
Anolis morazani, **219–20**
  dewlap of, *59*
  range of, 9
Anolis muralla, **220–21**
  dewlap of, *59*
  range of, 9
Anolis nasofrontalis, **221–22**
  dewlap of, *59*
  range of, 6
Anolis naufragus, **223**
  body of, *332*

dewlap of, *59*
range of, 9
Anolis neblininus, **224**
  body of, *345*
  dewlap of, *59*
  range of, 6
Anolis nebuloides, **225–26**
  body of, *332*
  dewlap of, *59*
  range of, 9
  synonyms of, 353
Anolis nebulosus, **226–27**
  body of, *332*
  dewlap of, 45, *59*
  history of research on, 12
  range of, 9
Anolis neglectus, **227–28**
  body of, *344*
  dewlap of, *59*
  range of, 6
Anolis nemonteae, **228–29**
  biogeographic barriers in, 363n20
  body of, *342*
  dewlap of, *59*
Anolis nicefori, **229–30**
  dewlap of, *59*
Anolis nietoi, 211, 353
Anolis nigrolineatus, 147, 350
Anolis notopholis, **230–31**
  body of, *341*
  dewlap of, *59*
  history of research on, 11
Anolis occultus, 30, 367n13
Anolis ocelloscapularis, **231–32**
  body of, *336*
  dewlap of, *60*
Anolis omiltemanus, **232–33**
  body of, *331*
  dewlap of, *60*
  range of, 9
Anolis onca, **233–34**
  anatomy of, 365n3
  dewlap of, *60*

Anolis orcesi, **234–35**
  body of, *345*
  collecting of, 32
  dewlap of, *60*
  eye color in identification of, *41*
  sleeping behavior of, 32
Anolis ortonii, **235–37**
  body of, *340*
  dewlap of, *60*
  range of, 6
Anolis osa, 257, 349
Anolis otongae, **237–38**
  body of, *342*
  dewlap of, *60*
Anolis oxylophus, **238–39**
  body of, *339*
  dewlap of, *60*
  range of, 9
Anolis pachypus, **239–40**
  body of, *335*
  dewlap of, *60*
  head scales in identification of, *38*
Anolis parilis, **240–41**
  body of, *342*
  dewlap of, *60*
  variation within, 46
Anolis parvauritus, **241–42**
  body of, *343*
  dewlap of, *60*
  range of, 6
Anolis parvicirculatus, **242–43**
  body of, *331*
  dewlap of, *60*
  range of, 9
  type locality of, 15
Anolis pentaprion, **243–44**
  body of, *338*
  dewlap of, 45, *60*
  range of, 7, 8
Anolis peraccae, **244–45**
  body of, *340*
  dewlap of, *60*
Anolis peruensis, **245–46**
  body of, *345*
  dewlap of, *60*

INDEX

Anolis petersii, **247**
  body of, *337*
  dewlap of, *60*, 362n5
Anolis peucephilus, **248**
  dewlap of, *61*
  range of, 9
Anolis phyllorhinus, **249**
  collecting of, 33
  dewlap of, *61*
  range of, 6
  sexual dimorphism in, 46
Anolis pijolense, **250**
  dewlap of, *61*
Anolis pinchoti, **251**
  dewlap of, *61*
Anolis planiceps, **252**
  dewlap of, *61*
Anolis podocarpus, **253**
  body of, *343*
  dewlap of, *61*
  identification of, 42, *43*
Anolis poecilopus, **254–55**
  body of, *346*
  dewlap of, *61*
Anolis poei, **255–56**
  body of, *342*
  dewlap of, *61*
Anolis polylepis, **256–57**
  body of, *336*
  dewlap of, *61*
  diet of, 23
  history of research on, 12
  synonyms of, 349
Anolis porcatus, 112
Anolis princeps, **257–58**
  body of, *344*
  dewlap of, *61*
  range of, 6
Anolis proboscis, **258–59**
  anatomy of, *18*
  collecting of, 33
  dewlap of, *61*
  habitat of, 26
  range of, 6
  sexual dimorphism in, 46, *47*

Anolis propinquus, **259–60**
  dewlap of, *61*
Anolis pseudokemptoni, **260–61**
  body of, *334*
  dewlap of, *61*
Anolis pseudopachypus, **261–62**
  dewlap of, *61*
Anolis pseudotigrinus, **262–63**
  dewlap of, *61*
  range of, 6
Anolis punctatus, **263–64**
  dewlap of, *62*
  genome of, 25
  history of research on, 10
  range of, 6
  type locality of, 14
  variation within, 46, *48*
Anolis purpurescens, **265–66**
  body of, *337*
  dewlap of, *62*
  range of, 6
  vocalization in, 20
Anolis purpurgularis, **266–67**
  dewlap of, *62*
Anolis purpuronectes, **267–68**
  body of, *339*
  dewlap of, *62*
  range of, 9
Anolis pygmaeus, **268–69**
  dewlap of, *62*
  range of, 9
Anolis quaggulus, 170, 350–51
Anolis quercorum, **269–70**
  body of, *332*
  dewlap of, *62*, 352
  range of, 9
  synonyms of, 352–53
Anolis quimbaya, **270–71**
  dewlap of, *62*
Anolis richteri, **272–73**
  dewlap of, *62*
Anolis riparius, **273–74**
  dewlap of, *62*
Anolis rivalis, **274–75**
  dewlap of, *62*

Anolis roatanensis, **275–76**
  dewlap of, *62*
  range of, 9
Anolis robinsoni, **276–77**
  dewlap of, *62*
Anolis rodriguezii, **277–78**
  body of, *333*
  dewlap of, *62*
  range of, 9
  synonyms of, 349
Anolis roosevelti, 366n25
Anolis roquet, 67, 350, 362n1
Anolis rubiginosus, **278–79**
  body of, *332*
  dewlap of, *62*
  range of, 9
Anolis rubribarbaris, **280–81**
  dewlap of, *62*
  range of, 9
Anolis ruizii, **281–82**
  dewlap of, *63*
Anolis sacamecatensis, 270, 352–53
Anolis sagrei, **282–83**
  abundances of, 21
  anatomy of, 20
  body of, *347*
  dewlap of, *63*
  habitat of, 366n30
  human dispersal of, 26, 366n34
  as naturalized species, 1
  range of, 9
Anolis salvini, **283–84**
  body of, *339*
  dewlap of, 46, *63*
  vocalization in, 20
Anolis santamartae, **284–85**
  body of, *345*
  dewlap of, *63*
  range of, 6
Anolis savagei, **285–86**
  body of, *338*
  dewlap of, *63*
  range of, 8

INDEX

Anolis schiedii, **286–87**
   body of, *332*
   dewlap of, *63*
   range of, *9*
Anolis scypheus, **287–88**
   body of, *340*
   dewlap of, *63*
Anolis sericeus, **289–90**
   body of, *333*
   dewlap of, *63*
   range of, *9*
Anolis serranoi, **290–91**
   body of, *336*
   dewlap of, *63*
   range of, *9*
Anolis sheplani, 368n27
Anolis sminthus, **291–92**
   body of, *335*
   dewlap of, *63*
Anolis soinii, **292–93**
   dewlap of, *63*
   head scales in identification of, *38*
   variation within, *48*
Anolis solitarius, **293–94**
   dewlap of, *63*
   identification of, 15
   range of, *6*
   sexual dimorphism in, *48*
Anolis spilorhipis, 314, 353–55
Anolis squamulatus, **294–95**
   dewlap of, *63*
Anolis stevepoei, 226, 353
Anolis subocularis, **295–96**
   body of, *332*
   dewlap of, *63*
   range of, *9*
Anolis sulcifrons, **296–97**
   body of, *341*
   dewlap of, *63*
   synonyms of, 350
Anolis tandai, **297–98**
   body of, *340*
   dewlap of, *64*
   distribution of, 21

Anolis taylori, **298–99**
   body of, *331*
   dewlap of, *64*
   distribution of, 21
   natural history of, 16
   range of, *9*
Anolis tenorioensis, **299–300**
   body of, *334*
   dewlap of, *64*
Anolis tequendama, **301–2**
   body of, *344*
   dewlap of, *64*
Anolis tetarii, **302–3**
   dewlap of, *64*
Anolis tigrinus, **303–4**
   dewlap of, *64*
Anolis tolimensis, **304–5**
   body of, *340*
   dewlap of, *64*
Anolis townsendi, **305–6**
   body of, *336*
   dewlap of, *64*
Anolis trachyderma, **306–7**
   body of, *340*
   dewlap of, *64*
Anolis transversalis, **307–8**
   body of, *341*
   dewlap of, *64*
   head scales in identification of, *38, 39*
   range of, *6*
Anolis trinitatis, **308–9**
   body of, *347*
   dewlap of, *64*
Anolis triumphalis, **309–10**
   body of, *339*
   dewlap of, *64*
Anolis tropidogaster, **310–11**
   body of, *341*
   as cryptic species, 362n12
   dewlap of, *64*
Anolis tropidolepis, **311–13**
   body of, *336*
   dewlap of, *64*, 351
   history of research on, 11

   population declines in, 26
   range of, *8*
   synonyms of, 351–52
Anolis tropidolepis species complex, 352
Anolis tropidonotus, **313–14**
   body of, *335*
   dewlap of, *64*
   natural history of, 16
   range of, *9*
   sleeping behavior of, 32, 368n21
   synonyms of, 353–55
   variation within, 46
Anolis uniformis, **314–15**
   body of, *334*
   dewlap of, *65*
   range of, *9*
   sleeping behavior of, 368n21
Anolis unilobatus, 290
Anolis urraoi, **315–16**
   dewlap of, *65*
Anolis ustus, **316–17**
   dewlap of, *65*
   range of, *9*
Anolis utilensis, **317–18**
   body of, *338*
   dewlap of, *65*
   range of, *9*
Anolis utowanae, 369n5
Anolis vanzolinii, **318–19**
   body of, *344*
   dewlap of, *65*
   as endangered species, 25
Anolis vaupesianus, **319–20**
   dewlap of, *65*
Anolis ventrimaculatus, **321–22**
   body of, *343*
   and crown-giant anoles, 362n6
   dewlap of, *65*
   variation within, 46
Anolis vermiculatus, 368n3
Anolis vicarius, **322–23**
   dewlap of, *65*
Anolis villai, **323–24**
   dewlap of, *65*
   range of, *8*

INDEX

Anolis vittigerus, **324–25**
  vs. *A. lyra*, 351, 362n15
  dewlap of, 45, 46, *65*, 351
  sexual dimorphism in, 45, *47*
Anolis vociferans, 29
Anolis wampuensis, 314, 353–55
Anolis wattsi, 11
Anolis wellbornae, 290, 363n34
Anolis wermuthi, **325–26**
  dewlap of, *65*
  range of, 8
Anolis williamsmittermeierorum, **326–27**
  body of, *345*
  dewlap of, *65*
Anolis wilsoni, 314, 353–55
Anolis woodi, **327–28**
  body of, *337*
  dewlap of, *65*
Anolis yoroensis, **328–29**
  body of, *336*
  dewlap of, *65*
Anolis zapotecorum, 226, 353
Anolis zeus, **330**
  body of, *333*
  dewlap of, *65*
anterior nasal scales, *41*, 42
Arteaga, A., 14
assemblages, definition of, 362n7
Atlantic coast, 6
Autoridad Nacional de Ambiente de Panama (ANAM), 28
Avila-Pires, Teresa, 12
axillary pockets, 41
Ayala, Fernando, 12
Ayala, Stephen, 12, 364nn45, 47

Barros, Tito, 12
behavior, 21–22
Belize, species lists for, 356
biking lights, 31, 368n16
biogeographic barriers, 5–6, 8, 363n20
biogeography of Latin America, 3–9

birds, 23, 367nn1, 2
bite force, 19
bodies, plates of, *331–48*
body color. *See also* dewlaps
  changes in, 18, 364n2
  intraspecific variation in, 45, 46, *47*
  in species identification, 41
body patterns
  intraspecific variation in, 46, 48, *48*
  in species identification, 41
body scales, in species identification, 39, *39*
body size
  sexual dimorphism in, 45, 369n13
  in species identification, 37
Bolaños, Federico, 12
Bolivia
  physiography of, 6
  species lists for, 360
boots, 28, 367n5
Boulenger, George, *10*, 10–11, 363n40
brain, anatomy of, 19
Brazil
  physiography of, 6
  species lists for, 360–61
Brazilian Highlands, 6
breeding, 22
brown anoles
  identification of, 43
  intraspecific variation in, 46
  listed by country and region, 356–61

Calderón-Espinosa, Martha, 12
cameras, 28
canthals, 40, *40*
Caribbean coast, 7–9
Carvalho, A. Leitão de, 368n1
Castro, Fernando, 12, 364n45, 364n47

catching anoles. *See* collecting
caudal autonomy, 365n5
caudal crest, 18
caudal vertebrae, 19, 365n6
CEB. *See* Comparative Evolutionary Biology
Central America. *See* Middle America
chameleons, 364n2
characters
  sources of data on, 13
  in species identification, 37–42
chemoreception, 20
Chiapas Central Highlands, 9
Chimalapas, 9
circulatory system, 20
clades, 1–2, 24, 362n2. *See also specific clades*
cladistics, 23, 365n13
claws
  anatomy of, 17–18
  in identification, 42, *42*
climate change, 26, 366n32
climates, 3, *5*, 363n17
climbing anatomy, 18
cloaca, 20–22
collecting anoles, 27–34
  equipment for, 28, *28*
  executing trips for, 33–34
  other approaches to, 33
  for pet trade, 25–26, 366n31
  planning trips for, 27–29
  with snares, *29*, 29–32, 367nn7–11
  with spotlights, 27, 30–33, *31*, *32*, 367–68nn15–23
Collette, Bruce, 364n46
Colombia
  endemic species in, 6, 7
  number of anole species in, 6, 363n19
  physiography of, 6, 7
  species lists for, 359–60
colon, 21
color, body. *See* body color

417

INDEX

color, dewlap. *See* dewlaps
color, eye, 41, *41*
color changing, 18, 364n2
comments sections of species accounts, 16
common species, criteria for characterizing, 16
Comparative Evolutionary Biology (CEB), 24
comparative research, 12, 24–25
computer identification key. *See* AnoleKey
computers, 28
conservation, 25–26
Continenteloa, 24, 362n1
convergence, 17, 25
Cope, Edward, *10*, 10–11, 19, 363n40, 364n42
copulation, 20, 22
Cordillera Central, 7–8
Cordillera de Guanacaste, 8
Cordillera Tilarán, 8
Costa Rica
    physiography of, 7–8
    population declines in, 26
    species lists for, 358
country, species lists by, 356–61
courtship behavior, 22
crests
    anatomy of, 18, *18*
    in identification, 39
crown-giant anoles, 362n6
cryptic species, 3, 12, 13, 362n13
Cuba
    dewlaps in, 368n3
    number of anole species in, 363n19

Dactyloa clade, 1–2, *24*, 24–25, 362n1
D'Angiolella, Annelise, 12
Darién range, 8
data deficiencies, 25–26
data sources, 13–14

Daudin, Francois, 10, *10*
daytime collecting, 29–30
de Queiroz, Kevin, 2
description sections of species accounts, 14
detectability, criteria for characterizing, 16
dewlaps, *50–65*
    anatomy of, 1, *2*, 18, 19
    functions of displays of, 1, 22
    intraspecific variation in, *45*, 45–46
    in species identification, 1, 35, 36, 41, 42, 369n14
dichotomous keys, 37, *37*, 369n6
diet, 22–23
digestive system, 20, 21
digits, 2, 18
Disney World, 368n30
distribution, *4*, 12, 21
disturbed habitat, 21, 26, 34
diurnal activity, 22
dorsal crest, 18, *18*
Draconura clade, 1–2, *24*, 24–25
Duméril, A. M. C., 10, 363n40
Dunn, E. R., 17

ears, 20, 40
ecology of anoles, 13, 21–23
ecomorphs, 12, 17
Ecuador
    number of anole species in, 363n19
    physiography of, 7
    species lists for, 360
edge habitats, 34, 368n29
egg production, 20, 22, 365n9
elevation
    in Latin America, 5–9, *6*
    sources of data on, 14
El Salvador
    physiography of, 8
    species lists for, 358
endangered species, 25–26

endemic species, 6–9, 363n23
endocrine system, 19
environmental ethics, 26, 366n33
equipment. *See* collecting anoles
ESC. *See* Evolutionary Species Concept
Etheridge, R. E., 19, 23–24, 365n6
Evolutionary Species Concept (ESC), 2–3, 362n9
evolution of anoles, 23–25
    convergence in, 17, 25
    history of research on, 12
    island vs. mainland, 1–2, 17
    radiations in, 1
    retrograde, 76
excretory system, 20, 21
extinct species, 25, 366n25
eye color, 41, *41*
eye contact, 29, 367n10

feeding, 22–23
female anoles
    dewlaps in, 1, 22, 36, 41, 45–46, *46*
    identifying sex of, 36, *36*
    identifying species of, 36, 368n2
    reproductive system of, 20
    sexual dimorphism in, 45–48, *46*, *47*, *48*
    territoriality in, 22
field notes, 28
field trips, planning and executing, 27–29, 33–34
fishing poles, 29, 31–32, 368n18
Fitch, Henry, 12
flashlights, 31, 368n17
foraging behavior, 22
French Guiana, species lists for, 361

GBIF. *See* Global Biodiversity Information Facility
geckos, 18, 364n2

INDEX

Geiger, Rudolf, 363n17
genomes, 17, 25, 366n24
gestalt
  categorization by, 43–44
  species lists by country, region, and, 356–61
giant anoles
  identification of, 44
  listed by country and region, 356–61
Global Biodiversity Information Facility (GBIF), 13, 364n48
global positioning system. *See* GPS
global warming, 26, 366n32
Goodman, D., 23
Google Earth, 28
Gorgona, 6
GPS coordinates, in species accounts, 14–15
GPS units, 28
grass snares, 367n9
green anoles
  identification of, 43–44
  intraspecific variation in, 46
  listed by country and region, 356–61
Grinnell, Joseph, 28
Guatemala
  physiography of, 9
  species lists for, 357
Guiana Highlands, 6
Gulf of Fonseca, 8
Guyana, species lists for, 361
Guyer, Craig, 12

habitats, 21–22
  in collecting strategies, 34
  disturbed, 21, 26, 34
  diversity of, 3, *4*, 21
  edge, 34, 368n29
  loss of, 25–26
  partitioning of, 21
Harvard University, 368n27, 369n16

headlamps, 28, 31
head scales
  anatomy of, 18
  in identification, 37, *38*, *39*, 40, *40*
hearing, 20
heart, 20
hemipenes
  anatomy of, 20
  in identification, 36
herpetology, definition and use of term, 27
herping, definition and use of term, 27. *See also* collecting
highlands, physiography of, 5–9
Hillis, David, 35
hindlimbs, in identification, 40–41, *41*
Histoire Naturelle (Daudin), 10, *10*
holotypes, 14
home range size, 22
Honduras
  physiography of, 7–9
  species lists for, 357
Hulebak, Erik, 368n19
Humphrey, Stephen, 12
hyoid bones, 19, 365n8

Ibañez, Roberto, 12
identification of sex, 36, *36*
identification of species, 35–48
  AnoleKey in, 36–37, 42–43, *43*
  approaches to, 35–37
  challenges of, 36
  characters in, 37–42
  dewlaps in, 1, 35
  dichotomous keys in, 37, *37*, 369n6
  intraspecific variation in, *45*, 45–48, *46*, *47*, *48*
  locality information in, 15, 36–37
  sex in, 36, *36*

species accounts in, 15, 36, 43–45, *44*
species concept in, 3
vetting of data on, 13–14
iNaturalist, *4*, 14, 28, 369n12
Instituto Nacional de Recursos Naturales (INRENA), 28
integument, 17–18
International Union for Conservation of Nature (IUCN), 25, 366nn26, 27
interparietal scale, 40, *40*
intraindividual variation, 46, *48*
intraspecific variation, *45*, 45–48, *46*, *47*, *48*
invasive species, 25–26, 366n31
iridophores, 18
Irschick, D. J., 12
island vs. mainland anoles
  ecology of, 21–22
  endemic, 6
  evolution of, 1–2, 24–25, 362n3
  research on, 1–2, 17
Isthmus of Téhuantepec, 9
IUCN. *See* International Union for Conservation of Nature

jaw, 19
Jenssen, Thomas, 12
Johnson, J. D., 25
Johnson, M. A., 20
Jubones River basin, 363n20
Jungurudó range, 8
juvenile anoles
  identification of, 36
  intraspecific variation in, 46, 48

Kahrl, A. F., 20
keeling
  in identification, 37–38, 369nn8–10
  intraspecific variation in, 46
keratin, 17

419

INDEX

keys
  AnoleKey, 36–37, 42–43, *43*, 369n11
  dichotomous, 37, *37*, 369n6
kidneys, 20, 21
Köhler, Gunther, *10*, 12, 14, 37, 351–55
Köppen, Vladmir, 363n17
Köppen-Geiger classification system, 3, 5, 363n17

Lacerta, 363n38
Lago de Managua, 8
Lago de Nicaragua, 8
lamellae. *See* toe lamellae
Langstroth, Robert, 368n23
large brown anoles
  identification of, 43
  listed by country and region, 356–61
large green anoles
  identification of, 43–44
  listed by country and region, 356–61
lassos, 367n7
Latin America
  physiography of, 3–9
  range of anoles in, 3–9, 21
  species lists for, 356–61
Lazell, James, 27
lectotypes, 14
Lee, J. C., 13, 14
life histories, 21–22, 365n12
lights, 28, 31, 368n17. *See also* spotlighting
limbs
  in identification, 40–41, *41*
  modularity of, 25, 365n19
Linnaeus, Carl, 10, 363n38, 367n3
live anoles, identification of, 36
lizard markets, 33
locality data. *See also* range; type localities
  in species identification, 15, 36–37
  vetting of, 13–15

Losos, Jonathan, 1
lowlands, physiography of, 7–9, 363n28
lungs, 20–21

Majé range, 8
male anoles
  dewlaps of (*see* dewlaps)
  identifying sex of, 36, *36*
  identifying species of, 35–36
  reproductive system of, 20
  sexual dimorphism in, 45–48, *46*, *47*, *48*
  territoriality in, 22
Malpelo, 6
maps
  elevation, *6*
  range, *4*, 13–14, *15*
  regional, *35*
  sources of data for, 13–14
markets, lizard, 33
mastication, 21
Mayer, Greg, 369n16
Mayr, Ernst, 369n16
McCranie, James, 12, 14, 354
melanin, 18
melanophores, 18
Mexico
  number of anole species in, 363n19
  physiography of, 7, 9
  species lists for, 356
Middle America
  elevation map of, *6*
  physiography of, 7–9
  range of anoles in, 7–9, 21
  regional map of, *35*
  species lists for, 356–59
middorsal crest, 18, 39
middorsal scale, 38–39, *39*, 46
middorsal stripe, 46, 48
Miyata, Ken, 30
model organisms, 17
monsoon climates, tropical, 3, 5

Moreno-Arias, Rafael, 12, 13
morphology
  intraspecific variation in, 45–48
  sources of data on, 13
  in species accounts, 15, 43
  musculoskeletal system, 19
museum acronyms, 14
Museum of Comparative Zoology, 368n27, 369n16
Myers, G. S., 368n1

nasal scales, *41*, 42
native species, 1
natural history sections of species accounts, 16
naturalized species, 1
neotypes, 14
nervous system, 19
Nicaragua
  physiography of, 8, 363n28
  species lists for, 357
niches, 2, 25, 365n18
Nicholson, K. E., 365n21
Nieto Montes de Oca, Adrian, 12
nighttime collecting, 30–34, *34*
nocturnal activity, 22
noosing, 367n7. *See also* snaring
Norops clade, 365n6
notebooks, field, 28
nuchal crest, 18, *18*

Oberhautchen, 17
olfaction, 20
Omar Torrijos park (Panama), *7*, 8, 16
Orinoco river basin, 5

Pacific coast, 5–9
Panama
  diversity of species in, *7*, 8
  number of anole species in, 363n19

420

INDEX

physiography of, 7–8, 363n31
population declines in, 26
species lists for, 358–59
Panama Canal, 7–8, 11
Paraguay, species lists for, 360
parapatric species, 6, 363n24
paratypes, 14
parking lots, 26, 366n30, 368n30
Paynter, R. A., 15
permits, 28–29
Peru
  physiography of, 7
  species lists for, 360
Peters, W., 10–11, 363n40
pet trade collecting, 25–26, 366n31
Phillips, J. G., 351
phylogeny, 23–24, *24*
physiography of Latin America, 3–9
Pisani, G. R., 28
plates
  of bodies, *331–48*
  of dewlaps, *50–65*
poles, 29, 31–32, 368n18
Pontificia Universidad Católica, 14
Popular Science (magazine), *27*
population status, 25–26
postcloacal scales, 36, *36*, 42
Pounds, J. A., 12, 26
Prates, Ivan, 12, 25
predation, 22–23, 30
preserved specimens, identification of, 36–37
proteins, 17

rainfall, 3
rainforest climates, tropical, 3, *5*
Rand, Stan, 12
range, 3–9. *See also specific locations and species*
  climates in, 3, *5*
  in Latin America, 3–9, 21
  maps of, *4*, 13–14, 15
  in species accounts, 15, 43

rare species. *See also specific species*
  collecting of, 28, 33
  criteria for characterizing, 16
  in South America, 6
references sections of species accounts, 16, 364n55
region, species lists by, 356–61
regional maps, *35*
Remsen Jr., J. V., 28
reproductive behavior, 22
reproductive system, 20
Reptile Database, 349
research on anoles
  future of, 12–13
  history of, 10–13, *11*
  island vs. mainland, 1–2, 17
  model organisms in, 17
  number of species in, 10, *11*, 12
respiratory system, 20–21
retrograde evolution, 76
Ribeiro-Junior, M. A., 14
ribs, 19
Rio Palenque field station, 6
Rivas, Gilson, 12
Rivero, J. A., 367n13
Rodrigues, M. T., 33

Savage, J. M., 13, 14
savannah climates, tropical, 3, *5*
scales
  anatomy of, 18
  intraspecific variation in, 46
  in sex identification, 36, *36*
  in species identification, 18, 37–42, *38*, *39*, *40*, *41*, 46
Schwartz, Albert, 33
semiaquatic species
  identification of, 44
  listed by country and region, 356–60
sensory system, 19–20
Serranía del Darién, 363n29

sexes. *See also* female anoles; male anoles
  identification of, 36, *36*
  variation between, 45–48, *46*, *47*, *48*
Sexton, Owen, 12
sexual dimorphism, 45–48, *46*, *47*, *48*
sexual maturity, 20, 22
shake and bake, 368n19
shoes, 28, 367n5
Sierra Madre of Chiapas, 9
Sierra Nevada de Santa Marta (SNSM), 6, 363n22
similar species sections of species accounts, 15, 36–37, 43, 45
Simpson, George Gaylord, 2
skeletal system, 19
skull, 19
sleeping behavior, 22, 30, *31*, 32, *32*
small brown anoles
  identification of, 43
  listed by country and region, 356–61
small green anoles
  identification of, 43–44
  listed by country and region, 356–61
Smith, Hobart, 11, 364n43
smooth ventral scales, 37, 369n8
snakes, 23, *23*, 30
snaring
  daytime, *29*, 29–30, 367nn7–11
  nighttime, 31–32
snout extensions, 18, *18*
snout to vent length (SVL), 37
SNSM. *See* Sierra Nevada de Santa Marta
solitary species, 6, 21, 363n25
South America
  elevation map of, *6*
  physiography of, 5–6
  range of anoles in, 5–6, 21
  regional map of, *35*
  species lists for, 359–61
Sparrman, Anders, 363n38

INDEX

specialization, 33
species accounts, 66–330
  components of, 1, 14–16
  sources of data for, 13–14
  using in identification, 15, 36, 43–45, *44*
species concept, 2–3, 362nn9, 10
species identification. *See* identification
specimens. *See also* collecting
  classification of, 14
  preserved, 36–37
sperm, 20
SpiderWire, 367n8
spotlighting, 27, 30–33, *31*, *32*, 367–68nn15–23
Stapley, J., 26
stomach, 21
stripes, middorsal, 46, 48
Stuart, L. C., 48, 368n1
subocular scales, 40, *40*, *41*
sunken species names, 3
superciliary scales, *41*
supralabial scales, 40, *40*, *41*
supraocular scales, 40, *40*, 46
supraorbital semicircles, 40, *40*
Suriname, species lists for, 361
survivorship, 22
SVL. *See* snout to vent length
sympatric species
  definition of, 362n4
  ecology of, 21
  in Middle America, 8
  in South America, 6
synonyms, 349–55
syntypes, 14

Tailor Rengifo, Jhon, 12
tails, 19, 365n5

targeting, 33–34
taste buds, 21
taxonomy, 10–12, 24, 363n39, 364n44
Taylor, Edward, 11
teeth, 21
temperatures, in range, 3
terrestrial species, 21
territoriality, 22
testes, 20
thermal biology, 22
Thomas, Richard, 30, 367n13
threatened species, 25
toe lamellae
  anatomy of, 17
  in identification, 39–40, *40*
toepads
  anatomy of, 1, *2*, 18
  functions of, 18, 365n3
  in identification, 1, 39, 42, *42*
  in other lizards, 18, 364n2
Tollis, M., 25
tongue, 21
topotypical specimens, 14
Torres, Omar, 12
trash anoles, 364n51
Trefaut Rodrigues, Miguel, 12
tropical climates, 3, *5*
trunk-crown anoles, 33
twig species
  identification of, *44*
  listed by country and region, 356–61
  range of, 6, 362n6
tympanum, 40
type localities
  definition of, 14, 362n14
  sources of data on, 13–15
  in species accounts, 14–15
  of sunken names, 3

United States, species lists for, 356
urinary bladder, 21

van der Waals adhesion, 18
Vanzolini, Paolo, 364n45
Velasco, Julián, 12, 13
Venezuela
  physiography of, 6
  species lists for, 361
ventral scales, 37, *38*, *39*, 46, 369n8
vertebrae, 19, 365nn5, 6
VertNet, 13, 28, 367n4
vicariance, 5, 363n18
vision, 19
Vitt, Laurie, 12, 13
vocalization, 20
vulnerable species, 25–26

Whitfield, S. M., 26
Whodini, 30
Wiley, Edward, 2
Williams, Ernest, *10*, 11–12
  on characters, 37
  on collecting, 33, 367n13
  Mayr and, 369n16
  significance of research of, 11–12, 364nn44, 45, 47
  on solitary species, 363n25
  synonymy in works of, 350
  on taxonomy, 24, 364n44

xanthophores, 18

Yucatan peninsula, 9